T0065829

Praise for *One River*

"A wild ride through one rapid after another . . . magnificent."
—Catherine Foster, *Boston Globe*

"An absolutely fascinating look at the field of ethnobotany."
—Richard Gehr, *Newsday*

"*One River* is a cross between Joseph Conrad's *Heart of Darkness* and *The Hitchhiker's Guide to the Galaxy.* For sheer adventure, it puts Indiana Jones to shame."
—James P. Lucier, *The Washington Times*

"Davis's riveting prose . . . brings the reader into oneness with the river, the jungle, and the people who explored it."
—Laurence A. Marschall, *The Sciences*

"Extraordinary. . . . A biographical tapestry rich in history, adventure, intrigue, and scholarship."
—Michael J. Balick, *Nature*

"A consistently enlightening and thought-provoking study."
—*Publishers Weekly*

"A fascinating narrative . . . an exceptional tale of 20th-century scientific exploration and a rousing travelogue to places both real and illusory."
—*Kirkus Reviews*

"Davis, a compelling writer and intrepid ethnobotanist, proves himself a master of synthesis in this engrossing history of plant exploration in the Amazon."
—Donna Seaman, *Booklist* (Editor's Choice for one of 1996's best books)

"Who would have thought it: reading a thick volume of botanical history can be an unexpectedly magical and riveting experience. *One River* manages to be entertaining, rich with meaning, and full of human biological drama."
—Sue Sutton, *The Financial Post*

"One of the richest books ever written about South America. Combining botanical lore, history, sensitive evocations of native cultures and a good deal of old-fashioned adventure, *One River* is as fascinating and densely varied as the rain forest itself."
—John Bremrose, *Maclean's*

"In *One River,* Davis has forged a rare combination of exploration and unobtrusive scholarship."
—Douglas Daly, *Audubon*

"*One River* has a hallucinogenic feel, in which science overlaps with myth, memory mingles with illusion, and time shifts unpredictably. Davis' outrageous brand of storytelling is well suited to his larger than life subjects, fusing traditional biography with one of Latin America's most potent legacies, magic realism."
—Miles Harvey, *Outside*

"Richard Schultes is the real Indiana Jones, *One River* is a crackling good story that reads like a combination of Sir Richard Burton, *Raiders of the Lost Ark,* and Richard Haliburton's *Book of Marvels.* . . . Buy the book and get ready for a great adventure."
—Jeff Baker, *Portland Oregonian*

"Although a botanist, Davis is a fine writer and schooled in history."
—Dean Sims, Tulsa *World*

"An exciting account of an unusual adventure."
—Broox Sledge, *The Neshoba Democrat*

"An epic—an absorbing combination of adventure, history, botany, and facts and impressions about South American cultures."
—David Bezanson, Austin *American-Statesman*

"Restores some rollicking adventure to the jungle, complete with snakebites and tropical illnesses."
—Will Nixon, *New Age Journal*

"This is a wonderful book about a great biologist, Richard Evans Schultes. It is a trip into a time fast disappearing, when biologists were often also explorers, trying to understand the rich biodiversity of our planet—a form of natural wealth that is rapidly disappearing."
—Paul R. Ehrlich, author of *The Population Bomb*

"Wade Davis, one of our most lyrical nature writers, has written the definitive book about the South American rain forest. . . . *One River* is a spellbinding account of magical places, plants, and people."
—Andrew Weil, M.D., author of *Spontaneous Healing*

"Wade Davis is a rare treasure—a professional scientist who writes like a poet. In tracing the adventures of Richard Evans Schultes, his remarkable mentor, Davis enthralls us with the mysteries of the plant kingdom. I couldn't help regretting that I became a zoologist rather than a botanist."
—Dr. David Suzuki, author and broadcaster

"Richard Evans Schultes is one of the last of those biologists and botanists who confronted a planet with vast unexplored tropical regions, and lived out the high adventure of a serious student of tropical nature. Wade Davis tells his story with humor and reverence. . . . *One River* is a must read."
—Terence McKenna, author of *Food of the Gods*

"Among twentieth-century ethnobotanical explorers, few can begin to approach the accomplishments and influence of Richard Evans Schultes . . . *One River* is wonderfully compelling, a sprawling account of discovery and high adventure that rivals those of such esteemed naturalist predecessors as Humboldt, Darwin, or Spruce."

—Peter Raven, Director, Missouri Botanical Garden

"A captivating tale of stalwart explorers, of mentors and students, drawn alike by the siren call of the tropical forest, its plants and its people. Wade Davis writes from the heart about an intellectual river as old as Socrates and as fresh as this morning's dew: teacher and student, forest and people bound together in a dream one hopes will never end . . . a staggeringly wonderful book."

—Thomas E. Lovejoy, Smithsonian Institution

"*One River* is a densely informative, vibrant book, a lovely labor—a fine homage to a scarcely describable web of phenomena that once seemed eternal."

—Edward Hoagland, author of *Notes from a Century Before*

"Wade Davis succeeds in weaving together a suitable tribute to Schultes and his most brilliant student, Tim Plowman, by combining myth, history, indigenous lore, exploration, travelogue, and the most adventurous kind of scientific discovery into a rich narrative that succeeds in conveying both the legendary fascination of the Amazonian rain forest and the importance of Schultes's work."

—David Maybury-Lewis
Professor of Anthropology, Harvard University
President, Cultural Survival

Also by Wade Davis

Light at the Edge of the World
Nomads of the Dawn
(with Ian MacKenzie and Shane Kennedy)
Shadows in the Sun
Penan: Voice for the Borneo Rainforest
(with Thom Henley)
Passage of Darkness
The Serpent and the Rainbow

SIMON & SCHUSTER PAPERBACKS
New York London Toronto Sydney

One River

Explorations and Discoveries in the Amazon Rain Forest

Wade Davis

For Timothy Plowman
1944–1989

SIMON & SCHUSTER PAPERBACKS
Rockefeller Center
1230 Avenue of the Americas
New York, NY 10020

For information about special discounts for bulk purchases, please contact Simon & Schuster Special Sales:
1-800-456-6798 or business@simonandschuster.com.

Picture Credits
Photographs 1–10 and 12–23 in the photo insert section and illustrations opening chapters 1, 3, 4, 6, 7, 10, 11, 12, and 14 are by Richard Evans Schultes. Photo 24 in the photo section and illustrations for chapters 2, 5, 8, 9, and 13 are by Wade Davis.

Designed by BARBARA M. BACHMAN
Maps by JEFFREY L. WARD

Manufactured in the United States of America

25 27 29 30 28 26 24

The Library of Congress has cataloged the hardcover edition as follows:
Davis, Wade.
One river: explorations and discoveries in the Amazon rain forest / Wade Davis.
p. cm.
Includes bibliographical references and index.
1. Schultes, Richard Evans. 2. Plowman, Timothy. 3. Ethnobotanists—United States—Biography. 4. Ethnobotanists—Amazon River Region—Biography. 5. Ethnobotany—Amazon River Region—Field work. 6. Hallucinogenic plants—Amazon River Region—Collection and preservation. 7. Medicinal plants—Amazon River Region—Collection and preservation. I. Title.
GN20.D38 1996
581.6'1'09811—dc20 96-21516 CIP
ISBN-13: 978-0-684-80886-4
ISBN-10: 0-684-80886-2
ISBN-13: 978-0-684-83496-2 (Pbk)
ISBN-10: 0-684-83496-0 (Pbk)

Contents

Preface

THE IDEA FOR this book emerged in a moment of great sadness. Timothy Plowman was a man of generosity, kindness, modesty, and honor, and his untimely death at the age of forty-five from AIDS on January 7, 1989, cut short a career of immense promise. A superb ethnobotanist with an uncanny ability to gain the trust and confidence of Indian people, he was a scholar of extraordinary depth and one of the finest Amazonian plant explorers of his generation. Of this there was little doubt, for he was the protégé of Richard Evans Schultes, the greatest ethnobotanist of all, a man whose own expeditions a generation earlier had earned him a place in the pantheon along with Charles Darwin, Alfred Russel Wallace, Henry Bates, and his own hero, the indefatigable English botanist and explorer Richard Spruce.

Twelve days after Tim's death a memorial service was held at the Field Museum of Natural History in Chicago. At the time Schultes himself was seriously ill, and though Tim had been like a son, the old professor was unable to attend the memorial. Instead he sent a tape, and it fell to me to deliver the eulogy. I, too, had been a student of Schultes, and Tim, though ten years older, had been a close friend. More than that. For well over a year we had traveled together, living among a dozen tribes, collecting medicinal plants, and studying coca, the source of cocaine. On expeditions that ranged the length and breadth of South America, Tim introduced me to the wonder of ethnobotany and a life of exploration and adventure that realized all my youthful dreams.

Tim's death was especially difficult for Schultes, who in his wisdom understood that the student is as important as the teacher in the lineage of knowledge. The people in the chapel, botanists and friends, sat quietly as his tired voice came over the speakers. He ended with the famous lines from Hamlet: "Now cracks a noble heart. Good night, sweet prince, and flights of angels sing thee to thy rest." It was then, as I stood at the podium, that I decided to write a book that would tell the story of these two remarkable men.

From the start I knew that I was incapable of producing a strict biography. A proper biographer, it is said, must be a conscientious enemy of the subject, and I was far too close to both men to qualify. The case of Schultes was especially complex. He was not a man who marked an age; he was an individual who slipped away from the confines of his own times to experience an exotic land on the cusp of change. His life as a plant explorer, the forest that sheltered him, the indigenous people and their transcendent knowledge of plants—these were the compelling themes. I was less interested in his formative years or his later experiences as a Harvard professor when, as the world's leading authority on hallucinogenic plants, he sparked the psychedelic era with his discoveries. I wanted to focus not on the man alone but on the people and places that made him great, and on the monumental changes that the Amazon had undergone in the decades since he first fell under its spell.

I decided to tell two stories. The narrative follows the travels that Tim and I pursued over a fifteen-month period in 1974–75, a journey not only inspired and made possible by Schultes but infused at all times with his spirit. Wrapped around this account are biographical chapters, identified by year, that describe the most extraordinary period of Schultes's life, a stretch of near continuous fieldwork between 1936 and 1953 that took him from the peyote cult of the Kiowa and the pursuit of *teonanacatl* and *ololiuqui,* the long-lost sacred plants of the Aztec, to the Northwest Amazon of Colombia. There, while searching for the identity of curare, he became engaged in one of the most important botanical quests of the twentieth century: the hunt for new sources of wild rubber, an investigation given urgency by the outbreak of World War II.

In the end the greatest achievements of both Plowman and Schultes were denied and even betrayed by the very government that had sponsored their work. Tim's elegant descriptions of coca as a mild and benign stimulant central to Amerindian culture and religion, and his discovery that the leaves play an essential role in the diet of the Andean farmers, did nothing to stop those committed to eradicating the crop with poisons that foul the many rivers that fall away into the Amazon. In the case of Schultes's work, the consequences of bureaucratic folly may prove even more serious. In destroying his work of a decade, along with that of so many other brave explorers attached to the rubber program, long-forgotten officials at the U.S. State Department left us a disturbing legacy. In the words of one of Schultes's colleagues, "A sword of Damocles hangs over the industrial world. We've created a scenario whereby a deliberate act of biological terrorism so simple it could be perpetrated by your grandmother could precipitate an economic crisis of unprecedented dimensions. And no one even knows about it. What's worse, it all could have been avoided."

Although the collapse of the rubber program was a great disappointment to Schultes and confirmed his long-standing conviction that most bureaucrats were fools, it did not make him bitter. He simply moved on, returning to Harvard, where, as the years passed, he devoted more and more of his time to those who would carry on his work. My experience as a student was not untypical. In early 1974 I walked into his office and explained that I had saved some money and wanted to go to the Amazon and collect plants. At the time I knew little about the Amazon and less about plants. He glanced up from a stack of specimens and said, "When do you want to go?"

Ten days later, equipped with two letters of introduction, I landed in Bogotá. Within a week I was invited to join a botanical expedition, intent on crossing the Gulf of Urabá to reach the rain forests of the Darien. The small fishing boat was serene and the passage calm until we hit open water. A cold northerly wind blew all night, and the boat lunged violently through the water. Then just before dawn the storm passed. The weather cleared, and beneath the shelter of a strange southern sky, crisscrossed by shooting stars, the night gradually gave way to day. Shapes emerged from the darkness, broad, rolling green waves and islands of coconut palms. It was like waking into a dream and coming upon an open sea and the pristine shoreline of a continent that stretched south to the horizon.

CHAPTER ONE

Juan's Farewell

WHEN I FIRST lived in Colombia, I used to stay from time to time on a farm just outside the city of Medellín. The land was owned by a *campesino,* Juan Evangelista Rojas, who was rich beyond his wildest imaginings, though he didn't know it. Juan and his twin sister, Rosa, neither of whom ever married, had lived on the farm most of their lives, and during that time—sixty or seventy years, no one really knew—the city had spread north following the new highway to Bogotá, and the barrios now lapped at the base of their land. Their property was worth millions of pesos, but for reasons of their own, both Rosa and Juan continued to work as they always had, she by gathering herbs and coaxing the odd egg from a flock of sad-looking chickens, he by making charcoal, which he sold by the bagful to the passing peasants on the Guarne road. I don't think either Juan or Rosa ever thought of selling off any land. They would never have been able to agree on which section to let go, and besides, from the main house at the edge of a pine forest at the top of the farm, it was easy to ignore the encroaching city.

The land ran in a narrow swath up and down a precipitous *quebrada,* and the gradient was so steep that the new highway, though but a mile away, lay more than two thousand feet below. Juan was forever scampering up and down the hillside gathering firewood or tending an astonishing variety of crops: potatoes and onions in the mist by the pine woods, coffee a thousand feet below near the waterfall, where the eagle had turned into a dove, and bananas, plantains, and cacao at the very bottom, where the hot tropical sun blistered the pavement of the highway. Juan's imagination breathed life and mystery into every rock and tree. Angels often appeared to him, and he claimed that the crosses set up to mark the route of his mother's funeral sometimes glowed red in the daytime and green by night. At one bend in the main trail, where the pallbearers had tripped and the corpse had tumbled out of the casket crushing the giant horsetails, he never failed to pause to say a prayer or at least to cross himself. Sometimes he brought manure to the spot to fertilize the ground so that the delicate plants need never again feel the weight of death.

There was a beautiful order to Juan's world. Everything had its place, and the land, though large enough to embrace all his dreams, was of a human scale. One could know all of it intimately. In a wild and ragged country the farm was safe. Sometimes late in the afternoon when the work was done, Juan and I would walk up the road to a nearby *estadero* and drink together on the verandah overlooking his fields. Fueled by a few glasses of *aguardiente* Juan would speak of the world beyond the farm, of the seasons in his youth when he, like everyone in rural Colombia, had to move to stay alive. He told of logging on the Río Magdalena when there were still forests to be cut, and of men eaten by black caiman in the swamps of the Chocó. The most horrific tales were of *La Violencia,* the civil war between liberals and conservatives, that racked the nation in the forties and early fifties, a time when entire villages were ravaged and populations destroyed. Juan had fought for the liberals, or at least had managed to be shot by the conservatives, who left him to die in a pile of dead children in the plaza of a small town in Cauca. Since then, he claimed, whenever he thought too much he felt pain. So he tried not to think, only to look and see.

Because of his own itinerant past Juan had no difficulty understanding what I was doing in Colombia. *"Buscando trabajo,"* he would explain to his incredulous neighbors. "The gringo is looking for work." In fact, just twenty at the time, I was in South America to study plants. A letter of introduction from my professor, Richard Evans Schultes, then director of the Botanical Museum at Harvard, had secured me a room at the Jardín Botánico Joaquin Antonio Uribe in Medellín. But I was never quite comfortable there. The garden, located in the north of the city, is a lavish complex of neocolonial buildings sequestered from the surrounding barrio by enormous white stucco walls crowned with barbed wire and shards of glass. Beyond the walls run acres of more modest structures of cinder block and mud, zinc roofs draped in wires which light the lamps of the hundreds of brothels that surround the garden. Inside the walls is the illusion of paradise, but the land used to belong to the poor, and according to Juan, the tranquil ponds with their lilies and papyrus once ran red with the blood of victims of *La Violencia.*

So while I kept my room at the garden, I preferred to live with Juan, and it was from the farm that most of my initial botanical forays originated. In the beginning these wanderings had a random quality. Colombia, with three great branches of the Andes fanning out northward toward the wide Caribbean coastal plain, the rich valleys of the Cauca and Magdalena, the sweeping grasslands of the eastern llanos and the endless forests of the Chocó and Amazonas, is ecologically and geographically the most diverse nation on earth. A naturalist need only spin the compass to discover plants and even animals unknown to science. In the past months I had gone away on several occasions to the rain forests of northern Antioquia, across the Gulf of Urabá to Acandi and the Darien, and south into the mountains of Huila. At the farm Juan became used to my comings and goings, and he always anticipated my return with some excitement. It was as if I had become his eyes and ears onto a world of his own past, a life of uncertainty and adventure, of magic and discovery.

As for life on the farm, it went along in its own casual way. It was an innocent period in Medellín, the spring of 1974. The Cartel was emerging, though no one knew quite how, and no one realized how sordid and murderous it would become. Most Americans had never heard of cocaine. For those who had, it was a sweet sister, incapable of harm. Just beyond the pine forest passed a country lane that led in a few hours to Río Negro and the farm where Carlos Ledher would eventually build his empire, complete with the bizarre statue of John Lennon, stuck like a hood ornament onto a hillside of the hacienda. No one could have imagined how rich he would become or that he would end up in a Miami jail, locked away for life. The cocaine trade, such as it was, still lay in the hands of the independent drifter, people like our neighbor Nancy, an elusive California surfer who lived alone and dazzled the locals with her beauty and the rainbows she painted each morning across her eyelids. Sometimes on Sundays, just as Juan and I were settling down to a day in the garden, musicians would appear with briefcases full of cocaine and simple requests to play their guitars into the night. Almost always Juan welcomed them and then slipped away into the forest that led to the waterfall, a place of tree ferns and mist, where plants dictated the mood and he felt free.

Juan had a brother, Roberto, who was a carpenter, the only one I ever knew who took wind into account before driving a nail. Roberto was explaining this to me one afternoon when Juan came by with the telegram I had been expecting. He found us hammering on the roof of the new pigsty. The telegram came from Tim Plowman, one of Professor Schultes's recent graduate students. For the past month Tim had been languishing in Barranquilla, a miserably hot and dismal industrial city on the north coast, haggling with customs officials who, by some curious sleight of hand, had determined that the pickup truck that he had shipped south from Miami was, in fact, theirs. Evidently, Tim had finally persuaded them otherwise and was at last ready to begin his botanical explorations. I was to meet him in three days at the Residencia Medellín in Santa Marta, a sun-bleached port on the Caribbean coast sixty miles east of Barranquilla.

Juan seized the opportunity like lightning. To get to the coast I would have to pass through a village he knew on the Río Magdalena where I could buy or capture an ocelot that he would train as the nucleus of a circus. It was an old idea of his. There hadn't been an ocelot in that part of the Magdalena Valley in forty years. Juan surely knew this, yet he clung to his dream like a limpet. Like all *campesinos* Juan had dozens of plans for getting rich, impossibly complicated schemes that had nothing whatsoever to do with the day-to-day reality of his life.

As I made ready for the coast, Juan prepared for my departure. For a Colombian there is no such thing as a casual leave-taking, and for Juan it was inconceivable that I might depart without a long afternoon at the *estadero.* Each of his brothers and sisters—and one would discover at such moments with stunning clarity just how many there were—would wax eloquent about the prospects of the journey, its promise and hazards, with each prognosis followed *en seguida* by a stiff shot of *trago* that brought the soft hues of sunset alive at midday. The afternoon grew charged with threatening earthquakes, impossible rapids, train wrecks, sorcery, volcanoes, floods of rain, horrendous unknown diseases, and sly deceitful soldiers with the demeanor of feral dogs.

Thieves lurked at every crossroads except on the north coast. There everyone was a thief.

"Life is an empty glass," Juan would say. "It's up to you how fast you fill it." Inevitably, Rosa would then begin to cry. That was the signal. One had to get out fast or make new plans for the night. This time, against all odds, I managed to get myself and Juan out of the *estadero* and onto the *flota* that bounced and rumbled down the dirt road to the city, leaving Rosa and a pair of sisters red-eyed and wailing in the dust.

In Colombia there are no train schedules, only rumors. The guide books maintain that there is no train service out of Medellín to the Caribbean coast. I was quite sure there was and did my best to get to the train station at an hour that made sense to Juan. His estimate left just enough time to swing by the Botanical Garden, pick up my mail and some extra equipment, and battle the traffic and crowds of the city. As we parted at the station, I promised Juan somewhat disingenuously that I would do my best to find his ocelot. I told him I expected to be back in a week or two. He was delighted and mentioned something about building a cage. I said it would be a good idea. Neither of us could have known that I would be away for many months and that this passage to the coast would be the first leg of an intermittent journey that would last eight years, taking me to some of the most remote and inaccessible reaches of the continent.

The train ran north past the sprawling outskirts of Medellín and into the rich farmland of Antioquia. In the fading light and through a cracked window coated in dust and fingerprints, I could just make out the flank of Juan's farm above Copacabana, and beyond, the mountains of the Cordillera Central looming jet black on the horizon. Inside the train the passengers—among them a dozen military conscripts, *campesinos* who still smelled of earth and smoke—settled in for the thirty-hour trip to the coast. Lulled by the rhythmic clacking of the wheels and the steady pounding of rain on the metal roof, many of them fell asleep almost as soon as the train left the station. Someone turned a radio on, and a fatuous song celebrating the construction of a highway bridge drowned out a dozen soft Spanish voices. I noticed a sign stuck to the back of the seat in front of me. It politely asked all passengers to be civilized enough to throw their garbage out the windows of the train.

I reached into my old canvas pack and pulled out a bundle of mail, which I quickly flipped through until I found the letter from Schultes that had been waiting for me at the Botanical Garden. Like everything about him, it was infused in sepia, the choice of language, the elegant handwriting, the tone of the letter itself, at once both intimate and formal like the correspondence of a Victorian gentleman. Schultes's hope was that Tim Plowman would one day take over as director of the Botanical Museum, just as he had inherited the post from his mentor, the famous orchid specialist Oakes Ames. Thus, when Tim received his degree, Schultes had appointed him research associate, and together they had secured $250,000 from the U.S. Department of Agriculture—an enormous sum in those days—to study coca, the sacred leaf of the Andes and the notorious source of cocaine. For an ethnobotanist, it was a dream assignment.

In three pages Schultes outlined the details of the proposed expedition. Though long the focus of public concern and hysteria, he wrote, surprisingly little was known about coca. The botanical origins of the domesticated species, the chemistry of the leaf, the pharmacology of coca chewing, the plant's role in nutrition, the geographical range of the cultivated varieties, the relationship between the wild and cultivated species—all these remained mysteries. He added that no concerted effort had been made to document the role of the coca in the religion and culture of Andean and Amazonian Indians since W. Golden Mortimer's classic *History of Coca,* published in 1901. The letter continued in Schultes's inimitable style—references to idiotic politicians and policies, an anecdote about one of the admixture plants used with coca that he had discovered in 1943, reflections on how much he had enjoyed chewing coca during the years he spent in the Amazon—but the essential point had been made. Plowman's mandate from the U.S. government, made deliberately vague by Schultes, was to travel the length of the Andean Cordillera, traversing the mountains whenever possible, to reach the flanks of the *montaña* to locate the source of a plant known to the Indians as the divine leaf of immortality.

The expedition was to begin at Santa Marta. Flanked by the fetid marshlands of the Magdalena delta to the west and the stark desert of La Guajira peninsula to the east, the city is and has always been a smuggler's paradise, the conduit through which flowed at one time perhaps a third of Colombia's drug traffic. It is by reputation a seedy place, noisy by night, somnolent by day, and saturated at all hours by the scent of corruption. Just beyond the city to the southeast, however, lies a world apart: the Sierra Nevada de Santa Marta, the highest coastal mountain range on earth. Separated from the northernmost extension of the Andean Cordillera that forms the frontier with Venezuela to the east, the Sierra Nevada is an isolated volcanic massif, roughly triangular in shape, with sides a hundred miles long and a base running along the Caribbean coast. The northern face of the mountains rises directly from the sea to a height of over nineteen thousand feet in a mere thirty miles, a gradient surpassed only by that of the Himalayas.

The people of the Sierra are the Kogi and Ika, descendants of the ancient Tairona civilization that flourished on the coastal plain of Colombia for five hundred years before the arrival of Europeans. Since the time of Columbus, who met them on his third voyage, these Indians have resisted invaders by retreating from the fertile coastal plains higher and higher into the inaccessible reaches of the Sierra Nevada. In a bloodstained continent they alone have never been conquered.

A profoundly religious people, the Kogi and Ika draw their strength from the Great Mother, a goddess of fertility whose sons and daughters formed a pantheon of lesser gods who founded the ancient lineages of the Indians. To this day the Great Mother dwells at the heart of the world in the snow fields and glaciers of the high Sierra, the destination of the dead, and the source of the rivers and streams that bring life to the fields of the living. Water is the Great Mother's blood, just as the stones are the tears of the ancestors. In a sacred landscape in which every plant is a manifestation of the divine, the chewing of *hayo,* a variety of coca found only in the mountains of Colombia, represents the most profound expression of culture. Distance in the mountains is not measured in miles but in coca chews. When two men meet, they do not shake hands, they exchange leaves. Their societal ideal is to abstain from sex, eating, and sleeping while staying up all night, chewing *hayo* and chanting the names of the ancestors. Each week the men chew about a pound of dry leaves, thus absorbing as much as a third of a gram of cocaine each day of their adult lives. In entering the Sierra to study coca, Tim was seeking a route into the very heart of Indian existence.

This letter from Professor Schultes, which I read in the yellow glow of dim lamps as the train shrieked and plunged in and out of tunnels through the Andean Cordillera, was like a map of dreams, a sketch of journeys he himself would be making were he still young and able. But nearly sixty, his body worn from years in the rain forest, it had been some time since he had been capable of active fieldwork. There was a certain tension in his writing, a sense of urgency that came from his awareness of both his own limitations and the speed with which the knowledge of the Indians was being lost and the forests destroyed. In this sense his letters were both gifts and challenges. It was impossible to read them without hearing his resonant voice, without feeling a surge of confidence and purpose, often strangely at odds with the esoteric character of the immediate assignments. This perhaps was the key to Schultes's hold over his students. He had a way of giving form and substance to the most unusual of ethnobotanical pursuits. At any one time he had students flung all across Latin America seeking new fruits from the forest, obscure oil palms from the swamps of the Orinoco, rare tuber crops in the high Andes. Under his guidance it somehow made perfect sense to be lost in the canyons of western Mexico hunting for new varieties of peyote, miserably wet and cold in the southern Andes searching for mutant forms of the tree of the evil eagle, or hidden away in the basement of the Botanical Museum figuring out the best way to ingest the venom of a toad rumored to have been used as a ritual intoxicant by the ancient Olmec.

His own botanical achievements were legendary. In 1941, after having identified *ololiuqui,* the long-lost Aztec hallucinogen, and having collected the first specimens of *teonanacatl,* the sacred mushrooms of Mexico, he took a semester's leave of absence from Harvard and disappeared into the Northwest Amazon of Colombia. Twelve years later he returned from South America having gone places no outsider had ever been, mapping uncharted rivers and living among two dozen Indian tribes while collecting some twenty thousand botanical specimens, including three hundred species new to science. The world's leading authority on plant hallucinogens and the medicinal plants of the Amazon, he was for his students a living link to the great natural historians of the nineteenth century and to a distant era when the tropical rain forests stood immense, inviolable, a mantle of green stretching across entire continents.

At a time when there was little public interest in the Amazon and virtually no recognition of the importance of ethnobotanical exploration, Schultes drew to Harvard an extraordinarily eclectic group of students. He held court on the fourth floor of the Botanical Museum in the Nash Lecture Hall, a wooden laboratory draped in bark cloth and cluttered with blowguns, spears, dance masks, and dozens of glass bottles that sparkled with fruits and flowers no longer found in the wild. Oak cabinets elegantly displayed every known narcotic or hallucinogenic plant together with exotic paraphernalia—opium pipes from Thailand, a sacred mescal bean necklace from the Kiowa, a kilogram bar of hashish that went on display following one of Schultes's expeditions to Afghanistan. In the midst of enough psychoactive drugs to keep the DEA busy for a year, Schultes would appear, tall and heavyset, dressed conservatively in gray flannels and thick oxfords, with a red Harvard tie habitually worn beneath a white lab coat. His face was round and kindly; his hair cut razor short, his rimless bifocals pressed tight to his face. He lectured from tattered pages, yellow with age, sometimes making amusing blunders that students jokingly dismissed as the side effects of his having ingested so many strange plants. Just the previous fall he had been discussing in class a drug that had been first isolated in 1943. "That was fourteen years ago," he added, "and we've come a long way since." Such oversights were easily forgiven, coming as they did from a fatherly professor who shot blowguns in class and at one time kept in his office a bucket of peyote buttons available to his students as an optional laboratory assignment.

Throughout the sixties, as America discovered the drugs that had fascinated Schultes for thirty years, his fame grew. Suddenly academic papers that had gathered dust in the library for decades were in fierce demand. His 1941 book on the psychoactive morning glory, *A Contribution to our Knowledge of Rivea Corymbosa, the Narcotic Ololiuqui of the Aztecs,* had been published on a hand-set printing press in the basement of the Botanical Museum. In the spring and summer of 1967 requests for copies poured into the museum, and florists across the nation experienced a run on packets of morning glory seeds, particularly the varieties named Pearly Gates and Heavenly Blue. Strange people began to make their way to the museum. A former graduate student tells of going to meet Schultes for the first time and finding two other visitors waiting outside his office, one of them passing the time by standing on his head like a yogi.

Schultes was an odd choice to become a sixties icon. His politics were exceedingly conservative. Neither a Democrat nor a Republican, he claimed to be a royalist who professed not to believe in the American Revolution. When the presidential election results are published in his local newspaper, *The Melrose Gazette,* there is always one vote for Queen Elizabeth II. A proud Bostonian, he will have nothing to do with one New England family. He will not use a Kennedy stamp, insists on calling New York City's Kennedy Airport by its original name, Idlewild, and will not walk on Boylston Avenue in Cambridge now that its name has been officially changed to John F. Kennedy Boulevard. When Jackie Kennedy visited the Botanical Museum, Schultes vanished. Rumor had it that he hid in his office closet to avoid having to guide her through the exhibits.

Schultes inherited his political views both from his conservative family and from Oakes Ames, head of the Botanical Museum when Schultes was a student. An aristocratic Bostonian whose family had made a fortune selling iron shovels on the American frontier, Ames was a scholar of the old school, securely positioned by class, instinctively elitist, inherently disdainful of the winds of democratic change sweeping America in the early years of this century. Schultes, whose father had a small family plumbing business, idolized Ames and absorbed his ideas and opinions. For the most part these political convictions were already archaic when forged by Ames himself almost a century ago. Though no doubt sincerely believed by Schultes, they in fact bear little resemblance to the profoundly democratic instincts by which he actually conducts his life. Like a boy stumbling around in his father's coat, Schultes awkwardly, sometimes foolishly, always amusingly, parrots the reactionary values of a ruling class that evaporated long ago.

These stubborn convictions, however rigidly held, belie the fundamental decency and kindness of the man. Schultes's disdain for liberal democrats and his contempt for government—he still calls Franklin Delano Roosevelt a socialist—are rooted in an intense dedication to individual freedom. On the issues that speak directly to personal choice —sexual orientation, abortion, use of drugs, freedom of religion—he is a complete libertarian. His devotion to struggling students is legendary. For years he used to travel around the country using an obscure taxonomic argument to obtain the release of dozens of young people charged with marijuana possession.

The argument went something like this: By law marijuana is illegal, but until recently, when the law was changed to defeat Schultes's crusade, the actual legal statute prohibited by name only *Cannabis sativa.* Schultes maintained that there were three species of marijuana, including *Cannabis indica* and *Cannabis ruderalis.* As an expert witness he would testify that there was no way of distinguishing the species with forensic material alone. That left the burden of proof on the prosecution to show beyond reasonable doubt that a bag of ground-up flower buds was *Cannabis sativa* and not one of its botanical relatives. Since even the botanists could not agree on how many species there were, it was by definition an impossible task. It made for great theater, of course, with Schultes and his entourage on the one side and arrayed against him a team of indignant botanists, often envious of his fame, infuriated by his stand on drugs, and openly contemptuous of his taxonomic position.

In truth, the evidence for Schultes's position was somewhat dubious. Marijuana is a multipurpose plant that has been used for over five thousand years as an oil, a food, a drug, medicine, and a source of fiber. The morphological variation that led him to recognize three distinct species may well be the result of artificial selection. In the heated passions of the times, however, when young students were being jailed for smoking an innocuous herb, none of these academic details were important. What did matter was Schultes's uncanny ability to break open the courtrooms and set the students free. This, as much as anything else he did, contributed to his mythic reputation on the Harvard campus.

Between the extremes of his personality, in the space created by what superficially appeared to be immense contradictions in his own character, there was room for anyone to flourish. His students ranged from quietly conservative, earnest scholars to a somewhat more unusual group drawn to his work on hallucinogens. By reputation the best of his students and his undoubted protégé was Tim Plowman. I had met Tim only once, very briefly, when I walked into his office in the basement of the Botanical Museum and found him hidden in a forest of living plants. He was tall and thin, strikingly handsome, with dark brown hair, a generous mustache, and warm smile. His office had the atmosphere of a gypsy tea room, oriental rugs on the floor, silk scarves coloring the lampshades, the musty scent of incense and patchouli oil. I can't remember why I went to see him. I do remember that in one corner of the room there was a beautiful woman naked to the waist typing a manuscript. Her name was Teza. She was an artist who lived with Tim. He later published her botanical illustrations of new species that he had discovered. They were the only drawings I ever saw that captured the feel of wind on paper.

Tim first arrived at the museum as a graduate student in 1966, and even before he was officially enrolled, Schultes sent him down to the Amazon. Two years later in Iquitos, a lowland town on the upper Amazon in Peru, he fell in with Dick Martin, another Schultes student, and together they cut a deal with a shady outfit called the Amazon Natural Drug Company, which gave them a boat and free rein to search for medicinal plants up and down the Río Napo. They collected together for several months, until the research director of the drug company arrived and displayed less interest in plants than in the whereabouts of Che Guevara. When it became clear that he didn't know the difference between a daisy and a palm, Martin and Plowman quit and left Iquitos.

Martin was the only botanist I ever heard of who took a saxophone into the field. By day he and Tim would collect furiously, then by night Dick would disappear into the bars and whorehouses of jungle settlements and blow sax till dawn. Sometimes when they were upriver, Martin would wander off and for much of the night Tim would hear soft plaintive music mingling with the haunting sounds of the rain forest. Schultes always referred to Martin as a genius. He often told of the indignant phone call he once received from a colleague complaining that one of his graduate students was doodling through every lecture of an advanced taxonomy class. Schultes looked into the matter and discovered that Martin was taking notes in Japanese.

Schultes's favorite story about Tim concerned a near fatal poisoning that occurred shortly after Plowman and Martin split up after Iquitos. By then Tim was on the trail of *chiric sanango, Brunfelsia grandiflora,* an important medicinal plant in the potato family used throughout the Northwest Amazon to treat fever. He had gone north to Colombia and in the valley of Sibundoy had contacted Pedro Juajibioy, a Kamsá Indian healer who as a boy guided Schultes into the upper Putumayo. Tim and Pedro retraced that journey to reach a group of Kofán Indians on the Río Guamués, visited by Schultes in 1942. The Kofán knew the plant as *tsontinba"k"á.* Another of Schultes's students, Homer Pinkley, who spent a year among the tribe in 1965, had written that their shaman occasionally ingested it in order to diagnose disease. This observation supported other reports going back to the nineteenth century that suggested the plant might be hallucinogenic. Plowman wanted to know. At Santa Rosa on the Río Guamués he found *chiric sanango* commonly cultivated in house yards, but he also met an old shaman who brought him from the forest a rare but related plant named for the tapir because of its inordinate strength as a drug. Plowman immediately recognized the plant as a new species, which he later named *Brunfelsia chiricaspi,* after the Quechua word meaning cold tree. He asked the old man to prepare it. The shaman refused. He described the plant as a dangerous messenger of the forest and disavowed any knowledge of taking it for visions. Tim persisted. Eventually the shaman agreed, though reluctantly and only on the condition that Pedro would also drink the preparation.

The drug, an extract from the bark, was murky brown and bitter to swallow. Tim felt the effects within ten minutes: a tingling sensation like one feels when the blood rushes back to a limb that has fallen asleep. Only in this case the sensation grew to a maddening intensity, spreading from the lips and fingertips toward the center of the body, progressing up the spine to the base of the skull in waves of cold that flooded over his consciousness. His breathing collapsed. Dizzy with vertigo, he lost all muscular control and fell to the mud floor of the shaman's hut. In horror he realized that he was frothing at the mouth. An hour passed. Paralyzed and tormented by an excruciating pain in his stomach, he remained only vaguely aware of where he was—on the earth, face-to-face with three snarling dogs fighting over the vomit that spread in a pool around his head.

The shaman, noticing his plight, did what shamans normally do under such circumstances: He went to bed. Desperate to escape the sensations, half-blinded by the drug and incapable of walking, Tim and Pedro stumbled and crawled through the forest for two hours until finally, toward dawn, they reached the village of San Antonio where they were staying in an abandoned jail. As light came to the forest, they crawled into their hammocks, where they remained, motionless, for two days. Pedro Juajibioy, whose experience as a folk healer had taken him on a thousand flights of the spirit, summed up the experience succinctly: "The world was spinning around me like a great blue wheel. I felt that I was going to die."

The train jolted along, stopping every so often, each time in what looked like empty countryside. But there were always voices in the dark and sometimes the faint light of shrines. At every station peddlers leaped aboard and pushed their way through crowded cars, prompting a babble of groans and protests from the crouched figures in the aisles. For a day and a night I lived on corn *arepas,* rounds of milk cheese sold wrapped in banana leaves, and small thimbles of *tinto,* a thick syrupy coffee that ought to be dispensed by syringe. It seemed like rich land, this northern flank of the Cordillera. The whitewashed houses that appeared out of the darkness had tile roofs draped in bougainvillea, and small patios where at all hours *campesinos* who looked like Juan gathered to drink and play cards. Always there seemed to be one who stood alone, still as a corpse, with vacant eyes protected from the night by a worn fedora and a blanket of wool.

There was a change of trains at Barrancabermeja, a small port on the Río Magdalena. The middle of the night, and the air was hot and humid. The first scent of the lowlands, moisture, and the slow flow of the river working its way around the trestles of the railroad, carrying pieces of the forest to the sea. On the platform passengers milled about in a confusion of bundles and boxes. Warm rain fell, and beneath the overhanging roof, three young barefoot boys slept beneath cardboard. Watching over them was a fourth, also barefoot, only he was standing in a pair of leather shoes several sizes too large. It looked as if the shoes belonged to them all. Ten hours to the coast, and already the silent explosions of lightning revealed the huge crowns of the ceiba trees.

CHAPTER TWO

Mountains of the Elder Brother

JUAN HAD BEEN right. Within minutes of arriving in Santa Marta I was fleeced by a pickpocket. I must have smiled, for moments later as I piled into a taxi, there was a commotion in the backseat. I turned to find my wallet on the floor. Only a few pesos had been taken. Colombia is not a nation of thieves, but those that there are have a certain flair difficult to begrudge. Once on Juan's farm, in the middle of one of the horrendous thunderstorms that periodically tumbled off the mountain, the power failed just as lightning struck a tall pine below the garden. Only the next day did we discover that a thief, perched on the electric pole and waiting for just the right moment, had cut the line and run off with four hundred feet of copper wire. Even Juan expressed a grim respect for the deed.

The taxi drove through a warren of modest houses that ran up to the public market and then turned back toward the sea, picking up passengers as it went along. On first glimpse Santa Marta seemed a pleasant, if somewhat shabby, seaside resort, a grid of white and pastel buildings laid out along a tranquil bay, backed by high shelving cliffs. Lighthouses overlook the harbor, and along the shore delicate palms shelter a promenade that separates the cafés and hotels of the waterfront from the black sand beaches that sparkle at night and stretch east a hundred miles to the Guajira. A dusty barrio spreads up one side of the city above the docks, and in the center of town, just four blocks from the sea, is a beautiful cathedral, said to be the oldest church in Colombia. Beneath the jumbled clutter typical of any Latin American town, the city has a rakish charm that makes it difficult to believe a man's life can be had for a hundred dollars.

It is a city, as one Colombian politician has discreetly put it, favored by geography. Midway between the coca plantations of Peru and Bolivia and the stateside market for cocaine, and serviced by dozens of clandestine airstrips scratched into the Guajira peninsula, Santa Marta exports the drug by the shovelful. Cocaine did not corrupt the city, however, it merely increased the take. Since the arrival of the Spanish, the people of Santa Marta have pursued a life of smuggling and trade. The place exists to make money. Besides drugs there are emeralds stolen from the mines of Boyacá, coffee lifted by the ton from the city docks, young women drawn from the interior to be bartered in the flophouses on Calle 10. From the lowliest peddler to the richest merchant, every native of the city shares a reverence for the deal.

In the early seventies, when no one could imagine a time when the continent would yield a million pounds of cocaine a year, marijuana was still the biggest game in town. The fertile lower slopes of the Sierra Nevada, blanketed by forest, well watered, and dissected by miles of ravines and hidden valleys accessible only to mules, produced the best harvest in Colombia. Before the development of domestic varieties slowed the demand for imported weed, and just after the Nixon administration blanketed the Mexican plantations with paraquat, ruining the market and sending enterprising smugglers to Colombia, Santa Marta was the dope capital of the world. On the waterfront loose joints of Blue Sky Blonde and Santa Marta Red, the Colombian cousin of Panama Red, could be had for a nickel, and the easy rhythms of the town drifted through a soporific cannabis haze.

• • •

The Residencia Medellín charged four dollars a night, twice the rate of the cheaper places on 12th Street, and though it was on the waterfront, the rooms were mostly empty. There was a reception desk of sorts, and behind it a *muchacho* named Lucho lounged in a hammock, his entire demeanor proclaiming that this was a day job. I asked for an American named Plowman. He placed a finger by the side of his nose. I shook my head. He pointed to a pair of doors that swung open onto a bright patio overlooking the sea. At one table a middle-aged foreigner in a fresh leisure suit sat nervously flipping the pages of a magazine, and just beyond two women were drinking beer. Tim Plowman was sitting alone, his back to the others, looking out toward the sea. There were topographical maps spread across his table.

"Dr. Plowman?"

"Yes," he said, looking up. I recognized his face. He seemed puzzled.

"My name's Willy. Actually, it's Wade, but no one down here can say it so I go by Willy. I just got in from Medellín. Schultes said you—"

"Wade Davis," he said, interrupting and rising out of his chair to greet me warmly. He was wearing a red T-shirt, jeans, and a single strand of white beads around his neck. I guessed he was about thirty, which at the time seemed old.

"I don't think we've met. I'm Tim Plowman. Forget about this 'doctor' business."

"Actually, we did meet once, in your office." I reminded him of the time I had visited him and Teza in the basement of the museum. At the mention of her name his face softened.

"Teza," he said in a manner that left me no choice but to ask what had become of her.

"She found Mr. Right"—he laughed—"and disappeared to the islands. She calls him the Texas Long Prong. They're out there somewhere. Last I heard they were sailing off the coast of Jamaica."

"So you split up?"

"Oh, no," he said without a touch of bitterness in his voice. "You don't ever lose a woman like her even if you want to. You just wait till she rolls back over your life."

"She always does?"

"Certain as a wave." He smiled. "But listen. I'm just glad you made it. Schultes has been calling."

"It took a while to get your telegram."

"You're in good time. We leave for the mountains in the morning.

Come have a look." I put down my rucksack and took a chair next to him. He gestured for a waiter.

"How about a beer?" he said. "Lucho, *por favor, dos Aguilas."* The kid from the hammock poked his face past the swinging doors and then slipped back into the shadow. "I spoke with Reichel-Dolmatoff in Bogotá. Do you know him?" Tim asked.

"Just by reputation." Gerardo Reichel-Dolmatoff was Colombia's foremost anthropologist and the leading authority on the peoples of the Sierra Nevada. A contemporary and close friend of Schultes, he and his wife had first lived among the Kogi in the early forties.

"He's an incredible man. And he warned me about the north slope. Yesterday I tried to get up beyond Masinga and Bonda. Right about here. I was turned back by military patrols. Day before yesterday I tried another road and ran into gunfire. Could have been anyone—grave robbers, the army, guerrillas. The forest is amazing, but it's dangerous. We'd be lucky just to *find* the Indians." He reached for a pack of cigarettes.

"Do you smoke?"

"No, thanks," I said. He was smoking Pielroja, a local brand. There was little breeze, and the heady, pungent scent lingered at the table.

"The Kogi live in the north." He pointed on the map to a number of river valleys draining directly into the Caribbean. "But Reichel suggested that we enter here, from the southeast."

Tim traced a route around the south of the sierra through the city of Valledupar and on to Atanquez, a small village where the road ended.

"In Atanquez we can get mules, and in a day we'll be able to reach the Ika settlements on the Río Donachuí. The Ika and Kogi are related, and there are a lot of similarities. They describe themselves as brothers, descended like two branches from the Tairona. Both tribes say they are the Elder Brothers, the ones charged with the protection of the earth, but the Ika are more willing to deal with the Younger Brothers."

"The Younger Brothers?" I asked.

"You and me and all of our kind who have screwed everything up. According to Reichel, the Ika and Kogi believe that the balance of life is maintained by their prayers and that as we cut into the earth, we are tearing at the heart of the Great Mother. He says they believe it literally. I don't know if he's just being romantic. You studied anthropology, didn't you?"

"For two years. Then I took time off and came down here."

"What do you make of what he says?" Tim asked.

"About the prayers?"

"Yeah."

"I don't know," I answered. "I suppose that's why the Kogi are up in those mountains, and the rest of us are down here in Santa Marta."

Lucho arrived and placed two large bottles of beer on the table. He took a cigarette from his mouth and snapped it into the sea.

"Señor Timoteo, *los gamines están aqui.*" The street kids are here.

"Bueno. Gracias, Lucho."

"What's going on¿" I asked.

"You'll see in a second. Just a little business."

"So what about the rains¿"

"They've already begun, but the worst of it falls in the north and west. There's a rain shadow in the south, and the slopes are a lot drier. There'll still be rain, and the rivers will be full. We'll be lucky to get to eight thousand feet. The Indians won't go much higher at this time of year."

"Where will we find coca¿"

"Below eight thousand feet," Tim said with a smile. There was some kind of noisy commotion behind us.

I turned to see Lucho sweeping four young urchins onto the patio and toward our table. They looked like the boys I'd seen sleeping on the train platform at Barrancabermeja, suitably stripped down for the coast. They were barefoot, dusty, dressed in rags, and smiling. Each lugged a flour sack. One of them had thick wavy black hair, the other three had only a thin stubble of hair.

"They shave their heads in jail," Tim said, turning to the tallest of the boys. "Hey, *flaco, qué pasa¿"*

The boy shrugged and lifted one of the bags onto the table. It was full of seashells. Tim felt the weight of each bag and then paid each boy, slipping a little something to Lucho who was standing by with an exaggerated air of indignation.

"Seashells¿" I said.

"Twenty kilos," Tim answered. I didn't pursue it. I needed a shower. The beer, the midday heat, the train trip, and now the prospect of this journey swept over my senses. I longed for sleep.

"It's getting on," he said. "Let's get you settled in a room. Later we'll walk around town, maybe get you a haircut. We have to look pretty straight. I know a barber down past the cathedral. He has only one arm, but he does a good job."

There was a knock at my door around five. I joined Tim in the lobby, and we headed along the waterfront for several blocks and then crossed into the shade of the *guayacán* trees in the Parque Bolívar. The sun was still hot, but there was a lot more activity than I had seen at midday. Shoeshine boys darted about, old women gathered to gossip, and beneath the shadow of an enormous statue of Simón Bolívar the photographers had set up their stations, pinhole cameras made of cardboard and tin perched precariously on wooden tripods. Each photographer displayed samples of his work, small black-and-white images, mostly of *campesinos,* young girls and sweethearts, all of them posed ramrod stiff as if terrified by the process.

The one-armed barber worked in a small shop with a red leather chair and a large mirror, cracked in several places. A friendly sort, he cut with his right hand, kept the comb beneath the stump of what had been his left arm, and traded comb and scissors back and forth with a dazzling dexterity that almost made you forget he was greasing your hair with his sweat. He snipped away for ten minutes and then pulled out a razor and shaved off the rest. I looked like a military recruit. I asked him why he hadn't sought other work after he'd lost his arm.

"I did," he answered. "That's why I became a barber."

Tim had left to run some errands, and after the haircut I decided to see the cathedral before heading back to the waterfront to meet him for dinner. Once inside it took a few moments for my eyes to adjust to the dark. The interior was an odd mixture of styles: opulent Baroque elements welded onto the starker features of an earlier period. To one side of the church doors I came upon the ashes of Rodrigo de Bastides, the Spaniard who founded Santa Marta in 1526. That was six years before Francisco Pizarro's rape of Peru, and seven years after Hernán Cortés and his men had stood stunned by the beauty and majesty of Tenochtitlán, the Aztec imperial capital then twice the size of Spain's largest city. Under orders from his king, Rodrigo de Bastides had reached the coast of South America in 1501. Moving westward from the Guajira and drawn to the snowcapped mountains that soared higher than any in the known world, he met the Tairona, the most elaborate civilization the Spaniards had encountered up to that time. A chiefdom descended from peoples who originally migrated to South America in the tenth century from the Atlantic slopes of what is today Costa Rica, the Tairona had transformed the slopes of the Sierra Nevada, establishing roads, architectural terraces, and irrigation systems of stunning complexity. Dazzled by their gold work, which was perhaps the most sophisticated and beautiful ever produced in the Americas, the Spaniards established a series of trading posts, including Santa Marta, which in time emerged as the dominant center.

For a hundred years, as the Conquest swept across the continent, an uneasy truce hung over the northern coast. There was conflict and rebellion, and death by enslavement and disease, but the Spaniards made no systematic attempt to destroy the Tairona. Few in numbers, they were initially content to control the coast, trading fish and salt, axes and metal tools for gold. The Tairona valued the peace even as they retreated further into the hinterland.

It was not until the end of the sixteenth century that the Spaniards launched a campaign of annihilation. Their excuse—and the Spanish, obsessed as they were with jurisprudence, always had an excuse—was completely bizarre. Hungry for gold, they were nevertheless scandalized by the phallic and sexual representations that formed a significant motif in Tairona ceramics and gold work. The chronicler Gonzalo Fernández de Oviedo described a gold piece weighing twenty pesos that depicted "one man mounted on another in that diabolical act of Sodom," a "jewel of the devil" that he righteously "smashed at the smeltering house at Darien." Such graphic depictions of sodomy confirmed their deepest suspicions. It was known that Tairona men gathered regularly in large ceremonial temples, often for nocturnal rituals that lasted until dawn and excluded women. From experience the Spaniards recognized that when their own sailors and soldiers spent long hours together, it was only the restraint of Christian virtue that kept them from "unnatural acts." Since the Tairona were not Christian, it was obvious, at least to the Spanish, what the Indians had been up to at those nightly assemblies. When in 1599 Santa Marta's new governor, Juan Guiral Velon, undertook the final destruction of the Tairona, he did so charged with the certainty that all of his enemies were homosexual.

The subsequent struggle was as violent and brutal as any recorded in the Americas. Tairona priests were drawn and quartered, their severed heads displayed in iron cages. Prisoners were crucified or hung from metal hooks stuck through the ribs. Those who escaped and were recaptured had their Achilles tendons sliced or a leg cut off. In Santa Marta, Indians absurdly accused of sodomy were disemboweled by fighting dogs in obscene public spectacles. Women were garroted, children branded and enslaved. Every village was destroyed, every field burned and sown with death. When the Spaniards took the Tairona settlement of Masinga, Velon ordered his troops to sever the noses, ears, and lips of every adult.

Marching inland, Velon attempted to vanquish an entire civilization. In the midst of the carnage, the Spaniards never forgot their ultimate mission. To ensure the legality of their deeds, before each military action Velon's captains read aloud in the presence of a notary public the famous Requirement, a standard legal document exhorting the heathen to accept the true faith. Recited in Spanish without translation, it was but a prelude to slaughter. "If you do not accept the faith," the text read, "or if you maliciously delay in doing so, I certify that with God's help I will advance powerfully against you and make war on you wherever and however I am able, and will subject you to the yoke and obedience of the Church and of their majesties and take your women and children as slaves, and as such I will sell and dispose of them as their majesties may order, and I will take your possessions and do all the harm and damage that I can."

The Spaniards were true to their word. In the end the entire Tairona population was either dead or given over as slaves to the soldiers as payment for their services. Those Indians who survived were expected to pay the costs of their own pacification. On pain of death they were prohibited from bearing arms or retiring into the Sierra Nevada. But flee they did—a tragic diaspora that brought thousands into the high mountains, leaving behind a desolate, empty coast of ruined settlements, shattered temples, and fields overgrown with thorn scrub and ultimately redeemed by the forest.

Knowing this history, one can almost too easily hate these men of Spain, too easily forget the kind of world in which they had lived. Rodrigo de Bastides and every other Spanish conquistador grew up in a land convulsed in triumph and terror. After eight hundred years of war the Spaniards had at last retaken Andalusia, driving out the Moors, expelling by edict the Jews. A Spaniard was Pope and Isabella, queen of Spain, was patroness of the Holy Inquisition. Christianity had swept Europe only to turn upon itself in bloody sectarian wars. Millions had died, by plague and war, and in the fires of the Inquisition that reduced to ashes anyone who failed to embrace the faith and power of the priests.

They were men of violent ideas. None among them doubted that the end of the world was at hand, that they would all at one time have to endure the purifying flames of a final judgment. One of their saints, Thomas Aquinas, declared that, next to contemplating God, the greatest pleasure in the afterlife would be watching the tortures of the eternally damned. Men who denied themselves by day the pleasures they so avidly sought by night drifted easily into perversion. From the pulpit celibate priests advised men to beat their wives regularly, not in rage but in charity for their souls. Midwives were burned because they eased the pain of childbirth, which the Church maintained was God's punishment for Eve's original sin. Young girls were shrouded in dark robes to do penance for sins they did not commit, joys of the flesh they would never know. Any woman who healed was a witch, and black-frocked inquisitors, having transformed the devil into the nemesis of their god, discovered him everywhere in bed with women. Twisted in torture, mothers and wives from all corners of Europe admitted that they had, in fact, mated with the devil, and they revealed that the priests had been right. The devil's member was as cold as ice.

The men who crossed an ocean to conquer America were the ones that Europe, for all its depravity, could not kill. Weather-beaten and torn by the wind, Columbus arrived off the shore of Santa Marta in 1494 still carrying with him a carefully annotated copy of Marco Polo's journal, still certain that he had reached the lands of Kubla Khan, still anxious to present to the Great Khan his credentials, written in Latin on parchment that he could not read. Those who followed him knew better. This land of demons, of birds with teeth and fish that flew, was one vast kingdom of the devil. To destroy was to serve God, a glorious mission made sweet by the presence of gold. Their Church embraced all their deeds. The rape of children, the violation of the earth, the destruction of all that was beautiful could be condoned by the halo of the faith. Men who had sex as if relieving themselves declared all native women to be whores, and branded the faces of children while the Pope debated whether or not they were human beings. Priests who exhaled disease declared pestilence to be the will of God. In their wake they left death. Three million Arawakans died between 1494 and 1508. Within 150 years of Columbus the aboriginal population of 70 million would be reduced to 3.5 million. In the southern Andes of Bolivia, on a mountain of silver once sacred to the Inca, an average of 75 Indians were to die every day for over 300 years.

From beneath the blue awning of the Pan-American bar you could survey the entire waterfront of Santa Marta. The sun was going down. There was a naval vessel at the dock, and in the bay just offshore a tanker with Liberian markings was flushing its hold into the harbor. I found Tim reading a newspaper at one of the tables close to a small bandstand. He had already finished a beer.

"Look at this." He passed me the paper. "It's a couple of years old. I noticed it when I went down to the depot to get newsprint for the plants." Beside a notice for an upcoming performance of *El Maestro de Obscuridad,* a magician from Bogotá who claimed to be able to predict the future, there was a small item about tomb robbers. They had formed an association and wanted the Ministry of Labor to recognize it as an official union. In Santa Marta alone the association had registered ten thousand members. Though it contradicted every law concerning the protection of archaeological sites, certain authorities within the Ministry had initially agreed, thus provoking a scandal.

"A union of looters?"

"The government finally killed it," Tim said, "It's crazy. But that's what this coast is all about. One decent artifact, and they can make it for a year. That's why everyone ends up here. There's a feel of easy money. In the rest of the country a handful of people own practically all the land. So if you're young and poor, and you don't want to work in some cane field, and you don't have a cousin who can get you a job on the trucks, you dream up some little business, buying and selling fruit, peddling fish or bread, and when you get tired of haggling with housewives, you drift down the Magdalena to Santa Marta."

"Like the guy who cut my hair?"

"He blew off his arm with a stick of dynamite," Tim explained. "He used to be a fisherman. He and Lucho at the *residencia* are cousins." At the next table two British sailors were rushing through their food. One of them kept looking at his watch. A night in Santa Marta and neither one of them wanted to waste time eating.

"Freaks come here when it dawns on them that for the price of a couple grams of coke back home they can fly to Colombia and pick it up by the kilo. Some figure out the scene. For them it's an easy life. For others it becomes a kind of hell. If they're lucky, they just get rolled on Calle 10. If they get behind on a deal, they can end up dead."

He glanced past the sailors toward the entrance to the patio where an agitated German tourist was arguing with a waiter. "You can always spot the ones just off the plane." He smiled, turning back to our table. "Walking down the beach beating off the kids. Fighting over every bill. Teeth grinding away. They have a kind of sweaty anxiety that comes with the first taste of decent cocaine."

A waiter brought our dinner, beer and two plates of fish and rice. For a few minutes we ate without speaking. A band started up, and a woman in a satin dress began to sing wearily. I looked up at Tim.

"It's hard to believe the Tairona were once here," I said.

"I know," Tim replied. "You think of this town and then try to imagine priests in cloaks woven with gold and jewels, feather headdresses. Beautiful fields of plants." He stopped eating, looked to the sea, and then turned back to me. "I'd like to know more about them—how they lived, what they thought. Have you ever paid attention to language?"

"In what way?" I asked.

"The choice of words. What they mean. There's a tribe in Uruguay, one of the Guaraní groups, whose word for soul was 'the sun that lies within.' They called a friend 'one's other heart.' To forgive was the same word as to forget. They had no writing, and when they first saw paper, they called it the skin of God—just because you could send messages."

"Like magic."

"It was magic," Tim said. "Did Schultes ever tell you about the Indians in the Amazon who couldn't tell blue from green? I forget the tribe. I asked him whether they saw the same color or whether they just considered the two colors to be one."

"What did he say?" I asked.

"He didn't know. I don't think he ever really thought about it."

"But you have," I said. Tim laughed.

"Reichel talks about all this. In one of his books he says the Tairona believed that gold was the blood of the Great Mother. He says the Kogi word for vagina is the word for dawn. Can you imagine what it means for a people to have such thoughts?"

"No," I said.

"I can't, either." He smiled. "Listen. Let's get the bill and get out of here. We've got an early start."

Tim's truck was magnificent, a bright red Dodge 4x4 pickup with a small camper on the back. Perhaps it was the early morning light or the novelty of leaving a town without having to deal with the rickety contraptions that pass as buses in Latin America—rusted-out chassis, rear seats that double as urinals, tires smooth as river stones—but as we drove along the Caribbean shore away from Santa Marta, I felt a perfect ease. There was nowhere else I wanted to be. The land was rich, the mountains in the distance a pale velvet on the horizon. As we turned south past the shallow waters of the Ciénaga Grande, a vast lagoon that seems as large as the sky, the light was still soft, salmon-colored, and the air and water were alive with egrets, herons, and doves the color of earth.

We drove south quickly, without stopping, past scruffy plantations of bananas, cotton, and oil palm. There was forest here a century ago, and the towns along the road—Tucurinca, Aracataca, Fundación—were lost villages which took their names from the rivers that drain the western flank of the Sierra Nevada. Today they all look more or less the same: whitewashed houses, dusty plazas, gray bridges over rivers of dirty water. But just beneath the surface is a history of betrayal and death. In the early 1920s the United Fruit Company brought in bananas, and with the railroad and telegraph, the roads, post offices, police stations, brothels, and bars came thousands of migrant workers to hack away at the land. Living in sheds, drinking water fouled by disease, they earned less than a dollar a day—money that quickly disappeared in company stores where shopkeepers with crooked scales robbed them shamelessly.

Then, in 1928, the field hands went on strike. Bananas rotted on the stem, trains ceased to run, and in the harbor of Santa Marta ships outbound for Boston hung empty on their anchors. The workers and their families were encamped at Ciénaga, awaiting the formal signing of the agreement that would end the strike. Meanwhile, in Aracataca the company and the army cut a deal. The next morning in place of an agreement, a general dictated an ultimatum. Before the children could be moved, before the old women were even awakened, the machine guns exploded. The banners fell and the dead blanketed the plaza. The army and company thugs worked throughout the night, washing the blood from the ground, tossing the dead into the sea. By dawn there was no sign of life, no sign of death.

Those who survived fled south to Aracataca. There they were hunted down, wounded men and terrified children. One hundred and twenty-five were shot in the graveyard before the eyes of a desperate priest. Just blocks away an infant boy slept. Fifty years later Gabriel García-Márquez transformed Aracataca into Macondo, the setting of *One Hundred Years of Solitude,* his novel of despair and hope, where life is scattered by the wind and people effervesce into angels. There is nothing today in Aracataca to remind you of its past, little to suggest it could inspire such a novel. Oil palms and mango trees baked by the sun, dirt paths leading off into plantations, schoolchildren in bright uniforms scurrying about. As we drove out of town a young boy stared at our shiny truck as if it were an apparition.

The land became drier as the road turned east around the southern side of the mountains. Plantations gave way to chaparral, acacias and tall, looming cacti, flocks of green parrots that seemed oddly out of place amidst the barbed wire, desert haze, and dusty horse tracks. Kapok trees soared over the homesteads, and on the bare hills tabebuia trees were in full flower, their crowns leafless and brilliant yellow. Mangos were in season, and we stopped every so often at roadside stands where children sold guanabana fruit, iguana eggs, and bread. Pineapples went for a dime, bananas a penny each. Once or twice we pulled over to look for plants, but we collected very little. The land was scrub, the vegetation scavenged by cattle and mules. There were always vultures flying overhead, and to the north the ascending ridges of the Sierra Nevada tearing at a violently blue sky.

Throughout a long day Tim and I traded stories about our lives. I spoke of growing up in British Columbia, working in logging camps and fighting forest fires to earn enough money to help pay for college. He told of his early years rambling through the woods of Pennsylvania, of his parents' love of gardening and the herbarium specimens he collected as a young boy and used to decorate the walls of his room. His father was a doctor, and Tim might have gone into medicine had it not been for his interest in plants. He had just one brother, but they were not close. Like so many families of that era they found themselves on opposite sides of a social divide created by the times. His older brother, John, had married his high school sweetheart, started a family, and found a job selling insurance in their hometown of Harrisburg. Tim went off to Cornell, took a heroic dose of acid in 1964 when it was still dispensed in sugar cubes, made love to a girlfriend in the university library, survived the official reprimand, and avoided Vietnam by heading directly to graduate school at Harvard in 1966. Along the way he took up music, painting, yoga, and, of course, Teza. For years they had lived with a group of friends in an old mansion in Roxbury, the inner-city ghetto of Boston. For thirty dollars a month everyone had his own apartment, though from the sounds of it Teza reigned over the place like Aphrodite. She was in love with both Tim and his friend Craig, and in the midst of their liaison had married a third friend, Aharon, an Israeli physicist. Everyone involved had attended the wedding, a simple affair held in the Dudley Street subway station.

As the afternoon wore on, the conversation turned to botany and in particular a new book that made a great fuss about house plants responding to music and human voices. For Tim the very idea was ridiculous.

"Why would a plant give a shit about Mozart?" I remember him saying. "And even if it did, why should *that* impress us? I mean, they can eat light. Isn't that enough?"

He went on to speak of photosynthesis the way an artist might describe color. He said that at dusk the process is reversed and that plants actually emit small amounts of light. He referred to sap as the green blood of plants, explaining that chlorophyll is structurally almost the same as the pigment of our blood, only the iron in hemoglobin is replaced by magnesium in plants. He spoke of the way plants grow, a seed of grass producing sixty miles of root hairs in a day, six thousand miles over the course of a season; a field of hay exhaling five hundred tons of water into the air each day; a flower pushing its blossom through three inches of pavement; a single catkin of a birch tree producing five million grains of pollen; a tree living for four thousand years. Unlike every other botanist I had known, he was not obsessed with classification. For him Latin names were like koans or lines of verse. He remembered them effortlessly, taking particular delight in their origins. "When you say the names of the plants," he said at one point, "you say the names of the gods."

Valledupar was a hot, dusty city of cowboys and pickup trucks and bars blasting *vallenatos,* the same raucous accordion music that had kept me awake on a dozen all-night Colombian bus trips. *Vallenatos* originated in Valledupar, which may be one reason the city seemed so dismal. We arrived at dusk and took a room at the Residencia Yavi, one of the few cheap hotels that was not a whorehouse. Rooms went for a dollar, $1.50 with a fan.

From Valledupar there are two routes into the Sierra. In either case, to enter land that as far as the Indians are concerned doesn't belong to Colombia, it is necessary to obtain a mound of permits, a letter from the Casa Indígena, the office in charge of Indian affairs, a permit from the conservation agency Inderena, and police clearance from the DAS, Departamento Administrativo de Seguridad, the national security forces. Even the mayor of Valledupar has a piece of the administrative action. Tim and I spent the better part of a day securing the necessary papers, and it was late afternoon by the time we finally reached Atanquez in the foothills of the Sierra some twenty-five miles northwest of Valledupar.

The town was small, again mostly whitewashed houses with thatch roofs, a bare plaza, and a tin-roofed church. A single road climbed away from the plaza and after a few blocks splintered into a series of mule tracks that headed up into the mountains. An Indian village but a generation before, Atanquez had become a *mestizo* settlement, with government officials and black-frocked priests, *campesinos,* and merchants dealing in coffee and maguey, the tough fiber extracted from the leaves of agave, the century plant.

By nightfall we had a place to stay and had met Aurelio Arias, a mule skinner and trader who for four dollars a day was willing to guide us into the mountains. He charged more for his mules—five dollars each a day. We hired two and told him we expected to be in the Sierra for two weeks, perhaps longer. To avoid the heat and ensure that we would reach the Ika settlement of Donachuí before the afternoon rains, we planned to walk by night, leaving Atanquez *de madrugada,* in the early hours before dawn on the following day.

• • •

The rains came, lasting well into the night. It was past three when I heard the clip-clopping of Aurelio's mules and his quiet profanities as he lashed our gear to the wooden pack saddles. By then the sky had cleared, and as we walked away from the village, the air tasted fresh and healthy. The track ran to the north and west, climbing gradually through scrub forest and crossing numerous streams swollen with the rains. At first it was quite dark, and it was all we could do to follow the mules. I became conscious of sounds and the whirls of odors that drifted quickly past, the smell of burned grass, dry stones washed by the rain, the sudden quelling scent of a dead animal in the brush. Once the moon had risen over the distant ridge of the mountains, the walking became much easier. The land was pale, and the moon cast long silvery shadows past tall columnar cacti. The tips of the acacia branches, still wet with rain, sparkled like sea spray.

An hour or so from Atanquez the track dropped onto the Río Guatapurí and turned south through a scattering of houses known as Chemesquemena. As we approached, dogs growled and roared and lunged at the mules. A single hissing curse from Aurelio sent them packing, tails curled miserably between their legs. One house had begun to stir and a few people moved about, their silhouettes showing past the kerosene lamps and through the mosquito netting that shut out the night.

From Chemesquemena we crossed the Río Guatapurí by footbridge and then climbed steadily through plantings of coffee and maguey. In time the frogs and cicadas and remnant forests of the valley bottoms yielded to a burning landscape of barren hillsides, treeless slopes covered with coarse grass and blackened boulders. We picked up a new trail, a hard and beaten track that led up a steep nose and thence onto a ridge. Another hour or more of steady climbing led to a high saddle. By then the sky had begun to lighten. The clouds took on a luminous appearance, and it was impossible to distinguish clouds from sky.

The rising sun touched the flank of the mountains, casting long shadows across almost imperceptible undulations on the earth. The shadows drew in toward dawn, quivered at the last moment, and gave way to a river of sunlight that poured impartially over every slope. A white sun and every color of the sunset returning in softer hues. There was mist in the valleys, and clouds enveloping the summits where the snow fields lay. To the northwest the mountains fell away to the sea, and shimmering on the horizon was an open expanse of the Caribbean. On the immediate hillsides, small puffs of dust scurried in the wind—foxes or perhaps brocket deer. There were hawks everywhere, and at one moment an enormous condor hovered before us, its cruel head ringed with white feathers, its fingerlike pinions outstretched on wings eight feet wide. Below and ahead of us lay the Río Donachuí and the lands of the Ika and Kogi.

"Look," said Tim. On a distant ridge there were two figures silhouetted against the sky. They looked like women, with long hair and robes, and they held something in their hands. But they were men.

"Se los lleva el sol," Aurelio said. They are being taken by the sun. He turned back to his mules. *"¡Mula! mula! Macho carajo!"* he yelled, slapping the animals' flanks with a switch and pushing them ahead on the trail to the valley below.

When Gerardo Reichel-Dolmatoff first went into the mountains, the Kogi told him a story about the birth of the world. In the beginning, they explained, all was darkness and water. There was no land, no sun or moon, and nothing alive. The water was the Great Mother. She was the mind within nature, the fountain of all possibilities. She was life becoming, emptiness, pure thought. She took many forms. As a maiden she sat on a black stone at the bottom of the sea. As a serpent she encircled the world. She was the daughter of the Lord of Thunder, the Spider Woman whose web embraced the heavens. As Mother of Ice she dwelt in a black lagoon in the high Sierra; as Mother of Fire she dwells by every hearth.

At the first dawning, the Great Mother began to spin her thoughts. In her serpent form she placed an egg into the void, and the egg became the universe. The universe was to have nine layers, four of the nether world and four of the upper world, with the plane of contact being the fifth, the central world of human beings. The four nether worlds were created first. Then the four upper worlds, each resplendent with the light of its own sun. The fifth layer, the plane that links the upper and lower halves of the universe, is sun-earth/night-earth, the land of human beings, the junction between the cosmic realms.

When the Great Mother conceived the nine-layered universe, she fertilized herself by anointing one of her pubic hairs with her menstrual blood and then by impregnating herself with a phallic lime-stick. She gave birth to Sintána, a black-faced jaguar, the prototypic human being. Then Sintána placed one of his mother's pubic hairs, a sliver of her fingernail, and a necklace of red stones on his mother's navel. With his coca stick he pushed them into her body, and thus she became pregnant with the Lords of the Universe, the four cardinal points, the zenith, nadir, and center. The Lord of the Zenith is the sun. The Lord of the

Nadir is the black sun, older brother of our sun. As soon as our sun sets below the horizon, this lord of darkness appears, a black sun that shivers like a darkling moon.

At the first dawning the universe was still soft. The Great Mother stabilized it by thrusting her enormous spindle into the center, penetrating the nine layers of the world axis. The Lords of the Universe, born of the Great Mother, pushed back the sea and lifted up the Sierra Nevada around the world axis, thrusting their pubic hairs into the soil to give it strength. Then the Great Mother placed potsherds on the surface and from her spindle uncoiled a length of cotton thread with which she traced a circle around the mountains, circumscribing the Sierra Nevada, which she declared to be the land of her children. Thus the spindle became a model of the cosmos. The disk is the earth, the whorl of yarn is the territory of the people, the individual strands of spun cotton are the thoughts of the sun. The white cone of yarn represents the four layers of the upper world, but below the disk the cotton is black and invisible. The sun in moving around the earth spins the yarn of life and gathers it about the axis of the cosmos, the mountains of the Sierra Nevada, the homeland of the Great Mother.

The trail met the river at a grove of fruit trees. The water was sweet and cold, the riverbed flush and full of massive white boulders that had been molded into beautiful shapes by the water. Some of them were larger than the deserted houses we had passed, wattle and daub structures with thatch roofs and small courtyards surrounded by walls made of river stones. From the ridge the valley had appeared unkempt and wild, but from within it seemed much gentler. There were remnants of the original vegetation, but the land had been worked and reworked for generations. Most of the trees had been planted for their fruit. There were mangos and avocados, *lucma,* guanabana, and beautiful *ingas* with their spreading delicate branches. The valley had all the splendid chaos of an Indian garden, but it was not a forest.

The trail up the Río Donachuí rose through plantings of corn and sugarcane, plantains, cotton, beans, squash, and peppers. Sweet manioc was grown, as well as native root crops such as *arracacha* and *xanthosoma.* It was still early. Smoke rose through the thatch of the houses, but the courtyards remained empty, save for small pods of children who laughed and giggled and scattered like chickens as we walked by. At one cluster of houses several men and women were working a large sugar press. It was made of wood, with three vertical rollers linked by a horizontal pole to the back of a mule. As the animal slowly circled the press, the rollers turned and ground the cane. The juice trickled into buckets, which the women carried to a pair of large iron cauldrons hanging over an open fire. Beneath a shelter two men were pouring the syrupy contents of a third cauldron into wooden molds. Beside them, stacked like bricks, were dozens of blocks of raw sugar. The men glanced at us as we passed, then returned to their work. They did not seem openly hostile, just fearful and suspicious. A few minutes later an old woman passed us coming down the trail. She wore a long frock of homespun cotton wrapped at the waist. There were dozens of coils of wine-red beads around her neck, and her head and black hair hung forward, burdened by an enormous load of firewood that she carried in a basket with a tumpline that ran across her forehead. She mumbled a greeting as she hurried past.

There was a portal at the entry to the village of Donachuí. It was unlocked, but a young Ika man blocked the trail. He was astonishingly handsome with fine bronzed features and black hair flowing down past his shoulders. He wore a white cotton *manta* pulled over the head so that it fell to the knees back and front like a tunic, held at the waist by a belt of fiber. His leggings were of the same rough cotton cloth; his sandals were cut from a rubber tire, and he wore on his head a fezlike hat of woven sisal. Across each shoulder hung a *mochila,* a woven bag decorated with brilliant geometric designs in crimson, the same color as the stripe that ran down his tunic. In his left hand he held a small bottle-shaped gourd. He had a thick quid of coca in his cheek.

"Buenos días, compadre," said Aurelio. The man returned the greeting. We all shook hands. For the Ika it was an exotic gesture. He touched our hands lightly, almost reluctantly, and then introduced himself as Adalberto Villafañe. He spoke in Spanish, with the carefully chosen words of one speaking a foreign language.

"These gates are to keep us apart," he said.

"They have their papers," Aurelio said, turning to Tim. "Show him your papers." Tim ignored him.

"We have come for only a few days," Tim said, "a very long way in order to know your plants." The Ika reached into a double calabash and scooped out a dab of a thick syrupy paste that he rubbed onto his teeth.

"To go higher you must have permission," he said.

"Of course," Tim replied.

"Then please come." He turned and began to walk up toward the village. "You are strangers," he said. "You must always enter through this gate."

Adalberto led us up a trail that passed through a small field of cotton and then disappeared into a dense planting of coca. The bushes were almost ten feet tall, scandent with small white flowers, red fruits like those of a barberry, and leaves of a brilliant yellowish green hue that set them apart from the other plants in the garden. Adalberto stopped for a moment and turned to Tim.

"This is *hayo,"* he said. He opened his *mochila* and lifted a small handful of the dry leaves, which sifted gently through his fingers.

Most of the village had gathered on a small patio outside the *kankúrua,* the ceremonial lodge that dominated the village. Tim took a small pack off one of the mules and told Aurelio to move the animals ahead toward a copse of eucalyptus just beyond the settlement. Adalberto then led us through a low door into the *kankúrua.* It was dark inside. A fire had burned down, leaving a thick pall of smoke that lingered above the ground. It took a moment for my eyes to adjust to the light. Several Ika men sat in a tight circle around the remains of the fire. Behind them were other men sitting on squat four-legged stools carved from single blocks of wood. Staffs lay at their feet. One old man lay in a hammock strung between two posts. Between the central fire and the doors were four other hearths. Smoke-stained feathers and animal skulls hung from the rafters. The roof was high and vaulted, but it was impossible to see beyond the beams. Everything was black with soot.

Adalberto walked slowly around the room and, without speaking, exchanged small handfuls of coca leaves with each man. Each in turn placed a few leaves into the open mouth of the other's *mochila.* After he had formally greeted everyone, Adalberto turned toward the hearth and tossed a small offering of coca into the fire. He introduced one of the elders as the *comisario* and then moved aside, taking his place on a small stool by the fire, carefully tucking the tail of his tunic between his legs as he sat. The *comisario* asked to see our papers. Tim withdrew from his pack a number of documents, including one embellished with ribbons and a wax seal. The old man read the letter aloud, paused, and then launched into a long monologue. Each man spoke in turn, and for thirty minutes the strange whispering sounds of the Ika language filled the room.

I watched Adalberto cast a spent quid into the fire and take more leaves from his *mochila.* From his poporo gourd he removed a lime-coated stick, which he placed in his mouth. He bit down gently and withdrew the stick, now wet with coca juice. Reflexively, he began to rub the head of his gourd with the tip of the stick.

The *comisario* began to speak in Spanish, reciting a litany of abuses suffered recently by the Ika: tourists who had trespassed on their land and used houses as latrines, Japanese mountain climbers who had broken up a door to kindle a fire. Tim assured him that we would be respectful of his community and wishes. There was mention of an administrative fee and various taxes, which we agreed to pay. Finally, the old man reached into a ceramic vessel and withdrew a lined notebook. With an air of resignation he wrote in elaborate script a letter that was duly signed by a number of his associates, granting us permission to spend three weeks in the mountains. Tim thanked him and, turning to leave, pulled from his pack a skein of red wool and a small bag of seashells, which he placed before the fire. A vague murmur of approval ran around the circle.

"To them you are the sowers of disease and misery," explained Aurelio, "so you must enter their land with patience."

It was difficult to be patient. Leaving Donachuí we had climbed to Sogrome, a second Ika settlement two miles upriver from Donachuí. For two mornings we had sat in and around an abandoned stone hut, waiting for an appropriate emissary from the Ika. The houses across the river were occupied—we could see smoke seeping out of the thatch roofs—but so far our only contact since leaving Donachuí had been a few inquisitive children and an old woman, her neck draped in beads, who had strayed into our camp looking for a missing chicken. We spent our days collecting plants and waiting for the rain to cool the air, and our afternoons and evenings preparing our specimens, waiting for the storms to pass.

Coca grew in abundance between Donachuí and Sogrome, but Tim deliberately showed only passing interest in the fields. He wanted to bide his time and not give the impression that we had arrived only to study this most revered of plants. He did explain the purpose of the seashells. To chew coca, or at least to absorb efficiently the cocaine in the leaves, one must modify human saliva by the addition of alkali. Any basic compound—baking soda, ash, limestone—will do, but the Ika and Kogi prefer burned seashells, which they acquire by trade or gather as part of elaborate pilgrimages to the ocean. They call their lime *impusi.* It is extremely caustic and must be applied to the moist quid of leaves with a stick, or *chukuna,* taking some care to avoid burning the mouth. To control the amount of lime, the tip of the *chukuna* is dried by vigorously rubbing it on top of the gourd. As the saliva evaporates, a layer of bright yellow lime is deposited around the mouth of the gourd, adding to the circumference of its ever enlarging head. The size, shape, and color of a man's *yoburu,* his lime gourd, is a matter of immense prestige and status. Symbolically the *yoburu* is a vagina, a "little mama." The stick or *chukuna* that penetrates the gourd and applies the lime to the coca quid is analogous to a penis. Just as a man fertilizes a woman, so the lime empowers the sacred leaf.

For Ika and Kogi men the chewing of coca is the purest activity of their lives. Women, by contrast, though expected to harvest and prepare the leaves, are forbidden to use the stimulant. The result is an axis of tension around which much of Ika and Kogi social life revolves. A man loses his virginity as a youth by copulating with the old widows of the tribe. His symbolic initiation into manhood does not occur, however, until he is initiated and ready to marry, for it is only then that he is allowed to taste the bittersweet leaves of coca. At that time the *máma,* the high priest, presents the bride with a carved wooden spindle with which to spin the thread to weave her husband's first coca bag, his *mochila* or *zijew.* The priest then selects a *yoburu* for the groom. At the marriage ceremony the priest perforates the *yoburu* and implants the lime into the bulbous base of the gourd. In the company of the groom, he then copulates with the bride, initiating her into womanhood. Just as marriage brings together man and woman, so it weds the groom to a life of chewing the sacred leaves.

Late in the afternoon of our third day at Sogrome I returned from gathering water to find Adalberto standing inside the door of our hut. At first he said nothing and made no movement. Aurelio, our mule skinner, sat to one side gnawing on a piece of sugarcane. Tim, who had been preparing our morning collections, stood before a half-dozen packets of pressed plants, wrapped in newspapers and soaked in a mixture of alcohol and formaldehyde. Crumpled newsprint and discarded foliage littered the courtyard. Adalberto moved close to Tim, his eyes somewhat scornfully passing over our specimens. In his right hand he held a *chukuna* that he kept rubbing around the head of his lime gourd. Leaning over the bags of preserved plants, he winced and fell back. Looking up at Tim, he spoke softly in Spanish. "These twigs will never grow. They are drunk, and besides, you need the seed."

The next morning and for two days after, Adalberto came with us as we foraged for plants. No money was exchanged, no official arrangement made, yet each day he arrived at our fire, shared breakfast, and then led us into the remnant patches of forest that remained on the steep side hills and in the moist ravines above and around Sogrome. In a landscape completely dominated by human beings, these small bits of forest had a wild beauty so powerful it seemed to annihilate memory. In the early morning the leaves still glittered with rain, and the wet trunks looked almost black under the green foliage. Then, with the sun rising over the mountains, the light turned the canopy a luminous green, a quivering fiery light that spread through the treetops but barely reached the forest floor. Subdued by the vegetation—the broad-leafed aroids and sprays of orchids, the ferns, epiphytic bromeliads, and hanging lycopodiums—the light fell in a golden hue, filling the lower layers of the forest with half-light and faint gray shadows.

The forest floor was an intricate maze of white roots, prayer plants, and herbaceous heliconias, lush and brilliant. On rocky outcrops grew wild begonias and delicate peperomias. There were anthuriums and a dozen species of climbing vines and lianas, morning glories, mandevilla, and philodendrons. Many of the trees grew gnarled, their branches draped in Spanish moss. Others had pale trunks dappled with lichens. The tallest trees rose stiffly from the ground, reached the height of the forest, and then exploded in dense bundles of branches. They had strange names, which Adalberto shared, peering over Tim's shoulder as he transcribed them phonetically into his notebook—*karaguara kaktil, ma müpusana, sarmósiya.* We knew the trees as *Buddleja, Chrysophyllum, Saurauia, Cassia*—names that in the moment appeared as arbitrary as the Ika words that, when spoken in the soft susurrating tones of the language, sounded like wind running over a forest.

The seasonal rains had left the plants flush with blossoms and fruit. In a silence broken only by the pure and expressive cries of distant *caracaras,* we moved from tree to tree making our collections. A slight breeze tempered the air and made it cool and pleasant. The plants were magnificent and rare. An orchid the size of a seed, ferns the height of small trees. In a ravine moist and overgrown with mosses and ferns, we found a new species of *Myrcia,* a large genus of trees in the myrtle family native to tropical America. On the riverbank between Donachuí and Sogrome I stumbled upon a new species of *Protium,* a lovely tree that, like its distant relatives that produce frankincense and myrrh, yields an aromatic resin, pungent and sweet. On the same day we found a third plant previously unknown to science, a large shrub of the genus *Psammisia* in the heather family.

Though somewhat confounded by our enthusiasm for things so obviously useless, Adalberto was nevertheless fascinated by our interest in wild plants. Gradually, however, like a father grown tired of indulging a child, he became impatient with our ignorance and began to show us plants that were worth something. He began with *Picramnia spruceana,* a tree known to the Ika as *urú,* the leaves of which yield a deep purple dye. The rhizomes and stem of a wild species of *Puya* could be eaten as a root vegetable. Exposure to the sap of *queraka, Toxicodendron striatum,* a tree in the poison ivy family, causes severe dermatitis, but Adalberto revealed how the leaves could be boiled and applied as an effective treatment. The sap of a different tree, *Mauria heterophylla,* was deadly poisonous. There were other plants that could kill, many that could heal. All of them, Adalberto said, were the gifts of the forest.

With Adalberto as our companion and plants the center of our attention, the purpose of our visit became clear to the rest of the Ika. Within just a few days our activities faded into an easy and predictable rhythm. The early mornings were tranquil, the air still and cold, the sky a tender lucid blue. By the time the sun became hot, the forest lay all around us. Then toward noon the clouds gathered, the forest darkened, and storms washed down from the mountains. Rain fell throughout the afternoon and often continued long into the night. Generally, it tapered off for a while around dusk, and the Ika would take the opportunity to slip in and out of the shelter of our house. Their visits, which began sporadically, increased to a steady stream. By the end of our first week we were never alone. The tension and suspicion that had marked our initial meeting had given way to curiosity.

Adalberto remained a constant. After three days he began to sleep by our fire, on the dirt floor of the hut on a bed of sheepskins and cowhide. Then he moved in his loom, which he leaned against the mud wall, close to the door. It was a simple rectangular loom with stout five-foot poles on each side. The two parallel bars that held the warp taut were lashed horizontally to the uprights, and the entire structure was reinforced by two intersecting cross poles. Often in the late afternoon after we had finished with the plants, or in the evening after dinner, Adalberto would turn to his weaving. For hours at a time he became lost in the fabric. On several nights, with the fire burning down, I fell asleep to the sounds of a heddle separating the strands of the warp, a shuttle bound with cotton being shot from side to side, a shed stick thumping the woof into place. Once in the middle of the night I woke in my hammock. Adalberto was at work, and beside him in the amber light of a kerosene lamp sat Tim, perched on a wooden stool, still as a stone.

Gradually, the ways of the village came into focus. The houses that we had thought abandoned were, in fact, only temporarily vacated. Ours belonged to Adalberto's eldest brother, Celso, who several years before had left the mountains to train as a dentist. Certificate in hand he had returned to set up a clinic in Nabusímake, the Ika ceremonial center two days' walk to the west. Most of the time Celso was on the go, traveling from settlement to settlement, his medical kit and foot-powered drill strapped to the back of a mule. Adalberto himself lived near his mother, Juanita, in a house just below ours, in the shade of a large mango tree. His other brother, Faustino, was the *comisario* of Sogrome and was one of the silent figures we had met that first morning at Donachuí. Yet another brother, Atilio, owned several mules and was working in the family cane fields two days beyond Nabusímake. Adalberto's uncle, Juan Bautista, was a *máma,* a priest, as was his father, José de Jesús. Both of them were living high in the mountains at Mamancanaca, where the family raised sheep and gathered medicinal plants that grew around the periphery of the sacred lakes. It was a place, Adalberto explained, far beyond the trees, where the plants had fur and the stones in the morning were sheathed in ice.

Like all the people of the Sierra, the Villafañe family was constantly on the move, even in the rainy season. Their narrow traffic on the mountain trails brought firewood to the grazing lands and grasses from the treeless alpine to the temperate settlements for use as thatch. The lowlands provided plantains and bananas, sweet manioc and maize, cash crops such as coffee, sugar, and pineapple. Lands a thousand feet above Sogrome grew potatoes and onions, peanuts and squash, broccoli and tomatoes. Coca grew at mid-elevation between Donachuí and Sogrome, and was carried to all parts of the Ika homeland. This movement seemed to fascinate Tim. He never failed to comment on the ceaseless parade that passed our door each day, mules loaded with blocks of sugar, stems of bananas, and burlap sacks bursting with wool. Men and women laden with goods hurrying along before the afternoon rains swelled the streams and ran down the trails.

"You know, Willy," he said one morning, "they're always heading somewhere. With food or something to trade. Visiting family, getting together for ceremonies or meetings. But there's something else going on."

I followed him out the door onto the dirt terrace that overlooked the valley. A chalky trail ran down the side hill and passed through a series of plantings of cotton, coca, and maize before reaching a wooden bridge that crossed the river just below Adalberto's house. On the far side grew plantains and sugar, interspersed with fallow fields, some of which rose high up the flank of the mountain, becoming fused with dense thickets of new growth.

"Reichel was the first to notice," Tim said. "He knew that the Ika and Kogi were not of our world. He knew it but he also felt it, and in the end it changed his life. That's why his insights were so profound."

Tim walked to the edge of the terrace. The key point, he said, was that the Indians did not have to move. There was plenty of land, good irrigation, all the means to survive and prosper without constantly climbing up and down the mountains. True, it allowed them access to a wider range of foods and resources, and from a material point of view, this may be all that matters. But Reichel understood that movement was in part metaphor, that in passing over the earth they wove a sacred cloak over the Great Mother, each journey like a thread, each seasonal migration becoming a prayer for the well-being of the people and the entire earth. The Kogi themselves refer to their wanderings as weavings.

A sudden gust of wind ran up the valley. I glanced down the slope and saw climbing toward us an old man twisted with age.

"Look at those fields. What do you see?" Tim asked.

"What do you mean?"

"Do you notice anything about them?"

"No. Nothing special."

"Reichel saw the same fields, the same gardens. But he stayed long enough to see them harvested and to see them planted. This is how they plant a field." He took a pencil and piece of paper from his pocket and sketched a rough rectangular figure, which he divided in two. The northern half of the field, he explained, was the domain of the men, the southern side that of the women. The men grow cotton and maize, the women coca and manioc. The women plant a field by beginning in the southeastern corner and working their way north until, reaching the center line, they turn again south, sowing lines of crops parallel to the sides of the field. The men, by contrast, begin on the center line on the western side of the field and move east, planting horizontal lines of crops until, having worked their way back and forth across the entire field, they finish their work at the northeastern corner.

"Now take these two tracings, and what do you have?" Tim asked.

"I don't follow."

"You can't *see.* Here, try this." He folded the paper in two, along the center line, and lifted it toward the sun.

"There. What is it?"

"A grid," I answered.

"No," he said. "A fabric. The garden is a piece of cloth. I have to show you something."

Tim turned and we went back into the hut. He rummaged through his gear and pulled out his journal.

"Everything begins and ends with the loom," Tim said, reaching into the fire for an ember to light a cigarette. He moved over toward the door. "For the Kogi, a person's thoughts are like threads. The act of spinning is the act of thinking. The cloth they weave and the clothes they wear become their thoughts. Listen to this." He opened his notebook and began to read.

I shall weave the fabric of my life,
I shall weave it white as a cloud;
I shall weave some black into it;
I shall weave dark maize stalks into it;
I shall weave maize stalks into the white cloth;
Thus I shall obey divine law.

"It's a Kogi prayer," Tim said. "I found it in one of Reichel's papers. You see, to us the loom is just a few poles, a simple piece of technology. But for these Indians it's something sacred—not the object but the act of weaving itself. Or at least the symbols that it invokes. In the simple act of making cloth, the weaver aligns himself with all the forces of creation."

According to Reichel-Dolmatoff, Tim explained, the loom is an image of the four corners of the world, with the point of intersection of the cross poles representing the sacred peaks of the Sierra Nevada. The loom is also the human body with the four corners representing the shoulders and hips, and the intersection of the poles being the human heart. Thus when a man crosses his arms, hands touching opposite shoulders, he embraces himself and becomes the loom of life. The earth itself, the surface of the land, is also a loom, an immense template on which the sun weaves the fabric of existence. In the four corners are the points of the solstices and equinoxes, the loci between which the divine weaver moves each day and night creating the worlds of light and darkness, of life and death.

This idea of the sacred infusing the material world informs every aspect of life in the Sierra. When the Great Mother conceived the nine-layered universe, she also dreamed into being the first temple, egg-shaped like the cosmos. The temple floor is the world of the living, the thatch roof a model of the upper worlds, mirrored beneath the ground by an inverted realm like that of the cosmos. To this day the Kogi build their temples around this cosmic model. They are simple structures with high conical roofs supported by four corner posts. On the dirt floors, positioned between the central axis of the temple and each of the four posts, is a ceremonial hearth representing one of the four lineages founded at the beginning of time by the Lords of the Universe. In the middle is the hearth of Lord Mulkuëxe, the representative of the sun.

When the Kogi build their temples, the alignment of these hearths is precise and critical. At the apex of the roof is a small hole, covered most of the year by a piece of pottery. The orientation of the temple is such that on the summer solstice, as the sun rises above the mountains in the morning, a narrow beam of sunlight falls on the hearth that lies in the southwest corner of the temple. During the day, as the sun moves across the sky, the beam of light moves across the floor until, just before dusk, it lands on the hearth in the southeast corner. Six months later, on the winter solstice, with the sun having shifted south, the beam of light passing through the roof touches the northwest hearth in the morning and in a similar fashion passes over the floor during the day, striking the northeast hearth at dusk. On both the fall and spring equinoxes, the beam of light passes through the roof and slices a path equidistant between north and south, such that at the meridian point, with the sun high in the sky, the central hearth, the most sacred of the five, is bathed in a thin vertical column of light. For that moment a *máma* or priest has been waiting. He lifts a mirror to the sun, and as the light of the Father fertilizes the womb of the living, the priest with his mirror creates a cosmic axis along which the prayers of the people may ascend to the heavens.

Thus over the course of a long year the sun passes over the earth and weaves the lives of the living on the loom of the temple floor. He weaves by day and night, in this world, and during the hours of darkness in the world that lies below, the inverted realm of the black sun. Above and below, the sun weaves two pieces of cloth each year, one for himself and one for his wife. The first strands of the warp are laid down on the solstice, the first cloth is completed on the equinox. At that time, according to Reichel-Dolmatoff, the Kogi priests begin to dance at the eastern door of the temple, moving slowly across to the western entrance, all the while in gesture and song acting as if they were drawing a rod behind them. Finally, reaching the western door, the priest pulls forth the imaginary rod and the fabric of the sun unfolds to the north and south. Within moments a new cloth is conceived, the divine weaver soars over the loom, and life continues.

Adalberto stood impassively to one side as Tim reached for the tip of a slender branch. He touched the leaves, examined the red fruits, lifted a small white flower to the glass of his hand lens. Then, turning from the coca plant, he glanced at Adalberto, who nodded in response. Tim plucked three leaves and crushed them between his fingers.

"Smell this," he said. I leaned forward and sensed the fresh scent of wintergreen. "Methyl salicylate," Tim explained. "You'll notice it even more with the dried leaves. That makes it certain that it's *novogranatense.* Bolivian coca has a grassy odor, almost like hay or Japanese tea."

After more than a week in Sogrome, Tim had finally asked permission to collect coca, and he approached the task with an intensity that Adalberto seemed to find appropriate. Small vials of alcohol contained flowers and fruits, soil samples filled a dozen plastic bags, tens of herbarium specimens left more than one plant looking frail and spindly. He gathered living material from several different plants, seeds, and cuttings, which he wrapped with moist sphagnum and placed in cloth bags, labeled and carefully cross-referenced to the numbered voucher specimens. Before strapping the vouchers into the plant press, he recorded all relevant information that would not be self-evident on a dried and mounted herbarium specimen: size and habit of the plant, color of leaves, flowers, fruits, and bark, as well as ecological notes including soil type, exposure, evidence of insect pests, and harvesting activity.

The cultivated coca of the Ika was, as expected, the Colombian species named in 1895 *Erythroxylum novogranatense* by the German botanist Hieronymous after the old colonial name for the country, Nueva Granada. This was the coca of the thirteenth century Muisica and Quimbaya goldsmiths, the stimulant of the unknown people who carved the monolithic jaguar statues and massive tombs at San Agustín in southern Colombia 1,500 years before Columbus, the plant that Amerigo Vespucci encountered on the Paria Peninsula of Venezuela in 1499 when he recorded the first European description of coca chewing. Once extensively grown along the Caribbean coast of South America, in adjacent parts of Central America, and in the interior of Colombia, it is now found in traditional context only in the rugged mountains of Cauca and Huila and in the Sierra Nevada de Santa Marta. Throughout the country it is known as *hayo,* the name used today by the Ika and Kogi.

The idea that different varieties of coca existed beyond the mountains of his homeland fascinated Adalberto. Was it possible to obtain seeds? How was the leaf harvested? Who was responsible for maintaining the fields? Tim answered these and a dozen other questions. Some of his information was relatively straightforward and readily accepted by Adalberto—the observation, for example, that Bolivian coca could be planted by cuttings whereas *hayo* always grew from seed. Other facts proved more difficult to accept. It appalled Adalberto to learn that in other regions both men and women chewed the leaf and that men took an active role in its propagation and harvest. Tim described coca as a sacred food, noting that to compare it to the pure alkaloid cocaine was as inappropriate as comparing a cup of coffee to the effects of ingesting pure caffeine. He spoke of endless mountainous lands where fields are blessed with offerings of the plant, and men divine the future by consulting the leaves, a gift of clairvoyance reserved only for those who have survived a lightning strike.

"They believe," Tim explained, "that as you move from one valley to the next, you must thank the mountain guardians for their protection. Every time they cross over a divide, they place a quid of coca on the rock cairns that mark the high passes and blow prayers into the wind."

"For everything there must be a payment," Adalberto said, his thoughts and Tim's finally achieving a certain symmetry. He lifted the *chukuna* stick from his gourd, placed it into his mouth, and bit down on the lime-coated end. A small trickle of green saliva ran out the corner of his lip.

"You are not Christians," he said.

"No," Tim agreed.

For both the Ika and Kogi the earth is alive. Every mountain sound is an element of a language of the spirit, every object a symbol of other possibilities. Thus a temple becomes a mountain, a cave a womb, a calabash of water the reflection of the sea. The sea is the memory of the Great Mother.

The life spun into being at the beginning of time is a fragile balance, with the equilibrium of the entire universe being completely dependent on the moral, spiritual, and ecological integrity of the Elder Brothers. The goal of life is knowledge. Everything else is secondary. Without knowledge there can be no understanding of good and evil, no appreciation of the sacred obligations that human beings have to the earth and the Great Mother. With knowledge comes wisdom and tolerance. Yet wisdom is an elusive goal, and in a world animated by solar energy, people invariably turn for guidance to the sun priests, the enlightened *mámas* who alone can control the cosmic forces through prayer and ritual, songs and incantations. Though they rule the living, the *mámas* have no evident privileges, no outward signs of prestige. They share the same simple food, live in identical stone houses, wear the same cloth, woven by their own hands. Yet their pursuit of wisdom entails an enormous burden, for the Kogi and Ika believe that the survival of the people and the entire earth depends on their labors.

One is called to the priesthood through divination. As soon as a child is born, a *máma* consults the Great Mother by reading the patterns that stones and beads make when they are dropped into water in ceremonial vessels. Those who are chosen are taken from their families as infants and carried high into the mountains to be raised by a *máma* and his wife. There the child lives a nocturnal life, completely shut away from the sun, forbidden even to know the light of the full moon. For eighteen years he is never allowed to meet a woman of reproductive age or to experience daylight. He spends his life in the ceremonial house, sleeping by day, waking after sunset to cross in the darkness to the *máma*'s house where he is fed. He eats twice more through the night, once at midnight and again shortly before dawn. His food is prepared only by the *máma*'s wife, and even she may see him only in the darkness. His diet is a simple one: boiled fish and snails, mushrooms, grasshoppers, manioc, squash, and white beans. He must never eat salt or foods unknown to his ancestors. Not until puberty is he permitted to eat meat.

The apprenticeship falls into two distinct phases, each lasting nine years and thus mimicking the nine months spent in a mother's womb. During the first years the apprentice is raised as a child of the *máma,* educated in the mysteries of the world. He learns songs and dances, mythological tales, the secrets of Creation, and the ritual language of the ancients known only to the priests. The second nine years are devoted to higher pursuits and even more esoteric knowledge—the art of divination, techniques of breathing and meditation that lift one into trance, prayers that give voice to the inner spirit. The apprentice learns nothing of the mundane tasks of the world, skills best left to others. But he does learn everything about the Great Mother, the secrets of the sky and the earth, the wonder of life itself in all its manifestations. Because the initiates know only the darkness, they acquire the gift of visions. They become clairvoyant, capable of seeing not only into the future and past but through all material illusions of the universe. In trance they can travel through the lands of the dead and into the hearts of the living. Finally the great moment of revelation arrives. After having learned for eighteen years of the beauty of the Great Mother, of the delicate balance of life, of the importance of ecological and cosmic harmony, the initiate is ready to shoulder his divine burden. On a clear morning, with the sun rising over the flank of the mountains, he is led into the light of dawn. Until that moment the world has existed only as a thought. Now for the first time he sees the world as it is, the transcendent beauty of the earth. In an instant everything he has learned is affirmed. Standing at his side, the *máma* sweeps an arm across the horizon as if to say, "You see, it is as I told you."

• • •

Adalberto carefully laid short lengths of reed side by side in a narrow hollow between two rocks. They weren't just any rocks; they were the ones we had searched for since leaving Sogrome: white river boulders that to my eye looked like any others. To Adalberto there was a difference. These were the stones meant to support the fire that would reduce the seashells into lime to fill his *yoburu.* He called the gourd his *mujercita,* his little woman.

On top of the reeds he placed nine shells and nine short lengths of yarn, which he covered with esparto grass and more reeds. Flanking either side of the reeds were two sticks, driven vertically into the ground. These he called guards.

"How did you get these shells?" he asked. Tim told of hiring the young boys in Santa Marta. Adalberto was not impressed. He recalled one of his own journeys to the sea, a five-day pilgrimage that had taken him to the land of frogs and spirit beings, past ice caves and lakes and other openings into the body of the Great Mother. Descending to the shore, he had waited for the dawn, walked onto the beach with his back to the sea, and begun slowly to spin, moving closer and closer to the water and the origin of the Divine Weaver.

"Your lime comes from these shells," Tim said, "but there are others who burn limestone from the earth or animal bones. There are some who use the ashes of certain leaves and stems mixed with urine and dew. There are plants that will sweeten the chew." Adalberto looked up. There was a rustle in the bushes, and we all turned to see an iguana frozen along a branch, his tough wrinkled skin looking a thousand years old. Adalberto flung a stone and missed.

"We use only these shells," he said, returning to his work. He struck a match and lit the dry grass, which quickly ignited, spreading a hot flame over the reeds. Adalberto fanned the fire. Within fifteen minutes the flames reduced the stalks to ashes, leaving nine pure white shells, purged of all organic matter, on a bed of dark ashes. Flipping them out of the fire with his hands, he let them cool for a few moments before dropping them one by one into a ceramic jar. With some care he poured onto the shells an infusion prepared earlier from flowering *moroche* stalks. A reaction occurred, and a faint chemical steam wisped out of the mouth of the jar. Having absorbed the liquid, the shells crumbled into a fine powder.

After two weeks at Sogrome our brief time in the Sierra Nevada was coming to an end. Aurelio had returned with the mules, and Tim each day was growing more anxious about the condition of our collections. The alcohol and formaldehyde we had bought in Santa Marta turned out to have been cut with water; mold had appeared on some of the packages, and several of the early collections already showed signs of rot. We had learned all that we could about the use of coca in the mountains. To discover more would have involved a commitment of time that was outside the scope of Tim's study. It is one of the frustrations of ethnobotanical exploration. At any place in the hinterland of South America, one could spend a lifetime and not come close to exhausting the reservoir of indigenous knowledge. Yet Tim's mandate was to study coca throughout its distribution, and a dozen other tribes scattered across the continent beckoned.

On the evening of our last night at Sogrome, Adalberto slipped into the light of our fire. There was an old man with him. He had a black mustache and wore a cloak of white cotton that dwarfed his frail body. His face had the sheen and warmth of burnished copper. Adalberto introduced him as his father, the *máma* José de Jesús. He had walked all the way from the family holdings high in the mountains at Mamancanaca in order to meet Tim.

For several minutes Tim and Adalberto spoke about this or that, but the words seemed to have no weight. They floated in the air, were purely decorative. Then Adalberto's father lifted his coca bag from around his neck and handed it to his son. Adalberto reached inside and pulled out a *yoburu,* which he gave to Tim. José de Jesús began to speak softly and deliberately.

"The *yoburu* is very important. It is the cradle of civilization. At night before you sleep you chew the leaves three, maybe four times. You think of this day. Then you think of the morning and the next night when once again you lie in your hammock. The leaves will make you think of this land."

Tim accepted the gift, reached into the bag, and took his first bittersweet taste of *hayo.*

An hour later we were once again alone. The fire was out. I fell back into my hammock and let the events of recent days run through my mind. I heard Tim get up and walk toward the door. I had no interest in sleep, so I joined him outside. The night was clear and the sky bright with stars. The air was surprisingly cold. Tim had the *yoburu* that Adalberto had given him in his hand.

"How are you doing?" he asked.

"I couldn't sleep."

"Me, neither."

"I was thinking about these people," I said, "and this work. How you could spend your life collecting plants—just going from place to place, different tribes, setting up and then moving on."

"And getting paid." I could sense Tim's smile in the darkness. "You can thank Schultes for it."

"How did he get started?" I asked.

"That's a long story. His was a different world."

Tim lit a cigarette and stepped to the edge of the terrace. The moon was rising and the wind moved through the branches of the tree that rose above Adalberto's house. Smoke was seeping out of the thatch. For a few minutes we remained quiet; then I heard Tim's voice once again.

"The Kogi have this word, *munse.* Do you remember?"

"No," I said.

"It means dawn and it means vagina. It is also a white light. The priests go to the highest peaks and sit with their backs to the mountain, their eyes ahead to the sea. They make offerings and they stay there until they feel a power surge through their bodies. That's when they see the light, *munse.* It comes on like a vision and then takes form. What they see is the vagina of the Great Mother, a cross that is shaped like a loom."

CHAPTER THREE

The Peyote Road, 1936

THERE IS A small photo album in the anthropological archives at the Smithsonian Institution that shows what it was like that summer nearly sixty years ago when Richard Evans Schultes, a young Harvard student, traveled west to Oklahoma to live among the Kiowa and participate in the solemn rites of the peyote cult. In one photograph the land appears as a blur of dust, the sky fading to gray, the air darkened by soil worked loose by the wind, the farmhouses on the horizon broken down and abandoned. Another image reveals the silhouette of a distant ridge of the Wichita Mountains, the place where the Kiowa elder Bert Crow Lance sought the medicine power. The caption, an almost illegible handwritten scrawl, explains that during his four-day vision quest, the Indian built a sweat lodge of willow and hides, fasted, cleansed himself with sage and cedar, and endured the heat of the fire until his spirit was released to soar over a field of snakes. His ordeal ended when a vision of his mother appeared and told him to go back home because he had forgotten his pipe.

Another photo is a portrait of an old woman, identified as Mary Buffalo, principal informant and wife of the Keeper of the Ten Medicines, the holy medicine bundles that the Kiowa say date back to the beginning of the world. She is granddaughter of Onaskyaptak, owner of the Tai-Me, the Sun Dance Image, the most venerated object of the Kiowa. Her medicine bundle has twelve scalps tied to it, seven of them taken from whites, including one from a long-haired woman killed in Texas in the last century. The photograph reveals a face formed by the open prairie, by winter blizzards and summer heat. It is dark, weathered, and stark. Her thin hair is drawn close to her head and hidden with a black net cap. A long dress and a blanket cover all of her frail body, save for her hands, which clasp the corners of the blanket to her chest. She has the strong, oversized hands of a woman who has spent her youth scraping meat and fat from hides. She appears proud, yet there is a deep sadness in her eyes that suggests the stoic indifference we have come to associate with Plains Indians is less a characteristic of a people than the result of a century of impossible grief. At eighty-eight her life has spanned the entire modern history of the Kiowa. As a child she was brought up to believe in the divinity of the sun. As a young girl she witnessed the return of war parties and made offerings to the Tai-Me at the Sun Dance. As a woman she discovered the affliction of defeat, endured famine and disease. She grew old listening to the brooding chants of broken warriors, the silence of a prairie without buffalo.

Yet another photograph shows a group of blurry-eyed peyote eaters lined up by a tepee at dawn. They are dressed in billowy cotton shirts, baggy trousers, and kerchiefs. Belo Kozad, the Roadman or leader of the ceremony, stands in front and wears traditional clothes: a buckskin shirt, moccasins, and an old trading blanket wrapped around his waist. His long hair is braided and wrapped in otter skins that hang well below his knees. He wears a fur hat, and on the front of it, just above his eyes, is a circle of beads. The red cross at the center is the Morning Star. Around the edge are eight triangles representing the vomit deposited on the earth by the ring of worshipers inside the tepee. The fringes of yellow beads symbolize the rays of the sun. A prairie falcon feather dangles over his left eye. Hanging from his left shoulder is a strand of mescal beans, the toxic scarlet seeds used as a hallucinogen and in ritual ordeal before the arrival of the peyote cult on the Great Plains. To his right stands Charlie Charcoal, nephew of Kicking Bear. In the Roadman's hand is the fan of eagle feathers which that night Charlie had seen turn into water, a river, the wing of a bird, and finally a ladder that had carried his prayers out of the tepee and into the heavens.

The most intriguing image of all is also the simplest. It shows the Roadman, Belo Kozad, flanked by two young white men standing in a field. On the Kiowa's left is Weston La Barre, a graduate student in anthropology at Yale who would go on to write the seminal book *The Peyote Cult.* His companion is the twenty-one-year-old Schultes. It is clear from the juxtaposition of the photographs in the album that all three men have just come out of an all-night peyote ceremony. La Barre looks like it. His eyes shy away from the glare, his hair is disheveled, his clothing loose. Schultes, by contrast, does not have a hair out of place. He is tall, dignified, and contained. In the heat of the morning and throughout a long night of chanting, prayers, and ritual vomiting, he has evidently not so much as loosened the red Harvard tie around his neck. One would never know that coursing through his blood is the residue of a sacred plant that has just sent a dozen Kiowa on a mystical journey to their gods.

The outline of Richard Evans Schultes's life appeared safe and predictable in the fall of 1933 when he first walked through the Johnson Gate as a freshman at Harvard College. He was born and raised in a staid, churchgoing family in East Boston at a time when the small, tightly knit community of Italians and a smattering of English, Irish, and Scandinavian immigrants was still on an island, cut off from the mainland by the ebb and flow of tides across the Chelsea Creek. The people of East Boston were hardworking, religious, and conservative. They supported eleven churches, and at the Sunday school attended by Schultes's mother there had been more than 1,300 children, so many that classes were held in shifts. Though the older residents could still vaguely remember the heady days when Donald MacKay's shipyards had sent clipper ships to all corners of the globe, the community had long since entered a period of economic decline. A few factories remained, but most men found work off the island in Boston, at the naval shipyards in Charlestown, or in the industrial plants that had sprung up beyond the mouth of the Mystic.

In his later years as a Harvard professor Schultes would often describe himself as a fourth-generation Bostonian. He was, in fact, the grandson of immigrants. His family on his mother's side had come from the

Midlands of England in 1860, arriving in East Boston on the Cunard Line. At dockside they were met by a Susan Damon, a social worker from the Unitarian Church, who offered to put up their three children while the parents found a place to live. This gesture was never forgotten by the Bagley family, who became Unitarians and staunch supporters of the Church. Schultes's grandfather, a master mechanic, landed a job across the harbor in Chelsea and commuted to work each day on foot, walking over the railway trestle that was the only link to the mainland. It was a dangerous, exposed crossing, and, tragically, one bitterly cold December day he tripped on a railway tie and fell one hundred feet to his death. His body was not found until the following March. His widow, left to raise five children, took in washing and sold wool from the flock of sheep that she grazed on the open slopes above the city.

The other side of Schultes's family were Germans who had left the country following the Prussian takeover under Bismarck. Settling in Hoboken, New Jersey, his paternal grandfather, a former military officer, worked as a teamster, delivering barrels of beer. Schultes's father, Otto, grew up around the beer industry, and shortly after the Boer War traveled to South Africa, where he spent eighteen months in Durban overseeing the installation of fermentation vats at the first brewery built in the country. He was by all accounts a painfully shy man, and this was the only occasion he ever left America. One photograph survives of that time. It shows a small man dressed in the standard white linens of a colonial planter and seated in a rickshaw being drawn by a hefty Bantu youth. Strapped to either side of the lad's head are the horns of an oxen.

Though certain members of the family prospered, including two uncles who successfully ran for public office as Republicans in thoroughly Democratic East Boston, Schultes's immediate family fell on more difficult times. Prohibition shut down the breweries and forced his father to find work as a household plumber. Later, during the Depression, his fledging plumbing business suffered. To survive he had to lay off men, and he ran up against the government, this time in the guise of Roosevelt's National Recovery Agency, which intervened on behalf of the workers and prevented the layoffs. Thus, twice within a decade what the family perceived as arbitrary meddling by the federal bureaucracy had jeopardized his father's livelihood. These events, embellished by the passage of time and filtered through the eyes of a family that considered Woodrow Wilson a radical socialist, would inform all of Schultes's political views. His lifelong dislike of the Kennedy family, however, was not based on political differences alone. Schultes's mother, Maude, had gone to school in East Boston with Joe Kennedy, Sr., the family patriarch. During Prohibition she had watched her husband's business disappear while Joe Kennedy built a fortune bootlegging whiskey. Later, during the Depression when so many suffered, Joe Kennedy increased his wealth by foreclosing on the mortgages of many of her friends.

Maude Schultes was a solid woman whose life had been tempered by personal loss. Her father drowned when she was two. Her own firstborn son died a month after his birth. Perhaps as a consequence Maude tended to be overprotective of her next two children, Richard and his younger sister, Clara. Her family, home, and church were her entire life, her children her closest companions. The yellow house on Lexington Street where Richard was raised had been her home since she was seven, and she would live there for fifty-two years. Her husband was a distant man, frugal, austere, and serious. He worked six days a week and spent much of his free time in small claims court trying to get people to pay their plumbing bills. He never took his wife out and had no social life save for his membership in the Independent Order of Oddfellows, a private club that his son preferred to call the "Peculiar Chaps." Though raised as a strict German Lutheran, Otto Schultes completely rejected his religious upbringing and wanted no part of any church, including the one that was the focal point of his wife's life.

Though tall and lanky as a teen, Richard Evans had been a sickly child, and when it came time to go to high school, the family doctor advised that he not commute to school by trolley through the damp and congested tunnel that linked East Boston with the mainland. Thus, rather than attending Boston Latin, the most prestigious public school in the city, he went to East Boston High School, where he excelled, particularly in Greek and Latin, chemistry, and foreign languages. In his spare time he read a great deal, raised rabbits, worked in the family garden, and ran errands, working for a nickel a day, which he put away toward his college expenses. Throughout his adolescence Richard dwelt largely in a world of his own, a brilliant student who managed somehow to be eccentric without suffering ridicule. In fact, though he had few friends, by all accounts his peers looked up to him. He was proud, confident, self-possessed, and charged with a burning ambition. His reference points were not totally of this world. He was, as his sister Clara would recall, "different."

When it came time for college, he applied only to Harvard, did well on his entrance exams, and became the first of his family to attend university. In many ways it was a leap of faith. His mother and father managed to put aside $400 to cover the first year's tuition, but after that he would be on his own. His plan was to become a doctor. During his first semester he studied hard, as he always had, taking chemistry, biology, and intensive German. To save money he commuted from home, and while his classmates dined in sumptuous halls served by women known as "biddies," he ate the same meal every day in a small shop on the corner of Harvard Square: soup and three pieces of rye bread with butter for fifteen cents.

Financial security came at the end of his first year when he received the Cudworth Scholarship, a small award given by the Unitarian Church of East Boston and named in honor of a former minister who had been with Lincoln at Gettysburg. The scholarship was endowed specifically to help Harvard students of good moral standing who came from either East Boston or neighboring Lowell. Those who recommended him for the award had watched Richard Evans Schultes grow up. They thought of him as a sober individual and no doubt expected that one day he might return to East Boston to set up a medical practice. His mother shared the same expectations, as did the young student himself. But no one knew what awaited him in Cambridge on the fourth floor of the Botanical Museum.

The Botanical Museum at Harvard dates to a letter sent in 1858 from Asa Gray, then America's most influential and famous naturalist, to Sir William Hooker, director of the Royal Botanical Gardens at Kew. In the letter Gray, who would soon emerge as Darwin's most vocal and distinguished opponent, announced Harvard's plan to establish "in humble imitation of Kew . . . a Museum of vegetable products." Hooker responded immediately by shipping to Cambridge the duplicate specimens—vegetable ivory from Ecuador, palm trunks from Southeast Asia, rubber from the Amazon, narcotics from Turkey—that became the nucleus of the museum's economic botany collections. Unfortunately, even as the museum was being established, Gray's energy was increasingly diverted and consumed by his lifelong effort to prevent the acceptance of Darwin's evolutionary theory, an intellectual struggle that would set back biology at Harvard for a generation. It was not until George Goodale was named first director in 1888 that the museum grew and became, in his words, "a place where rare drugs could be identified or unusual fibers compared."

By 1933, the year Schultes entered Harvard, the Botanical Museum was to a great extent the personal creation of its second director, Oakes Ames. The world's foremost orchid specialist, a leading economic botanist, and a man who in his time held more academic appointments at Harvard than anyone in the history of the university, Ames was a gentleman scholar who viewed science as an avocation, a refuge from the pedestrian affairs of ordinary people.

Ames devoted his life to Harvard and the Botanical Museum. A millionaire many times over, he subsidized virtually every aspect of the museum's operation. Drawing on his extensive contacts in Europe and the Americas, and spending his own money whenever necessary, he built up a herbarium of more than fourteen thousand specimens of economic plants, a library of some thirty thousand volumes, and a collection of amber, lacquerware, and various plant products unequaled in the country. In 1906 he donated his collection of living orchids, then the country's largest, to the New York Botanical Garden. His orchid herbarium, some sixty-four thousand dried and mounted specimens including more than a thousand new species that he had described, went to the Botanical Museum. Ames personally paid all the salaries of the museum staff: librarian, research assistants, and secretaries. When he decided that "a botanist's research should be a jewel worthy of a proper setting," he bought a printing press and hired a young printer who worked at the museum for over sixty years. When Harvard proposed constructing a vast complex of biological laboratories, the president of the university turned to Ames to raise the money.

In person, Oakes Ames was aloof, shy, and retiring, a man who, as his own son would recall, found it easier to be with plants than with people. He had absolutely no interest in the politics of science. He hated faculty meetings and in his entire career never once attended a national botanical congress. When the highly regarded and influential American Association for the Advancement of Science held an annual meeting in Cambridge, Ames avoided legions of admirers who came to pay their respects by escaping to Harvard Square, where he spent the day watching silent films in the old University Theatre.

By his own admission Ames was a poor teacher who found it impossible to deliver formal lectures. In class he preferred to sit quietly on the edge of a table in the front of the lecture hall and chat with the students, who rarely numbered more than a half-dozen. He always wore a buff-colored jacket and vest, and had a mesmerizing habit of twirling his glasses, which were connected to his vest by a length of thin black cord. While his students quietly placed bets on how long the glasses would remain attached, Ames's intellect would sweep over the entire field of economic botany, picking and choosing the subject matter in a manner that was completely idiosyncratic. In a course that ostensibly dealt with "Plants and Human Affairs," the aristocratic Ames scarcely mentioned wheat, said nothing of lumber, paid lip service to rice, but spent a month on arrow poisons, discussed fish toxins and narcotics for several weeks, and devoted one entire lecture to the study of amber.

There was a perfect logic to his choice of subject matter. Ames had no interest in the mundane plant products already established in the world economy. His concerns lay with the unknown, the obscure ethnobotanical mysteries that were to be found amid the remnants of ancient traditions. In this regard his instincts verged on the clairvoyant. His emphasis on arrow poisons came years before the isolation of d-tubocurarine, the muscle relaxant that would revolutionize modern surgery. He directed his students to study fish poisons a decade or more before derris root went into mass production in the tropics as a source of rotenone, the most important biodegradable insecticide. His focus on medicinal plants came half a century before scientists extracted from the rosy periwinkle compounds that achieved 99 percent remission rates in the treatment of lymphocytic leukemia, thus saving the lives of thousands of children.

As a man Ames was firmly rooted in the past, yet as a botanist he was curiously ahead of his time. A profoundly original thinker, Ames was one of the few scholars in the country seriously concerned about the origins of cultivated plants. At a time when anthropologists maintained that man was a relatively recent arrival in the New World, Ames published a book that, on the basis of botanical evidence alone, shattered the dogma. Ames noted that in five thousand years of recorded history not a single major crop had been added to the list of cultivated plants. With the origins of maize and beans, peanuts and tobacco lost in the shadows of prehistory, it was simply unrealistic to assume that agriculture had emerged in the New World within the past ten thousand years. The antiquity of agriculture alone suggested that humans had reached the New World far earlier than anthropologists then believed. He was right, but it would be twenty years or more before his ideas became generally accepted.

Schultes fell into Ames's orbit during his second year at Harvard when, in order to earn a little money—thirty-five cents an hour—he took a job at the Botanical Museum filing cards and stacking books in the Economic Botany library. Young and curious, he found himself in the midst of one of the most eclectic libraries in the country. Between the folios of Linnaeus and the herbals of Fuchs and Brunfels, there were volumes on African ordeal poisons, monographs on distant tribes, travel accounts of plant explorers, and entire shelves of books devoted to narcotics and stimulants, arrow poisons, tropical fruits, fibers, sugars, essential oils, and spices from places in the world about which he had never heard. Intrigued by this material, he decided in the spring of his junior year to enroll in Ames's course, Biology 16, Plants and Human Affairs.

There were six students in the class, and in the introductory lecture Ames outlined the course requirements. In addition to the basic readings, examinations, and written assignments, there would be a practical laboratory during which the students would experiment with various plant products. They would make paper and ink, mix essential oils to create perfumes, extract sugar to produce molasses, turn fatty oils into soap, dye clothing with leaves and roots, sample rare and exotic spices, and practice the art of herbal medicine. And naturally, Ames continued, they would brew beer and distill alcohol, just as his students had done every year throughout Prohibition.

Several weeks into the semester Ames announced a slight change of routine. In the laboratory devoted to stimulant and narcotic plants there would be bourbon, brandy, gin, rum, tequila, crème de cacao, sake, and vodka to sample. Both species of tobacco were on hand, *Nicotiana tabacum* and its older and considerably more potent relative *N. rustica.* There were five plant sources of caffeine to become familiar with, including *yerba maté* from Argentina and *yaupon,* a powerful purgative native to the Carolinas. Naturally, betel nuts, cola nuts, and other masticatories were available, as was a fresh root of *kava-kava,* a mild soporific and ritual beverage from the South Pacific. Regrettably, time and prudence did not permit the sampling of certain more exotic fare: opium from Turkey, Moroccan hashish, and coca from Peru. Also out of the question were the curious plants classified in the lab manual as *Phantastica,* those capable of causing "excitation in the form of visions and hallucinations, often in color." There were, however, specimens to examine and literature to consult. On a table at the back of the laboratory, Ames noted, were six books. Each student was expected to read and prepare a report on a volume of his choice. Hard-pressed by other assignments, Schultes at the first opportunity rushed to the back of the room and grabbed the thinnest book. The title was *Mescal: The Divine Plant and Its Psychological Effects.* Published in 1928 and written by the German psychiatrist Heinrich Klüver, it was the only monograph then available in English that described the astonishing pharmacological effects of peyote.

Late that evening, in his room back in East Boston, Schultes opened the book and read the first short chapter, which outlined what little was then known about the plant. It was, Klüver noted, a small spineless cactus, native to northern Mexico and commonly found on both sides of the Rio Grande. Blue-green in color and shaped not unlike a thick carrot, without branches or leaves, peyote grew alone or in dense clusters, in the open sun or more frequently in the shadow of the tall yucca trees that thrive amid the tar bushes and agaves of the Chihuahuan desert. The top of the plant, a rounded crown that alone appears above ground, is divided radially by a number of ribs that bear small tufts of whitish gray hairs. It is for these that the plant is scientifically named *Lophophora,* meaning "I bear crests." According to Klüver the word "peyote" also refers to these hairs and is derived from *peyotl,* meaning cocoon in Nahuatl, the language of the Aztecs. Schultes would later disagree with this derivation and suggest instead the Nahuatl words *pi yautli,* meaning a small herb of narcotic power. For him the source of this power was mescaline, one of more than thirty alkaloids eventually isolated from the plant. For the Indians, he would soon discover, the strength of peyote lay in an altogether different realm. According to legend the plant had been born in the hoofprints of the Sacred Deer, and the songs that heralded its visions were composed at the moment when shamans first heard the sound of the rising sun.

In 1928 the botany and chemistry of peyote was, in Klüver's words, "a matter of dispute." All that was certain about the plant was that it could induce visionary experiences that were as startling as they were indescribable. For the rest of that memorable night, as he worked through the book, Schultes was enchanted by page after page of exquisite accounts of those who had taken the drug. One person reported seeing "stars, delicate floating films of color . . . then an abrupt rush of countless points of white light swept across the field of view, as if the unseen millions of the Milky Way were to flow a sparkling river before the eye . . . the wonderful loveliness of swelling clouds of color. All the colors I have ever beheld are dull compared to these. Here were miles of rippled purples, half transparent, and of ineffable beauty." Another found himself beneath "a dome of the most beautiful mosaics, a vision of all that is most gorgeous and harmonious in color. The prevailing tint is blue, but the multitude of shades, each of such wonderful individuality, makes me feel that hitherto I have been totally ignorant of what the word color really means. The color is intensely beautiful, rich, deep, deep, deep, wonderfully deep blue . . . [like] the blue of the mosque of Omar in Jerusalem. . . . The dome has absolutely no discernible pattern. But circles are becoming sharp and elongated . . . figures wildly chasing one another across the roof."

For many, the senses became confused, sounds became visions, colors became taste, touch became rhythm. "Each audible stroke of the pendulum produced an explosion of color. The beat of a drum increased the beauty of the visions, the low notes of the piano produced an hallucination of violet, while high notes give rise to rose and white." Another musician reported that "the effect of the sound of the piano was most curious and delightful . . . the whole air being filled with music, each note of which seemed to arrange around itself a medley of other notes which appeared to me to be surrounded by a halo of color pulsating to the music." Yet a third was more succinct: "I hear what I am seeing. I think what I am smelling, I am music, I am climbing into the music." Finally one mescaline user put into plain English what all the others, despite their lyrical descriptions, knew to be true: "The display which for an enchanted two hours followed was such as I find it hopeless to describe in language which shall convey to others the beauty and splendour of what I saw."

These descriptions of peyote visions stunned Schultes, who almost sixty years later would still recall his amazement. "That a plant could do such things! It was wonderful. I had to know about it." The day after reading Klüver's little book, Schultes approached Professor Ames and asked if he might write about peyote for his undergraduate thesis. Ames agreed on one condition: It would not be enough to research the subject in the literature. Schultes would have to travel west to Indian country in Oklahoma and see the plant in use. Peyote, Ames explained, originated in Mexico but in the mid-nineteenth century spread north, reaching the Great Plains around 1870. Carried from the Mescalero and Lipan Apache to the Kiowa, and from the Kiowa to the Comanche, the cactus had become the basis of a visionary nativistic religion, legally organized as the Native American Church. From the Kiowa, the Peyote Cult had passed to the Arapaho and Cheyenne, the Shawnee, Wichita, and Pawnee, eventually touching not only the peoples of the northern Plains, the Crow, Sioux, and Blackfoot, but going beyond to the Seneca and Creek, the Cherokee, Blood, Chippewa, and reaching even into northern Canada, where it was taken up by the Cree. Despite violent opposition and anti-peyote laws enacted by nine states, peyote had within seventy years reached almost eighty tribes, a phenomenal rate of diffusion of better than a tribe a year.

Those in the "damnable government" who opposed the use of peyote by the Indians, Ames suggested to Schultes, knew nothing about its history and were completely ignorant of its importance as both a medicinal plant and a ritual sacrament. A proper ethnobotanical study was long overdue. "For such a purpose," Ames said, "we just might be able to find a small grant." A month later Schultes received his grant, which he would discover years later came right out of Ames's pocket.

Over the next weeks Schultes read everything available on peyote, from the chronicles of the Spanish conquistadors to the turn-of-the-century experiments of Philadelphia psychiatrist S. Weir Mitchell, who ate the plant at home and had visions of luminous gems floating in a sea of limpid light. From the journals of the Danish explorer Carl Lumholtz he learned of the Tarahumara, Indians of the Sierra Madre Occidental in Mexico who were the best runners in the world. Traveling steadily and carrying a peyote button and the dried head of an eagle under their girdles as protection from sorcery, Tarahumara men could run 170 miles without stopping. Employed by the Mexican postal service, one Tarahumara man had delivered a letter 600 miles in five days.

For the Tarahumara peyote was *hikuli,* the spirit being that sits next to Father Sun. It was a plant so powerful that it bore four faces, perceived life in seven dimensions, and could never be allowed to rest in the homes of the living. To gather *hikuli* the Tarahumara traveled far to the east and south, beyond the foothills of the Sierra and into the desert. There they would find the plant by listening for its song. *Hikuli* never stops singing, even after being harvested. One man told Lumholtz of returning from the desert and trying to use his bag of *hikuli* as a pillow at night. It had been impossible. The singing was so loud that he could not get to sleep.

Once safely home, the Tarahumara would spread the *hikuli* on blankets, sprinkle blood over the top, and then carefully store the dried plants until the women were ready to grind them on a *metate* into a thick brown liquid. A large fire would be built, with the logs oriented to the east and west. Sitting to the west of the fire a shaman would trace a circle on the ground and within it draw the symbol of the world. A peyote button was placed on the cross and covered with an inverted gourd, which would amplify the music and please the spirit of the plant. The shaman wore a headdress of feathers, which imparted the wisdom of birds and prevented evil winds from entering the ring of fire. Following the prayers, peyote was passed around and barefooted men and women wrapped in white cloth began a slow clockwise dance that lasted until dawn. Then, with the first sign of the sun, the shaman and the people would face east and wave farewell to the *hikuli,* a spirit that had descended on the wings of green doves, only to depart in the company of an owl.

Needless to say, when Schultes turned to the early Spanish chronicles, he found a quite different perspective on this remarkable plant. For the Spaniards peyote was the "diabolical root," yet another sign of "satanic trickery" in the New World. The Franciscan friar Bernardino de Sahagún, who first wrote of the plant in 1560, noted that *peyotl* was "common food of the Chichimeca, for it sustains them and gives them courage to fight and not feel fear nor hunger nor thirst. They say it protects them from all danger. . . . They lose their senses, see visions of terrifying sights like the devil." Francisco Hernández, personal physician to Philip II and the first to describe the plant botanically, wrote of the "miraculous properties attributed to this root" that allowed "those devouring it to be able to foresee and predict things." This gift of clairvoyance was explained in 1737 by a Padre Arlegui, who lived among the Zacatecans: "It intoxicates them with a paroxysm of madness, and all the fantastic hallucinations that come over them with this horrible drink they seize upon as omens of their future." This priest also reported that upon the birth of a first son, a Zacatec father refused food for twenty-four hours and then imbibed a "brew concocted of a root called peyot." With the onset of visions the man took his place upon a ceremonial stag horn. One by one his people, armed with sharpened bones and animal teeth, stepped forward and cut into his flesh, wounding him mercilessly. Thus a father would be tested that his son might display similar courage in his own life. It was, the Zacatecs told the priest, the least one could do for a child.

Ritual practices such as these appalled the Spaniards. Shortly after the Conquest, and in a gesture indicative of the times, Juan de Zumarraga, the first archbishop of Mexico, combed the land for any manuscripts or artifacts that contained information about the vanquished civilizations, any heretics who still practiced the ancient religions. Then in a final orgy of destruction, on a pyre fueled both by human beings and thousands of religious texts, he attempted to eradicate the memory of all that had gone before. Such violent acts were common after the introduction of the Inquisition to Mexico in 1571, and Indians who used peyote were among those who suffered. By 1620 the plant was officially declared the work of the devil. In 1760 a priest near San Antonio, Texas, published a religious manual containing questions to be asked of potential converts. "Have you eaten the flesh of man? Have you eaten peyote?" Another priest, Nicolás de León, asked, "Art thou a soothsayer? Dost thou foretell events by reading omens, interpreting dreams and figures on water? Dost thou garnish with flower garlands the places where idols are kept? Dost thou know certain words with which to conjure for success in hunting, or to bring rain? Dost thou suck the blood of others, or dost thou wander about at night, calling upon the demon to help thee? Hast thou drunk peyote? Dost thou know how to speak to vipers in such words that they obey thee?"

This plant that so concerned the Spaniards grew in a relatively small area of their colonial domain. Native to the Chihuahan desert, peyote grew from the Rio Grande Valley in Texas south across the flank of the Sierra Madre Oriental into the high central plateau of northern Mexico.

A land of thorn scrub and mesquite, creosote bushes, yucca, and dozens of species of cacti, the desert was at the time of the Conquest home to nomadic hunters and gatherers, the Teochichimeca and the Guachichil, peoples who, according to Sahagún, had used peyote for almost two thousand years. In fact, we now know, based on recent archaeological discoveries, that the native people of Mexico have eaten peyote for seven thousand years. The Conquest brought disease and death to the high deserts of northern Mexico, and in its wake the native inhabitants scattered, with many fleeing south and west into the isolated valleys and *barrancas* of the Sierra Madre Occidental. One population established itself in a particularly remote and inaccessible region some four hundred miles south of the lands of the Tarahumara. Their descendants in all likelihood gave rise to the Huichol, the people of the Sacred Deer, an extraordinary tribe that in its isolation managed to maintain its traditional way of life well into the twentieth century.

Schultes first read about the Huichol in the second volume of Lumholtz's journals, and he realized immediately that their reverence for peyote surpassed even that of the Tarahumara. What's more, in reading Lumholtz's description of their peyote ceremonies, the whirling dancers, the weeping of the supplicants, the prayers and the pure exhaltation of the wondrous plant, he recognized the ritual activities of the Teochichimeca, first described by Sahagún in 1561. Though the Spanish had taken nominal control of the Huichol lands in 1722 and established five missions, their presence and influence had been so ephemeral that the peyote ceremony endured almost unchanged for close to four hundred years. What Schultes could not possibly have understood, given his age, lack of experience, and the information available at the time, was the astonishing fact that the Huichol peyote hunt was, in essence, a profound evocation of the ancient impulse that first gave birth to religion.

Shamanism is arguably the oldest of spiritual endeavors, born as it was at the dawn of human awareness. For our Paleolithic ancestors, death was the first teacher, the first pain, the edge beyond which life as they knew it ended and wonder began. Religion was nursed by mystery, but it was born of the hunt, from the need on the part of humans to rationalize the fact that to live they had to kill what they most revered, the animals that gave them life. Rich and complex rituals and myths evolved as an expression of the covenant between the animals and humans, a means of containing within manageable bounds the fear and violence of the hunt and maintaining a certain essential balance between the consciousness of man and the unreasoning impulses of the natural world. It is precisely this balance that the Huichol seek when they undertake their holy pilgrimage to Wirikúta, the mythical homeland of their ancestors, the ones who first tasted the bitter flesh of *hikuri,* or peyote.

For the Huichol, peyote, deer, and corn are as one. At the beginning of time the water that made life in the desert possible sprang from the forehead of a deer. In the tracks of that first deer grew peyote, which in turn became the first ear of corn as well as the drinking bowl of the greatest and oldest of the Huichol gods, Tatewari, Grandfather Fire. It was Tatewari who first led the Huichol to Wirikúta and introduced them to peyote. Unless this memory is honored by the living, Tatewari will stop the rain, there will be no corn, and the deer will die of thirst. Thus each year the Huichol leave their mountain home to travel through a sacred geography two hundred miles to the northeast. Guided by shamans, the pilgrims take on the identity of the ancestral gods. By completing the peyote hunt and eating the bitter flesh of Elder Brother Deer, they find their way to the center of the world and thus become whole.

Every moment of the pilgrimage is charged with ritual meaning. The shaman, or *mara'akame,* wears a broad straw hat stuck with eagle and hawk feathers that allow him to see and hear everything, to transform the dead, heal the living, call down the sun. The feathers are peyote, corn, deer, and the flames of Grandfather Fire. The pilgrimage begins with a ritual confession, the naming of all sexual partners, a guiltless and cathartic purification intended to return the pilgrims to a pure state of innocence. For each transgression the shaman ties a knot in a string that he will later burn. He then takes a new piece of yarn, ties one knot for each pilgrim, and spirals the twine around his bow. The spiral is the journey, the twine the symbol of the umbilicus, the knot the memory of that moment at birth when one is severed from one's mother to become a child of the earth.

Thus purified, the pilgrims assume the identity of spirits, for it is only by becoming a god that one may pass through the Gateway of Clashing Clouds which separates the realm of the living from the world of the ancestors. This dangerous passage occurs shortly before the pilgrims reach the mountains and deserts of Wirikúta. Once inside the land of the ancestors, once safely through the opening of the Clouds, the pilgrims walk in single file, and the shaman, transformed into the image of Tatewari, begins to chant. A fire is lit, and as the pilgrims feed small branches into the flames as offerings to Tatewari, they form a circle and begin to weep for *hikuri,* the Elder Brother Deer. "Do not be angry," they pray, "we have come from afar to greet you."

Sometime later the pilgrims begin the silent stalk. Bows and arrows at the ready, the men move cautiously through the chaparral until one of the hunters finds the first tracks, in a thicket of mesquite or beneath the shadows of a creosote bush. The shaman takes aim and shoots an arrow into the earth, just to one side of the peyote. A second arrow, then a third and a fourth, flies until the plant is surrounded on four sides. Laying ritual power objects and food offerings before the cactus, the shaman weeps and in song beseeches the Elder Brother Deer not to be angry, to know that his spirit will rise again. Then, after lifting his prayer arrow to the sky and the four directions, the shaman presses it down until the hawk feathers touch the surface of the plant. Chanting that the deer brothers have sprung up to give life, the shaman carefully slices off the crown of the cactus, leaving the root to thrive. He then offers a small piece to each pilgrim. Chew well, he cautions, that you will find your life.

For a week the pilgrims eat peyote by day and night, until finally, extinguishing the fire of Tatewari, they gather their full baskets and set off on the long journey home. As they approach their village and begin the metamorphosis from spirit to human being, they pause for three days to hunt deer. Senses heightened by hunger and fatigue, they kill several animals, enough to ensure that rain will fall and that people at home will share not only the gifts of *hikuri,* the tracks of the Sacred Deer, but the actual meat of animals. For the Huichol, they are one and the same.

On June 24, 1936, the *Anadarko Daily News* announced the arrival in Oklahoma of the Harvard–Yale–American Museum Indian Expedition. It consisted, the local paper noted, of two men: Weston La Barre, whose "special interest is in talking with individual Indians learning from them first hand information on history which many of them had an active part in making," and Richard Evans Schultes, "an ethnobotanist interested in Indian plants of various varieties." The short article mentioned that La Barre, "Ethnographer is his technical title," had spent the previous summer working among the Indians as part of a team headed by Professor Alexander Lesser of Columbia University and that he had returned to Oklahoma to "select a subject for his doctors dissertation" at Yale University. It said nothing about their trip west from Pennsylvania in the broken-down 1928 Studebaker that La Barre had traded a train ticket for, nothing about the eight flat tires or the day they made ninety-six miles in twelve hours through the back roads of Tennessee, clouds of dust boiling out behind the car, windows shut tight because the hot wind was impossible to bear. The newspaper also failed to note that the younger of the two had never been west of the Hudson and that for the rest of the summer he and his new friend would be eating peyote twice and sometimes three times a week.

Weston La Barre's main contact in Anadarko was Charlie Apekaum, or Charlie Charcoal, one of the few remaining full-blooded Kiowa. Charlie was the nephew of Kicking Bear and Mary Buffalo. His father was one of the first to bring peyote to the Kiowa. Charlie himself had eaten peyote as a medicine at the age of two and attended a ceremony when he was twelve. As a child he celebrated his first kill with a family feast and vision quest. He had known Fort Sill when it was still manned by cavalry troops, when stagecoaches came through with money destined for the chiefs. He had met Quanah Parker, the great Comanche leader who, assured by medicine men that the bullets could do no harm, had led the disastrous attack against the buffalo hunters at the Battle of Adobe Walls. In the dying years of the Kiowa, he attended the Ghost Dance with Kicking Bear, who told him that a new earth was about to slide over the world from the west, bearing upon it buffalo and elk, and that the whites would die as the holy dance feathers of the Kiowa lifted the faithful up onto the new world.

Instead, as he grew older, Charlie saw only more evidence of the whites—schooner wagons arriving thirty at a time, ranchers and farmers dividing up the land, missionaries setting up their churches and schools. In 1901, at the age of thirteen, he was thrown out of school after the Baptist teacher discovered that during the hymn "Hallelujah, Thine the Glory," Charlie was singing a verse in Kiowa that roughly translated, "Howdy do, shit on you." Charlie spent the rest of his youth breaking horses, visiting friends among the Arapaho and Cheyenne, and hunting bear and wild turkeys in the Wichita Mountains. He knew of peyote ceremonies; his family had hosted the great Smithsonian ethnographer James Mooney, who coined the term Native American Church, but it was not until after World War I, when he returned from service in the navy, that Charlie became an active member of the peyote cult.

With Charlie Charcoal as companion and guide, La Barre and Schultes together visited fifteen tribes over the course of the summer and attended peyote sessions whenever possible. They lived out of the old Studebaker and slept wherever they found themselves at night—in cool willow arbors, in cheap dives where they lay on the floor to avoid bedbugs, or in the simple houses that the Indians, who still preferred to live in tepees, built specifically for white guests. At county fairs they dodged Baptist preachers and spoke softly with elders in great black hats, bright shirts, and long braids greased with fat and bound in brightly colored cloth. Their informants had names like James Sun Eagle, Heap o' Bears, Old Man Horse, White Fox, and Little Henry.

Schultes spent much of his time with Mary Buffalo, who introduced him to the medicinal and ritual plants of the Kiowa. She mixed nightshade berries with animal brains to tan buckskin, smoked sumac leaves to purify body and mind before taking peyote, chewed willow bark for toothaches and wood sorrel to provide salt. As a young wife and mother she had washed her children with soap made from yucca roots, scented their bodies with the smoke of sweet grass. She gathered the fruits of a creeper to yield dyes and war paints, larkspur seeds for her husband's peyote rattle. Osage orange provided the best wood for bows and the ceremonial staffs of the peyote Roadmen. She even showed him the leaves of grass worn into battle by warriors who had killed an enemy with a lance.

Mary Buffalo had a special fondness for both Schultes and La Barre. Most white people, Charlie Charcoal explained, sound like a pack of coyotes. Schultes and La Barre were different, and Mary Buffalo appreciated their ability to listen. Indeed, Weston was so cautious and quiet that Mary's granddaughter, Lily Jean, fell in love with him, for the etiquette of Kiowa courtship demands that lovers spend days together without speaking. With the magnanimity that often comes to those who have lived long and survived untold hardships, Mary Buffalo shared every aspect of her past—not only secrets of the plants and stories from the wars and buffalo hunts, but also intimate details of the way the Kiowa give birth, the character of love and religion, even the contents of the sacrosanct Ten Medicine Bundles that had been in her family for generations. Often in the evenings when the interviews had been completed and the Kiowa had returned to their tepees, Mary would build a sweat lodge with willow saplings and hides, a bed of sage, and a rimmed hole in the earth to receive the red hot stones. Arms folded in an old trading blanket, she would stand outside while Schultes and La Barre, accompanied by Bert Crow Lance or Charlie, endured the heat as the elders prayed and poured water over the stones. In such moments both young students established the trust that was the foundation of their work. They were the last generation of scholars actually to know Kiowa men and women who had lived the culture of the Plains, a way of life that withered and died within a century of its birth.

Unlike the Cheyenne and Arapaho, whose legends still speak of a time when they lived to the east and grew corn, the Kiowa have no tribal memory of ever having been anything but hunters. In the beginning, their elders say, they emerged from the hollow of a cottonwood log that lay in a dark forest far beyond the mountainous headwaters of the Missouri River. They flung their children into the heavens, their daughters becoming the Dipper, the sons fusing with the night sky. Their language was related to no other. It was spoken only by the storm spirit, an animal molded long ago by the Kiowa from clay, a terrible creature with the breath of lightning and a tail that beat the hot tornado winds into being.

Slowly, beginning around 1700, the Kiowa moved to the south and east, leaving the mountains for the grasslands of the Dakotas. There, beneath an immense sky, they met the Crow, who gave them the religion and culture of the Plains. They acquired the horse and the Tai-me, the sacred image of the Sun Dance. They learned to hunt buffalo and for the first time in their history were able to secure in a single day enough meat to feed their families for a year. Liberated from the crude struggle for survival, they became a society of warriors whose raiding parties ranged for a thousand miles across the open prairie.

Sometime during the last years of the eighteenth century, shortly after the American Revolution, the combined strength of the Cheyenne and Dakota Sioux forced the Kiowa to abandon the Black Hills and move south. At the headwaters of the Arkansas they ran up against the Comanche. The tribes fought, made peace, and around 1790 forged an alliance that gave them complete control of the southern Plains. Though never numbering more than 1,500, the Kiowa became known as the most predatory of all the Plains Indians. Their warriors brought back slaves and women to swell their ranks, scalps to empower their future, and horses in such numbers that the Kiowa owned more per capita than any other tribe. They were also the tribe that, in time, would kill more whites in proportion to their population than any other. War was their calling. Painted and adorned with eagle down, their warriors rode into combat like the wind. Among the finest horsemen the world has ever known, they came upon their enemies at a full gallop, flung themselves to one side of their horses, and shot their arrows from beneath the creatures' necks. In the entire tribe there were only ten men who held membership in the Ka-itsenko, the warrior society of "Real Dogs." Each carried into battle a long sash and a sacred arrow. When attacked, the warrior would impale his sash to the earth with the arrow and thus stand his ground until victory or death.

Once each year, at the height of the summer, when the grass cracked beneath one's feet and down appeared on the cottonwoods, the Kiowa came together for the Sun Dance, by far the most significant religious event in their lives. It was a celebration of war, a time of spiritual renewal, a moment in which the entire tribe partook of the divinity of the sun. It began with a buffalo hunt and the ceremonial construction of the medicine lodge. The tepees went up in a wide circle, their entrances facing inward, the encampment itself oriented to open to the rising sun. The medicine lodge was the focal point, for within it, on a stick planted on the western side, hung the Tai-me, the sacred image of the sun. It was a simple fetish, a small human figure with a face of green stone, a robe of white feathers, and a headdress of ermine skin and a single erect feather. Around its neck were strands of blue beads and painted on its face, neck, and back were the symbols of the sun and moon. For the Kiowa the Tai-me was the source of life itself. Kept in a rawhide box under the protection of a hereditary Keeper, it was never exposed to light save for the four days of the Sun Dance. At that time its power spread into all and everything present: the children and the warrior dancers, the buffalo skull that lay at its base as the animal representative of the sun, the Ten Medicine Bundles displayed before it, the men who for four days and nights slowly turned their shields to follow the passing sun, the young dancer who stared at the sun all day every day, sacrificing his vision that the people might come to see.

On the night of November 13, 1833, a meteor shower flashed across the prairie sky, awakening the entire tribe. The elders said it was the beginning of the end. The first contact with American soldiers occurred the following summer. A treaty of friendship was signed in 1837. In the winter of 1839 what became known as the evil medicine of the black robes swept the entire plains, killing thousands of Indians and nearly exterminating tribes such as the Mandan. Smallpox struck again in 1841. Eight years later California emigrants brought cholera, which killed hundreds of Kiowa and left the tribe so demoralized that dozens of them committed suicide. By 1850 settlement pressure had driven several of the eastern tribes west into Kiowa territory. War broke out, and in 1854 a party of over a thousand Kiowa, Comanche, Apache, and Cheyenne warriors was defeated by a handful of Sauk and Fox Indians armed by the whites with long-range rifles. In 1861 smallpox struck yet again. By the fall of 1863 the Kiowa had had enough. Joined by the Dakota, Cheyenne, Arapaho, Comanche, and Apache, they called for a general uprising. In response the U.S. military went to war, issuing orders to "kill every Indian in the country." One government agent wrote, "Lead, and plenty of it, is what the Kiowa want and must have before they will behave."

Even after the Civil War had ended, and thousands of veteran cavalry troops were available for service in the West, defeating the Indian insurrection proved a difficult task. The tribes were highly mobile, intimately familiar with the terrain, and increasingly well armed. With pressure on the frontier building and with settlers demanding safe passage and railroad magnates calling for the pacification of the tribes, the U.S. military set in motion a campaign of biological warfare that, in its deliberateness and destruction, was unparalleled in the history of the Americas.

As late as 1871 buffalo outnumbered people in North America. In that year one could stand on a bluff in the Dakotas and see buffalo in every direction for thirty miles. Herds were so large that it took days for them to pass by. Wyatt Earp described one herd of a million animals stretched over a grazing area the size of Rhode Island. Within nine years of that sighting, the buffalo had vanished from the Plains. U.S. government policy was explicit. As the Civil War hero General Philip Sheridan wrote at the time, "The buffalo hunters have done in the past two years more to settle the vexed Indian Question than the regular army has accomplished in the last thirty years. They are destroying the Indians' commissary. Send them powder and lead, and let them kill until they have exterminated the buffalo." Between 1850 and 1880 more than 75 million hides were sold to American dealers. No one knows how many more animals were slaughtered and left on the prairie. A decade after native resistance had collapsed, Sheridan advised Congress to mint a commemorative medal, with a dead buffalo on one side and a dead Indian on the other.

The Kiowa resisted encroachment until their war leaders were dead or imprisoned and the buffalo eliminated from the Plains. With the herds gone, the people were finally forced to settle on reservations. In 1887 the last Sun Dance was held on the Washita River, just above Rainy Mountain Creek. To consummate the sacrifice, to secure the buffalo head to place at the foot of the Tai-me, the Kiowa had to send a delegation to Texas to beg for an animal. Three years later they could find no buffalo anywhere and had to make do with an old and weathered hide. But before they could begin the dances, a company of troops arrived to disperse the gathering. On July 20, 1890, the Sun Dance was officially outlawed, and on pain of imprisonment the Kiowa were denied their essential act of faith.

Like the Ghost Dance, with its messianic hope that the world might once again be free of whites and replenished with the spirit of the old ways, the Peyote Cult flourished in the wake of the collapse of Indian culture. The dried plant first appeared among the Kiowa around 1850, acquired no doubt by war parties that commonly raided deep into Mexico, striking settlements as far south as Durango and reaching southwest to the Gulf of California. Away from their homeland for months at a time, the Kiowa found respite in the mountains of the Sierra Madre, among the Mescalero Apache and other tribes that, in turn, had learned of peyote from the Tarahumara. At a time when the entire world order of the Kiowa and Comanche was collapsing, when the Tai-me no longer protected the weak and the Sun Dance itself had faded in memory, peyote offered the Kiowa and Comanche an astonishing affirmation of their fundamental religious ideas. Unlike the formal religions of the Southwest, where the cult of the seed had overthrown the hunt and a hierarchical priesthood had turned the shaman's poetry into prose, the Kiowa still placed immense value on the individual visionary experience. Peyote was a pharmacological shortcut to distant mystical realms traditionally reached by the pain of ordeal, the rhythm of ceremonial drums, the hunger and burning thirst of the vision quest. The first Kiowa to use peyote was Big Horse, who ingested the plant alone, became an eagle, and soared over the land to locate the war parties of his enemies. Half Moon, a Caddo Indian known to the whites as John Wilson, ate peyote every day and night for a month until his spirit fused with the sky and the ritual outline of the ceremony of the peyote church came to him in a vision.

Charlie Charcoal sat on a small wooden bench at the edge of a willow arbor open to the cool evening breeze. Heavyset, with a warm round face and dark hair, cut short and parted on one side, he wore as he always did a white shirt and a wide tie that in the fashion of the times reached just halfway to his waist. Schultes sat on the dirt floor beside La Barre, who was taking notes as Charlie described a vision.

"Once in the fire I saw a young lady," he said, "waist high, hair down her back, small braids drawn back from her temples and tied with a ribbon. Wind was blowing her hair, and she was dancing and keeping perfect time with the singing. She was looking toward the peyote. She was wearing a buckskin dress. And she was smiling, too. After the music stopped the picture disappeared. Now how would you explain this?" Charlie asked. Schultes looked over toward La Barre.

"There is a woman," Charlie continued, "who comes to meetings. You can't see her, but sometimes when two men sing, the tones throw out a higher voice, a beautiful woman's voice that sings together with the men. Some call her Peyote Woman. They say she is the daughter of Buffalo Woman."

"And both are children of the sun," La Barre said. Charlie nodded. Schultes didn't understand. "Peyote is the incarnation of the sun," La Barre explained, "as are buffalo. One is a plant, the other an animal.

The visions are like dreams. They come and go, and for the old people dreams and visions are the same—revelations that must be obeyed."

"I've never had a dream," Schultes said. La Barre looked at him incredulously.

"You must have."

"No, really. Nothing that I can remember."

"But if you haven't ever experienced a dream," La Barre asked, "then how can you be so sure you haven't dreamed? How do you know what a dream is? How do you know what you've been missing?"

"I've just never had a dream," Schultes repeated.

"Then you can never have a vision," La Barre replied.

"I've never had a vision, either, just colors."

"You know," Charlie said, smiling, "it's kind of different for you and me. I mean, like once I was going down into a canyon with large pine trees. As I looked back, there was this steep bluff. I had to keep going downward until I was lost. And then a voice spoke to me and guided me out of the canyon. Then that voice became a squirrel. And I thought that was interesting. What do you think, Wes?"

He looked at La Barre, who smiled but said nothing. Charlie continued.

"All these white men come to study peyote, and they don't see no animals. Animals don't mean nothing to them. They just see people, people playing music. Now, Indians don't see no piano playing, but they catch songs from the wind blowing and they make music from the birds singing.

"One time I saw a man wearing a flowered cloak and gold braids with flower designs like oak, with a white cloth hanging down his chest. An elderly man, not old. He had his arms folded and was smiling toward the peyote, nodding his head very slowly as if it— I did not hear him speak, but it seemed to me every time he bowed his head he said in my mind, 'It is good, it is good.' "

There was a rustle in the darkness, and Mary Buffalo slipped quietly beneath the arbor. In her hands she carried three folded blankets and a bundle of sage. She nodded to Charlie, who rose slowly, beckoning Schultes and La Barre to follow.

"The Roadman is waiting," Charlie said, taking the blankets from Mary. They walked swiftly through dry grass to a small grove of blackjack oaks and hickory that grew at the back of Mary's land. A dozen or more men stood around a fire, and behind them the white canvas of a large tepee shone in the moonlight. As Schultes approached, a grasshopper sprung out of the grass and struck his face. He swatted it impulsively and sent a murmur of laughter around the fire.

The men greeted Charlie and his young companions. Schultes worked his way through the crowd, offering his hand to each bemused person. In the darkness it was impossible to tell whether he had met them before or not. They were all dressed in cowboy clothes, dark trousers and open shirts. Several carried folded blankets over their shoulders. Most were young. The one elder wore face paint and had his hair long in braids. As he stepped from the entrance of the tepee into the firelight, Schultes could see a band of yellow on his forehead and thin red lines running up and down his cheekbones. A mescal bean necklace and a leather satchel hung over his shoulders. He wore a blanket around his waist and held in one hand a beaded staff, in the other a fan of eagle feathers, a rattle, and a whistle made from the wing bone of a large bird.

"I am going into my place of worship," he said quietly. "Be with us tonight."

The Roadman turned and began to walk. Schultes glanced at Charlie, who nodded. The other men stopped talking and one by one fell in behind the Roadman until the entire gathering had been reduced to a single file shuffling slowly around the tepee. Reaching the entrance once more, the Roadman paused for a moment to pray and then slipped beneath the deerhide flap that covered the entrance. The men followed and once inside continued to move around the periphery in a clockwise direction until each had taken his place on a bed of sage that encircled an altar and small fire. To the west of the fire the Roadman sat alone.

As Schultes settled in somewhat awkwardly, Charlie Charcoal quietly pointed out the officers of the church and the ritual objects that lay just behind the Roadman—a gourd, a sprig of sage, a ceremonial staff, and a large leather bag full of peyote. To Schultes's immediate left sat the Drummer, and beyond him was the Roadman. To the Roadman's left sat the Cedarman, and then a row of worshipers around the tepee led to the Fireman, whose position was to the east, beside the entrance of the tepee and the source of the rising sun. The Water Bearer sat just in front of the Fireman. Beside him offerings of corn and water lay in a row, in line with the fire and perpendicular to the raised crescent altar, that curved to shelter the flames. The symbolic effect was that of an arrow, its shaft running from the east and its head piercing the Roadman's heart.

"The altar is the moon," Charlie said quietly. "It is the mountain where Peyote Woman found the first plants. Do you see the track running across it?" he asked, indicating a narrow groove along the length of the altar. "That is the road you must follow to obtain peyote knowledge."

The fire flared momentarily and cast bright shadows across the opposite side of the tepee.

"The peyote spirit is like a little hummingbird," Charlie said. "When you are quiet and nothing is disturbing it, it will come to a flower and get the sweet flavor. But if it is disturbed, it goes quickly."

The Fireman moved close to the fire and stirred the flames. He added a few small pieces of wood and then drew the ashes away from the base of the fire and mounded them between the fire and the altar. With his hands he spread the ashes in an arc parallel to the altar and then slowly worked them into the shape of a wing.

Schultes glanced over at La Barre, who sat beside the door, just across the opening from the Fireman's position. It was the spot Weston always managed to sit in so that in the morning, when the ritual offerings of food were passed around, he would have the first taste and not have to share the bowls with everyone present. Schultes smiled. Since leaving Boston he had found himself in a new world, a place where the rules he had been brought up with no longer had much relevance, where his own eccentricities went unnoticed by people content that he was willing to sleep beside them on the ground, eat their food, and respect their ways.

"The Drummer," Charlie said. Schultes turned to his side and watched as the man on his left poured water into an iron kettle and then added hot coals and a handful of fresh leaves. He took a wet piece of buckskin and tied it tightly over the kettle. The steam smelled of the prairie. As the buckskin tightened with the heat, the drummer tested the sound, tuning it by twisting the small round stones that held the rawhide cord in place on the side. He blew air across the surface, rubbed it with his thumb, and tapped it twice with a beaded drumstick decorated with a tuft of red horsehair.

"The drum is thunder," Charlie said, "the coals lightning, the water rain. Now, follow the Roadman."

On a dirt floor the color of pipestone the Roadman had spread a velvet cloth on which he placed a sheath of tobacco leaves, a leather bag, a whistle, a rattle, and a stack of corn husks and leaves. Reaching toward the fire, he spread a small rosette of sage on the center of the altar. Then with a gesture that silenced all casual talk, he laid a large dried peyote plant on top of the sage.

"Father Peyote," Charlie whispered.

A pouch of tobacco and oak leaves passed around the circle of worshipers. Each man rolled a smoke, and then the Fireman took a ceremonial smokestick from the fire, blew its tip into flame, and handed it to the Roadman, who lit his cigarette and then passed the smokestick to his left. With the ember slowly making its way around the circle, the Roadman made an offering of tobacco to the altar.

"Look down on us, Father Peyote," he prayed, "and guide us for we are ignorant." Schultes was the next to last to receive the smokestick. He lit his cigarette and inhaled deeply. He dug a fingernail into the wood. It was soft and white, probably cottonwood. Carved into the handle was the figure of the waterbird, the same image that was slowly taking form in the ashes before the fire. For several minutes he and the men smoked and prayed in silence. When they had finished, each in turn stood up and placed his cigarette butt on the end of the altar. The Roadman cast juniper leaves into the fire, and the smoke rose like incense. Some of the men had fans of eagle feathers. Others drew the smoke to them with their hands. The Roadman lifted his bag of peyote to the altar and described four circles with it in the air. Charlie Charcoal nudged Schultes.

"The ones with eagle feathers," he whispered, "are those who have seen the eagle in their visions."

The Roadman took four peyote buttons from the leather bag and placed them before the altar. The Cedarman repeated the gesture and then circulated the bag in a clockwise direction. Taking a bundle of sage in his hands, the Roadman knelt before the fire and rubbed his entire body with the leaves. Others did the same, and the scent of crushed sage and juniper mingled with the blue veil of tobacco smoke that hovered just above the worshipers. As the men reached into the smoke with their fans, the Roadman added more juniper to the fire. He took his staff in one hand, his rattle in the other, and slowly described four circles in the smoke.

"Eat these and remember that they are sweet," Charlie said as he passed the leather bag to his left.

Schultes withdrew four peyote buttons and placed the bag on the ground in front of the Drummer. The texture of the dried plant was that of brittle leather. Once in his mouth, it was several minutes before the flesh softened and the first waves of nauseating bitterness swept over his senses. He swallowed hard.

Charlie spat a mouthful of peyote into his hand, turned it over with his finger, and picked off the small tufts of hairs.

"These will make you blind," he said.

"Wonderful," Schultes whispered.

The Roadman held a sprig of sage and his staff before the altar. He closed his eyes and, shaking his rattle, began in a high nasal tone to sing the Hayätinayo, the Opening Song that heralds the spirit of Father Peyote.

Hei yana hi ya na nei
Hei yana hi ya na nei
Hei yana hi yoi na nei, hei yana ha yoi na hei nei
Hei yana hi nei na dok' igo ba ko onta
Hei nei yo wah.

May the gods bless me,
help me, and give me power and understanding.

The Drummer motioned to the altar with his drumstick, pulling smoke from the fire over the drum. Then began the sound that would continue intermittently until dawn, a single drumstick struck on wet buckskin. The beats came so fast that they merged into a steady, almost electronic hum, exactly the tone of the auditory hallucinations that soon would be running around Schultes's mind.

It was forty minutes or more before Schultes felt the first effects of the plant. An unpleasant nausea was soon overcome by a whimsical sensation in the periphery of his field of view, an intuitive sense of space, a cushion between himself and everything else in the tepee. There followed a fleeting moment of pure clarity, an almost crystal awareness of the appropriateness of where he was and what he was doing. He looked at Charlie, whose eyes had changed to sparkling beads, whose face flushed in gentle undulations that were slowly growing in intensity. Schultes blinked. The music had stopped. The Roadman was handing his staff to the man on his left; the Drummer was passing his drum to the Roadman. Each man in turn was drumming for the one singing. The songs ran together. Schultes found himself paying the utmost attention to every sound, and each one burst into another thought that washed over him with immense feelings of goodwill. All around him there were old men praying, tears running down their cheeks, voices tender with emotion, bodies swaying in adoration, hands reaching out for Father Peyote. He tried to move. Charlie stopped him.

"Never step between the fire and a man praying," he cautioned.

Schultes began quietly to laugh. The shadows on the tepee wall were so much larger than the men beneath them. It was as if a gallery of spirits were dancing.

A pouch of tobacco came his way. His fingers fumbled over the moist brown strands. The scent was as rich as memory. He crushed the oak leaves between his fingers. He took a peyote button in his hand. He shut his eyes and experienced warm flushing sensations and a sound that seemed to link his body to the earth. There was the sense of the feel of the earth, the dry desert soil passing through his fingers, the stars at midday, the smell of cactus and sage, the feel of dry leaves through hands. When he once again opened his eyes, the men were slowly fusing one into another, and every movement shot pulsating bolts of color in orblike brilliance—diamonds turning into tunnels, windows into waves, oceans into rain. He lifted his hand before his eyes and was amazed to see a trail of light from his thigh to his fingers. Everything was reduced to sensation. His heart. His hair like straw. His jaw moving up and down chewing more peyote. The men around him ate it constantly. He watched their faces, and the taste in his mouth changed. It was no longer bitter and sour. If the desert itself had a flavor, this was it.

Time turned into color. Every thought unleashed a sound, every gesture a rainbow of light. Schultes tried to concentrate, to follow a single train of thought, but found it was impossible. Resisting the wild flow of his thoughts was physically painful. There was another cloud of smoke as the Roadman placed more juniper on the fire. He seemed so calm, so outwardly untouched by the peyote. The Fireman went about his work, sweeping the floor of the tepee, tossing the cigarette butts into the fire, tending the flames, drawing the ashes into the shape of the Waterbird. Sometime close to midnight the Fireman placed a bucket of water close to the fire while the Roadman sang the Yáhiyano, the first of the midnight water songs. After a second song the Roadman left the tepee, and moments later Schultes heard four high, piercing whistles. When the Roadman returned, he prayed over the water and gathered the Cedarman, Drummer, and Fireman to the sides of the bucket so that together they might form a cross, an image of the four directions. The Fireman blew four puffs of smoke over the water and thanked all those present for the honor of having tended the fire that carried their prayers to the heavens. The bucket was then passed around the circle and taken outside.

At the Roadman's invitation several of the men left the tepee. Schultes remained behind, uncertain whether his legs would carry him. Time passed, and the singing and drumming continued almost constantly, for there were twenty or more worshipers and each had four songs to chant. The amount of peyote consumed varied, with some men eating as many as forty buttons. Schultes had eaten ten or twelve; it angered him that he didn't know. In the middle of the night, well after the Midnight Water ritual, there was a healing. The patient was an old woman, a cousin of Mary Buffalo, and thus it fell to Charlie Charcoal to chew the peyote for her. The Roadman placed the partially masticated buttons in her mouth and with his hands slowly worked her jaw. He bared the upper part of her chest, and taking a burning ember between his teeth, blew sparks over her skin four times. Then accepting the bucket from the Water Bearer he sprayed water onto her face and chest, all the while tracing the form of a cross in the air and whispering soft invocations to the spirits.

A sudden wave of nausea took Schultes out of the tepee. Once clear of a small group of women huddled around the fire, he stumbled against a tree and retched as he never had, deep spasms that were less painful and disturbing than peculiar. Though the peyote had begun to run its course and the wild imagery was already receding into memory, his mind and body remained apart. He was watching his body vomit. He looked up. The dawn was slipping through a pearl gray sky. The branches of the oak trees spread like veins. Birds were beginning to stir. Behind him in the tepee he heard the Roadman singing the first words of the Wakahó, the daylight song for the morning water. The voice seemed terribly far away.

"You okay?"

He turned. Charlie Charcoal was walking toward him. He had a blanket over his shoulders, and his words seemed to come out of a perfect stillness, as if the air itself was held in time, suspended between the past and this new morning. A branch snapped, and the sound was that of ice cracking from an eave and falling into snow. He tried to answer and found his words falling away from his thoughts. His jaw was sore. His mouth felt rubbery, and the words that finally formed emerged into the moment as if spoken by someone else. He could listen to their deep and resonant tone.

"I'm fine," he said.

The air was cold. A sharp and distant birdcall held his attention for an instant and then was gone. The morning songs became louder. Schultes listened and discovered in the prayers a weary melancholy that he hadn't noticed during the night. And in the midst of the sounds he heard the unmistakable yelp of a coyote.

He-na-wa-ya, he-na-we-yo
He na we yo, he ha we wo wo wo
He-na we ne
Ya-na he-na-we-ne
Ya-na he na we ne yo wah.

"Roadman's doctoring," Charlie said, "blowing away the sickness." Schultes looked and saw smoke rising past the tepee poles.

"I hear coyotes," he said.

"That's how it happened. Some Comanche medicine man got his healing power from the coyote. So he made the song and now it's dedicated to the coyote. That's why he sings. Because it's morning and he's got the power. He's made everyone well." Charlie reached into his pocket for a cigarette.

"Seems like old Wes had a pretty good night," Charlie said.

"What?" Schultes asked.

"He was just there all the time. Then his eyes kind of started to grow on him, and pretty soon the Roadman's head turned into some kind of duck, and the drum, too. Something like a gila monster."

"He said that?" Schultes asked. Charlie nodded. Behind him Schultes could see Mary Buffalo and her granddaughter passing small bowls of food into the tepee. Once again he realized that the night was over. The sweeping of the tepee floor, the ashes drawn into the final form of the Waterbird, and the ritual food offerings of sweetened meat, parched corn, fruit, and water had for almost a month now marked the beginning of his days. How strange to have had so little sleep, to have experienced such hallucinations, to have known nothing that was normal and yet feel so little fatigue.

"Gonna be winding up pretty soon," Charlie said, leading Schultes quietly back toward the tepee. Once inside it took a moment for his eyes to adjust to the light. The singing had stopped. The worshipers sat as they had all night, cross-legged and upright, eyes lost in prayers. La Barre, too, remained in place. The circle shifted to the left, allowing Schultes and Charlie Charcoal to sit beside him by the door. Weston's eyes, Schultes noted, were wildly dilated, and there were streaks of sweat and dust on his face. He had a small pile of food on his lap, and his smile suggested that eating was just about the last thing he wanted to do.

Food and water passed around the circle, and for half an hour or more the worshipers remained silent. Then in a final ceremonial gesture the Roadman sang the Gayatina, the round of four Quitting Songs that marked the end of the ritual.

Ki da bw da ya-na hai yoi no,
He ne yo wah
Dok'i ki-da bw-da ya-na hai yoi no
De k ä on ki-da bw-da ya-na hai yoi no
He ne yo wah.

As with most peyote songs, the verses were composed of words scattered among sounds that had no literal meaning. The effect was that of a single human voice surrounded and echoed by the syllables of nature.

"Day is coming," the Roadman sang. "Creator, it is good. Creator, it is good."

When the last cycle of songs was complete, the Drummer unlaced the buckskin from the surface of the drum, and the kettle slowly made its way around the circle. Each man dipped his fingers into the scented water and touched his lips.

"The charcoal and water are sacred. They are the thoughts of our grandfathers. You must do as the Old Man says. He has lived. He is good."

With this final message and blessing the Roadman removed Father Peyote from the altar, put away the last of his paraphernalia, and instructed the Fireman to lead the worshipers out of the tepee. As they filed out into the morning light, Schultes and La Barre spoke for the first time since the previous night.

"Well?" La Barre said.

"Charlie thinks the meeting went through over four hundred peyote buttons. It's no wonder they purchase them by the thousand lot down in Laredo."

La Barre laughed. A thousand peyote buttons for $2.50. It was incredible. All of that for the price of a pair of blue jeans. He turned to Schultes.

"I saw something so beautiful: the Roadman standing before a half-moon altar as he pierced the night sky with arrows of light. The sky collapsed over my head, bringing down heavenly dust that mixed with the smoke of the fire and left me alone on the ground, floating between a velvet cloth and the warmth of my mother's womb."

"I just saw colors," Schultes said. Charlie Charcoal grinned.

"Come on," he said, leading the two young men away from the clearing and toward the trail that led to Mary Buffalo's house.

"We must speak with the Roadman," Schultes said. Behind them the Kiowa were removing the canvas from the tepee. The tall poles cast harsh shadows across the ground, and the morning wind had already begun to scatter the ashes of the fire. The Roadman graciously accepted their thanks, encouraged them in their work, and even suggested that a photograph be taken as a souvenir. La Barre handed his camera to Charlie, who lined up the Roadman and the two students for the shot.

"Just a moment," Schultes interrupted, reaching into his pocket for a comb, which he ran quickly through his hair. He tightened his tie, dusted off his trousers, and adjusted his belt.

"Okay," he said, looking grimly at the camera. A week later he was, as planned, riding a Greyhound bus out of Tulsa on his way back to Boston.

Weston La Barre to this day cannot understand why it was he and not Schultes who experienced the full onslaught of the peyote visions. At first he felt, as he put it, that Schultes was a prisoner of reality. "Certainly," he would explain almost sixty years later, "there was no chance of Schultes going native. He was born grown up. The only way for him to go native would be to go to England." Yet La Barre remembers his time with Schultes with great fondness and pride. It was, after all, the six weeks among the Kiowa that forged Schultes's career in ethnobotany. Schultes, too, never understood why he did not experience true hallucinations.

"I get colors," he would later write, "lightninglike flashes, little stars like when you break a glass, sometimes colored smoke going by like clouds. I wish I could see visions. La Barre has tried to explain it to me, but I don't understand what he is talking about." It was not until the release of the film *Fantasia* in 1941 that Schultes finally found a reference point for what he had seen.

"The 'Toccata and Fugue' of Bach," he would exclaim to anyone who would listen. "That's what I saw!"

Though his own experiences with peyote never made much sense to him, Schultes did, at a remarkably young age, understand and respect the role that the cactus played in the lives of the Kiowa. Beginning in 1916 there had been no fewer than nine attempts to introduce legislation in Congress prohibiting the traditional use of the plant. Schultes abhorred such initiatives. In February 1937, six months after returning to Harvard from Oklahoma, he testified against Senate Bill 1399, the latest attempt to interfere with the religious practices of the Kiowa. He said it was a cruel and violent obstruction of religious freedom. The Chavez Bill failed, in good measure because of the testimony of Schultes and others who, unlike the proponents of the legislation, had firsthand experience with the Peyote Cult. It was pretty heady stuff for a young student who had never before challenged authority. In his undergraduate thesis he wrote that through the use of peyote the Indians are "able to absorb God's Spirit in the same way that the white Christian absorbs the Spirit by means of sacramental bread and wine." This, too, was a bold idea in the spring of 1937. What he did not yet fully realize, however, was that his studies of peyote were about to reveal to him the missing clue that would allow him to solve one of the most enigmatic mysteries in the entire history of ethnobotany: the long-lost identity of *teonanacatl* and *ololiuqui,* the most sacred plants of the Aztec Empire.

CHAPTER FOUR

Flesh of the Gods,

1938–39

IN THE *Codex Vindobonensis,* one of a handful of pre-Columbian documents to escape the fires of the Inquisition, there is an image of the Aztec feathered serpent god Quetzalcoatl, resplendent in precious stones and wearing a mask with the face of a bird. On his back he carries a woman, just as a bridegroom would carry a bride in ancient Mexico. The woman, too, has a mask, and from her headdress emerge four mushrooms. She is the incarnation of the spirit that dwells within these sacred plants. The next painted figure in the sequence shows Quetzalcoatl singing, beating time with a drum made from a human skull, and addressing one of his divine princes, who holds aloft two other mushrooms. The prince, enchanted by the song, sheds a tear of delight and awe. Above and to the left of this scene there are seven different gods and goddesses, each with a pair of mushrooms. There can be little doubt what this gathering of the gods represented in the imagination of the Mixtec priest who painted the text. It was a memory of fire, a reflection of the first encounter between human divinities and the power of *teonanacatl,* the mushrooms known in the language of the Aztec as "the Flesh of the Gods."

As a young student Schultes first came across references to *teonanacatl* in the early Spanish chronicles. Francisco Hernández, physician to Philip II, wrote of the miraculous properties of peyote, but he also described four mushrooms, including one called *teyhuinti,* "the inebriating one." In his monumental *Historia de las Cosas de Nueva España* the Franciscan friar Bernardino de Sahagún described "a small black mushroom which they called *nanacatl.* They ate these before dawn with honey, and when they began to be excited by them, they began to dance, some singing, some weeping. . . . Some sat down as if in a meditative mood. Some saw themselves dying; some saw themselves being eaten by a wild beast; others imagined that they were capturing prisoners in battle; some believed that they had committed adultery and were to have their heads crushed for the offense."

In the *Florentine Codex* of Sahagún's work, there is an illustration of a demonic figure dancing in a field atop a cluster of mushrooms identified as *teonanacatl.* This image of the devil, clothed in fur and replete with a beaked face, claw hands, and a cloven foot, linked the mushrooms with every savage instinct presumed by the Spaniards to be dwelling in the heart of the New World. In 1524, Cortés, whose soldiers dressed their wounds with the fat of dead Indians and fed the remains to dogs, invited twelve Franciscan priests to introduce Christianity to Mexico. Motolinía, a member of this first contingent, wrote that mushrooms were the flesh of the "devil that they worshiped, and . . . with this bitter food they received their cruel god in communion." Hernando Ruiz de Alarcón employed torture to tear from the Indians the secrets of their faith, information that later figured in a guide to native idolatries written for missionaries by Padre Jacinto de la Serna in 1656.

Yet try as they might, the Spaniards could not veil the sublime character of *teonanacatl.* In Serna's strange book Schultes read that "priests and all the men . . . went to the hills and remained almost the whole night praying. At dawn, when a certain breeze that they know began to blow, they would gather them, attributing to them deity." According to the Dominican friar Diego Duran, intoxicating mushrooms were served at the coronation of the Aztec emperor Ahuitzotl in 1486. Tezozómoc, writing in Spanish in 1598, described the same ritual occurring at the crowning in 1502 of his grandfather Moctezuma, who ruled until overthrown and murdered by the Spaniards in 1520. The true character of *teonanacatl* and its profound role in the religious life of pre-Columbian Mexico are perhaps best revealed by an illustration that appears in the *Codex Magliabechiano,* another early-sixteenth-century source. Unlike the *Florentine Codex,* which was drawn by a Spaniard, the *Codex Magliabechiano* comes from the pen of an unknown Indian, a person evidently still imbued with the spirit of his past. The illustration, labeled simply *teonanacatl,* reveals a seated man eating mushrooms. Hovering over his shoulder is an image of his god. At his feet three mushrooms emerge from the earth. They are painted green, the color of jade, the Aztec symbol of the sacred.

The significance of *teonanacatl* was not lost on young Schultes. But no sooner did he learn of the mushrooms than he read that they did not exist. In a series of academic articles beginning in 1915, William E. Safford, a highly respected economic botanist at the U.S. Department of Agriculture, wrote that the early Spaniards had been deceived. Three centuries of field research, he argued, had failed to uncover any evidence of a "narcotic Mexican fungus." His own surveys of the literature and herbaria had been equally disappointing. What's more, he suggested that "a knowledge of botany has been attributed to the Aztecs which they were far from possessing. . . . The botanical knowledge of the early Spanish writers Sahagún, Hernández, Ortega, and Jacinto de la Serna was perhaps not much more extensive." They had all been fooled, Safford maintained, by Indians intent on concealing the true identity of their sacred plant. *Teonanacatl* was not a mushroom but rather an Aztec name for the dried tops of peyote, which resemble dry mushrooms "so remarkably that, at first glance, it will deceive a mycologist."

Schultes knew that Safford's argument made no sense. True, when dried, peyote and mushrooms generally assume a drab olive green color, but they share this characteristic with about half a million species, roughly as many plants as there are in nature. There, however, the similarities end. Even a cursory glance at dried peyote reveals closely packed silky hairs on top and the remains of vascular structure below, traits lacking in a fungus. *Teonanacatl* and peyote grow in completely different habitats, the former in wet mountain pastures and the latter in hot, dry deserts. The Spanish recognized this obvious ecological difference just as they explicitly distinguished "the root which they call *peiotl* and *nanacatl* which are harmful mushrooms." Unlike Safford, Schultes did not question the botanical knowledge of the Aztecs or the early Spaniards. He was versed in the classics and read Latin and Greek. He knew that in the sixteenth century physician and botanist were one and that their training in botany was precise, their insights consistent. His work with Mary Buffalo and the Kiowa had revealed the depth and complexity of indigenous plant lore. Indeed, he had already committed himself to a life of unveiling its mysteries. For Schultes, Safford's argument was a classic case of scientific hubris. In a nutshell, Safford was saying that because he couldn't find contemporary evidence of *teonanacatl,* the mushroom could not exist—this notwithstanding the fact that no American scientist, least of all Safford, had gone to Mexico to have a look.

Had anyone but Safford proposed such a theory, it would never have become enshrined in the literature, as indeed it was for more than twenty years. The problem was that Safford was a good botanist and his professional prestige was considerable. The recognized authority on New World narcotics, he had published an important monograph on datura, the toxic hallucinogen known as the Holy Flower of the North Star. In 1916 he had successfully shown that the botanical source of the psychoactive snuff *cohoba* was not tobacco but rather the seed of a leguminous tree identified now as *Anadenanthera peregrina.* It was an important discovery that had eluded botanists for four hundred years, ever since the use of the drug was first observed among the Taino Indians in Hispaniola in 1496.

Discoveries such as these lent credence to Safford's ideas about *teonanacatl,* and by 1936, when Schultes came onto the scene, there had been no published objection to his theory. Those few who cared about such matters had long concluded that intoxicating mushrooms had never existed in the Americas—everyone, that is, except two young scholars who spent the summer that year in Oklahoma eating peyote. Both La Barre and Schultes had the audacity to go on record as saying that Safford was wrong. Their interpretation of the early Spanish sources left them certain that *teonanacatl* was indeed a mushroom and that somewhere in the distant hills of Mexico there might still be a contemporary cult employing the plants as a sacrament.

An unexpected breakthrough came several months after Schultes returned from the West. While writing his undergraduate thesis he traveled to Washington, D.C., to study peyote specimens preserved at the Smithsonian in the collections of the United States National Herbarium. On herbarium sheet number 1745713 he found a letter dated July 18, 1923, addressed to the director of the herbarium, J. N. Rose, and written by one Blas Pablo Reko of Guadalajara, Mexico. Safford, who had died several years before, had had the foresight and generosity of spirit to pin the letter to a voucher specimen of peyote. Schultes was stunned by an offhand comment that came at the end of the letter. "By the way," Reko wrote, "I see in your description of *Lophophora* that Dr. Safford believes this plant to be the *teo-nanacatl* of Sahagún, which is surely wrong. It is actually, as Sahagún states, a fungus that grows on dung heaps, and which is still used under the same name by the Indians of the Sierra Juarez in Oaxaca in their religious feasts."

Schultes wrote immediately to Reko and received by return post several mushrooms said to have been collected from among the Otomi Indians of Puebla. The specimens were poorly preserved and impossible to identify. They appeared, however, to be in the genus *Panaeolus*. This was all the incentive that Schultes required. He had always intended to carry his ethnobotanical studies into the rain forests of Latin America and had already discussed with Oakes Ames the possibility of studying the economic botany of Oaxaca for his dissertation. A chance to solve the mystery of *teonanacatl* was an added bonus. In the early summer of 1938, equipped again with a grant from Ames's pocket, he was once more aboard a Greyhound bus, this time making his way south to Mexico City.

The first thing Schultes discovered about travel in Mexico in the thirties had to do with timing. The best way to get to Oaxaca from the capital was to take the train; there was no road. If a train was scheduled to depart at seven in the morning, you would phone the station at ten and ask when the seven o'clock train was leaving. "At one," might be the response. That meant you began to pack at three and make your way to the train station around half past four. At six, with tremendous fanfare and not the slightest indication of concern or embarrassment, the dispatcher would ring a bell, signaling the conductor, whose high, piercing whistle left the platform whirling with commotion. At seven in the evening or slightly before, the train would make its way slowly out of the station.

The second thing Schultes learned was that the track, which ran south from Mexico City through Puebla and into the dry intermontane valley of Oaxaca, had been repaired—quite often, as it turned out—with the wooden ties replaced by lengths of cactus. Anticipating problems, the crew carried massive hydraulic jacks. Between derailments—there were two on his first trip south, one every 125 miles—he practiced Spanish with the Indians and German with his new traveling companion, Blas Pablo Reko, an Austrian by birth who spoke only broken English. Schultes's initial attempts at Spanish failed miserably. Turning to an elderly Zapotec woman, he politely asked her age.

"¿Quantos anos tiene, señora?" he said, inadvertently leaving off the tilde in *años.* Her face blanched.

"Solamente uno, señor," she answered, recovering her composure. Reko explained to Schultes that he had just asked her how many assholes she had.

Schultes did much better in German, at least until the conversation took an unexpected turn. Dr. Reko was a pleasant if somewhat strange and eccentric little man in his late sixties with a passion for anthropology and botany. As a young physician he had worked in the mining camps and on the railways that had opened up southern Oaxaca and the neighboring state of Chiapas. A tireless field-worker, he later explored much of southern Mexico on foot or horseback, returning to the capital with wild theories about Indian astronomy, linguistics, religion, and race that were dutifully printed in *El México Antiquo,* a small journal published by the German community in Mexico City. Infused as they were with his own peculiar ideology, few of his ideas were taken seriously by Mesoamerican ethnologists. His ethnobotany, however, was remarkably sound.

As early as 1919, four years before he wrote to the Smithsonian, Reko had reported in an article on Aztec plant names that *"nanacate* was a black mushroom that produces narcotic effects." It was not until 1936 that he obtained actual samples of the mushrooms; the specimens he sent to Harvard arrived in such poor condition as to be unrecognizable. These had come, he informed Schultes as their train struggled south, courtesy of Robert J. Weitlaner, an anthropologist who had been working among the Mazatec in the town of Huautla de Jiménez. During Easter week of 1936, while engaged in studies of the Mazatec calendar, Weitlaner had become the first outsider in four hundred years to see *teonanacatl.* His informant was a shopkeeper named José Dorantes, a Mazatec merchant living in Huautla who claimed to have witnessed the mushrooms being used in divinatory rituals.

This information was a stunning revelation for Schultes, and it almost made up for the diatribe he was forced to endure for the rest of the train trip. It was July 1938, only three months after the German rape of Austria and little more than a year before the outbreak of war in Poland. Mexico was thick with German sympathizers, and Reko, Schultes discovered to his dismay, was one of them. As the train bounced and clattered along the dry riverbed of the Río Salado, this Mexican citizen of Slavic blood and Austrian birth boasted of racial purity and spoke confidently of the imminent takeover of the world by the Germans. When Schultes begged to differ, suggesting that the moment Germany went after Britain, Hitler would be destroyed, Reko was shocked.

"Schultes is a good German name," he said. "You must be Jewish."

"Unitarian," Schultes responded.

"With Jewish blood," Reko decreed. Schultes suggested that they restrict their conversations to matters botanical. Thus in a world on the edge of a precipice, with madness as the backdrop of the age, Schultes found himself about to enter the mountains of Oaxaca to search for the long lost *teonanacatl* with an ardent Nazi as a companion.

In 1938, long before roads sliced into the rain forests of Chinantla in the east and dams flooded immense valleys to the north, before thousands of itinerant travelers clogged the marketplaces of Huautla and Ixtlán, there was a mystique that touched all scholars swept into the remote and inaccessible valleys of northeastern Oaxaca. On a map the land was not large: a hundred miles or more in length and half that in width. Yet within that condensed space it could take four days to travel twenty miles. Entire Indian nations were compressed into astonishingly small territories. The Mazatec land encompassed perhaps eight hundred square miles; the Chinantec and Mixe, Cuicatecs and Popolocas, controlled far less. The region was poorly known. Of the fifty-five thousand Mazatec, 80 percent spoke no Spanish. Over 70 percent of the Chinantec remained monolingual. The figure for the Mixe, a people never conquered by Aztec or Spaniard, was even higher. There were tribes—such as the mysterious Guatinicamame, mentioned in colonial documents and not seen since—still believed by some to be living in the midst of the Chinantla rain forest. Only a few years before, an anthropologist had rediscovered the Ocuiltecos, an indigenous group thought to have been extinct for centuries, living within fifty miles of Mexico City. In remote Oaxaca, anything seemed possible.

Topography made for an extraordinary diversity of landscape and habitat. In the east, the lowland forests of the coastal plain swept like an immense wave up against the slopes of the Sierra Madre. There the torrid air coming from the Caribbean condenses in soft billowy clouds that soak the mountains. Winter rains last until those of spring and summer, months pass without a sign of the sun, and there is no dry season. Ten miles to the west, the cloud forests with their elfin vegetation, dangling epiphytes and cascades of orchids, bromeliads, and aroids give way abruptly to rain shadows and forests composed of willow and a dozen species of oak. The summits of the highest peaks are rocky outcrops, barren save for scrub pines and the odd raptor overhead. The transitions are abrupt and dramatic. Mountains rise to ten thousand feet, valleys drop close to sea level, and the interior is dissected by deep *barrancas* that fall away from the heights and ultimately run into the desert of the valley of Oaxaca.

Before the arrival of the Spaniards, these mountains were known as the Land of the Deer. The Mazatec venerated the animals as gods; hunting was forbidden, and as a result thousands of tame deer wandered over the land. When the Spaniards came, shamans stole their souls and hid them inside the deer, thus ensuring that as the creatures were slaughtered, the killers would doom themselves. In the immediate wake of the Conquest, missions spread, Dominicans mostly, but they were soon abandoned. By the early part of this century Christian and Indian beliefs had found a certain balance. Though every village had its little church, high-peaked thatch roofs, adobe walls, lovely baroque façades coated with whitewash—which time and rain faded to a golden yellow—and iron bells strung up beneath more thatch, priests were few and far between. Ancient ideas survived remarkably intact. Indian *cantores* recited the liturgy of the saints, but they also whispered the names of the malevolent dwarfs with childlike faces and withered bodies who stole the souls of the living. Blood sacrifices were offered to the guardians of rocks and rivers, mountains, stars, sun, and moon. In markets old women sold curing bundles made up of eggs for strength, cacao for wealth, and copal resin for the spirit. Bright feathers and paper tree bark recorded messages of the gods, as they had since the time of the Aztec.

Language is the filter through which the soul of a people reaches into the material world, and for the Mazatec, in particular, communication was not limited to the spoken word. The language had four different tones. Every conversation had its own key, which was determined by the one who initiated the exchange. In solemn moments words broke into meaningless syllables, songs became chants, sounds resonated in low humming prayers that might go on for hours. On less propitious occasions the Mazatec conversed by whistling. The whistles were not just sounds with generally recognized significance; they comprised an entire lexicon, like a vocabulary based on the wind.

Precise and literal interpretations of whistle speech could be rendered by any Mazatec. Detailed messages, extended conversations, and urgent requests with substantive information could be expressed simply by whistling. It was possible, for example, for two men to meet on a trail, discuss the weather, argue about the worth of a commodity, settle on a price, and continue on their way without a single word having been exchanged. Travelers separated by a quarter mile or field hands working on opposite sides of a valley could communicate across distances that would muffle ordinary voices. The secret lay in an ability to replicate the tonal and rhythmic features of the spoken language, but in a manner that few outsiders, however fluent in Mazatec, could understand.

Schultes and Reko got off the train at a small town called Teotitlán del Camino, stayed long enough to buy four mules and find a blacksmith willing to equip them with horseshoes, and then began the long climb into Mazatec country. The rock-strewn trail rose over broken ground through a desert of frail acacias and cacti, which slowly gave way as the air became cooler and small wisps of cloud appeared in the sky. At the Aztec village of San Bernardino a beautiful woman whose black hair shone with oil fed them tortillas and a frothy drink made from roots and cacao. Beyond the village the road narrowed and climbed gradually to a high summit flanked on two sides by deep valleys. A long descent followed, then a river crossing, another slow ponderous climb, and finally the trail opened up into a high rolling landscape increasingly dominated by lush fields of green corn. Emerging from the midst of the fields were small clusters of houses, white adobe with long, shaggy ears of thatch protruding at the end of each roof gable.

"That is Mazatec," Reko said, pointing to the houses. Schultes was surprised to see a row of men with digging sticks working across a new field.

"They plant a little even now in the middle of the rainy season," Reko explained, "but most of what you see growing was sown in April. They call it *hno-htsee,* rain-corn. The corn they prefer is planted in October at the beginning of the dry season. It's sun-corn, *hno-ndwa.* You will find that this plant teaches you a great deal about the Mazatec."

Schultes drew in the reins of his mule and let Reko ride ahead. Once still, he realized that there was activity and color all around. The brilliant red and white *huipiles* of the women, all ribbons and embroidery, the cotton cloth and straw hats of the men, the flowers, which grew in every doorway, and the red earth itself combined to give the landscape a richness and depth unlike anything he had ever known. Then, in a moment, the sun passed into cloud, and the fields changed hue; the corn became blue-green, and the sky took on the tone of a Japanese print. Such a strange quality of light, he thought, such a sense of a new land.

Ten hours after starting out, Schultes and Reko reached Huautla de Jiménez, the Mazatec capital. It was a small town spread out along the side of a mountain, alone and detached from the world. The streets were dirt tracks, more like dry creek beds than roads, that tumbled through a warren of thatched adobe onto a stone plaza dominated at one end by the bell tower of a large tin-roofed church. By the time they arrived, the sun was going down and the air was damp and cold. The people they met on the way into the center of town greeted them in half-audible, crushed tones and swerved away from their mules. The Mazatec were short, especially the women, who scurried along barefoot, hands clasped tightly to shawls that held their babies close to their backs.

At a small cantina on the corner of the plaza, Reko asked an old woman for directions to the shop of José Dorantes, the merchant they had come to meet. With a wave of her hand that could have meant anything she said, *"Por allá."* Over there. Reko turned and promptly ran into a plump Mazatec wreathed in smiles and reeking of mescal.

"Un poco de fiebre, señor," the Indian explained. Just a little fever.

"Borracho y sucio," Reko said with disgust, his hands dusting off the lapels of his field coat.

"No, no, señor. Solamente un fiebre de Dios." Only a fever of God. Schultes watched as the man stumbled away, his smooth brown back and the muscles of his legs showing through the tears in his tattered clothes.

Señor Dorantes owned a large *tienda* that carried all the goods worth having in Mazatec country. The store was located in the one part of town that had electric lights, a short row of tin-roofed wooden shops facing onto Huautla's main street. Cut into the side of the mountain, the street was the only paved and level surface in town. Every evening between six and ten when the electricity was on, a crowd of children would gather to chase Señor Dorantes as he rode up and down the cobblestones on his prized possession, a red bicycle that had been brought in over the trail from Teotitlán. It was the only bicycle in Huautla and a source of endless fascination and mystery for the Mazatec. The rest of Dorantes's life was rather ordinary. Three days a week he checked his inventory of ribbons and thread, cigarettes, Epsom salts, aspirin, castor oil, canned fish, matches, and macaroni, and then prepared his orders, which went out with the evening pack trains bound for Teotitlán. On Tuesdays, Thursdays, and Saturdays, when the mules and donkeys returned and crowded the street in front of his store, he spent his time sorting through the supplies, making sure that nothing was missing. Thieves and bandits on the mountain slopes beyond Mazatec country to the west were a favorite subject of conversation among the regulars in his shop.

Like most of the town's leading citizens, the *gente de razón* or *civilizados,* Dorantes basked in the company of foreigners. He spoke Spanish and had been educated in the local school. When in the midst of one of his evening rides he ran into an unexpected pack train loaded down with strange-looking equipment and two tired botanists, he welcomed the visitors warmly and invited them to stay in a room above his shop. Later that night, after he had closed the shutters and the crowd of townspeople had finally dispersed, Dorantes sat down with Reko and Schultes and listened as they outlined their plan of study.

Perhaps prompted by Reko's connection with Robert Weitlaner, the anthropologist who had worked previously in Huautla, Dorantes was from the start remarkably candid about the Mazatec use of *teonanacatl.* He confirmed, for example, that he had witnessed the adoration of the mushrooms. He had been to the nocturnal ceremonies, seen the *curanderos* cleanse the mushrooms in a veil of copal incense and smoke, and listened as these remarkable healers invoked through prayer the power of the spirit. Apparently he had not ingested the mushrooms himself, but he knew exactly why the Mazatec used them.

"We are a poor people," he explained, "and we have no doctors or medicine. That is why Jesus gave us the mushrooms. Because He could not be here forever."

"And can we find these mushrooms?" Schultes asked.

"*¡Cómo no!* Why, of course. They grow wherever the earth is alive. In all the fields, all the best fields, I have seen them many times." According to Dorantes, when Jesus walked, mushrooms sprang up in his wake, wherever his blood or saliva touched the ground.

"So Jesus and the mushrooms—"

"This is not for me to say," Dorantes interrupted.

"Can we meet these healers?"

"We can try."

"What about attending a ceremony?" Reko asked.

"Perhaps."

"When?"

Dorantes ignored Reko's question and directed Schultes to a small adobe house three doors down from his *tienda.* "There is a woman," he said,"an American, maybe she will know."

Unlike many missionaries with their uncomfortable zeal, Eunice Pike seemed at peace with herself and her place in the lives of her Mazatec neighbors. A woman of simple grace, she lived in a one-room thatch house with a mud floor and a single window cut into the adobe wall.

Like the Mazatec, she ate corn and beans, drew her water from the one narrow pipe that served the entire town, and when necessary washed herself with the dew that gathered on leaves. Her clothes were plain, but she was tall and strikingly handsome, with long dark hair that she wore pulled back into a bun at the nape of her neck. The daughter of a Connecticut country doctor and only twenty-four, she had been living in Huautla for two years and intended to stay for however long it took to master the Mazatec language. Her goal was to translate the New Testament, a task that she addressed with all her time and energy. She had no interest in buying converts with aluminum pots and modern trinkets. She was honest enough to know that most conversions were shallow and ephemeral, less transformations of the spirit than triumphs of expediency.

"Once I tried to explain heaven to a young woman," she said, smiling, as she poured Schultes a cup of tea. "I said it was a beautiful place, a place where there are no tears. She asked me whether I had been there. I said no. I explained that only the dead know heaven. Then she looked at me with the saddest face. She said she was so sorry for me. And she left almost in tears."

"How strange," Schultes said.

"It was only later I realized that most Mazatec actually claim to have been to heaven."

"With the mushrooms?"

"Yes. They believe Jesus speaks through the mushrooms, that their visions are messages from God. What was it you called them?"

"Teonanacatl," Schultes said. "Some believe it means 'flesh of the gods.' "

"In Mazatec the mushrooms have several names. One translates roughly as 'the little holy ones.' "

"Have you ever seen them?"

"No," she said.

"What about the effects? What do the people say?"

She held his eyes and for a moment said nothing. Then with a sigh of resignation she explained, "There are things we know that we cannot know. Christianity is a thin veneer over the lives of these people. I've heard them singing at night. They always begin with the Lord's Prayer. The leader will say she has the heart of Christ and is the daughter of the Virgin Mary. But then in the next moment she is the daughter of the moon and the stars, snake woman, bird woman, whatever." She smiled and began to laugh softly.

"It doesn't disturb you?" Schultes asked.

"Yes, of course," she said. "But then, no. I mean, how can it, really?

When I first came here I complained about the use of mushrooms to an old man. Do you know what he told me?"

"No." Schultes smiled.

"He said, 'But what else could I do? I needed to know God's will, and I don't know how to read.' "

They both laughed.

"So how does one get the message of God to a people who seem to have something far more spectacular and immediate than anything we have to offer?" She asked the question he had wanted to but hadn't dared.

"With difficulty, I suspect," he said. "What do the padres say?"

"Oh, the Catholics have it even worse. It's hard enough to translate the meaning of the Last Supper, but the Eucharist! Compared to the mushrooms, bread and wine must seem rather tame."

Schultes laughed once more. What an extraordinary woman, he thought—a missionary who could laugh, one who could love God without hating people.

"I once was waiting for an airplane, and I started to sing a hymn. It was one no Mazatec knew. I had just translated it. Two of the women said, 'Isn't it beautiful! How lovely! It's just like the mushroom.' I turned and rather piously told them that it wasn't like the mushroom. That God and Jesus were different. But they wouldn't listen. Can you imagine what they said?"

"No," said Schultes, ready for anything.

"They said, 'We mean, wasn't it gracious for the mushroom to teach you that song.' "

For the rest of July and the early part of August, Schultes and Reko pursued their ethnobotanical studies in and around Huautla and the neighboring town of San Antonio Eloxochitlán. Wherever they went they heard vague rumors about the mushroom cult, but at no point did they see *teonanacatl* or meet a *curandero* willing to describe the rite. Part of the problem was language. Schultes still struggled with Spanish, and neither he nor Reko spoke adequate Mazatec. The Mazatec themselves were reticent. Thus he found himself filling his notebooks in English with information gleaned from somewhat reluctant informants and filtered through not one but two languages that neither he nor the informants understood fluently. To make matters worse, he was completely dependent on Reko for the Mazatec-Spanish-German part of the chain.

In good conscience he could not turn to Eunice Pike for help. The very issues that interested him involved possibilities that she could not acknowledge, publicly or privately. He could sense that the missionary had come to a certain balance with the Mazatec, based in part on her quiet respect for the very religious practices she had been taught to abhor. She did not condone what they did, but her reverence for all things sacred prevented her from condemning them for doing it. She could not expose herself to the mushroom cult, nor would she be willing or able to expose the tradition to the outside world. Thus for the time being Schultes remained content to botanize in the hills around Huautla, focusing more on the plants than the people.

While Schultes was off in the hills collecting plants and searching for specimens of the mushrooms, the mystery was being approached from a different angle by a different group of explorers. The same month that Schultes and Reko journeyed to Huautla, a team of anthropologists arrived led by a tall, strapping Englishman named Bernard Bevan, a man rumored, Schultes would later discover, to be affiliated with the British Secret Service. With Bevan was Louise Lacaud, Jean Bassett Johnson, a young anthropology student from Berkeley, and Irmgard Weitlaner, Johnson's fiancée and the daughter of Robert Weitlaner, the man who two years earlier had obtained the first samples of the mushrooms, the specimens Reko had forwarded to Schultes at Harvard.

On the night of Saturday, July 16, 1938, and using contacts provided by, of all people, José Dorantes, this small party became the first outsiders to attend a vigil in which *teonanacatl* was ingested. It was a healing ceremony, convened in the home of an old *curandero* who spoke no Spanish. Johnson later described the experience in a paper on Mazatec witchcraft published in an anthropological journal in Sweden. The *curandero* began by taking his place before a low table on which rested a large red-and-blue feather, a candle, a red paper package, a square of black bark, a parcel of copal, forty-eight kernels of maize, and a Mixtec basket containing six eggs. On a shelf to one side of the table was another candle, and beside it, wrapped in fresh banana leaves, were the mushrooms.

The *curandero* ate three, chewing them slowly as he invoked the saints and the Holy Trinity. He then instructed the relatives of the patient to place copal in an incense burner. He added copal dust, which ignited, sending up a narrow stream of smoke that hovered at eye level and slowly dissipated throughout the room. Making the sign of the cross and invoking the spirit of the patient, he demanded to know the circumstances of the patient's birth, the position of the stars, and the location where the afterbirth had been buried. Then he prayed a long rambling liturgy that called forth both the saints and all the *dueños* of the rocks and rivers, the mountains, thunder, earth, stars, plants, sun, and moon. A prayer of supplication followed, a direct communication with God that empowered the *curandero* for the critical act of divination. With a final invocation of the Holy Trinity, he picked up the kernels of maize and scattered them over the table. In their pattern lay the future, and with each successive throw came further insights that together formed the prognosis. The patient's fate rose and fell with each toss until on the seventh and final throw the *curandero* announced that the patient would live. As a final healing gesture he told the relatives that he would prepare six small packages made up of power objects from his table—copal, cacao, eggs, feathers, bark paper—which they were to bury in an east-to-west direction beneath the patio of the patient's home. At two in the morning the ceremony ended. No one but the *curandero* had ingested the mushrooms.

A week or so later, when Schultes ran into Bevan's party in Huautla, Johnson shared his discovery. The mushrooms, he explained, were clearly the vehicle of transformation that gave legitimacy to the divinatory rite. All the prayers and chants, each ritual gesture, were the voice of the mushrooms speaking through the body of the *curandero.* Yet it had been difficult to tell what effect they had had on the old man. He had not seemed drunk. His movements had been slow and methodical, as deliberate as those of a priest giving Holy Communion. Moreover, only three mushrooms had been eaten. To take as many as six, according to the *curandero,* was to risk insanity and the certain demise of the patient. Finally, Johnson told Schultes that at least three different kinds of mushrooms could be used. One was known in Spanish as *los hongitos de San Ysidro,* the little mushrooms of San Isidro. The other two were much smaller. One was called *tsamikíndi,* the other *tsamikíshu,* the landslide mushroom.

Ten days after Johnson and the others attended the ceremony, a tremendous rainstorm swept over Huautla. By morning the town had been washed clean. A cool breeze blew from the east, carrying the sound of the market to Schultes's small window in the loft above Dorantes's *tienda.* Waking up in his hammock he could hear the soft voices, a sound not unlike that of palm leaves scraping in the wind. The clouds overhead moved in circles, and beyond the bell tower of the church, high against the flank of the mountain, he could just make out a hawk of some kind, wing tipped into the wind, rising with the air. Across the street in the courtyard of a small house, a young boy knelt as his sister poured a bucket of cold water over his gleaming black hair. How strange, he thought, how much the Indians have taught us, and how little we have given in return. He was not thinking of maize, potatoes, tomatoes, chocolate, pineapples, tapioca, papaya, and a host of other foods, or of the medicines that had changed his world—cocaine, quinine, aspirin. It was their vision of life itself, something he was only beginning to grasp.

He walked down the wooden stairs to the *tienda,* past the hanging pots and machetes and the counters stacked with bolts of fabric, and out a narrow door that led to a small open yard where he had set up his plant dryer. The air had the scent of morning in Oaxaca, roses and carnations, resinous smoke from the kitchen fires, fresh coffee, and the faint odor of leaves mingling with dust and animal droppings, human urine and sweat, and the earth itself, still moist from the evening rain. He washed in a small basin, relieved himself in a corner of the yard, and then turned his attention to his specimens.

The dryer was a simple portable affair, four detachable legs beneath a metal frame that supported a plant press placed horizontally on its side. A canvas skirting around the base of the dryer directed the heat from a small kerosene lamp up through the specimens, which were separated in the press by corrugated aluminum dividers. Depending on the species, it generally took twelve to twenty-four hours to dry a plant. On rainy nights he covered the dryer with a sheet of rubberized canvas, which he now removed. He felt the top of the dividers, tightened the straps that held the press together, and knelt to light the lamp. He wished he could keep it going all night, but with the pigs and chickens the risk of fire was too great.

Perhaps only a botanist can know the quiet pleasure he felt as he worked through his specimens. Between the sheets of newsprint there was a wild iris with a bulb that had been roasted and eaten since before the time of the Aztec; an orchid the Mazatec dried and ground to a powder used to stanch wounds; a tree that yielded the bark paper employed by the *curanderos* to wrap the power objects of their rituals. There were at least two species new to science, a small orchid the size of a silver dollar, and the inflorescence of a tree, the relatives of which spread south through the rain forests of lower Central America and beyond to the Amazon. He held in his hand the dried specimens of a climbing vine and a small shrub, both used by the Mazatec to treat snakebite. One had never before been reported. The other had been used for at least six hundred years and was among the medicinal plants heralded by Hernández in his report to the Spanish king. This simple ethnobotanical collection at once aligned him with the past, the unknown, and a living culture that survived in no small measure because of its ability to understand the very plants that had drawn him to their land.

"Doctor! Doctor!" He turned back to the *tienda* and saw Dorantes walking toward him. There was a Mazatec with him, a thin middle-aged man in threadbare clothes with a face that was all bones and eye sockets. Dorantes was dressed, as he always was, in pressed khakis and a white cotton shirt.

"Doctor! *Por fin. Por fin.*" Dorantes repeated. He walked with an odd shuffle and kept rubbing his hands together.

"Buenos días," Schultes said.

"Buenos días, buenos días."

"What's up?" Schultes asked. He had never seen Dorantes so triumphant.

"Dr. Reko. *¿Dónde está?"*

"I don't know. Probably out walking or in the market. Why? What's wrong?"

"Nada, nada. It's just as well. He will be surprised." Dorantes turned and said something in Mazatec. The man beside him reached into a small basket and carefully pulled out a package wrapped in newsprint. Schultes recognized the newsprint as having come from his supplies. He frowned at Dorantes, who lifted his hands and shrugged. The Mazatec slowly pulled back the fold of the package. Inside were a dozen fresh mushrooms.

"The Saint Children," Dorantes said, "the Little Ones Who Spring Forth."

Schultes reached for the package with both hands. In a moment he became lost in the collection. With one finger he gently separated the individual mushrooms. There were two, no, three different fungi. The freshest he recognized immediately, at least to the level of genus. It was a species of *Panaeolus,* a small mushroom with a slender stipe and a conical pileus or cap. It resembled the *Panaeolus* mushrooms that grew in his mother's lawn back in New England. The other two mushrooms were in poorer condition. One was quite large, with a copper tone to the cap and a dark ring around the stipe or stem. It was stained dark purplish black where the flesh had been injured. The third species had a whitish stipe, a brown cap, and a purplish color to the gills. He looked up.

"Thank you," he said. "You have been most kind." The Mazatec nodded. Schultes turned to Dorantes.

"May I ask a question?"

"Of course."

"Ask him if these are all the mushrooms that are used."

Dorantes spoke with the Mazatec and then looked to Schultes. "He says he doesn't know."

"How many has he eaten?"

Again Dorantes turned to the Mazatec, who said nothing. "They eat twelve or fifteen," Dorantes answered. "Some eat as many as sixty, but they go insane."

"What do they see?" Schultes asked. Dorantes spoke and the Mazatec answered.

" 'Colors,' he says," Dorantes replied. Schultes reached into his pocket and pulled out a small bundle of pesos, which he handed to the Mazatec.

"Que Dios le pague," the man said, speaking perfect Spanish. God will pay you. Schultes took the mushrooms, placed them on top of the press, and carefully began to separate the species. In his notebook he recorded the date, description, and collection number of the specimens. It was July 27, 1938, Schultes and Reko number 231, the first identifiable botanical collection of *teonanacatl,* the Flesh of the Gods. Later that afternoon, a fortnight before his scheduled return to Boston, Schultes wandered through the moist fields outside of Huautla. To his amazement a mushroom that he had failed to find for a month appeared to be everywhere. He collected more than a dozen specimens, which are preserved to this day in the Farlow Herbarium at Harvard.

On February 21, 1939, Schultes reported his discovery in the *Botanical Museum Leaflets,* a journal established seven years earlier by Oakes Ames and printed privately on a hand-set press located in the basement of the museum on Oxford Street in Cambridge. Needless to say, in a world moving toward war, a paper entitled *Plantae Mexicanae II: The Identification of Teonanacatl, a Narcotic Basidiomycete of the Aztecs* did not receive wide circulation. It was, nonetheless, a major ethnobotanical breakthrough. With the assistance of Harvard mycologist David Linder, Schultes had positively identified *teonanacatl* as *Panaeolus campanulatus* var. *sphinctrinus,* a variety now recognized as the distinct species *Panaeolus sphinctrinus.* By securing a proper voucher specimen with an accurate botanical determination, Schultes had distilled ethnographic reports and rumors into scientific fact. Together with Jean Johnson, whose contribution is cited in the paper, Schultes had provided the first irrefutable evidence of a psychoactive mushroom used by Indians. Not only had he resolved the controversy initiated by Safford, and in a sense vindicated the early Spanish botanists, he had laid the foundation for further research. With a proper botanical identification, chemical work could begin and steps could be taken toward isolating the active compound responsible for the bizarre effects of the mushrooms. Perhaps most important, by suggesting that *teonanacatl* was a generic term for a number of different species of psychoactive mushrooms, Schultes left the door wide open for future exploration. It would take many years, but eventually others would follow in his wake; a series of expeditions in the 1950s would result in, among other things, the birth of the psychedelic era.

But all that was to come. For the moment, buoyed by his success with *teonanacatl* and encouraged by Oakes Ames, Schultes decided to take on *ololiuqui,* the vine of the serpent, the second of the mysterious and missing Aztec hallucinogens. In the fall of 1938 he began once more to comb the early Spanish records. To his astonishment he discovered that the Aztec held this plant in even higher regard than *teonanacatl.* "It is remarkable," Hernando Ruíz de Alcarón wrote in 1629, "how much faith these natives have in the seed. . . . They consult it as an oracle to learn . . . things beyond the power of the human mind to penetrate. . . . They venerate *ololiuqui* so much that they do all in their power so that the plant does not come to the attention of the ecclesiastical authorities." In Jacinto de la Serna's infamous book on Mexican idolatries, Schultes read that the Indians "venerate these plants as though they were divine. . . . They place offerings to the seeds and place them in the idols of their ancestors." When arrested and questioned, the author noted, the Aztec denied knowledge of *ololiuqui,* not because of fear of Spanish law but out of reverence for the plant itself.

An extraordinary description of the use of this sacred plant comes from the wanderings of Friar Francisco Clavigero. "The Aztec priests," he wrote, "went to make sacrifices on the tops of the mountains, or in the dark caverns of the earth. They took a large quantity of poisonous insects, burned them over a stove of the temple, and beat their ashes in a mortar together with the foot of the *ocotl,* tobacco, the herb *ololiuqui,* and some live insects. . . . They presented this diabolical mixture in small vessels to their gods, and afterwards rubbed their bodies with it. When thus appointed, they became fearless to every danger. . . . They called it *Teopatli,* the divine medicament." The place that *ololiuqui* once held in the lives of the Aztec is perhaps most perfectly distilled in a single understated line that Schultes found in a document published in 1634 by the Spanish priest Bartolomeo de Alua. Responding to questions put to him during confession, an anonymous Indian penitent explains very simply, "I have believed in dreams, in magic herbs, in peyote, in *ololiuqui,* and in the owl."

As he worked through this early literature, Schultes was not at all surprised to come across precise and detailed botanical descriptions of the plant. Indeed, by now he had almost come to expect it of Sahagún and Hernández, his most dependable sources. A 1651 edition of Hernández included a carefully drawn figure and a Latin description that in its wording seemed positively modern: "*Ololiuqui,* which some call *coaxihuitl,* or snake plant, is a twining herb with thin, green cordate leaves, slender green terete stems, and long white flowers. The seed is round and very much like coriander." It was taken orally, "when priests wanted to commune with their gods. A thousand visions and satanic hallucinations appeared to them." In a 1905 edition of Sahagún, Schultes found an even more remarkable illustration which showed quite clearly that *ololiuqui* had heart-shaped leaves, a swollen root, and a twining habit. That this unknown Aztec hallucinogen had to be a climbing vine was confirmed by the writings of Alarcón, who described *ololiuqui* as "a kind of seed like the lentil that is produced by a species of ivy of this land."

As early as 1854 botanists had suggested, based on the Spanish descriptions and illustrations, that *ololiuqui* was a plant in the morning glory family. In 1897 the Mexican botanist Manuel Urbina identified it as *Ipomoea sidaefolia,* a plant now known as *Turbina corymbosa.* There the matter would have stood had it not been, once again, for William Safford at the U.S. Department of Agriculture. This time Safford suggested that not only were the early Spaniards lousy botanists but so were the contemporary Mexicans, including Urbina. *Ololiuqui,* he argued, could not possibly be a morning glory because to date no member of that family had been shown to have narcotic or toxic properties. Urbina and the early Spaniards had been fooled by Indians intent on concealing the true identity of their sacred plant. *Ololiuqui,* Safford pronounced, was in fact *Datura meteloides,* a well-known and highly toxic hallucinogen belonging to a group of plants that, perhaps not entirely coincidentally, he had just monographed. In making his argument Safford stressed superficial similarities in the shape and color of the flowers —both are tubular and white—while overlooking the obvious: Every datura is a shrub or erect herb, but every account of *ololiuqui* describes a climbing vine. Though his argument had no merit, his professional stature assured that it would result in immense confusion. As in the case of *teonanacatl,* Safford's interpretation triumphed, and most botanists and anthropologists accepted his view on faith.

Schultes did not. Nor did Reko. In the same letter that Schultes had found pinned to the specimen of peyote at the Smithsonian that had alerted him to the true identity of *teonanacatl,* Reko had written that the

Zapotec of the Sierra Juárez of Oaxaca use in their religious feasts "*ololiuqui* which is doubtless *Ipomoea sidaefolia.*" *Ololiuqui* had to be a morning glory. In the spring of 1939, Schultes decided to return to Oaxaca to find out.

In the first week of April, Schultes began his second expedition to Oaxaca, this time entering from the east through Tuxtepec on the Río Papaloapam in the state of Veracruz. He traveled first to the town of Chiltepec, where he established a base, bought mules and supplies, and hired a field assistant, Guadalupe Martínez-Calderón, a young Chinantec youth who would remain with him for the entire expedition. With Guadalupe's help and guidance he proceeded into the mountains, following the traditional trade routes that penetrated the Chinantla rain forest and linked the land of the Chinantec with that of the Zapotec and Mixe to the south, and the Mazatec to the north.

In places the trails were wide and clearly marked, the river crossings shallow. For the most part, however, they were little more than vague etchings passing through dense tangles of undergrowth or scratched into the sides of tropical mountain walls. The villages Schultes passed hung suspended in clouds, cut off from all outside contact. In some of them the people had never seen a mule. In others the children peered from behind reed walls or scuttled for shelter as Schultes and Guadalupe approached. Sometimes an eerie silence greeted them, a silence broken only by whispers coming out of dark houses and the sound of hands patting tortillas, of clothing being beaten on river rocks. As their expedition pressed on, they came upon plantings of coffee, bananas, tobacco, and cotton, but nowhere had the forest been transformed. It lay over all the land, a soft mantle of green.

It was impossible to travel more than a few miles without coming upon a river or the steep banks of a forest stream. Over many of these crossings the Chinantecs had built ingenious suspension bridges, some more than a hundred feet in length. Called *hamacas,* they were made entirely from vines cut from the forest, laid lengthwise and bound together for strength, precisely like the metal cables of a modern bridge. The surface of the footpath consisted of a dozen or more bundles tied together to form a rope four to six inches thick, which was suspended from stout poles driven into the banks on either side of the crossing. Two narrower ropes served as hand rails, and the entire bridge was stabilized by an intricate network of smaller vines.

For Schultes and his party these narrow footbridges were a curse. The Chinantecs did not use mules. Schultes had two, and at every crossing it was necessary to hack a route down the steep banks to the water's edge, with the hope that the animals would be able to negotiate the rushing streams. The best one could expect was a delay of several hours while the packs were unloaded, the gear carried over the bridge, and the mules coaxed through the water. If the crossing was especially hazardous, Schultes would have to wait until Guadalupe obtained help. Usually it took at least three men to drag an animal across: two braced on the opposite bank and hanging on to a rope tied to the mule's neck, and one cursing and throwing stones from the near side. In almost every case the mules disappeared beneath the water, and for a moment no one knew whether they would ever surface.

To a great extent logistics defined the expedition. A year or two later, while working in the wet tropics of South America, Schultes would pioneer a method of preserving plant specimens by dipping them in alcohol or formaldehyde before laying them between sheets of newspapers. Two or three bundles of specimens could then be sealed in waterproof bags and packed into *costales,* the large burlap bags available in any marketplace in Latin America. The advantages were enormous. Treated collections could be stored for a month or more, and because the specimens remained pliable, they could be transported without risk of damage. By monitoring the bundles and adding more preservative when necessary, it was possible to remain on expedition for several months and then dry the entire collection upon return from the field. This simple technique represented a revolution in field methodology and was later adopted by virtually every tropical botanist.

In Oaxaca, unfortunately, Schultes was still drying his plants in the field as botanists had since the time of Alexander von Humboldt and Aimé Bonpland. To do so, an expedition had to carry cumbersome materials—plant presses, dryer frame, canvas skirting, stoves, kerosene, corrugated dividers—and find a suitable place to set up shop every two or three days. This was especially critical in the tropics, where fresh specimens begin to rot within forty-eight hours. While the plants dried, one waited, perhaps collecting in the immediate vicinity, always aware of the danger of fire, which in the history of ethnobotany had consumed many a collection and no doubt an equal number of Indian houses. Once the plants were dry, the specimens were brittle, extremely fragile, and a nightmare to transport.

Consequently, as Schultes and Guadalupe explored the Chinantla, their movements were to a great extent defined by expediency. Over the course of the fourteen-week expedition they described a series of circuits, each lasting roughly ten days and ending up at a base, one of the two or three small settlements where supplies could be replenished and specimens shipped out by post. Thus the first foray took Schultes and Guadalupe south, then west over a mountain range and back to the lowlands at Chiltepec where they had first met. The second phase led them south once again and then southeast into the heart of the Chinantla rain forest. Passing through a number of Chinantec villages, they finally reached Yaveo, where they stayed with an old Austrian couple, Mr. and Mrs. Wilhelm Barth, who owned a coffee plantation.

At Yaveo the expedition remained for several days as Schultes consolidated his specimens and nursed several open sores on his head that had begun to fester. Three days previously, upon reaching San Juan Lalana late in the evening, he had been surprised to notice the Indians bedding down beneath mosquito netting. It was cold, mountainous country, hardly the habitat normally associated with malaria. Too exhausted to bother with his net, Schultes had collapsed in his hammock, only to wake in the morning with his hair encrusted with congealed blood. He had been bitten by five vampire bats. Not actually bitten, as he had explained to the Chinantec. The bats feed at night, and to ensure that their victims do not awaken, they hover over the point of attack, beating their wings vigorously to create a flow of air that partially numbs the surface of the skin. Each bat then makes a quick incision with a single razor-sharp tooth. Their saliva contains an anticoagulant. Vampire bats do not exactly suck blood, they lap it up as it flows unimpeded from wounds. His Chinantec hosts were less than pleased by this explanation. How, they asked Guadalupe, could a human being know so much about a bat?

From Yaveo, Schultes made a week-long excursion into Mixe territory, climbing Cerro Zempoaltepetl, the highest mountain in southern Mexico, before returning through Yaveo and Yahuivé to Choapam. There he paused to ship off specimens before continuing into Zapotec territory where, a week later, he mailed more specimens out of Villa Alta. From there he proceeded north to Ixtlán, took a side trip to Oaxaca City for a week of rest and resupply, and then continued north, climbing two more high peaks, Cerro Cuasimulco and Cerro Zacate, before moving through a number of Zapotec villages and back into Mazatec country. In the middle of July, three months after starting out, he reached Huautla, where he remained for ten days before traversing the Cerro de los Frailes, a high and difficult mountain pass that led to the desert and Teotitlán del Camino, the railhead where he had begun his investigations with Reko the year before.

Throughout all of these travels Schultes had been searching for both *teonanacatl* and *ololiuqui,* and any evidence of rituals associated with the plants. On rocky outcrops he came upon evidence of animal sacrifice.

At Usila he spent a day in the hills with a *curandero* who spoke of the use of both plants in divinatory ceremonies. At San Juan Lalana, where he was injured by the bats, an elderly healer escorted him into the forest, where he made a valuable collection of medicinal plants. Near San Pedro Sochiapam he observed old men gathering mushrooms in open pastures. In the *pueblito* of Santa Cruz Tepetotutla he exchanged malaria tablets for five mushrooms known to the Chinantec as *nañ-tau-ga.* Informants in San Juan Zautla told him how the mushrooms were used, and he realized for the first time that most of the traditional healers were women. The man who officiated at the mushroom vigil Jean Johnson attended in Huautla had been the exception, not the rule.

By far his most significant discovery occurred in the Chinantec-Zapotec town of Santo Domingo Latani in the District of Choapam. There, with the help of Guadalupe, who knew the plant well, he encountered a single immense vine covering the entire house of an old *curandero.* Heavily laden with fruit, it was the only such plant in the village, and selling its seeds was the *curandero*'s only source of revenue. Guadalupe called it *a-muk-ia,* medicine for divination. The Zapotecs said it was *kwan-la-si.* Schultes knew it as *ololiuqui.* Without doubt it was *Turbina corymbosa,* the morning glory identified forty years before by the Mexican botanist Urbina. Once again Schultes had proved Safford wrong. With this field identification, together with subsequent collections made among the Zapotec, Mixtec, and Chinantec, he settled for good the vexing mystery of *ololiuqui.*

It was at this point that history intervened in the story of both *teonanacatl* and *ololiuqui.* In August 1939, a month before the Germans marched into Poland, Schultes returned to Harvard, where, after a near fatal bout with blood poisoning, he finished his doctoral degree. Uncertain about the future and eager to fulfill a dream to travel in the forests of South America before the United States was dragged into war, he accepted a Guggenheim fellowship to study arrow poisons in the Northwest Amazon. Before leaving he published or had in press two books and twenty-seven academic papers based on his work among the Kiowa and in Oaxaca, but none of these would be widely read. He was only twenty-six, and his career as a plant explorer was just beginning.

One by one the other players in the Oaxacan adventure disappeared from the scene. José Dorantes, the Huautla merchant, was shot and killed on the streets of Mexico City. A Swedish chemist, Professor C. G. Santesson, who had begun an analysis of *teonanacatl,* died suddenly in 1939. Jean Johnson, the anthropologist who had brought out the first exquisite account of the mushroom ceremony, was killed in the Allied landings in North Africa in 1942. As for Reko, the outbreak of war and the ultimate Nazi debacle silenced him in Mexico City, where he would die fourteen years after parting ways with Schultes in Oaxaca.

Despite all that had been accomplished, much remained unknown, including the nature of the intoxication and the identity of the active compounds in each of the sacred plants. The thread of the mystery was not picked up until a highly unusual series of events began to unfold in the early 1950s. For over twenty-five years Gordon Wasson, a banker and vice president of J. P. Morgan & Co. in New York, and his Russian-born wife, Valentina Pavlovna, had been studying the role of mushrooms in European and Asian cultures. One thing they noticed was that human societies could be readily divided into those that revered mushrooms and those that despised them. In an early publication they coined the terms "mycophiles" and "mycophobes" to describe the two alternatives. The Wassons were serious scholars, and this puzzling observation, together with their analysis of linguistic data, led them to suggest that our primitive ancestors had worshiped certain mushrooms. They did not know which kind or why any human might have such a reverential attitude toward these plants. They were simply certain that it had occurred.

In September 1952, just as the Wassons were struggling to prove their assertion, they received a letter from the poet Robert Graves in Majorca who somehow had stumbled upon Schultes's 1939 paper identifying *teonanacatl.* On October 27, Wasson wrote to the Botanical Museum at Harvard requesting a reprint. A reply from Schultes did not come until December 31, a week after his return from an expedition to the Colombian Amazon. The letter informed Wasson that the article was out of print but might be obtained by writing directly to the editor of the journal. Schultes thanked the banker for his interest, encouraged him to visit the Botanical Museum, and closed by noting that he would presently be returning to the Amazon to "complete twelve years of botanical explorations." Such was the beginning of one of the most significant friendships in the history of ethnobotany.

Wasson obtained a copy of the *teonanacatl* paper and after reading it, made immediate plans for Oaxaca. In the summer of 1953 he and his wife traveled for the first time to Mexico, where they were warmly received by the anthropologist Robert Weitlaner, who directed them to Huautla. Though enchanted by the experience, they were unable to penetrate the guarded world of the Mazatec healers. Nevertheless, Wasson persisted, returning to Huautla alone or in the company of his wife and others the following two summers. Finally, two years after his first visit, he met a *curandera* who invited him to attend a midnight vigil. Her name was María Sabina. Thus on June 29, 1955, Wasson and his photographer, Allen Richardson, became the first outsiders actually to ingest the mushrooms in a sacred context.

In his extraordinary account of the experience Wasson wrote of mushrooms gathered before sunrise in places where mountains are caressed by mysterious winds. Under the influence of María Sabina's prayers, Wasson heard infinitely sweet voices hovering in the dark, saw music assume form, and felt his spirit soar out of his body. With his imagination awash in color, Wasson lay beneath a blanket on the mud floor of her home as María Sabina sang in the beautiful tonal language of the Mazatec:

> *Woman who thunders am I, woman who sounds am I.*
> *Spider woman am I, hummingbird woman am I. . . .*
> *Eagle woman am I, important eagle woman am I.*
> *Whirling woman of the whirlwind am I, woman of a*
> *sacred enchanted place am I.*
> *Woman of the shooting stars am I.*

In the stillness of the night, with the rain falling softly on the thatch roof, this humbled New York banker struggled to find words to describe his "soul-shattering experience." They did not exist. Months later he would write, "We are all confined within the prison walls of our everyday vocabulary. With skill in our choice of words, we may stretch accepted meanings to cover slightly new feelings, but when a state of mind is utterly distinct, then our words fail. How can you tell a man who has been born blind what it is like to see?" For the Mazatec, too, the experience was impossible to describe. They called the mushrooms the "Little Ones That Leap Forth" because, as Wasson's muleteer informed him, "the little mushroom comes of itself, no one knows whence, like the wind that comes we know not whence or why."

When on May 13, 1957, Wasson published an account of his Huautla expedition in *Life* magazine, a young editor attempted to capture the ineffable quality of the experience in a snappy title, "Seeking the Magic Mushrooms." Neither the editor nor Wasson could have anticipated that the name would stick or that the article would mark a certain watershed in the social history of the United States, the beginning of the psychedelic era. Before its publication, the general public was completely unaware of the existence of hallucinogenic mushrooms. Among those who read it, and whose interest was sufficiently piqued that he sought out the more sober academic articles of Richard Evans Schultes, was a young Harvard lecturer named Timothy Leary.

In the summer of 1960, Leary would travel to Mexico and ingest the magic mushrooms in Cuernavaca. "Like almost everyone else who has had the veil drawn," he would later write, "I came back a changed man." Shortly thereafter Leary returned to Harvard, where he began the controversial experiments that eventually culminated in his dismissal from the university. Naturally he consulted Schultes. One of their conversations touched on the use of the word "psychedelic," a term that had been recently coined by the psychiatrist Humphrey Osmund. Schultes cautioned Leary that the word, meaning "mind manifester," was appropriate, but the spelling was incorrect. The proper Greek was "psych*o*delic," and Schultes was concerned lest a Harvard man be associated with the bastardization of a classical language. Leary suggested that "psychedelic" sounded better. An unverified account of the meeting indicates that long before Harvard fired Leary, Schultes had disowned him for using improper Greek.

There remained only one missing link in the chronicle of *teonanacatl* and *ololiuqui:* the isolation of the chemicals responsible for the hallucinogenic effects of the plants. By the late 1950s Wasson and his co-workers, notably a French mycologist named Roger Heim, had found in Mexico no fewer than twenty-four different kinds of psychoactive mushrooms, thus verifying Schultes's assertion that the word *teonanacatl* was a generic term. In Paris, Heim identified the species and managed to cultivate several of them in quantity. But all attempts in France and the United States to identify the active ingredients failed. Finally, somewhat in desperation, Heim forwarded a one-hundred-gram sample of *Psilocybe mexicana* to Albert Hofmann, then director of the natural products department of the Sandoz research laboratories in Basel, Switzerland.

Hofmann began his investigations by feeding the mushrooms to mice and dogs. Nothing appeared to happen. So Hofmann himself ate thirty-two mushrooms. Something did happen. The landscape outside his laboratory window began to look like Mexico, the face of his colleague overseeing the experiment turned into that of an Aztec priest, the pencil in his hand became an obsidian blade. After ninety minutes "the rush of abstract motifs . . . reached such an alarming degree that I feared I would be torn into this whirlpool of form and color and would dissolve."

Such an experience might have unnerved an ordinary scientist, but Hofmann was not of that sort. He had spent much of his career investigating the chemicals responsible for mass poisonings that had periodically convulsed European towns since the Middle Ages. Known as Saint Anthony's fire, these outbreaks killed thousands and left hundreds scarred for life. Many victims lost their fingers and toes or had their noses literally rot on their faces. Others experienced horrific hallucinations and went mad. The source of the affliction was a parasite on rye crops, a fungus known as ergot, which contains a series of compounds that among other characteristics causes the blood vessels to contract—hence the gangrene in the extremities. Sandoz and Hofmann hoped that this very quality might prove useful in modern medicine, particularly as a means of stanching excessive bleeding after childbirth. The challenge was to find out which of the many chemicals in the fungus was responsible for this unique property.

In a series of experiments in the early 1930s, chemists at Sandoz managed to isolate the drug, which they called *ergobasine.* Hofmann's task was to learn how to synthesize it. He began by manipulating its basic nucleus, a compound known as lysergic acid, the fundamental building block of all the ergot alkaloids. In 1938 he produced in his lab the twenty-fifth substance in a series of lysergic acid derivatives. The drug was tested for its effect on the uterus, the results duly recorded, and the compound promptly shelved and forgotten. Five years later, in the spring of 1943, a combination of intuition and serendipity led Hofmann to make this very compound once again. It was a Friday, and during the final step in the synthesis, something exceedingly odd occurred. He began to feel dizzy and restless, and was compelled to leave his laboratory for home. Due to wartime shortages of gas, he had no car, so he set off on what turned out to be one of the most momentous bicycle rides in history. The compound he was making, a trace of which he had accidentally absorbed through his skin, turned out to be the most potent hallucinogenic agent ever discovered: lysergic acid diethylamide-25, LSD for short. On his ride home Dr. Hofmann went on the world's first acid trip.

So Hofmann was quite prepared for the visual onslaught that resulted from eating the mushrooms, and no sooner had the intoxication passed than he set about identifying the active ingredients. He succeeded in a remarkably short time. In March 1958 he announced the discovery of psilocybin and psilocin, two new substances that turned out to be very close in structure to serotonin, a compound that plays an important role in brain chemistry. By November, Hofmann had managed to synthesize both drugs and was ready to move on to *ololiuqui.* He notified Wasson, who, with the assistance of Irmgard Weitlaner and a host of Zapotec and Mazatec collectors, managed to send him twenty-six kilograms of morning glory seeds. Once again the analytical work proceeded smoothly. What Hofmann found, however, was scarcely believable. The active principles of *ololiuqui* were two indole alkaloids, lysergic acid amide and lysergic acid hydroxyethylamide, compounds that he already had sitting on the shelves of his lab. They differed from LSD only by the replacement of two hydrogen atoms for two ethyl groups. Four years before Hofmann discovered LSD, Richard Evans Schultes had found its analog in nature, in the seeds of a humble morning glory that was worshiped as a god incarnate by the ancient peoples of central Mexico.

CHAPTER FIVE

The Red Hotel

THE WIND IN the evening runs away from the shores of Panama. An hour or two after sunset, when the sparkling cruise ships waiting at the mouth of the canal light up like carnival tents, the fishermen at the settlement of Veracruz drag their skiffs across the beach and into the sea. Those with small motors disappear quickly into the night. The others have to row, struggling through the surf and then pulling away from the shore and each other with long, smooth strokes of the oars. Their destination is the edge of a coastal shelf where the shallow seabed falls away and the deep, cold waters of the Pacific rise, bringing schools of fish to the surface. They know they are there when they can no longer

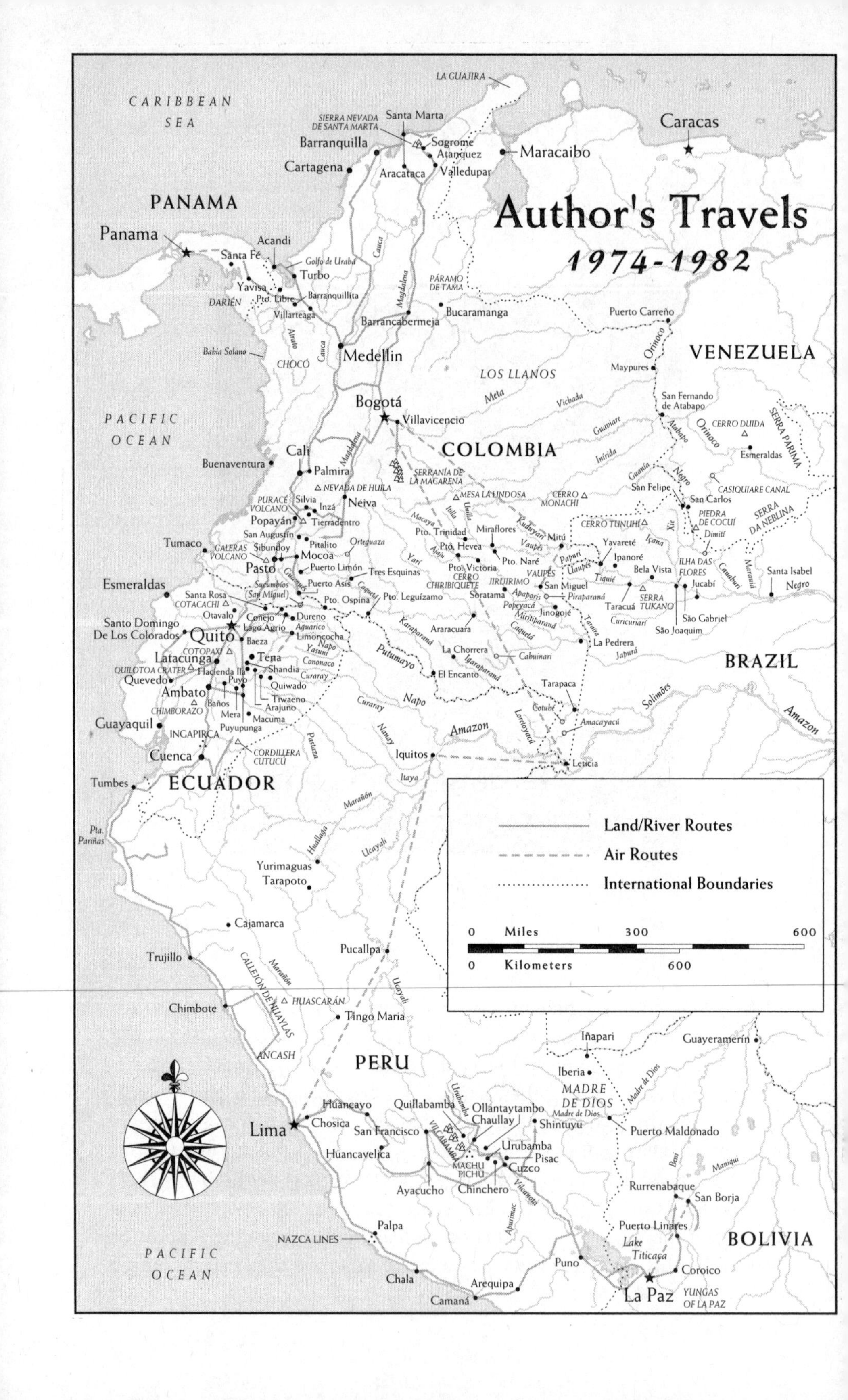

Author's Travels
1974-1982
Land/River Routes
Air Routes
International Boundaries
0 Miles 300 600
0 Kilometers 600
CARIBBEAN SEA
PACIFIC OCEAN
PACIFIC OCEAN
PANAMA
COLOMBIA
VENEZUELA
ECUADOR
PERU
BRAZIL
BOLIVIA
LA GUAJIRA
SIERRA NEVADA DE SANTA MARTA
Santa Marta
Sogrome
Atanquez
Barranquilla
Cartagena
Aracataca
Valledupar
Maracaibo
Caracas
Panama
Acandi
Santa Fé
Golfo de Urabá
Turbo
Yavisa
DARIÉN
Pto. Libre
Barranquillita
Villarteaga
Cauca
Magdalena
PÁRAMO DE TAMA
Bucaramanga
Barrancabermeja
Puerto Carreño
Orinoco
Atrato
Bahía Solano
CHOCÓ
Medellín
LOS LLANOS
Maypures
Meta
Vichada
Bogotá
Villavicencio
San Fernando de Atabapo
PACIFIC OCEAN
Guaviare
CERRO DUIDA
SERRA PARIMA
Esmeraldas
Cali
Buenaventura
Palmira
SERRANÍA DE LA MACARENA
Inírida
Guanía
Negro
CASIQUIARE CANAL
NEVADA DE HUILA
San Felipe
San Carlos
PURACÉ VOLCANO
Silvia
Inzá
Neiva
MESA LA LINDOSA
CERRO MONACHI
PIEDRA DE COCUÍ
SERRA DA NEBLINA
Popayán
Tierradentro
CERRO TUNUHÍ
Dimití
San Augustín
Pto. Trinidad
Miraflores
Mitú
Tumaco
GALERAS VOLCANO
Sibundoy
Pitalito
Orteguaza
Pto. Hevea
Vaupés
Yavareté
Içana
Mocoa
Pto. Naré
Ipanoré
ILHA DAS FLORES
Pasto
Puerto Limón
Tres Esquinas
Pto. Victoria
CERRO CHIRIBIQUETE
VAUPÉS
Bela Vista
Santa Isabel
Esmeraldas
Puerto Asís
San Miguel
Jucabí
Negro
Santa Rosa
COTACACHI
Pto. Ospina
Pto. Leguízamo
JIRUIRIMO
Soratama
Apaporis
Piraparaná
Taracuá
SERRA TUKANO
Otavalo
Conejo
Dureno
Jinogojé
São Gabriel
Santo Domingo De Los Colorados
Quito
Lago Agrio
Aguarico
Limoncocha
Araracuara
Curicuriarí
São Joaquim
Baeza
COTOPAXI
Yasuní
Napo
Putumayo
La Pedrera
Latacunga
Tena
Cononaco
La Chorrera
Cabuinari
Japurá
QUILOTOA CRATER
Hacienda Ila
Shandia
Curaray
El Encanto
Quevedo
Puyo
Quiwado
Ambato
Tarapaca
Baños
Tiwaeno
Arajuno
Napo
Solimões
CHIMBORAZO
Mera
Macuma
Curaray
Cotuhé
Amazon
Guayaquil
Puyupunga
Amacayacú
INGAPIRCA
Pastaza
Nanay
Amazon
Loretoyacú
CORDILLERA CUTUCÚ
Cuenca
Iquitos
Leticia
Tumbes
Itaya
Marañón
Pta. Pariñas
Huallaga
Ucayali
Yurimaguas
Tarapoto
Cajamarca
Pucallpa
Trujillo
CALLEJÓN DE HUAYLAS
Marañón
HUASCARÁN
Ucayali
Chimbote
Tingo María
Iñapari
Guayeramerín
ANCASH
Iberia
MADRE DE DIOS
Madre de Dios
Huancayo
Quillabamba
Urubamba
Ollantaytambo
Chaullay
Shintuyu
Lima
Chosica
San Francisco
VILCABAMBA
Puerto Maldonado
Urubamba
Pisac
Huancavelica
MACHU PICHU
Cuzco
Ayacucho
Chinchero
Vilcanota
Beni
Manique
Apurímac
Rurrenabaque
San Borja
Palpa
Puerto Linares
NAZCA LINES
Lake Titicaca
Puno
Coroico
Chala
Arequipa
La Paz
YUNGAS OF LA PAZ
Camaná

smell the land or distinguish the lights on the horizon from the stars in the sky.

In the short time that I stayed in Veracruz I often went out at night with one of the fishermen, a young man named Ohilio. He was a gentle person, a mixture of a dozen races, short and thin, with the rough hands of one who had worked the nets for years. Unable to hear or speak since birth, Ohilio had found in fishing the perfect vocation. On land he had the look of someone who had spent his life shying away from people. On the water at night, with quiet all around and the darkness broken only by the dazzling phosphorescence in the sea, he was completely at ease.

Ohilio rowed by choice, not necessity, for he always caught fish. He believed that the fish found him and that it was God's way of compensating for his misfortune. And so it seemed. For while the others labored at their nets, pulling them in and letting them out many times a night, shifting locations, chasing the fish, Ohilio would set a single net and then, as his boat drifted in the current, curl up in a bundle in the bow and go to sleep. Generally he woke up only once during the night to check his net. Sometimes in the cold hours before dawn he got up a second time to relieve himself, casually straddling the forward thwart, pissing over the bow as the skiff pitched and rolled precariously in the waves. He was completely fearless. Once at dawn as we pulled in the net, killing the fish we could sell with a swift blow to the head and throwing the others back into the sea, an enormous shark suddenly broke the surface just beside the boat. In an instant I found myself tossing a large fish past rows of teeth into a gaping jaw. The shark swallowed the fish sideways and then fell back into the water with a crash that nearly swamped us. Its tail struck the side of the skiff, the planks by the keel shuddered, and the entire boat spun violently. Stunned, I looked to Ohilio, whose eyes were gleaming and whose mouth was shaking with the laughter of the mute.

Most of our outings were less eventful, of course, and the quiet hours of the night were rare moments of peace and forgetfulness. Free of language, with no friction to our thoughts, we shared a strange solitude, a life momentarily drained of volition, as elemental as the sea. At the time I had little energy for new sensations. For a month or more I had been in the forests of the Darien, a difficult passage that had begun, perhaps predictably, with a contact provided by Professor Schultes. Eight weeks before, shortly after we left the Sierra Nevada and Tim returned for a month to Harvard, an old crony of Schultes, a geographer and explorer attached to the Botanical Garden in Medellín, invited me to join a British expedition intent on traversing the Darien Gap, a broad expanse of roadless swamp and rain forest that separates Colombia from Panama. The expedition turned out to be one man, Sebastian Snow, an English adventurer who, having just walked from Tierra del Fuego at the tip of South America, intended to continue north as far as Alaska. It was June, the height of the rainy season, and the Darien was said to be impassable.

Our route from Colombia had taken us on foot from Barranquillita, a ramshackle settlement just off the Medellín-Turbo road, sixty miles west across the Tumaradó swamp to Puerto Libre, a row of huts strung out along the banks of the Río Atrato. The Atrato runs four hundred miles south to north, draining the Chocó, the wettest region of South America, thirty thousand square miles of forgotten rain forest cut off from the Amazon millions of years ago by the rise of the Andean Cordillera. Downstream from Puerto Libre is the Gulf of Urabá and the Caribbean. Upstream is more swamp and forest, and a land that for Colombians is synonymous with disease and disappointment.

Like so many lowland settlements Puerto Libre was a place of lassitude strangely at odds with the intensity of life that surrounded it in the forest. It consisted of ten sun-bleached shacks and three floating outhouses, each with three holes in the floor—one to relieve oneself in, one to wash in, and a third to draw water. The lives of the local women revolved around these riverfront latrines. They were there with the children at dawn, and they remained for much of the day, washing clothes or idly gossiping. In the evening, when the night air finally offered some relief from the heat, mosquitoes rose from the river like a miasma, driving everyone indoors to the isolation of their netted hammocks. Once it was dark, caiman came ashore by the score and for the rest of the night sprawled on the grassy slopes leading up to the shacks or lay about on the wooden landings where so few hours before children and infants had bathed.

After three miserable days, including a morning when I awoke on the floor to discover that a dog had given birth on my foot, a local skin trader ferried our party upriver to a place called La Loma. There we hired mules to carry our gear beyond the Atrato up a narrow track that crisscrossed the Río Cacarica and rose toward the Darien. Three days later, having abandoned the mules and engaged three Emberá Indians as guides, we reached the height of land at Palo de Letras, the border between Colombia and Panama that, as the name suggests, is marked only by a pair of letters carved into the bark of a tree. Once beyond the frontier we entered a world of plants, water, and silence. For the next ten days we moved from one Emberá or Kuna village to the next, soliciting new guides and obtaining provisions as we went along. Nowhere did we stop long enough to understand the lives that we drifted through, but each day became part of a veil that gradually enveloped us as the forest closed in, absorbing our party as the ocean swallows a diver.

It was during those days that I first experienced the overwhelming grandeur of the tropical rain forest. It is a subtle thing. There are no herds of ungulates as on the Serengeti plain, no cascades of orchids—just a thousand shades of green, an infinitude of shape, form, and texture that so clearly mocks the terminology of temperate botany. It is almost as if you have to close your eyes to behold the constant hum of biological activity—evolution, if you will—working in overdrive. From the edge of trails creepers lash at the base of trees, and herbaceous heliconias and calatheas give way to broad-leafed aroids that climb into the shadows. Overhead, lianas drape from immense trees binding the canopy of the forest into a single interwoven fabric of life. There are no flowers, at least few that can be readily seen, and with the blazing sun hovering motionless at midday there are few sounds. In the air is a fluid heaviness, a weight of centuries, of years without seasons, of life without rebirth. One can walk for hours yet remain convinced that not a mile has been gained.

Then toward dusk everything changes: The air cools, the light becomes amber, and the open sky above the rivers and swamps fills with darting swallows and swifts, kiskadees and flycatchers. The hawks, herons, jacanas, and kingfishers of the river margins give way to flights of cackling parrots, sungrebes, and nunbirds, and spectacular displays of toucans and scarlet macaws. Squirrel monkeys appear, and from the riverbanks emerge caiman, eyes poking out of the water, tails and bodies as still and dull as driftwood. In the light of dusk one can finally discern shapes in the forest, sloths clinging to the limbs of cecropia trees, vipers entwined in branches, tapir wallowing in distant sloughs. For a brief moment at twilight the forest seems of a human scale and somehow manageable. But then with the night comes the rain and later the sound of insects running wild through the trees until, with the dawn, once again silence: The air becomes still, and steam rises from the cool earth. White fog lies all about like something solid, all-consuming.

After just a fortnight on the trail, our passage began to take on the tone of a dream. In part this was because we rarely slept. With the rain sleep was not often possible. At the end of long days we simply lay in our hammocks in an unnatural rest, like a state of trance, dulled by exhaustion and insulated from the night by mosquito netting and the smoke of a smoldering fire. But mostly we became infected by the spirit of the place. The Darien turned out to be less a piece of terrain than a state of mind, a wild frontier utterly divorced from the moral inhibitions of ordinary human society.

In each of the small villages—Paya, Capeti, Yape, El Común—that marked the route from the frontier to the main settlement at Yavisa, there was a recent story of murder or death. On the Río Cacarica five men had fought and wounded one another with machetes. In Capeti a black Colombian thief known as Mentiroso Serio, the Serious Liar, killed a woman and was himself hunted down, shot, and strung up in the forest beyond Paya, just short of the Colombian frontier. Seven Indians were murdered on the Río Chico; a Colombian was killed for his cooking utensils on the trail to Tigre; a man and his wife were fatally tortured near Yavisa. Those investigating these crimes, or at least responsible for recording them in their moldy logbooks, were the Guardia Civil, a clumsy and corrupt paramilitary force then under the command of a young Manuel Noriega.

Midway through our journey trouble with the Guardia Civil at Yavisa forced us to change our route. Stripped of most of our gear and driven away from the settlement, we followed three Kuna guides up a series of rock falls and cascades, a serpentine route designed to evade pursuit. In the process the Kuna themselves became disoriented, and for the next week we wandered through the forest lost or, at best, only vaguely aware of where we were. Free of distractions, one became honed by life in the forest—the howler monkeys overhead, the incessant streams of ants, chance encounters with snakes and jaguar, the haunting cries of harpy eagles; iridescent butterflies, teasing with their delicate beauty, while at one's feet bronze and purple frogs, poisonous to the touch. In my journal I noted the simple luxuries of forest life; "the smoke of a fire that chases away the insects, a rainless night, a thatch hut found in the woods, a banana almost gone bad found lying in a trough, abandoned plantings of manioc, a fresh kill, whatever it may be, water deep enough to bathe in, a hint of a solid shit, a full night's sleep, a lemon tree found in the forest."

By the time we came upon the road head some twenty miles east of Santa Fe, the cumulative effects of two years on the road had physically broken my English companion. In all he had lost fifty pounds. Leaving him with one of the Kuna, the remainder of our party went ahead to seek help. Several miles on, we came upon the right-of-way of the Pan American Highway, a cleared and flattened corridor that stretched to the horizon. We hesitated, momentarily confounded by so much space. Then we began to walk past the charred silhouettes of trees and onto a beaten track that meandered through the slash. It was several hours before we heard the sound of machinery—chain saws at first and then the dull roar of diesel trucks. We walked for another mile or two before coming upon a D-9 cat, the largest bulldozer made, buried up to its cab in mud. A second bulldozer, snorting and belching smoke, tore into the ground with its blade while two others, attached to the trapped machine with thick cables, attempted to haul it out of the mud. None of the workmen noticed us. The sound was deafening: the hiss and moan of hydraulics, the iron cables snapping like strings, and the smell of grease and oil.

The Kuna had never seen machinery of such a scale. Clinging to their rifles, struggling through the cloying mud, they walked past the bulldozers and gravitated toward a small work gang clustered at a bridge site half a mile beyond. It was dusk and the crew had broken for dinner, served by Kuna kitchen boys from a mobile canteen. The foreman asked where we had come from. When I said Colombia, the workers in a single gesture leaned forward to offer us their plates of food. I looked about, invited my companions to eat, and then glanced past the foreman to the road ahead, a scene of desolation that ran north as far as one could see. He followed my gaze. "The civilization of nature," he said, "is never pretty."

Four days later, having successfully crossed the Darien, I abandoned Sebastian to his walk and at Santa Fe climbed aboard a small plane for the short hop to Panama City. A last-minute addition to the passenger list, I was squeezed into the rear seat, my knees pushed up to my chest, unable to move and scarcely free to breathe. The pilot led us immediately into a tremendous tropical storm. Visibility dropped to nil. The woman beside me threw up on my lap. Her mother, a corpulent black merchant, turned to offer consolation and promptly threw up herself. For an anxious few moments, as the winds buffeted the plane, I feared that having survived the Darien, I was about to die ignominiously. When finally we landed at Panama City, I walked off the plane drenched in vomit, with only two dollars to my name. As for Sebastian, the last I heard he made it as far as Costa Rica before being admitted to a hospital. In the middle of the night he awoke and left the hospital in his pajamas, starting to walk north. He was arrested and spent a delirious week in jail before being rescued by a staff member at the British embassy who bundled him back to England.

When I returned from Panama to Medellín, flying in an hour across lands that had taken weeks to traverse on foot, I was relieved to find a telegram from Tim at the Botanical Garden. Dispatched from Boston ten days earlier, it left me only twenty-four hours to reach Bogotá. I took the overnight train and by dawn was rolling toward the outskirts of the capital across a pale green savannah beneath an intense and fragile blue sky. For once it was not raining in Bogotá, and the raw light of the high Andes burst through tattered shreds of clouds above the mountains that form the backdrop of the city to the east. From the base of the mountains the city sprawls endlessly in three directions, swallowing up the rich farmland of the savannah, erasing rivers and forests that but a generation ago were distant landmarks. Six million people now live in Bogotá, and thousands of new immigrants arrive each year from every corner of the country. Approaching the hard edges of the city, the train runs for a full hour through dozens of forgotten barrios built of concrete and cinder blocks where even new buildings look like ruins.

This was not the city that Schultes knew when he first arrived in Colombia in the fall of 1941. Then the population was just over 300,000 and Bogotá still belonged to a handful of families whose surnames went back to the time of the Spanish. They lived in the north, in the Chico and Cabrera neighborhoods, in a world that was more English than Spanish. They had modern houses with flower gardens and well-tended lawns guarded by watchdogs that looked less like the mangy animals of the provinces than the fine creatures of English country portraits. The rich wore clothes tailored in London and shoes that shone like mirrors. With vast haciendas on the savannah, estates in Tolima and Cundinamarca, grand houses on the Calle Real, those born to privilege ruled over a small provincial capital where talent, service, and even money had nothing to do with social advancement. Lineage and manners counted for everything. At the racetrack or by the warm fires of the Jockey Club, men with pink healthy faces and hair highlighted by brilliant flashes of silver drank whisky while they traded power and gossip with their cousins.

To the south lived the poor, the maids and laborers, bootblacks and newspaper boys, flower peddlers and taxi drivers, those who sold lottery tickets and all those who ran the hundreds of minor rackets that somehow kept hunger at bay. For the most part they appeared meek and obedient, a faceless urban peasantry, quaint in their woolen *ruanas* and straw hats, hidden away by night in barrios named for the saints or squatter shacks that clung to the edge of the mountains. The small middle class—shopkeepers newly arrived from the provinces, lawyers and civil servants who filled the ministries—lived between rich and poor in Chapinero, a dark and gloomy midtown neighborhood where women over forty wore black and the rain never seemed to quit. The men also dressed in black: dark flannel suits, felt hats, and umbrellas.

They rode to work in open streetcars, were obsessed by chills and kidney ailments, had a passion for funerals, and lived in constant fear of losing their jobs. Like the *campesinos* who drove mules across the cobblestoned streets, the avocado vendors by the National Palace, the factory workers, and the whores, they accepted the social order as it was, a hierarchy of Church and state, propped up by the army and dominated by the powerful few.

Schultes arrived in Bogotá from Mexico where he had spent the summer of 1941 working as a translator for a team of scientists surveying the agricultural potential of that nation for the Rockefeller Foundation. In eight weeks the commission traveled six thousand miles by road, visiting every farming state in the country. When the work ended, he had two job offers. A private boys' school just outside of Boston needed a biology master. The National Research Council wanted someone to travel to the Northwest Amazon and hunt for arrow poisons. Schultes chose the latter.

He landed in Bogotá early on a Sunday when the light was soft and translucent, the dust on the road in from the old airport settled from the night's rain, and all the city idle and peaceful in anticipation of midday Mass and empty and endless *paseos* in the countryside. He checked into the Hotel Andina on Avenida Jimenez and tried to contact the Institute of Natural Sciences, which was expecting him. Naturally, the Institute was closed, so he left his hotel to explore the city that would be his home for the next twelve years. Drifting up and down the broad avenues, he walked past the fountains that then graced the Plaza Bolivar, beneath the austere façade of the Granada Hotel overlooking Santander Park, the balconies and ornate doorways on Calle Real, and along the deserted alleys that ran up to the foot of Monserrate, the mountaintop shrine that even then was the symbol of the city. A veil of innocence hung over the morning. He took in a band concert, watched a parade of military cadets, drank fresh juices from street stands that sold cherimoyas, guavas, zapotes, lulos, and passion fruits of a dozen colors. Bogotá seemed to him a city of priests and organ grinders, bird sellers, gypsies, and the innocently insane living and thriving happily, everyone dressed in black.

In the afternoon of his first day in the city he hopped aboard an open-sided trolley, paid a penny, and sat back to see where it would take him. It went south, winding its way toward the outskirts of the city to the munitions factory where the line ended at the base of a steep hill lush with vegetation. Schultes got off and followed a group of schoolchildren and a nun up a stone stairway that climbed into a beautiful forest. As he walked through the trees, he noticed a small orchid partially concealed by a cluster of ferns. It was no more than an inch high and unlike anything he had ever seen. He carefully made a collection, which he pressed between the pages of his passport. He later sent it by post to Oakes Ames, who described it as a new species, *Pachiphyllum schultesii.* Thus on his first day in Colombia, at the edge of the national capital, Schultes had discovered an orchid unknown to science. It was his first botanical collection in Colombia, the first of more than twenty-five thousand he would make.

I took a room at the Hotel Paloma, a modest *residencia* on Calle 14 in La Candelaria, the colonial quarter that runs from Plaza Bolivar to the base of Monserrate. Tim was late. I fell asleep, and when I woke there was still no sign of him. I walked out into a weak drizzle and decided despite the rain to ride the gondola up Monserrate. Perched one thousand feet above the savannah, it is a quiet refuge and the only vantage point that allows one to take in the entire city. On a clear morning, before the smoke and exhaust fumes cloud the sky, you can see west one hundred miles to the Nevado del Tolima, a beautiful volcanic cone rising above the snowcapped peaks of the Cordillera Central.

On my way up the narrow streets of the quarter toward the base of the mountain, I passed by the house of an old man I used to stay with when visiting the capital. He had been a politician, a member of the Liberal Party, a communist to some. I had met him soon after arriving in Colombia, and he had insisted that to understand his country I would have to read *La Vorágine,* a novel about the rubber era written by José Eustásio Rivera. He bought me a copy, and I remember him standing beside a tibouchina tree in his patio, rain running down the red tiles of the roof, as he read aloud from the book. "I have been a rubber worker," the hero says at one point, "I have lived in the muddy swamps in the solitude of the jungles with my crew of malaria-ridden men cutting the bark of the trees that have white blood like that of the gods. I am a rubber worker. And what my hand has done to trees, it can also do to men." Before my friend passed away, we used to walk through the streets of the city as he pointed out the places where his life and beliefs had been forged. He had been present at the very moment when the Bogotá that Schultes had known died.

It all began in 1928 with the massacre by the army of several hundred striking banana workers and their families on the north coast. The Colombian president had charged the workers with treason, promoted the officers responsible for the slaughter, and accused the victims of having "pierced the loving heart of the Fatherland." In the Congress a lone voice had begged to differ. Jorge Gaitán, then a young liberal legislator, noted that the striking workers had had their wages slashed by the United Fruit Company, which paid no taxes and operated on lands donated by the state. He accused the Colombian army of committing murder at the request of a foreign company. Then he pointed out the obvious: Colombia was a country where the rich only became richer as the poor grew poorer. It was a simple idea, self-evident to anyone who thought about it. But few had. By giving expression to the misery of the poor, Gaitán shook the fragile foundations of the old order, and over the following years, as his oratory soared and his fame grew, the structure began to topple. In the neighborhoods of the north, matrons began to complain about uppity maids. Business leaders downtown encountered the scorn of shoeshine boys. Beggars spat into the paths of priests. Around the tables of La Cigarra Café, one of the centers of political intrigue, it was said that the poor cried out through Gaitán's mouth and that his words could silence the wind.

On April 9, 1948, U.S. Secretary of State George Marshall was in Bogotá to address the Ninth Pan-American Conference. At 2:00 P.M. Gaitán, then the head of the Liberal Party and heavily favored to win the next presidential election, was scheduled to speak before a group of students elsewhere in the city. At one o'clock a man loitering in front of the Black Cat Café crossed Carrera Séptima, pulled out a gun, and fired three bullets into Gaitán as he stood in front of the Agustin Nieto building. Gaitán died immediately. Within minutes a whirlwind of pain and anger ripped through Bogotá. Market women abandoned their stalls, factory workers burst into the streets, the barrios emptied, and students fled from their classrooms. Within an hour thousands had surged into the heart of the city: men with red flags and machetes, women with children and gasoline. Everything was looted and destroyed. Burning streetcars rolled empty down the avenues, churches and monasteries went up in smoke, machetes severed fire hoses, and the sky turned red with flames that gutted the core of the city. The air filled with the scent of burned metal and stone, spilled liquor, and in time the bodies of the six thousand who were killed. For three days Bogotá burned.

At the height of the *Bogotazo* a mob besieged Plaza Bolivar and flung the body of the assassin against the doors of the National Palace. The president, Mariano Ospina Pérez, declared that it was better for the nation that the president die than flee in the face of disorder. His wife, Doña Bertha, made arrangements for her family to be taken to the United States Embassy. Three tanks, festooned in red banners and swarming with students shouting Gaitán's name, arrived at the plaza.

Hysterical with hope, the people believed the army had sided with the uprising. The tanks rolled forward, reached the palace steps, and then slowly the turrets turned away from the building, pointed toward the crowd, and opened fire. In that instant the quiet provincial capital, where presidents in top hats and frock coats had mingled fearlessly with the people, disappeared forever.

It was late in the evening, well after I had returned from Monserrate, that I stepped out of my hotel and nearly ran into Tim's pickup truck as he pulled up in front of the curb. I could tell it was Tim from the strength of the headlights alone. In the narrow streets of La Candelaria, the power wagon, all shiny and bright after a month in a local garage, looked almost absurdly oversized. The traffic piled up immediately, and a string of *busetas* clamored for Tim to move on. I leaped into the dark cab, turned to greet him, and was promptly licked across the face by a dog.

"Willy, meet Pogo," Tim said.

"Pogo?"

"Sorry." Tim flipped on the inside light. Sitting between us, nervously surveying the scene beyond the windshield, thinking no doubt about territory, was a large handsome dog with a white face, long nose, and short brown fur. He looked like a cross between a red fox and Rin Tin Tin.

"He's a little uptight," Tim said.

Tim pulled onto Carrera Tercero and headed north. I turned on the radio and spun the dial until I found a station that played something besides salsa. It was called Radio Folklorica, or some such thing, and there was a woman singing over and over, "I want to kill you by burying you in my breast."

"I think he has altitude sickness," Tim said.

"A dog?"

"We could get a tincture of coca together. Put it in his food."

I looked at Tim. He wasn't kidding. A bus backfired. Pogo barked like mad and tried to climb over the steering wheel to reach the window. Tim pushed him away. Then the dog tried my window, his front paw landing right between my legs. I gasped. He climbed off the passenger seat and curled up on the floor on top of my feet. He looked up at me until I moved them.

"So what have you been up to, mate?" Tim asked.

"Well, I walked to Panama."

"You what?" Tim said. I told him the story. By then we had left downtown and were approaching Chapinero.

"So I take it Pogo—"

"I couldn't leave him behind," Tim said. "He'll be okay. He's always wanted to see Bolivia."

"Bolivia?"

"Eventually." Tim smiled. "In the meantime, how'd you like to cross the Andes?"

"Where?"

"You name it."

We stopped at a restaurant where we could eat outside and keep an eye on the truck. Before a waiter could bring us a drink, Tim had spread out a map of South America on the table. His plan was to move south slowly to Bolivia, crossing the Andean Cordillera at a dozen or more points, exploring the eastern flank of the mountains for wild and cultivated species of coca. His challenge was to trace the lineage of the domesticated plants. He knew that the genus *Erythroxylum* contained some 250 species, most of them scattered throughout the American tropics. Although minute amounts of cocaine had been detected in several wild species—seventeen to be exact—only the two cultivated species were widely exploited by Indians. We had already seen one in the Sierra Nevada: *Erythroxylum novagranatense,* the coca of Colombia. Adapted to hot, seasonally dry habitats and highly resistant to drought, it produced small, narrow leaves of a bright yellowish green hue. Bolivian coca, by contrast, thrives in the moist steamy climate of the *montaña.* There we would find it growing in immense plantations carved into the lowland forests. Along the way we would search for wild relatives. By mapping the range and distribution of the plants, and by studying them in the field, Tim hoped to reconstruct their evolutionary relationships and thus understand the history of the domesticated species. It was a hunt for the point of origin of the most sacred plant of the Andes, a search that promised to lead us to some extraordinary places.

"There's been traffic on the lower Amazon for hundreds of years," Tim said. "People always forget that. The wildest areas are here, close to the mountains at the headwaters." His hand swept over the map in an arc that ran from southern Venezuela through Colombia and south to Bolivia and northern Argentina. He pointed to southern Colombia.

"The Putumayo is navigable, and the main trunk of the Amazon in Peru. But almost every other major river is blocked by rapids, especially in Colombia and Ecuador. Even today boats can't go upriver. And until thirty years ago access from the west was limited to the old trading routes. From Colombia to Bolivia there was barely a decent road across the mountains. When they first punched the roads through in Ecuador, they discovered uncontacted tribes living within one hundred miles of Quito."

Tim explained one other important feature of the eastern *montaña.* During the Pleistocene epoch, as global climate change gradually transformed much of the Amazon into a vast grassland, the original vegetation retreated to the west toward the wetter slopes of the Andes. There, in remote pockets in the upper reaches of the river valleys, remnant patches of lowland tropical forest became isolated. Natural selection modified many of the species. When the climate changed once more, and rain returned to the Amazon basin, these biological refuges served as repositories from which plants, animals, birds, and insects radiated to all parts of the Amazon. Thus to this day many of the major river valleys of the *montaña* have astonishingly high numbers of endemic species and remain critical centers of biodiversity. It is not unusual to find genera of plants or butterflies with ten or more distinct species, each localized to a particular valley along the eastern flank of the Andes, precisely the areas now being pierced by roads and pipelines, and destroyed by deforestation.

"It's something you have to think about," Tim said. "We're both the first and probably the last generation of botanists to have a chance to explore these forests."

We began by traveling east from Bogotá, across the Cordillera Oriental to Villavicencio, a sprawling cowboy town that is the main commercial center of the Llanos, the vast lowland plains that cover an area of eastern Colombia larger than Great Britain. From Villavicencio we drove south through the rain and across the savannahs to the Serranía de la Macarena, an ancient and isolated upland massif that predates the rise of the Andes by millions of years. A hundred miles long and soaring seventy-five hundred feet above the plains, the Macarena is an island of astonishing diversity, one of the richest biological preserves in the world, a lost world of mist and rain that to this day remains largely unexplored. Schultes was there in 1951, on the eastern slope and summit of Cerro Renjifo and on the Mesa del Río Sansa. Accompanied by Jesús Idrobo, a colleague from Bogotá, he collected for a month before being urged to leave the region by the local military commander. Soon after his departure a battle took place between the army and a band of armed guerrillas, insurgents who had fled to the Macarena after the assassination of Gaitán.

Tim and I camped for a week close to where Schultes had stayed on January 23, 1951, the day he discovered a new species of an extremely rare genus of trees found only in Colombia. Perched on the side of a high cliff, it had a dense crown of compound leaves, a long inflorescence, and a striking appearance. He named it *Rhytidanthera regalis.* Intermediate between related species found in the northern Andes and one known from the sandstone hills of the Vaupés and Caquetá five hundred miles to the east, it was the missing link that verified his theory that there had been a major migration of Andean plants eastward toward the ancient mountains of the Guiana Highlands. With uncharacteristic pride Schultes called his find "one of the most significant phytographical discoveries of the last two decades."

Rain and the threat of guerrillas, who after nearly thirty years were still active in the mountains, limited our own movements in the Macarena. Nevertheless, within a week Tim found four species of wild coca. We stayed on a farm at the base of the mountains and collected mostly along the flank of a great uplifted slab of stone, several miles in length. It rose mysteriously from the savannah, only to fall away in a dramatic escarpment that plunged to the Río Guejar, the most beautiful river valley I had ever seen. Our host was a simple and generous man with two sons and a beautiful young daughter who knew the mountains far better than her brothers. She accompanied her father as he guided us through the forests and displayed an astonishing sensitivity to the natural world. Once as we rested on a high sandstone bluff, the meandering river below and a cool breeze blowing off the valley, she said that when she died she wanted to become the wind. Her father sighed.

"But your spirit can take many forms," Tim told her.

"That may be," she answered, "but nobody can kill the wind."

A few days later I came upon her father trembling in his garden as he tried to dispatch a large snake that had been eating his chickens. His shotgun wouldn't work, and the snake hung menacingly in a coffee bush. I impulsively struck it with a machete, slicing open the head. The young girl cried out. She walked over to the creature and, before it was limp, dipped her fingers into its blood. She turned, and there were tears in her eyes. She was weeping for a snake.

When Tim and I returned to Villavicencio, passing through the same endless savannahs with their herds of zebu cattle and scattered islands of moriche palms, we learned that a landslide had wiped out the Bogotá road two days after we had come down. The *derrumbe* occurred at Quebrada Blanca, a notoriously unstable creek draw where the mountains had given way a dozen times, most recently three months earlier when more than two hundred people had perished, including two busloads of children. These narrow dirt roads, overhung with vegetation and cut into the edge of the Andes, are perilous at the best of times. In the rainy season, when over twelve feet of precipitation fall on the Llanos and even more on the mountains, the danger is compounded by the constant threat of landslides. There is almost no way to engineer a safe road through the Andes. When the forest is disturbed and the ground exposed by excavation, the land can give way at virtually any point. If it wasn't for the cloud forest, all of the Andes would have fallen into the Amazon long ago.

Since the road went through in the late 1930s, Villavicencio has grown into the largest urban center in the eastern half of Colombia. Its only attraction is the tremendous flurry of traffic itself, the constant movement of trucks, which like clipper ships from a landlocked port carry the products of the Llanos into the mountains. But with the landslide blocking the shipment of petrol from the capital and local reserves virtually exhausted, the trucks had ground to a halt. Those with enough gas in their tanks to reach Bogotá sat idly in a long line at the western edge of town, waiting for word from Quebrada Blanca. The others were parked all over town, their cargoes of meat, fruit, and vegetables slowly decaying in the hot sun. Without gas, we, too, were stranded. Tim managed to secure a few gallons from the mayor, but it was only enough to get back to Bogotá. Unable to collect, we passed the time in the local bars, drinking beer and listening to the unique flamenco rhythms of *la música llanera.*

After three days, word passed through the streets that the Bogotá road would reopen the following afternoon. We left town immediately and joined a procession of trucks slowly moving into the Andes. Approaching Quebrada Blanca, the road from the lowlands runs across a steep mountain slope that drops away into a deep gorge. At the hairpin turn where it crosses the river and climbs abruptly up the flank of the opposite side of the valley, successive landslides had carved away both sides of the draw, depositing thousands of tons of rock and debris into the river and exposing a dangerously unstable cutbank rising more than one thousand feet above the road. It was a scene of desolation; you could still see at the bottom of the gorge the rusting hulks of the schoolbuses, yet for the moment the site of the disaster had a festive air. Landslides at Quebrada Blanca were so dependable that *campesinos* had built a string of shacks and soup kitchens on both sides of the river to cater to the stalled traffic. Children ran up and down the line of trucks hawking *empanadas* and corn, *tinto* and soft drinks. A primitive cable car ran high above the gorge, carrying supplies and the odd daring passenger. On all sides truck drivers and their lovers slugged back *aguardiente* and beer as they toasted the imminent opening of the road.

Finally, around four in the afternoon, with much fanfare and accompanied by a chorus of air horns that reverberated throughout the valley, government officials from Bogotá cut the ribbon on the new bridge. It was a one-lane Bailey bridge, invented by the British during World War I; the Colombian engineers had managed to bolt it into place in less than a week. A hundred or more gasoline tankers coming from the capital slowly rumbled across. By the time we finally began to move, it was almost dark. Rain was falling, and we climbed cautiously toward Bogotá, pulling over frequently to allow trucks bound for the Llanos to pass. For men who believe that horns are brakes and headlights too precious to be used, the drivers seemed unusually determined.

"*¿Hay paso?*" they shouted grimly as they went by. Is it open?

We stopped to eat at the first town, an hour or more up the road. Before we were finished, word came that the bridge had collapsed. Some driver had remembered seeing our truck cross just before the accident, and a local policeman asked us for information. We had none. There was no way to confirm the rumor until the next morning in Bogotá when the story made the front page of *El Espectador.* Above a grainy photograph of the opening ceremony was a banner headline, SE CAYÓ EL PUENTE! The bridge falls! A second photograph showed the twisted bridge at the bottom of the gorge, and beside it an enormous truck, remarkably intact. The driver, after having tried to cross with a load of rice weighing seven tons more than the capacity of the span, had walked away unscathed. In the confusion, no one had yet managed to identify him. No doubt, Tim remarked, he had already fled to the hinterland and would resurface within a week at the wheel of another truck, tumbling down some distant road, drinking *aguardiente* and laughing in the night. Only one thing was certain: He wouldn't be making the Villavicencio run anytime soon. There wasn't another Bailey bridge to be had anywhere in the country. Tim and I had come within ten minutes of being trapped in the Llanos for a month.

Ten dreary days in Bogotá waiting for our plants to dry at the Instituto de Ciencias Naturales left us anxious for the open road. Our next destination was southern Colombia and ultimately the valley of Sibundoy, home of the Kamsa and Inga Indians, headwaters of the Putumayo River, and site of the highest concentration of hallucinogenic plants on earth. For students of Schultes, collecting in Sibundoy is a virtual rite of passage, an opportunity to walk the same hills that he explored in 1941 on his first ethnobotanical expedition in Colombia. "One cannot truly call oneself a botanist," one of his old friends at the Instituto told us, "until you have worked in Sibundoy."

The valley hangs in a high basin, an ancient lake bed surrounded on all sides by mountains that rise two thousand feet above the plain. To the west a road runs beyond the Río Atriz 40 miles to Pasto, the commercial center of southernmost Colombia, an old colonial town perched on the side of the immense Galeras volcano. To the east, another road traverses the Cordillera Portachuelo, the last ridge of the Andes, beyond which the mountains fall away precipitously to the lowland rain forests of the Putumayo. The location of Sibundoy is significant. Less than 60 miles to the north the Andes split into three distinct ranges that fan out across the face of Colombia. To the south in Ecuador the mountains coalesce into a single chain more than 150 miles wide, studded with volcanoes that soar to nearly twenty thousand feet. At the latitude of Sibundoy the Andes are only 70 miles across, and no ridge reaches higher than eight thousand feet. In all of South America there is no other point where it is shorter to cross from the Pacific lowlands to the Amazon. Thus, despite its relative isolation, Sibundoy has for thousands of years been a pathway for ideas and goods moving across the heart of South America.

From Bogotá we drove south swiftly, reaching Cali in a day and moving on the next morning for Popayán, a picturesque and unspoiled university town with cobblestoned streets and old colonial buildings, white and dazzling against the soft green hillsides of the upper Cauca Valley. By air Popayán is only one hundred miles north of Sibundoy, but in between lies the Colombian Massif, a rugged knot of mountains that give rise, within a twenty-mile radius, to the Río Cauca, the Río Magdalena, Colombia's largest river, as well as the Río Patía, which flows into the Pacific, and the Río Caquetá, one of the major Colombian affluents of the Amazon. To the north and east of Popayán the patchwork fields of barley and wheat yield within miles to dense elfin cloud forests, beyond which the volcanic summits of the Cordillera Central rise to over seventeen thousand feet. It is a wild, solitary land, remote and inhospitable, one of the few isolated pockets in highland Colombia where Indian traditions flourish. It is also the only place in the Andes, north of Peru and south of the Sierra Nevada de Santa Marta, where coca is still revered.

There are two major surviving Indian societies. The Guambianos number perhaps eight thousand and live close to Silvia, a small and beautiful town nestled in a gentle valley an hour by road northeast of Popayán. Higher in the mountains, living on both sides of the Cordillera north as far as the Nevado de Huila, one of the highest peaks in southern Colombia, are the far more numerous Paez, a defiant people who, unlike the Guambianos to the south and the fierce cannibal Pijaos to their north, resisted Spanish domination to a remarkable degree. Throughout the seventeenth and eighteenth centuries, as Spanish wrath fell on the Guambianos, and the Pijao were virtually exterminated in a campaign of pitiless cruelty, the Paez were subjected only to the incursions of missionaries who, by all accounts, met with little success. The terrain was difficult, the climate forbidding, the shamanic traditions deeply entrenched in the culture. One Jesuit priest was rendered mute and catatonic by the Paez habit of laughing uncontrollably at his every attempt to convert them.

Before continuing south for Sibundoy, Tim and I spent a fortnight in these mountains, traveling from market to market, purchasing coca leaves, and learning what we could about the local use of the plant. The Indians for the most part were reticent, and with good reason. Though commonly grown in household gardens and used as a medicine and tonic by both Indians and *campesinos,* coca is officially illegal in Colombia. Its cultivation and distribution has been a criminal offense since 1947. Needless to say, the law is selectively enforced. Some of our most robust collections of Colombian coca came from the best neighborhoods of Cali, where the plant is a popular ornamental shrub. In the villages of Cauca, where local police can be bought for a few hundred pesos and most are already on the payroll of the *narcotraficantes,* prohibition is but a pretext for harassment, as we ourselves discovered one morning in Silvia.

Tuesday is market day, and well before dawn the streets of Silvia begin to clatter with the din of traders setting up their stalls. Pogo and I were out early, exploring the town and, in his case, marking turf. Sunrise found us sitting beside a small white church perched on a hill, watching the horse carts and trucks slowly wind their way down the valley roads to the confusion of dust and noise that marked the edge of the plaza. Around seven the thought of breakfast drew us toward the market. At a small stall nestled between an herb dealer and an Indian merchant from Otavalo in Ecuador, I ordered coffee and *sancocho.* The soup came as expected with a grizzled piece of pig, which I discreetly slipped to Pogo. As he ate, I struck up a conversation with the herbalist, a small and wrinkled old man. The Guambianos who live around Silvia abandoned the ritual use of coca nearly a century ago. Still, with the Paez coming down from the mountains to the north and local *campesinos* using the plant as a treatment for stomach ailments and lassitude, it was a sure bet he had a stash of coca beneath his table.

We spoke about kidneys for fifteen minutes, moved on to magical ailments including *susto* and *malaires,* and were back to the liver before I finally chanced the question. I asked for a pound of leaves. He sold me three small packets, each containing an ounce or two, for seven pesos apiece. The leaves were dark greenish brown from toasting, fairly moist, and far larger than those we had collected among the Ika in the Sierra Nevada. I let him know that I'd be interested in buying some more. His grunt seemed noncommittal.

Pogo and I dropped the coca off where we were staying, went out once more to look for Tim, and returned just in time to see the mattresses of our room flying into the patio of the *residencia.* It was a standard police sweep: an officer barking orders at scuttling Indians in ill-suited uniforms, panicky travelers trying to remember what they'd been told about bribing cops, a few penniless local hippies being hauled away, all the while the owner feigning hysteria, flinging his arms about as if the place had never before been busted. The *carabineros* seemed to be paying a lot of attention to our room. I walked in and found Tim standing cross-armed in a corner as three of Colombia's finest rummaged through our gear. With stacks of specimens, bottles of ginseng oil and herbal remedies, a closet jammed with machetes, chemicals, and plant presses, a bathtub overflowing with roots and cuttings, a metal balance that we used to weigh shipments of plants destined for the herbarium in Bogotá, dozens of seeds, shipping envelopes, vials of pickled flowers, cloth bags, and boxes of small plastic bags scattered all about, they were having a field day. My mind did a quick inventory. I had just finished assuring myself that anything incriminating was safely stashed in the truck and had begun to relax somewhat when I noticed a small glass vial of white powder sitting on the bureau by the bed. It was too late.

"¡Aha!" the *teniente* wheezed. With a look of triumph he grabbed the vial, spun open the top, and lifted it to his nose. He sniffed. His face showed disappointment.

"¿Qué es eso?" he demanded of Tim.

"Eso es el polvo de yohimbina," Tim replied calmly. Powder of yohimbine. I had no idea what it was. Tim sidled up to the cop and spoke quietly for a few moments, gesturing more than once to the man's groin. I couldn't hear what he was saying, but whatever it was, the officer's demeanor changed and he seemed visibly impressed. He was just beginning to soften up when one of his men found the three packets

of coca leaves. He quickly regained his professionalism. Now, he harrumphed, he had no choice but to take us in. Just then two little old Indian ladies carrying large bags of coca shuffled into the room, did a quick about-face, and disappeared out the door.

We dutifully filed out and made our way to a nearby police station. It was not too serious. Tim had permits for coca. It just meant another hour or two down at the station as the *teniente*'s three superiors scanned our documents, admired Pogo, and traded jokes with Tim. They kept us there until noon, when the party broke up for lunch. Just before we left, the fellow in charge of the whole place, the one Tim and everyone else had been calling *mi jefe,* pulled the glass vial of yohimbine powder out of a desk drawer.

"It's yours," Tim said graciously.

"Muchisimas gracias, y que tengas un buen viaje," said the *jefe.* Many, many thanks, and have a great trip. He was a short man, somewhat pudgy, with a very large smile stretched across his face.

"What's all this about 'yohimbine'?" I asked as soon as we were clear of the station.

"It's from the bark of an African tree," Tim explained.

"So what's it do?"

"You might say it's an aphrodisiac." Tim smiled.

Not wanting to push our luck, Tim and I beat a hasty retreat from Silvia. We headed south toward a town where we could pick up the road that climbed over the Cordillera Central to the heartland of the Paez Indians. A mile or two from Silvia, a tire blew and the power wagon spun off the road, coming to rest along a deep cutbank crowned with a hedgerow. While I got us back on the road, Tim went botanizing. He returned with a handful of bright tubular flowers that perfectly matched in color the red cloth he was wearing around his head.

"Flor de quinde," he said, beaming. "The hummingbird's flower, *Iochroma fuchsiodes.* It's a shrub, sometimes a small tree, in the potato family. Schultes collected it first, in 1942 in Sibundoy. He's been writing for years that it's hallucinogenic, but no one's had the nerve to try it."

"Tim," I said.

"Don't worry. We're not going to eat it. I don't fool around with the *Solanaceae."* I was glad that he had some limits.

"But what about that time you and Pedro Juajibioy ate *brunfelsia?*"

"How do you know about that?"

"Everyone does."

"Well," Tim said, smiling, "that was a medicinal plant. Besides, I wouldn't want to do it again." He knelt on the ground and carefully separated the specimens while I dug out a press from the camper at the back of the truck.

"Still," he said, "this must be some plant. They rasp the bark and boil it with the leaves. This old Kamsa *curandero* Chindoy—if he's still alive, you'll meet him—told Schultes he took this as a last resort when he couldn't figure out what was wrong with someone. Apparently you get pretty sick."

There was a hint of temptation in his voice that had me worried. Not twenty minutes up the road we pulled over once more, this time by choice.

"What's up?" I asked. Tim was already out the door. Pogo and I caught up with him at the top of a steep road bank; he was standing in the midst of a thicket of low shrubs, the tallest just brushing his face. It was a tree datura—I could tell by the very large and distinctive pendulous blossoms—but I didn't know which one. The flowers were deep salmon red, the lobes creamy yellow, and there were distinct yellowish striations running parallel along the length of the corolla. Tim seemed more interested in the woody fruits, roughly the size of small mangos, hanging awkwardly from the plant's thin branches. He examined their stout woody stems and then took a knife and cut into the tissue at the base of the shrub. His eyes had that glazed look botanists always get when something important is at hand. I could see in an instant that he had identified the plant. Even he was impressed.

"It's only the third time it's been collected," he said. "It's a *borrachero,* but this is the one known to the Guambianos as the tree of the evil eagle. *Brugmansia sanguinea,* subspecies *vulcanicola.* Schultes found it on the northern slopes of Puracé. Three days later the volcano blew up. That's why he called it *vulcanicola.*"

"What happened to him?"

"Nothing," Tim answered. "But he found a book in Popayán with an incredible account of the plant. There's a drawing in it of a woman sitting beneath a flowering tree with an eagle diving from the sky. I have a photocopy of it. Remind me to dig it out for you."

There were ten individual shrubs, and it took almost an hour for us to make a complete collection. By then a driving rain had begun, and we were wet and cold by the time we returned to the truck. I took care of the specimens and got into the cab with Pogo while Tim rummaged through the back. The camper, which I had dubbed the Red Hotel, was no higher than the roof of the truck, but the interior had been cleverly outfitted in Boston by a carpenter friend of Tim's. There was sleeping room for three, storage bins beneath the bunks designed specifically to hold plant specimens, a small built-in file cabinet, and a set of shelves that Tim had stocked with some fifty books, including all the classics of South American natural history—Charles Waterton, Richard Spruce, Henry Bates, Alfred Russel Wallace, the journals of Hipólito Ruiz and José Antonio Pavón. The front half of the file contained clippings and reprints of botanical and ethnobotanical articles, many of them written by Schultes, all of them related to some plant or idea that Tim expected to stumble upon. The back half of the file always held a bottle of Black and White scotch, which I was glad to see in Tim's hand when he got back behind the wheel.

"Here," he said as I poured. "Have a look at this."

It was a naive illustration drawn by a Guambiano artist, and though highly stylized, there was no missing the resemblance to the plant we had just collected. The title identified the tree as Borrachero the Intoxicator. The text, translated from Guambiano, read:

> How pleasant is the perfume of the long, bell-like flowers of the Yas. . . . But the tree has a spirit in the form of an eagle which has been seen to come flying through the air, and then to disappear; it vanishes completely in the leaves, between the branches, between the flowers. The spirit is so evil that if a weak person stations himself at the foot of the tree, he will forget everything, travelling up in the air as if on the wings of the Yas.

The document went on to say that if a young Guambiano girl sat beneath a *borrachero,* she would dream only about the Paez, "those men who never stop chewing coca," and that six months later she would give birth not to a child but to the intoxicating seeds of the tree. This same spirit that so cruelly impregnates maidens also punishes those who in clearing their fields uproot the wild plants without leaving even one to set seed for the next generation.

"So it's both light and dark," I said.

"It's neither," Tim replied. "It's in a realm by itself. Like all the *Solanaceae.* Just think of the names: mandrake, henbane, belladonna, the holy flowers of the north star. Strange, isn't it? Yet the same family gives us potatoes and tomatoes. My grandmother would never eat tomatoes. She said they were the devil's fruit, that we only thought they could be eaten, and that eventually everyone who ate them would be cursed.

"The *borracheros* are real mysteries," he continued as we pulled back onto the road. "They're always found growing near people, in fields, by houses, often in cemeteries, never in the wild. The seeds are infertile. The Indians grow them just by sticking a cutting in the ground. They're native to the Andes, but no one really knows how they came into being. It's the one hallucinogen that Schultes has never tried."

"Have you?"

"Once."

"What was it like?"

"Maybe you know. You probably had it before you ever breathed." He paused for a moment. "The main drug is scopolamine, a tropane alkaloid, same as in belladonna. If you take a big whack, it brings on a wild, crazed state, total disorientation, delirium, foaming at the mouth, a wicked thirst, terrifying visions that fuse into a dreamless sleep, followed by complete amnesia. You don't remember anything. The drug used to be injected into women during childbirth. They called it 'twilight sleep.' Supposedly it caused them to forget the pain of childbirth. It just made them deranged. And of course they didn't forget the experience. It was seared forever into their subconscious—and yours. It was probably in your blood when you were born."

I tried to imagine my mother on belladonna but couldn't.

"But did you eat it?" I asked.

"No, I smoked it, and drank it once as a tea."

"What happened?"

"For the Indians it's always associated with death. The Chibcha used to feed it to the slaves and wives of dead kings and then bury them alive in the royal tomb. In Peru the priests eat it to communicate with the ancestors. It's always considered the most frightening plant, the one you turn to when all else fails. Just like Chindoy told Schultes."

"But what happened when you took it?"

"I don't know. Don't you see? The only way you can know is to take the tea. But then you won't remember. So there will be no story to tell. Just the raw experience. Pure, like madness." Tim sighed in the way he always did before drifting off on some introspective journey. We drove on for a while without speaking.

His attraction to this potent group of plants was telling. There was a wildness in him, a willingness to court fate, a fascination with almost anything that stretched the limits of normalcy. For those who knew plants and knew Tim, his love of the Solanaceae made perfect sense. I often teased him about smoking, knowing full well that he would never quit, simply because of the allure of the source plant and its botanical cousins. For a man who had been obsessed for six years with brunfelsia, a dose of which had nearly killed him in the Putumayo, tobacco was tame.

"Willy," he asked suddenly, "have you ever heard of the Jivaro?"

"Yeah, I have. Actually, I was with them once." He was talking about a rain forest people of southeast Ecuador, a tribe notorious for their ritual preparation of shrunken heads. I had met them at the summit of the Cordillera Cutucú on a previous botanical expedition.

"They call themselves Shuar," I said.

"That's right. They believe that ordinary life is an illusion: everything you see—that mountain, this truck, your own body. The true determinants of life and death are invisible forces that can be perceived only with the aid of hallucinogenic plants. When a boy is six he must obtain an external soul, one that will protect him and allow him to communicate with his ancestors. He and his father go off to a sacred waterfall. The boy bathes, fasts, and drinks infusions of tobacco and other drugs. If the soul does not appear after all that, then father and son will take *borrachero.* You see, they are always plants of desperation."

The road was climbing steeply, through dense fog and past fields carved unnaturally into the cloud forest. Homesteads were scattered at great distances, simple mud houses nestled low to the ground, their thatch roofs blackened by smoke. Gradually even these few signs of human presence faded, and the land opened onto a vast treeless expanse of vegetation, which Tim identified as *páramo,* an exotic and mysterious ecological formation unique to the northern Andes. Farther south in Peru and Bolivia the plateaus and valleys lying above eleven thousand feet are arid, barren, windswept, and cold, a highland desert or *puna,* useful only for the grazing of alpaca and vicuña. Nearer the equator, the same elevation is equally forbidding, only it is constantly wet. The result is an otherworldly landscape that seems on first impression eerily like an English moor grafted onto the spine of the Andes. In the mist and the blowing rain, there are only the *espeletias,* tall and whimsical relatives of the daisy, spreading in waves to remind you that you are still in South America. With bright yellow flowers that burst from a rosette crown of long, furry, silver leaves, *espeletias* look like plants belonging in a children's book. The Colombians call it *frailejón,* the friar, because seen from a distance it can be mistaken for the silhouette of a man, a wandering monk lost in the swirling clouds and fog.

For the next thirty miles, as the road climbed past ten thousand feet, reached the crest of the *páramo,* and then fell away to the east, Tim was in and out of the truck like a jackrabbit, grabbing specimens and regaling me with anecdotes about the plants, all the infinite stories that kept him sane. How the Indians tie the soft *espletia* leaves around their foreheads to relieve headaches, how the *campesinos* use the leaves to stuff pillows and mattresses, or place them inside their jackets for protection against the cold. He collected a moss with a natural antibiotic that was used to make battle dressings during World War I, a gelatinous alga considered a culinary delicacy by the Inca, an insectivorous plant the size of a pea. We made five collections of *Gunnera,* an astounding genus of plants in which some species have leaves an inch across, whereas others, including the one from the Andes, have leaves six feet wide. Inside its roots, sheltered from the elements, lives another plant, a blue-green alga that pays rent by fixing nitrogen from the air and thus feeding its giant host. Many of the plants of the *páramo,* the lobelias and fuchsias, bomareas and gesneriads, have long tubular flowers pollinated by hummingbirds—which, as it turned out, was the reason that Tim was wearing a red bandanna. The color was that of the flowers, and as he moved across the *páramo,* hummingbirds appeared out of nowhere to pollinate his head.

We stayed on the *páramo* as long as there was light and then began the long descent along the Río Ullucos toward the small town of Inzá, which lay in a temperate valley almost five thousand feet below. Arriving well after dark, we found that the hotels listed in the guidebook did not exist. Nor did any others. Not relishing a night in the back of the Red Hotel with a wet dog and a dozen sacks of specimens, we went house to house looking for a place to stay. On our fourth attempt we were met at the door of a *tienda* by a toothless, retarded boy who tried immediately to sell us a pound of marijuana. An old Indian servant intervened, led us past a room where a couple were making love, and into a deserted living room where, for an outrageous price, the shopkeeper invited us to sleep on the floor. We accepted, spread out some blankets over the dirt, and only then discovered that we were sharing the room with a stentorian rooster that the family kept in a carpet bag tied to a rafter just above our heads.

The morning was miserable—rainy and cold, and the town seemingly deserted. We left early but not before Tim had a chance to poke around and ask a few questions. From a local schoolteacher we learned that the Paez call coca *esh,* and that they chew the leaves whole, mixing in as an alkali a fine white powder prepared from raw limestone. The *campesinos* refer to this lime as *mambe* and buy it in small round balls, which they wrap in banana leaves and bury in the ground for several weeks to develop the flavor. The Paez consider the mass-produced *mambe* of the whites crude and caustic. For them the character of the lime and the care with which it is produced are signs of culture. Black limestone yields *kuétan ch'ijmé,* a product that is perfectly white. From a dark reddish stone they make *kuétan kútchi,* their sweetest and most effective powder. Whatever the source of the lime, the preparation is the same. The stone is heated until red hot, then subjected to water, which causes a chemical reaction. Heat is given off, and the calcium carbonate of the limestone is transformed to calcium hydroxide as the rock is reduced to a fine powder. For the Paez the act of making lime is itself a ritual discipline: the gathering of the stones, the kindling of the fire, the rhythmic blowing of the cane flutes that, like a bellows, lifts the temperature of the flames. As if to acknowledge the significance of the lime catalyst, they call the woolen pouches in which they carry the sacred leaves *kuétan yáha,* not coca but lime bags.

Following up on a tip from the teacher, Tim and I left Inzá and drove east to a fork where a road turned off for San Andrés de Pisimbalá, a small Paez village close to the archaeological site at Tierradentro. In Pisimbalá we bought samples of lime and spoke for some time with a group of Paez men gathered in front of an old and beautiful thatched church. They were short and barefoot, and all of them wore felt hats, coca bags, and woolen *ruanas,* or ponchos. Tim bought one of the bags, and then we left for the ruins. The site included more than one hundred ancient burial chambers carved into the soft rock and decorated with stunning geometric motifs. But Tim was far more interested in a series of quite unrelated stone sculptures created one thousand years after the tombs. Though no one knows what relationship, if any, these prehistoric monoliths have to the contemporary Paez, it was nevertheless astonishing to come upon one of the statues at the site of El Tablón. The head and limbs were missing, but hanging from its side was a coca bag, carved in a literal style with distinct geometric designs precisely like those that appear on the *kuétan yáha* of contemporary Paez. On the other side of the statue, again at waist level, was a lime gourd, so faithfully reproduced that we could guess the species.

Later, at the small museum adjacent to the site, we learned that the statue was known as *El Coquero,* the coca chewer. According to the local guardian, the Paez believe that at the beginning of time a young woman was raped by a jaguar, and from this terrible event the Thunder-Jaguar was born. Today it is the Thunder spirit that calls the apprentice to the shamanic path. The initiation occurs at a sacred lake, but the process of attaining the supernatural power and the authority to heal is ongoing. It involves, more than anything, the gradual attunement of the physical body, which in Paez shamanism is the actual medium for the spirit.

The Paez believe that in a healthy body energy flows continually from the earth into the right foot, rises through the leg and the right side of the chest to the head, and then falls away down the left side of the body to the ground. Any disruptions to this flow bring imbalance and thus misfortune. Diagnosis is a form of divination. The shaman remains alone at night, seated in the open air, chewing copious amounts of coca. The leaves stimulate the motionless body, provoking involuntary muscular twitches, which the Paez identify as *senas.* By interpreting the location and direction of the *senas,* the shaman foretells the fate of the patient. An impulse on the cheek suggests a tear, and hence sadness. A spasm running with the flow indicates improvement; one running against it spells misfortune. To select the appropriate herbal remedy, the shaman rubs various medicinal plants along his skin until a *sena* reveals the one that is correct. Thus the body of the shaman, infused with coca, becomes the template on which his own spirit works for the betterment of his patients.

Inspired by the ruins at Tierradentro, Tim decided to continue east to the Río Magdalena, the Valley of Sorrows, south to Pitalito, and then back into the mountains to San Agustín, where we spent several days wandering the hillsides overlooking the ravine of the upper Magdalena. There, arrayed at a series of archaeological sites on either side of the river, standing over burial sites and sarcophagi, are some five hundred anthropomorphic statues. In aspect and scale they rival those of Easter Island, but their symbolism is rooted in the forests of the Amazon. Carved sometime during the first millennium A.D., though possibly much earlier, by a people we know little about, they depict animals and demons: eagles with fangs, felines copulating with men, faces emerging from the tails of snakes. In many instances the figures have cheeks bulging with stylized quids of coca. These are some of the oldest representations of the plant and the earliest evidence of its sacred role in the lost civilizations of the northern Andes.

San Agustín is one of a handful of towns in South America—Santa Marta and Cuzco are two others—where itinerant travelers, particularly those interested in drugs, invariably end up. The land is stunningly beautiful, the ruins mysterious, the living cheap. An added attraction is the San Isidro mushroom, *Psilocybe cubensis,* a powerfully hallucinogenic species that grows throughout subtropical South America but does especially well in the mountain pastures of Colombia, including those around San Agustín. Found always in association with cattle manure, it has a light tan cap that may be several inches across, dark gills, and a distinctive black veil around the stem. When the flesh is bruised, it invariably turns bluish purple within minutes. The largest of all the psychoactive fungi, and by far the easiest to identify, it is the mushroom of choice of every hippie who has ever dragged a backpack across South America. Schultes was the first to find it, in Oaxaca in 1938, growing on cattle dung in the mountains above Huautla.

On the evening of our last night in San Agustín, we ended up eating in the vegetarian restaurant of a hostel that catered to low-budget travelers. Tim sat across from me at a long table shared with four or five others. Pogo slipped between my legs and curled up on the floor. Beside me was a young woman named Sky, and beside her a pair of Germans. A Colombian hippie named Alejandro and his Swedish girlfriend sat alongside Tim, and beyond them was a strange character dressed in saffron robes. He had wooden beads around his neck, a long red beard, and the sort of crazed eyes normally seen on a deer through the sights of a rifle. He introduced himself as Prem Das, but his accent gave him away as Australian. The conversation ran from traveler's checks to cheap hotels but always came back to how long they had been on the road and the wonders of the San Isidro mushroom.

Prem Das clearly had been traveling for some time: Bali, Katmandu, Kabul, Benares, Goa, Marrakesh, and now San Agustín.

"Years ago I gave up shoes," he said. "I never travel anywhere if I need shoes."

"Far out." Sky sighed. I looked under the table. It was true. He had a ring on his toe. I glanced at Pogo, who was eating a bowl of tofu and not liking it. A waiter brought a couple of beers, and then Tim ordered a round for the table.

Prem Das began to describe some of his more exotic drug experiences. Mostly he rambled on about spaceships and mushrooms, but it became interesting when he started to talk about *borrachero.* Even Tim, who normally stayed away from such conversations, paid attention. Apparently, while passing through Barranquilla on the north coast, Prem Das had eaten several handfuls of leaves that he had gathered from a tree datura found growing in the patio of his hotel. Not knowing what to expect, he had decided to take a walk and see some of the city. That was the last thing he could remember. From what he was later able to piece together he had ended up wandering around the main marketplace in Barranquilla stark naked for three days. So mad was his demeanor that even the police had left him alone.

"Incredible, man," Sky said, stretching her vocabulary.

"I know that market," Alejandro remarked. "I wouldn't even buy a mango there."

"What happened to you next?" I asked. Prem Das explained that in the end he had been arrested after all. Somebody had given his clothes to the police, and they had found a joint in the pocket of his pants.

"They were even afraid to steal your clothes," one of the awestruck Germans said.

"What was jail like?" the other asked.

"Not cool," Prem Das answered. By this time they were fawning over him—everyone except Tim, who kept looking at Prem Das with an odd inquisitive stare. Suddenly he perked up.

"Howard," Tim said. "Howard Ziegler. That's it!" The man who called himself Prem Das looked momentarily stunned.

"Howard Ziegler," Tim repeated. "Bogotá in '66, up in the eucalyptus forest." Prem Das broke into a broad smile.

"Yeah, yeah, man." He nodded. "Tim Plowman. Now I remember. We ate acid together. It was far out." I and everyone else looked at Tim, who smiled benignly. It turned out that almost a decade earlier on the slopes of Monserrate he had introduced a skinny Australian tourist named Howard Ziegler, alias Prem Das, to the wonders of LSD.

Tim and Howard reminisced for a few moments but soon ran out of things to say. It was slightly awkward, so I asked Howard where he intended to go next.

"There's this really far-out place, kind of a lost world nobody knows about." He looked as if he was about to share a state secret.

"What are you talking about?" I asked.

"It's called Sibundoy," he said. It was my turn to be stunned. I looked at Tim, who displayed no outward emotion.

"You'll need shoes there, Howard," Tim said.

"Really?" Obviously disappointed, Prem Das said that rather than deal with that hassle, he'd head back to the coast and Santa Marta. I glanced again at Tim. I was glad that Prem Das had changed his plans. Nevertheless, things did not look good.

"What did Schultes tell you before you left?" Tim asked later that night back in our room at the Residencia Luis Tello. It was a modest place run by a schoolteacher—hot water and decent beds for a dollar a night.

"What do you mean?"

"Did he offer any advice?" I thought for a moment. Just before leaving Cambridge I had dropped by the professor's office at Harvard with the vague sense that I might pick up a few useful pointers before heading off for a year or more in South America. Three things had leaped immediately to his mind.

"He told me not to bother with heavy boots because all the snakes bite at the neck. He also said that I should take a pith helmet. He said that in twelve years he had never lost his bifocals."

"Anything else?" Tim asked.

"He said not to come back from Colombia without trying *ayahuasca.*"

Tim laughed. *Ayahuasca,* known also as *yagé* or *caapi,* is the vision vine, the vine of the soul, the most curious and celebrated hallucinogenic plant of the Amazon. The drug is prepared by pounding the woody stem of a liana and boiling it with various admixtures. The Indians see the plant as a magical intoxicant that can free the soul, allowing it to wander in mystical encounters with ancestors and animal spirits. Some users maintain that collective visions occur and that under the influence it is possible to communicate across great distances in the forest. When the active ingredient harmaline was first isolated, Colombian scientists called it telepathine.

"For forty years he's been giving the same advice, and to a lot of people. That's why Howard was on his way to Sibundoy. It's the closest place to the Pan American Highway where you can score *yagé.* The Indians bring it up from Mocoa and the lowlands."

Tim took a small book out of his pack and tossed it over to me. It hit the edge of my bed, bounced off, and landed on Pogo's head. He grumbled and fell back to sleep. The book was *The Yagé Letters,* a thin volume of correspondence between William Burroughs and Allen Ginsberg.

"It's mostly letters from Burroughs written in the early fifties while he was running around South America looking for the ultimate mind-bending high," Tim explained with a smile.

I turned to the first page and found Burroughs in Panama City, a place of "whores and pimps and hustlers," he wrote, "inhabited by the crummiest people in the hemisphere. I ran into my old friend Jones the cab driver, and bought some C off him that was cut to hell and back. I nearly suffocated trying to sniff enough of this crap to get a lift. That's Panama. Wouldn't surprise me if they cut the whores with sponge rubber."

"He's got Panama down."

"Keep going," Tim said as he headed into the bathroom to wash up. I lay back on my bed and thumbed through the book.

The last week of January 1953 finds Burroughs in Bogotá, riding a trolley toward the university and thanking God that he hadn't arrived junk sick in the cold and gloomy capital. He wants information on *yagé* and hopes to find it at the Instituto de Ciencias Naturales. He describes the place to Ginsberg:

> This is a red brick building, dusty corridors, unlabelled offices mostly locked. I climbed over crates and stuffed animals and botanical presses. These articles are continually being moved from one room to another for no discernible reason. The porters sit around on crates smoking and greeting everybody as "Doctor."
>
> In a vast dusty room full of plant specimens and the smell of formaldehyde, I saw a man with an air of refined annoyance. He caught my eye.
>
> "Now what have they done with my cocoa [*sic*] specimens? It was a new species of wild cocoa. And what is this stuffed condor doing here on my table?"
>
> The man had a thin refined face, steel rimmed glasses, tweed coat and dark flannel trousers. Boston and Harvard unmistakably. He introduced himself as Dr. Schindler.
>
> I asked about yagé. "Oh yes," he said, "we have specimens here. Come along and I'll show you," he said taking one last look for his cocoa. He showed me a dried specimen of the yagé vine which looked like a very undistinguished sort of plant. Yes he had taken it.
>
> He told me exactly what I would need for the trip, where to go and who to contact. He suggested the Putumayo as being the most readily accessible area where I could find yagé.

"Dr. Schindler?" I said, looking up from the book.

"They were in the same class at Harvard," Tim said, coming back into the room and flopping on his bed. "No, wait. I think Schultes was a year later. Burroughs was class of '36."

I read on. At the end of January, following Schultes's advice, Burroughs heads off for the Putumayo. A month later, "with nothing accomplished," he is back in Bogotá. Having been conned by medicine men, jailed by the police, rolled by a local hustler, and laid prostrate by malaria, he decides to stick close to "Doc Schindler" on his next foray into the jungle. He writes Ginsberg on March 3, "I have attached myself to an expedition, in a somewhat vague capacity to be sure, consisting of Doc Schindler, two Colombian botanists and two English Broom rot specialists from the Cocoa Commission."

Six weeks later Burroughs writes once more from Bogotá. This time, thanks to Schultes, he's had better luck. Within a day of reaching the lowlands Schultes has introduced him to an old friend, a German farmer and former gold prospector who within half an hour provides Burroughs with twenty pounds of *yagé* and an appointment to take the drug with a local *brujo,* or witch doctor. That night finds him on the dirt floor of a hut, sitting before a crude altar as the *brujo* chants over a red plastic bowl containing a brown liquid, oily and phosphorescent. Burroughs drank it "straight down." It had, he noted wryly, the "bitter foretaste of nausea."

Two minutes later a wave of dizziness swept over him, and the hut began to spin. Hit by a sudden, violent urge to throw up, he stumbled outside, flung himself against a tree, vomited six times, and fell down to the ground "in helpless misery." His numb body swathed in imaginary layers of cotton, his feet transformed into blocks of wood, his eyes lost in a blue haze of larval beings, this veteran of a thousand strange scenes had one cardinal thought: "All I want," he said to himself again and again, "is out of here." Fumbling with an emergency bottle of downers, he managed to pop six Nembutals. He spent the rest of the night on the floor of the hut, fighting off malaria-like chills and the crazed, obsessive thought that this *brujo,* alone among all *brujos,* made a specialty out of poisoning gringos. In the morning he attempted to compare notes with Schultes, who by this time in his career had taken *yagé* on more than twenty occasions.

"I never get sick," Schultes told him. Burroughs mentioned that at one point he felt himself change into a black woman, then a black man, then a man and a woman at the same time, with everything writhing as in a Van Gogh painting. He had achieved pure bisexuality, becoming a man or a woman at will, awash with wild convulsions of lust.

"I only get colors, no visions," Schultes replied.

A month later Burroughs and Schultes found themselves stranded at Puerto Ospina, a military post on the upper Putumayo. Burroughs recounted:

> The agent doesn't have a radio or any other way of finding out when the plane gets there if it gets there . . . so I says to Doc Schindler, "We could grow old and simple minded sitting around playing dominoes . . . and the river getting higher every day and every motor in Puerto Espina [*sic*] broke. Doc, I'll float down to the Atlantic before I start back up that fuckin river." And he says, "Bill, I haven't been 15 years in this sonofabitch country and lost all my teeth without picking up a few angles. Now down yonder in Puerto Leguisomo [*sic*]—they got military planes and I happen to know the commandante is latah. . . .
>
> So Schindler went on down to Puerto Leguisomo while I stayed at Puerto Espina. Everyday I saw the plane agent and he came on with

the same bullshit. He showed me a horrible scar on the back of his neck. "Machete," he said. No doubt some exasperated citizen who went berserk waiting on one of his planes.

Several days later Burroughs turned up in Puerto Leguízamo after all, where the commandante allowed him to bunk with Schultes on the *Santa Maria,* a naval gunboat anchored on the Putumayo. For a week they shared a cabin, anxious for no doubt different reasons to get out of Puerto Leguízamo. The town, Burroughs wrote,

> looks like it was left over from a receding flood. Rusty abandoned machinery scattered here and there. Swamps in the middle of town. Unlighted streets you sink up to your knees in.
>
> There are five whores in town sitting out in front of blue walled cantinas. The young kids of Puerto Leguisomo cluster around the whores with the immobile concentration of tom cats. The whores sit there in the muggy night under one naked electric bulb in the blare of the juke box music, waiting.

I finished the book and snapped off the light, my head spinning with the thought of these two unlikely Harvard characters hanging out in such a place.

"Now at least you know why Howard was going to Sibundoy," Tim said quietly.

"I thought you were asleep."

"Good night, Willy."

" 'Night, Tim."

In the early morning before leaving San Agustín, Tim and I took one last walk through the ruins at the site known as Las Mesitas. Located close to the village, it consists of a series of dome-shaped earthen mounds, roughly eighty feet wide and twelve feet high, scattered over a flat man-made plain between two affluents of a small tributary of the Magdalena. At the center of each of the barrows is an underground chamber built of huge vertical slabs of stone, and within each tomb are the large and imposing statues that once served as guardians of the dead. Other monoliths stand like sentinels around the site. Some have fallen over; others look as if they had been cast aside by the treasure hunters who long ago plundered the tombs.

The images on the statues, worn and beaten into the stone by a people who had no metal tools, are lurid, monstrous, even frightening.

Though captured in time and removed from the cultural context that once gave them meaning, they retain a brooding ferocity, a taut, aggressive power that seems at every moment ready to burst out of the confines of the stone. Some of the monoliths are surprisingly naturalistic: a stout eagle clasping the head of a snake in its beak, toads emerging from enormous boulders. But the majority are phantasmagoric visions of transformation, with the jaguar as the dominant iconographic symbol. There are felines overpowering women, men mutating into cats, and jaguar-toothed rodents dominating men whose genitals are bound by ropes to their waists.

The statues that Tim and I had come to photograph were relatively benign. Both stood alongside the northwest barrow at Mesitas B, the name given to one of the two major groupings of tombs at the Mesitas site. One was a sort of San Agustín happy face, a massive triangular head over seven feet high with enormous bright eyes, a broad smile accentuated with fangs, and two highly stylized protuberances in the cheeks. The other monolith of note was a six-foot columnar guardian statue, a warrior bearing a club across his chest in one hand and clasping a stone in the other. Above his head loomed a spirit being, protective and domineering. Again in each of the cheeks was a prominent bulge. Though more realistic than the "happy face," the carving nevertheless shared the essence of the jaguar. The nostrils were flared, the eyes glaring.

"It has to be coca," I said as I ran my hand over the surface of the statue. There was no mistaking the resemblance of the stone cheeks to the face of a modern *coquero.* We looked around and soon found yet a third monolith with an obvious representation of coca chewing, this time a warrior guardian with a single quid held in the left cheek.

"What you're looking at is the jungle coming into the mountains," Tim said, "the place of fear and the place of healing lifted into the highlands by the imagination of these people."

"It looks like the whole place was tripping," I replied foolishly.

"Reichel-Dolmatoff sort of thinks they were," Tim said, referring to the Colombian anthropologist. "The jaguar was sent to the world as a test of the will and integrity of the first humans. Like people, it is both good and evil. It can create and it can destroy. The jaguar is the force the shaman must confront. To do that he takes *yagé.* That's when things get interesting."

"How do you mean?" I asked.

"If the shaman can tame the jaguar, the energy may be directed for the good. But if the dark aspect of the wild overcomes, the jaguar is transformed into a devouring monster, the image of our darkest selves.

Either way the shaman and the jaguar become one and the same. Reichel-Dolmatoff would say that the jaguar spirit must be mastered by everyone if the moral and social order is to be preserved. The wildest of instincts, like the impulses of the natural world, must be curbed if any society is to survive. That may be what these stones are all about."

"You mean in guarding the dead, the statues reveal what it means to be alive?"

"Right. They also show the consequences of failure."

We continued to wander over the site and came upon a large trough-shaped sarcophagus carved in stone. Beside it was a short but dramatic statue. The figure held a trophy skull in its hands and had a ferocious expression that suggested it had most definitely enjoyed ripping the head from the torso of its enemy.

"Whoever lived here didn't have a lot of time or patience for compromise," Tim said. "They knew what they believed, and they knew it was true because the plant revealed it to them. That's the key. And I think that's what Burroughs came looking for; that's what he wanted to find. Conviction. But he thought it would be somehow pleasant, like another kick."

"You mean taking *yagé.*"

"Yeah. *Yagé* is many things, but pleasant isn't one of them."

CHAPTER SIX

The Jaguar's Nectar

TIM WAS NEVER one to hurry a situation that was enjoyable, and thus we found ourselves late on a Sunday afternoon high above Sibundoy on the *páramo* of San Antonio, chewing coca and watching the wind tear apart the clouds that lay over the valley. We had arrived in Pasto, the state capital of Nariño and the commercial center of southernmost Colombia, around noon. The city had been unnaturally quiet. The government's decision to double the price of gasoline had provoked a general strike, and a demonstration the day before had ended in violence. Burned-out shells of cars and trucks still smoldered along the main approach to the city center, and in the plaza tattered remnants of banners hung from the tree branches. There was a tank stationed by the steps of the old church, and military patrols posted at the major intersections. The weather was cold and wet. Those few who ventured into the streets shuffled along, shoulders bent into the wind, faces hidden beneath woolen ponchos and broad felt hats. Everything was gray—the people, the stone buildings, the dark glistening streets, the clouds that fell off the slopes of the Galeras volcano and smothered the city. The market was deserted, and it was only with a good deal of luck that we managed to come upon a small but well-stocked *tienda* on the way out of town, east on the road to Sibundoy. While Pogo guarded the truck, Tim and I bought supplies—old newspapers and alcohol, fresh fruit and bread, cigarettes for his old friend Pedro Juajibioy, a dozen bottles of *aguardiente* for the *curanderos* of Sibundoy and the upper Putumayo. As we pulled away from the city we passed a small field where a young Indian girl lay alone on the wet grass, sucking on the udder of a cow.

The road to Sibundoy climbed through rich farmland to a high ridge shrouded with fog, and then fell away to a valley that held a shimmering lake, the Laguna La Cocha, source of the Río Guamués, a tributary of the Putumayo. Skirting the northern end of the lake, the road rose once more to the *páramo* of San Antonio. There we stopped. Though the rain that had followed us from Pasto increased from a drizzle to a downpour, Tim insisted it was an ideal time to botanize.

"In some years they get twenty-five feet of rain," he remarked as we left the truck and tramped through the sodden ground. "It's one of the wettest places in the world."

"So I'm supposed to get used to this?"

"Willy, you're from Vancouver."

Actually, I was delighted to be walking. The coca had numbed my throat and mouth, producing a warm glow, a pleasant mingling of physical well-being and mental alertness that seemed in the moment to be the most desirable of all sensations.

The rain came in great gusts, bending the tall and thin *espletias.* The local farmers had recently set fire to the *páramo,* killing dozens of the plants, blackening the trunks of those that survived. In the fog they seemed like shadows heaving in the wind. Wrapped in a dark poncho and wearing a corduroy jacket, faded jeans, and high leather boots, Tim moved among the plants, clipping a specimen or two from each one.

"This is the one named for Schultes," he said, handing me one of the silver leaves. *"Espletia schultesiana.* He found it in December of '41 on his first trip. He made a century set."

"What's that?" I asked.

"A hundred specimens from the type locality of a new species. It's the best you can do, a hundred collections of an unknown plant dispersed to herbaria throughout the world. You've never heard of it because no one does it anymore. It's too much trouble."

"How did he know it was new?"

"He didn't. It was only the second time he had ever seen an *espletia.* But he had been on the *páramos* above Bogotá with Cuatrecasas." José Cuatrecasas was a Spaniard who had fled Franco and settled in Colombia. The leading authority on the flora of the Andes, he was one of the few botanists alive who was in Schultes's league.

"They got lost for a while and ended up sleeping in a cave—or what Schultes thought was a cave. Turned out to be an abandoned coal mine. It rained like hell, and they woke up black as night, completely covered in coal dust and mud. There was some American with them who was furious. Schultes just laughed about it."

"But what about the *espletia?"* I asked.

"Cuatrecasas was the expert, and he had never been to Sibundoy. So Schultes took an educated guess that no botanist had ever seen the plant. He was right. At least no one had ever collected it. That's why Cuatrecasas named it for him."

"It was that easy."

"The country was wide open."

I smiled in amazement. In North America and Europe the plants are so well known that the discovery of a single new species marks the highlight of a botanist's career. Schultes found over three hundred. Dozens of plants are named for him. Even genera. Panama hats, which are actually made in Ecuador, are woven from the fibers of *Schultesiophytum palmata. Schultesianthus* is a genus of nightshades. *Marasmius schultesii* is a mushroom used by Taiwano Indians to treat ear infections. The Makuna use *Justicia schultesii* for sores, *Hiraea schultesii* for conjunctivitis, *Pourouma schultesii* for ulcers and wounds. The Karijonas relieve coughs and chest infections with a tea brewed from the stems and leaves of *Piper schultesii.* The list goes on. So many botanists wanted to name plants for him that they ran out of ways of using his name and had to use his initials. On a cliff in the Vaupés he found an extremely rare and beautiful plant, a new genus in the African violet family. *Schultesia* had already been used so the specialist named it *Resia,* for Richard Evans Schultes.

Schultes himself took all of this rather casually. The only creature named for him that he ever spoke about was a lowly insect. The story was one of his favorites. He disliked traveling with large scientific expeditions, but in 1967 he went up the Río Negro in Brazil with a dozen entomologists, among them the world's cockroach expert, a fellow who worked for the U.S. Army. The river was as high as it had been in twenty-five years. Walking on shore was impossible. The expedition was equipped with four outboard motors, but only one worked. The scientists were at each other's throats. The cockroach man, a New Yorker who had never been out of the city, had sailed all the way up the Amazon without seeing a cockroach. He was going mad on the ship, so Schultes flagged down a dugout and hired an Indian to take them into the flood forest. Schultes doesn't know a beetle from a bat, but he knows the jungle, and as they paddled along he noticed some oropendola nests hanging over the river. He looked behind him and said, "Wouldn't your damned roaches love all that bird shit in those nests?" The nests, as it turned out, were crawling with some of the biggest cockroaches ever found. Every species of oropendola had a different species of cockroach living in its nest. There were three new genera, and the specialist was so happy he named one of them *Schultesia.* It was an ugly thing, but for years Schultes carried a photo of it in his wallet.

By the time Tim and I had stuffed our collecting bags with *espletias,* the rain had stopped and the wind had picked up, clearing the fog from the *páramo.* Thousands of feet below us, Sibundoy emerged from the clouds, a strange and beautiful world hidden in the midst of mountains. The floor of the valley was emerald green, lush and fertile, and seen from a distance the towns and hamlets that lay scattered along the flanks of the mountains appeared toylike and fragile. The entire valley covers just one hundred square miles, and you could see in an instant that but for a chance of geography the place would still be a lake. The land is almost perfectly flat, its surface disturbed only by the glint of unknown streams and the many rivulets that drain into the basin from the surrounding mountains. To the north and east the San Pedro and San Francisco rivers cut deeply into the mountains, extending the bottom lands. Other rivers flow in from the west and south, coalescing in a shallow lake surrounded by marsh and a tangle of vegetation, remnants of the forest that once covered the valley. Out of the marsh flows a small river, little more than a brook, that crosses the plain and then plunges through an abrupt notch in the southeast corner of the valley. This is the headwaters of the Río Putumayo. The water that soaked the *páramo* at our feet, the clouds that burst over the valley to the south, the veil of rain that obscured the high ridges of the mountains on all sides would eventually join with tens of thousands of narrow streams falling out of the Andes and into the Amazon.

The road reached the valley at the town of Santiago and then turned north through Colón and San Pedro before heading for Sibundoy, the largest of the four communities. Santiago is the center of the Inga, a Quechua-speaking tribe that some believe are descended from the Inca. If true, they, like their close relatives, the Ingano, of the adjacent lowlands, would have been relatively recent arrivals in the valley. A more likely scenario, given that the Inca never fully controlled southern Colombia, is that the people who became the Inga and the Ingano were forced to adopt the Quechua language by the early Spanish missionaries who used it as a *lingua franca* during the Conquest. Sharing the valley with the Inga and living in the northern half around the town of Sibundoy are the Kamsá, an isolated culture with a language related to no other. The Kamsá maintain that their ancestors were born in the valley, that the land has always been theirs.

"It's peculiar," Tim remarked as we drove through the stone streets of Santiago. "The Kamsá speak Inga, but none of the Inga speak Kamsá. And all the plant names are Inga here and in the lowlands."

"Wouldn't that make you think the Spaniards imposed the language on them?"

"Probably. I don't think the Inca were here. Look at those fields." I glanced over the valley. The land was incredibly rich, the volcanic soil dark and moist. On all sides maize grew taller than the horses and mules that grazed by the roadside. Just beyond a small thatched house a row of men and women worked their way across an open field sowing seeds with primitive digging sticks. The women wore red blouses and black skirts, brilliantly colored waistbands, and bright woolen shawls. Their hair was black and long, tied with ribbons at the nape of the neck. Several carried babies beneath their shawls. The men wore long white ponchos decorated with vertical black stripes. Both men and women were barefoot and around their necks hung heavy necklaces of green and white beads.

"What do you notice about them?"

"They're growing corn."

"Exactly," Tim said. "This land and climate is perfect for potatoes. Yet they live on corn and beans. If the Inca had been here, they'd be planting potatoes and chewing coca."

"There's no coca?"

"No, at least not now. There might be in the lowlands. That's what we want to find out."

"Do the tribes get along?" I asked.

"They have to. The way of life is the same. Small plots of land, everyone just trying to get by. A hundred years ago there were just sixty settlers in the valley. Now the mestizos outnumber the Indians. The

Kamsá and Inga control only twenty percent of the valley. The rest of the land has been taken from them."

"What about Pedro?"

"He's Kamsá. Schultes met his family in 1941. By then the Capuchins owned the valley. Pedro went to their schools and became a devout Catholic. They made him an altar boy. He went to church twice a day and might have become a priest if Indians had been welcome. Then Schultes came along. Pedro's father was worried about malaria, but he let Schultes take his son into the lowlands as far as Mocoa. That's how Pedro got turned on to plants. Now he's a botanist. He's worked with everyone, all of Schultes's students, anyone who comes to Colombia. His feet are in both worlds. His dream is to build a herbarium right in Sibundoy. He already has a garden of medicinal plants all labeled with scientific names. Among the Kamsá he is known as a household healer, an expert at treating the ailments of the ordinary world. He has even trained among the shamans of the lowlands."

"Can he make *yagé*?"

"He can but he won't. That's the domain of the medicine men, the ones who have mastered the visions. They alone can deal with sorcery. Pedro wouldn't dare try."

"But he uses *yagé*."

"Naturally."

Pedro Juajibioy lived with his family in a small house on a back street of Sibundoy, not far from the enormous church and seminary that dominate the town. It was late, well past dark, when we knocked at his door.

"¿Quien es?"

"Don Pedro! Timoteo *a sus órdenes.*"

"¿Timoteo Plowman?" the voice asked carefully.

"Sí, sí, estamos aquí."

Another voice inside the house, that of a young girl, squealed, and the door flung open to reveal a short man holding candles in both hands. Behind him two teenage girls were bouncing up and down. Out of the shadows came an older woman, whom I took to be Pedro's wife. When she saw Tim, she flung her hands in the air in surprise. The girls leaped past their mother and wrapped their arms around Tim. They were wearing cotton blouses with lace collars. Both were very pretty, and the youngest had a flash of gold in her smile. All of them kept saying *"Dios mío"* over and over, except for Pedro, who just smiled as he deliberately worked his way up one side of Tim and down the other with his candles. Finally he broke into a broad grin.

"Pues," he said warmly, *"por fin el botánico loco vuelve."* At last, the crazy botanist returns.

Tim and Pedro hugged and slapped each other on the back for about five minutes. I shook hands twice with everyone. We had woken up half the neighborhood before the welcoming was complete. By then Pedro's wife had food on the table: baskets of roasted maize, a steaming cauldron of soup, and several flasks of chicha—an effervescent drink made by chewing up corn or yucca, spitting the mush into a vat, adding water, and waiting for the natural fermentation process to turn it into a frothy delight. Enzymes in saliva convert the starch into sugar, and yeasts in the air convert the sugar into alcohol. Generally one makes chicha in the morning, knowing that the fermentation will continue all day so that by nightfall the beverage will have the desired potency. Pedro's blend was wicked, and after each of us had drained a couple of large calabashes, Tim insisted on cracking open a bottle of *aguardiente.*

"No point risking a hangover," he explained as he poured shots of the cheap anise-flavored cane whisky.

"Pedro, *tomese un tragito."* Have a little drink.

"Gracias."

"Salud."

"Don Guillermo," Pedro said, turning to offer me a toast. *"Salud. Paz, amor, y amistad."* Peace, love, and friendship. The formalities at an end, Pedro asked how I had managed to get hooked up with a lunatic. Tim said something in return, and for a while the banter went back and forth—their trip together on the Río Guamués, who was to blame for the mistake about dosage, why they hadn't gone back and murdered the shaman, which plants had yielded new drugs. Tim asked about Pedro's family, Pedro asked about Schultes. As Tim brought him up to date, Pedro punctuated each new bit of information by saying softly in Spanish, "That was a man. That was a man."

It was difficult to guess Pedro's age. His body had been molded by the damp and cold, his face perfumed by smoke. His unmarked skin, the warmth and light in his eyes, and the tinge of gray in his hair gave him a look that might have been that of either a young man prematurely old or an old man remarkably well preserved. He smelled of rain, smoke, and sweat.

"So, my friends," he said as Tim drained the last of the bottle. "Where are we going¿"

"Mocoa," Tim answered.

"El camino a la muerte," Pedro replied. "It won't be good. There's been a lot of rain."

"The road to death?" I said, looking toward Tim.

"That's what they call the route across the *cordillera* to the lowlands. It's one of the most dangerous roads in the country."

"Terrific."

"It won't be what you imagine," Tim said.

"What do you mean?"

"You'll see."

The church bells woke me before dawn. I told Pogo to get off my pillow. He didn't move. I eased out of my bedroll, stumbled into the kitchen, and managed to stand just long enough to drink a bubbly herbal potion that Pedro's wife had waiting for me by the hearth. There were two other glasses on the table. Both were full.

"What is it?" I asked in English, amazed that I had managed to say anything in any language. It was only with tremendous willpower that I repeated the question in Spanish.

"Don Guillermo *tiene un ratón,"* she said with a laugh, not a tinge of sympathy in her voice. "You're hung over. So drink up." The potion had nothing to do with jaguars. It was crushed cabbage mixed with vitamins, hot chilies, and a splash of rum. I drained all three glasses, and wrapping a blanket around my shoulders, lay by the fire and went back to sleep. An hour or two later Pogo tugged on my foot. No one was around. I went out to the garden, poured a bucket of cold water over my head, got dressed, and slipped out to find some coffee.

Perhaps foolishly I had anticipated in Sibundoy a land lost in time and somehow miraculously preserved; the image I held of the place was so strong. More than anywhere else in South America I associated Sibundoy with Schultes. The town itself was not disappointing. It was a small village still, and with the exception of the new electric street lights, probably much as it had been in his day. What little commerce there was—small shops and a few *residencias,* a gas station, and several mechanics set up to service the trucks and buses, an Indian cooperative selling curios—straddled the through road to Mocoa. The main plaza lay several blocks away at the other end of a grid of muddy streets. The people still lived in houses built of adobe or wooden planks, some roofed in tin, others in old ceramic tiles. Each house had a garden. Milk cows grazed in backyards, and small plantings of maize spread continuously behind and between the houses. As in most Andean communities, there was no separation between the town and the fields and the mountains that surrounded it.

Yet as I walked through the streets I felt the disappointment that invariably arises when one comes face-to-face with a place that cannot match expectations. Like all of Schultes's students I had my own memory of Sibundoy drawn from his stories, and especially from the old black-and-white photographs that hung in his office at Harvard. Schultes was a naive photographer. For him a beautiful image was one of something beautiful. But he was technically adept at using his Rolleiflex camera, and he approached photography with the same meticulous attention to detail that characterized his work with plants. His photographs have a timeless, ethereal quality, especially those taken on his first long expedition to Sibundoy and the Putumayo. His favorite is a portrait of a young Kamsá boy holding the leaves and blossom of a tree known as the jaguar's intoxicant. The boy is dressed in a white woolen poncho with broad stripes. His skin appears soft, unblemished, and his thick black hair has been cut with a bowl. His only adornment is a mound of necklaces of small white and dark glass beads. His expression is completely natural. He neither fears the camera nor is concerned about its disapproval. He has the freshness and ease of a photographic subject who has never seen himself in a photograph. Though neither sentimental nor condescending, the image is touched with pathos. It is as if in taking the photograph, in freezing that moment of the boy's life, Schultes was both testifying to the youth's vulnerability and mortality, and bearing witness to the relentless corrosion of time.

As I reached the main plaza, I realized that in a sense I was looking for that young boy, just as I was searching for his other photographs: the Kamsá girls dancing at Carnival, medicine men with feather coronas and necklaces of jaguar teeth, barefoot Indian schoolboys packed ten to a bench, scribbling in their primers beneath the zealous gaze of priests. Naturally I found none of these. The buildings for the most part remained, and the priests still wore their dark robes. But the faces in the plaza had changed. The elderly women coming out of the church, the workers repairing the walls of the monastery, the schoolgirls in their black tunics—all of them were *mestizos.* Sibundoy had become an Indian town without Indians.

"Don Guillermo!" I turned and saw Pedro coming across the plaza. He was wearing a poncho and had a small bouquet of Easter lilies in his hand.

"Buenos días, Pedro."

"¿Cómo está, Don Guillermo? *¿Qué haces?"*

"Paseando, no mas." A priest wandered by. Pedro said good morning. The priest nodded and walked on.

"When I was a boy," Pedro said, smiling, "they used to laugh at our *cusmas.* They called them dresses. They with those robes called us women for what we wore! They called our language *coche,* the language of pigs."

Pedro took me by the arm and led me past the church and into a cemetery. The plots nearest the church had elaborate brick and masonry tombs, decorated with plastic flowers and color photographs of the dead. The names were Spanish. On the other side of the cemetery were unmarked graves—mounds of fresh dirt, a few with crumbling wooden crosses etched faintly with the names Chindoy and Juajibioy.

"Graveyards always tell the truth," Pedro said. "For every one of their children who dies, four of ours are lost. The Church owns the land and cattle. They send cheese and butter to Pasto, while our children go hungry. The government pays for schools, but the bishop decides how to spend the money. Everyone must go to school, that's the law. So they make rules. The children must have shoes, books, and uniforms. Who can pay for these things?"

Pedro paused in front a small grave. He knelt, placed the lilies on the dirt, whispered a prayer, and then crossed himself. When he was finished, I asked him if it was true that he had once wanted to be a priest.

"Yes. I was a believer," he answered. As we walked out of the cemetery, he hesitated for a moment at the gate. "What the padres don't realize," he said, "is that we have many lives, only one of which may be claimed by death."

That afternoon Pedro learned that one of his relatives living in Pasto had been injured in an accident. The city was three hours away, so Tim and Pedro left immediately in the truck. I remained behind with Pogo and spent the rest of the day and all of the next in the archives of the church, looking through old books and photographs, trying to get a feel for a history that clearly had left Pedro a tired and wary man. Initially the priests were somewhat defensive, but once they understood that I was a student of Schultes and that he had been the guest of Gaspar de Pinell, a monumental figure in the early days of the mission, they were gracious and forthcoming. They sat me at a large desk in a bright room in a corner of the monastery. The whitewashed walls were thick adobe, the floorboards whipsawed lumber of various widths, polished and oiled by hand. The stone lintels over the doorways, the carved balustrades overlooking the courtyard, and the massive walls that held the cold and damp lent a medieval air to a building that was not yet a century old. To understand Sibundoy and the impact of the Church on the Indians, you have to begin with the Conquest and a legend that lured hundreds of Spaniards to their deaths.

In the spring of 1541 word reached Bogotá that Gonzalo Pizarro had left Quito and crossed the Andes on the ill-fated expedition that would result in Francisco Orellana's epic voyage down the Amazon. Pizarro's goal was the fabled land of El Dorado. Eager to beat him to the prize, Hernán Pérez de Quesada, brother of the founder of Bogotá, recruited 260 Spaniards and, equipped with two hundred horses and six thousand Muisca men, women, and children, he marched east the following September. The expedition crossed the frozen mountains above Bogotá, reached the llanos, and then turned south along the eastern flank of the Andes. To find his way, Pérez de Quesada relied on the confessions of tortured Indians. He was a man of notorious cruelty. When Indians failed to deliver gold, he sacked their villages, killing the men, slicing the breasts off women, and the noses and ears from young children. Girls of seven were raped and sodomized. Noting that the Indians had little fear of hanging, he instructed his men to impale victims by inserting a stake between the legs, thrusting it out the top of the head. Other of his documented atrocities included hanging captives upside down over fires, burning them alive in boiling oil, killing infants so that mothers could carry heavier loads, and setting hunting dogs loose to disembowel children. Thus he was led further and further into the unknown.

At the Río Guaviare, the Spaniards encountered for the first time the forests of the Amazon. They hesitated and then plunged ahead; the men lived on roots, and the highland Indian porters died by the score. Sending search parties in all directions, the expedition turned west across the headwaters of the Caquetá, and entered the valley of Mocoa. The resistance of the native tribes increased. At one point five Spaniards were captured and quartered in full view of the remainder of the expedition. Pérez de Quesada took this as a good sign, evidence that the Indians were defending the approaches to El Dorado. Rumors spread of a golden land known as Achibichi, a fertile paradise lost beyond the horizon of the clouds. Certain that this was their destination, the Spaniards pushed on, climbing the flank of the mountains, slashing their way through the dense cloud forest. Finally they reached their goal. It was Sibundoy. There was no gold, and to make matters worse, there were Spaniards already in the valley, two soldiers left behind by Sebastien de Benalcázar, the founder of Pasto and archrival of Pérez de Quesada. Half the horses were dead, and a third of the Spaniards. To deliver the motley band of survivors to Sibundoy had cost the lives of all six thousand Muisca Indians.

Within four years of this debacle, Franciscans based in Quito established a mission in Sibundoy. Their work produced mixed results. Christianized Indians proved useful in enslaving the tribes of the lowlands and forcing them into the gold mines. Attempts to purge the valley of ancient religious ideas were less successful. More than a century after the arrival of missionaries, the Franciscan bishop Peña Montenegro, frustrated by the persistence of traditional beliefs, noted that "this evil seed planted such deep roots in the Indians that it appeared to become their very flesh and blood so that their descendants acquired the same beings as their parents, inherited in the blood and stamped in their souls." Though the priests outlawed dances and ceremonies, and did all they could to ferret out and destroy the ritual paraphernalia, the drums and deer heads, feathers and other "instruments of evil," the power of the shaman remained the principal obstacle to the spreading of the Gospel. These "witches," the Bishop wrote, "resist with diabolical fervor so that the light of truth shall not discredit their fabulous arts. Experience teaches us that to try to subdue them is like trying to subjugate jaguars."

The shamans prevailed. In 1767 the Franciscans were expelled from Colombia, and for well over a century the Indians of Sibundoy and the Putumayo had limited contact with the outside world. In the lowlands Mocoa remained a commercial outpost, and a modest trade in forest products, skins and vegetable ivory, gums and resins, herbal medicines, balata rubber, cinchona bark, beeswax, dyes, and lacquer moved through Sibundoy en route to Pasto. But the impact of the Church was minimal. From 1846 to 1899 there was no resident priest in the valley of Sibundoy.

Interest in the region grew considerably in the 1860s and 1870s. British demand for quinine bark to treat soldiers afflicted with malaria in India sparked an economic boom that was but a taste of what would come at the end of the century with the commercialization of rubber. For the first time the forests of the Putumayo and the unsurveyed frontiers of Colombia, Ecuador, Peru, and Brazil became of national concern. In 1893 at the invitation of the bishop of Pasto, Capuchin priests from Barcelona visited Mocoa for a year. In 1896 they returned to stay. In 1900 the government granted the Capuchins complete and total control of the Colombian Amazon. Their mandate was to evangelize the natives; their purpose was to establish their presence and secure the economic and political interests of the nation. The Capuchins chose

Sibundoy as their administrative and spiritual base, and for the next forty years, from the homeland of the Kamsá, they ran a colonial theocracy unlike anything seen in the Americas since the heyday of the Jesuits. Their power was absolute.

The first order of business was the construction of a road that would link the Putumayo with Pasto and the rest of Colombia. As late as 1900 most cargo moving across the Cordillera Portachuella to and from the lowlands was carried on the backs of Indians, including missionaries. The Indian porters who specialized in transporting people were known as *silleros.* They hauled loads of up to 250 pounds, had life spans of forty years, and fully expected to be spurred in the flank as they labored across the mountains. In a sermon delivered in Bogotá, the Capuchin priest Father Carrasquilla described the indignity of being exposed to death and nature on the back of an Indian. "In order to traverse the trail between the capital of Nariño and the residence of the missionaries of the Putumayo took me an entire week, carried on the back of an Indian crawling on hands and feet over terrifying cliffs along the edges of vertiginous abysses, descending precipices like those portrayed by Dante in the descent to hell."

Road construction began in 1909. Indians did the work, received no pay, and were forced by law to submit to the priests. Those who defied the mission edicts were placed in stocks or exposed to the whip. In three years a narrow track, little more than a horse trail, was punched through to Mocoa. Over the next twenty years, in fits and starts, the road pushed deeper into the Putumayo. The promise of free land, together with the seduction of a modest gold strike at Umbria, the head of canoe navigation on the Putumayo, attracted considerable settlement. In 1906 there were just over two thousand white colonists in all the Colombian Amazon. By 1938 there were over thirty thousand in the Putumayo alone. At the vanguard of the influx were the Capuchin missionaries. By 1927 they had built twenty-nine churches, sixty-one schools, and twenty-nine cemeteries in their domain. Their strategy for converting the natives ran from the coldly logical to the ludicrous. They forged an alliance first with the Colombian rubber traders. In exchange for permission to "conquer" Indians and use them to gather rubber, the rubber tappers agreed to teach the padres how to track down the Indians. Wherever they went the priests drew the Indians into small settlements dominated by a church and a mission school. Children were taken away from their parents, separated by sex, dressed in western clothing, and forbidden to speak their native language. In the most famous evangelical excursion, Gaspar de Pinell traveled among the "savage tribes" for over a year in the 1920s. Wearing heavy dark woolen robes tied at the waist with a cord and carrying a large painting of the Divine Shepherdess, the beloved mother and patron saint of the Capuchins, Father Gaspar went from village to village using magic incantations that had been provided by Pope Leo XII to exorcise the demons presumed to have lived inside the Indians since the beginning of time.

It was as a result of these efforts that civilization came to the Putumayo. As a contemporary news item in a Pasto newspaper put it, "If the Caquetá is no longer the asylum for wild beasts and half-animal people, . . . if the light of intelligence has penetrated into these uncultured brains, it is not the Indians who are responsible. . . . All this is owed to the fecund labour of the missionaries, with their assiduous and constant abnegation." The Colombian army took note of the glorious achievements of the Spanish priests. In a letter to the president of the republic, General Benjamin Guerrero commented: "Whereas before they used to flee into the jungle like wild beasts on seeing someone civilized, there are now elements of science. To hear the little Indians of the woods intone the songs of the Creator and the National anthem is most moving. The missionaries raise their spirits to immortality. They teach them to know and adore God, just as they do the Fatherland."

This fusion of Church and state soon paid dividends. In 1928, by international treaty, Colombia gained sovereignty over the Caquetá and the north bank of the Putumayo. Four years later Peruvian forces, angered by the stipulations of the treaty, attacked Leticia, Colombia's only port on the Amazon. War broke out, and Colombian troops advancing into the Putumayo marched in mud up to their waists over the Capuchins' road. In Puerto Asís, at the end of the road, the soldiers found shelter with the padres before embarking on the gunboats that carried them down the Putumayo.

The war lasted for just over a year, and casualties were light; far more troops succumbed to disease than to enemy fire. But by its end mission posts had become garrisons and the Colombian presence in the Putumayo was firmly established. There were gunboats at Puerto Ospina at the mouth of the San Miguel, a naval base on the lower Putumayo at Puerto Leguízamo at the mouth of the Caucaya, and a military garrison at Tarapaca just above the Brazilian border. With their southern flank well protected the Capuchins were free to consolidate their hold on a territory eight times the size of Switzerland.

Following their return from Pasto and before setting off for Mocoa, Tim, Pedro, and I spent a few days wandering around the valley. We had no particular purpose. The flora had already been well studied by Schultes and his close friend Hernando García Barriga, a Colombian botanist who traveled by horse all the way from Popayán to visit Sibundoy in the 1930s. The ethnobotany of the Kamsá had been documented by Mel Bristol, a graduate student of Schultes who lived in the valley for a year in 1962. Tommie Lockwood, another Schultes student who later died on a collecting trip in Mexico, had worked out the complex biology of the *borracheros,* the tree daturas. Tim knew that the wild plants, the orchids and bromeliads, grasses and mints, the mountainside forests of *Podocarpus, Weinmannia, Clusia,* and *Cecropia* were novel and rare. What he wanted to do, I suspect, for he never said so directly, was take the pulse of the valley, measure the changes that had occurred, and try to understand them, at least for himself.

Pedro guided us to many of the localities where Schultes had made botanical history. It was in Sibundoy in 1941, on his very first expedition in Colombia, that he had come upon the greatest concentration of hallucinogenic plants ever discovered. In a valley that can be crossed on foot in a morning, there were over 1,600 individual hallucinogenic trees —and that was just in one genus of the Solanaceae. At the approaches to the valley, near Laguna de la Cocha, Schultes made the second collection of the Tree of the Evil Eagle. In the foothills beyond the town of San Pedro he first found the hummingbird's flower, *Iochroma fuchsiodes.* In the gardens of *curanderos* he documented 12 cultivated varieties of tree datura, including an aberrant form that he later described as a new genus, *Methysticodendron amesianum,* named in honor of his mentor, Oakes Ames. On the Páramo of Tambillo, two thousand feet above the valley to the northeast, he found yet another magic plant, a beautiful shrub with dark glossy leaves like those of a holly, tubular red flowers tipped in yellow, and white berries, bright and lustrous. This was *Desfontainia spinosa,* known to the Kamsá as the "intoxicator," a source of dreams and visions employed by shamans from Colombia south as far as Chile. In his first month in the field, before he had even begun to explore the lowlands, Schultes discovered no fewer than four psychoactive plants new to science.

It was not only the hallucinogens that drew his attention. In the fields he found unfamiliar foods: tree tomatoes, taro, and arracacha roots, new varieties of beans and maize. Working with the healers he collected flowers used to treat fever, roots employed to kill parasites, herbal treatments for infections, tonics for nerves, and infusions taken to ease childbirth. To dress wounds the Kamsá chewed selaginella and tobacco, mixed in urine, and plastered the paste over the injury. Ant bites were soothed with a poultice of peperomia, and ulcers were relieved with the red resin of a plant from the lowlands known as *sangre de drago,* the blood of the dragon.

Schultes's main informant in Sibundoy was a famous shaman, Salvador Chindoy. In the photographs I had seen of Chindoy he was almost always singing or leaning over a patient as he swept away the illness with a fan of jungle leaves. He wore a black *cusma* tied at the waist, a necklace of jaguar teeth, pounds of glass beads, a magnificent corona with a halo of erect macaw feathers, and a cape of parrot feathers that hung down his back to the waist. His ears were pierced by the tail feathers of a scarlet macaw; his wrists were decorated with leaves. The entire costume, Schultes once told me, was a walking vision. The beads and feathers, the sweet leaves on his arms, and the delicate motifs painted on his face were a conscious and deliberate attempt to emulate the elegant dress of the spirit people that the shaman met when he ingested *yagé.* When Tim and I learned from Pedro that Salvador was still alive, we delayed our departure for a day so that we could visit him.

There are moments in life that open wide the soul. Meeting Salvador Chindoy, as it turned out, was not one of them. He lived in a small thatched house, in the midst of a garden of tree daturas not far from the town of Sibundoy. We arrived in the early afternoon and learned from his wife that he was asleep, recovering apparently from a long night of work—a ceremony, she said, put on for the benefit of three gringos. Pedro said something to her in Kamsá, and she slipped inside the house. When she came out, she was wearing beads and a red shawl.

"Uno momento," she said. Through a crack between the boards of the wall I could just make out a figure propped up against a post, slowly beginning to stir. A few moments later the door opened. A frail and suffering face appeared, mumbled something to Pedro, and fell back into the dark interior. When the old man finally emerged, he was in full regalia, complete with necklaces, corona, and a feathered cape that had seen better days. It wasn't clear whether he was dying of a hangover or simply dying.

Salvador knew Pedro, of course, remembered Tim, and claimed to remember me. He offered to sell us *yagé.* He said he was a great *médico,* pestered Tim to bring him a certificate saying as much, and then asked us if we had brought him *aguardiente.* Unfortunately, we had. Tim handed a load of groceries to Salvador's wife, who immediately reached for the bottle, twisted off the tin cap, and began to pour shots. We drank in turn, with Salvador managing to drink twice as much as the rest of us. It was a scene depressing beyond words. Finally, after enduring several unhappy bowls of an especially pungent chicha, we escaped.

It was well past dark and the moon had risen, touching the mountains with a silvery light that glittered against the blue-black sky. It was the first clear night since we had arrived in Sibundoy. The path through the fields was easy to follow, chalky white and dappled with shadows cast by the cornstalks. When we came near the truck and were about to negotiate a slippery log over a creek, Pedro stopped walking and turned to Tim.

"In the first years of your life," Pedro said, "you live beneath the shadow of the past, too young to know what to do. In your last years you find that you are too old to understand the world coming at you from behind. In between there is a small and narrow beam of light that illuminates your life. That is where Salvador became blind."

The next morning Pedro took us to his garden. On a small parcel of land he had built a modest house, raised off the ground and entered by means of a notched log ladder. Around the structure he had planted hundreds of medicinal plants, including all varieties of tree daturas distinguished by the Kamsá. Most were cultivars of *Brugmansia aurea.* The most common was *buyé,* known as the water intoxicant. Others were named for the boa, the hummingbird, the deer, and the snake. The variety associated with tapir is fed to hunting dogs on the night of a waxen moon to empower their spirits. Others are used medicinally. Botanists believe that the grotesque forms of many of the *borracheros* are caused by viral infections. The Indians note that the varieties breed true and that each has quite specific pharmacological properties that can be manipulated by the shaman.

"This is the one that *el doctor* Schultes liked," Pedro said, pointing to a variety with narrow leaves and a deeply divided corolla. "We call it *mits-kway,* the jaguar's intoxicant."

"What do you mean he liked it?" I asked.

"It's *Methysticodendron,"* Tim explained, "the one that he described as a new genus."

"But did he try it?" I asked Pedro.

"No, no, never. He planted it. In the garden of the church and around the seminary."

"The trip lasts four days," Tim said. "Not something to take casually."

"I know this plant," Pedro said. "The first time when I was young I drank a tea made from six leaves. I saw forests full of people and animals, pastures crawling with green snakes. They coiled to strike me. Then, as the intoxication took greater hold, the house I was in started to revolve. Everything inside began to spin as well. But not the snakes. They stretched out for the kill."

Pedro stepped between two rows of low-planted shrubs and herbs and reached to snap a flower from a spindly branch of another of his *borrachero* trees. The trumpet-shaped flower was almost a foot long.

"Smell it," he said. The aroma was sweet and heavy. The corolla was as wide as a child's face. "The scent grows all day long until at dusk it fills the air. Old women place it under their pillows at night so they will dream. That something so beautiful can—" Pedro hesitated for a moment and then continued to speak.

"I once ate six leaves of this tree. I became drunk and my vision went dim. During the day I saw unknown people splitting in half and becoming two. I felt crazy. I started running. I took off my clothes and ran around naked in the garden, showering myself with dirt and weeds left by the hired men who were working at the time. I insulted them. Then I started to kiss trees, thinking they were my fiancée."

"But you can remember?" Tim asked.

"Yes, just a few things. And the others told me what I had done," Pedro explained. "Later I went out to the pasture with a rope to catch a horse to ride, but it turned out to be a dog."

Tim laughed. So did Pedro, who continued along the edge of the garden, leading us to a sheltered corner where the ground was covered by the dense foliage of a scandent vine.

"Look at this," he said as he knelt beside the plant. Tim glanced at the foliage and then quickly squatted beside Pedro. He reached down and examined the underside of a leaf, running his finger down the midrib until it rested at the base of the blade.

"Ayahuasca," Tim said softly. *"Yagé."* He looked at Pedro, who was smiling.

"But I thought it was found only in the lowlands," I said.

"So did I," said Tim.

"Banisteriopsis caapi," Pedro said proudly. "I got tired of buying it from those people."

"But how did you get the plant to grow here?" Tim asked.

"It wasn't easy. The Ingano say there are seven different kinds. To me it's just one species. What do you think, Timoteo?" For the first time since I had known him, Tim was speechless.

"How do they tell them apart?" I asked Pedro.

"They say that you must prepare the plant at the right time of month. Then, once you come under its influence, you can distinguish the varieties based on the tone of the songs that each one sings to you on the night of a full moon."

"Do you think it's possible?" I looked at Tim.

"I don't know."

"I'm growing them all so that I can find out."

"Should be a whole lot more interesting than counting stamens," Tim said. Pedro smiled and nodded in agreement.

The distance by road from Sibundoy to the foothills of Mocoa is perhaps sixty miles, and with any luck the journey takes no more than three hours. That is, assuming there are no accidents or breakdowns, that the bulldozers positioned along the road to clear away the inevitable landslides do not run out of gas, and that the attendants stationed at checkpoints to control the traffic on the single-lane road do not make any mistakes. All of these are very large assumptions, particularly in the rainy season. We left Sibundoy at noon and arrived in Mocoa in the early evening, four days later.

Initially, all went well. The road crossed the valley to San Francisco and then climbed gradually along the western flank of the Cordillera Portachuelo, the last range of mountains before the Amazon. At the first checkpoint a young Indian in a hard hat warned us of a possible landslide and then waved us through. Thus it came as something of a surprise to come around a blind corner and confront a huge Mercedes truck just barely in control careening down the middle of the road. Tim cranked the wheel to the right, and the Red Hotel skimmed along the side of the grade, a foot or two from the edge of a precipice that dropped off into a cloud. A blast of air pounded us as the truck screamed by.

"Holy fuck," Tim yelled as he drove back onto the road. "Why didn't they warn us there'd be traffic?" Pedro, who was with us, had no idea. He didn't own a car, and like everyone, he traveled the road only by night. That way, he said, you didn't have to spend all day exposed to the abyss. You could sleep, comforted by the protective wall of darkness beyond the reach of the headlights.

After the episode with the Mercedes, we crawled along for more than an hour without encountering any other traffic. At several points where the road widened slightly we pulled off and did a little collecting in the cloud forest. Orchids, begonias, calceolarias, bomareas, four different species of fuchsias, and dozens of delicate ferns and lycopodiums. After several trips across the Andes, the pattern of the flora was gradually coming into focus. This to me was the great revelation of botany. When I knew nothing of plants, I experienced a forest only as a tangle of forms, shapes, and colors without meaning or depth, beautiful when taken as a whole but ultimately incomprehensible and exotic. Now the components of the mosaic had names, the names implied relationships, and the relationships resonated with significance.

Tim gunned the truck up through the gears, and we climbed higher into the mountains. By late afternoon the clouds swept across the road, the cold mist becoming so thick you could make out the beam of the headlights in the failing light. It wasn't raining. Water just hung suspended in the air. As we went around a bend Tim leaned on the horn, but the sound faded, absorbed by the fog. If at any moment we had come upon a truck driven as recklessly as that Mercedes, we would not have stood a chance. Instead, the absence of traffic suggested another possibility. In all likelihood there was a landslide ahead, blocking the traffic coming up from Mocoa. It was at this point that the Red Hotel let us down.

As the truck roared up a grade, bouncing in and out of potholes, there was a sudden wrenching clank followed by the terrible grating sound of some part of the undercarriage dragging along the road. Tim pulled over, and I crawled under the truck to see what was wrong. It is amazing how quickly a new pickup can be reduced to a useless hunk of metal. A few thousand miles of dirt roads had loosened the bolts holding the rear transfer box to the chassis. It had fallen away, putting a tremendous strain on the drive shaft. The universal joint linking the drive shaft to the transmission was twisted like a pretzel, its yoke ripped clear off. The only hope was to cannibalize the front axle yoke, replace the universal joint with the spare Tim had brought from the States, bind the transfer box into place with bolts or wire, and limp into Mocoa using the rear wheel drive. It was a messy job that would have to wait until morning. With night coming on there was nothing to do but crawl into the back and listen to the wind as it buffeted the mountainside.

Morning revealed a hell of a campsite. On one side the mountain fell away one thousand feet into a gorge that carried a raging river. On the other the cutbank rose straight up from the road, the slope so steep that the dense vegetation seemed ready at any moment to tumble away from its roots. The road itself swept around a bend and beneath an overhanging ledge that poured a veil of water across the grade. There was water everywhere. It dripped from every leaf and branch, filled every depression in the soil, turned every tire track into a running stream. Within immediate sight of our truck were seven waterfalls tumbling off the mountain and disappearing into the canopy of the cloud forest.

Though we worked through the morning, there was no way to repair the truck with the tools that we had. After a hot breakfast and a couple of stiff shots of scotch, Pedro and I headed off in the rain for Mocoa to find a mechanic. We walked for about two hours and then heard the distant rumble of an engine coming up the road behind us. It was the Flota Cootransmayo, making the run from Pasto to Puerto Asís. The bus strike evidently was over. Pedro flagged it down and we squeezed aboard. Apparently there had been more demonstrations in Pasto and more violence. Several people had been killed, and the government had finally backed off from its proposal to hike the price of gasoline.

"Nothing happens," Pedro said, "until there are a few dead."

Ten minutes later the road reached the crest of the summit ridge and the *flota* rolled into line at one of the last checkpoints, a local landmark known as Mirador, the Lookout. A landslide had closed the road. The driver announced the news casually, as if it was just another stop on the line. He predicted a delay of an hour. There was a collective groan from the passengers, and then one by one they filed off the bus, many of them heading for a roadside shack where an old woman was selling soup and coffee. Pedro and I drank *tintos* and then walked up a path that climbed to a large cross marking the summit. Just beyond the cross the *cordillera* dropped away to the east, and through the clouds I could make out a ribbon of road falling thousands of feet in narrow switchbacks carved into the sheer face of the mountain. At the base of the mountain the forests of the Amazon basin spread to the horizon.

"Aquí mismo. Right here!" Pedro said as we stopped by the cross. "This is where I brought Schultes. We stood there, by that bush. It was raining. He was soaking wet, but he waited for a long time, long after I had dropped back down to the road. He just stared to the east."

"Did he say anything¿" I asked.

"No, no. Claro que no. Estaba fluyendo en sus piensamientos la ceja de la montaña." No, of course not. He was touching the eyebrow of the forest with his thoughts.

I left Pedro to his memories and wandered farther along the trail that ran across the summit. The plants were somewhat different from what we had been seeing, a great profusion of heathers and filmy ferns, unusual orchids and gingers amid the more familiar *tibouchinas* with their purple flowers and furry leaves. I made a mental note to return with Tim to make a collection. Then as the cold wind blew up the slope, I found my thoughts turning from the plants back to Professor Schultes and the moment years before when he had stood here above the clouds and peered for the first time into the forest that would become his laboratory.

It is difficult to know what it was like for him during his first months in the Amazon. In twelve years he rarely kept a journal. He had no time. His collecting notebooks record where he went and when, but they give no insight into his thoughts or emotions. He generally traveled alone or with one native companion, learning early to eschew the cumbersome gear that dragged down so many expeditions of his era. High boots, complicated tents, stools, and portable kitchens were not for him. He wore a pith helmet, khaki trousers and shirt, a kerchief, and, in the low country, leather moccasins saturated with oil purchased by mail order from L. L. Bean back in New England. Rarely did he carry a gun. Besides a machete, hammock, and his plant collecting gear, he brought a camera, a spare set of clothes, and a small medical kit complete with hypodermic syringe and snakebite serum. For food he lived off the land, carrying as emergency rations only a few cans of his beloved Boston baked beans, less for sustenance than to boost his morale when things got rough. For reading he took Virgil, Ovid, Homer, and a Latin dictionary, as well as the eighteenth-century journals of the Spanish explorers Ruiz and Pavón, which he intended to translate in his spare time.

The region he entered, the Northwest Amazon of Colombia, was and remains the wildest area in South America. The Amazon has fifty thousand miles of navigable rivers and one thousand major tributaries, twenty of which are larger than the Rhine; eleven of these flow more than one thousand miles without a rapid. The Amazonian lowlands of Colombia, by contrast, contain only one major navigable river, the Putumayo; all others are interrupted by rapids and waterfalls. The riverboats that 150 years ago transformed the Amazon in Brazil and Peru into a highway have never been able to penetrate the heart of Colombia.

The land explored by Schultes covers over half a million square miles. On a map it is roughly triangular in shape, with a base running from Sibundoy through Iquitos in Peru and then along the Amazon to the Brazilian city of Manaus, with the apex at Puerto Carreño, the point at which Colombia projects into Venezuela and touches the Orinoco River. In Colombia the region is traversed by five major rivers that run east to west on more or less parallel courses before fusing into one massive stream at Manaus, the center of the Amazon basin. Farthest to the south is a short section of the Amazon that, rising in the Peruvian Andes, touches Colombia for only eighty miles above the port of Leticia. The one important north bank tributary is the Río Loretoyacu, home of the Tikuna Indians.

North of the Amazon is the Río Putumayo, with its two major north bank affluents, the Karaparaná and Igaraparaná, home of the Witoto and Bora Indians. Next is the Río Caquetá, formed by several important rivers, including the Río Miritiparaná, home of the Yukuna and Tanimukas; the Río Yarí, with its unexplored branch, the Mesaí; and the poorly known Río Cahuinarí, homeland of several scattered populations of Bora and Witoto. To the north of the Caquetá is the Río Apaporis, a black water river 1,350 miles long dissected by spectacular rapids and gorges that have long isolated several groups of Indians; Taiwanos on the Río Kananarí, and farther downstream on the Río Piraparaná, Makuna, Barasana, and the curious Makú, a nomadic people once employed as slaves by their sedentary and more powerful neighbors. North of the Apaporis is another black water river, the Río Vaupés. Along the banks of its major affluents, the Papurí and Kuduyarí, live the Desana and Cubeo. Finally, northeast of the Vaupés is yet another black water river, the Guainía, homeland of the Arawakan-speaking Kuripakos. The Guainía is the headwaters of the major north bank tributary of the Amazon, the Río Negro, a massive river that carries more water than either the Congo or the Mississippi.

This, then, was the world into which Schultes disappeared: hundreds of thousands of square miles of undisturbed rain forest, traversed by thousands of miles of unexplored rivers, the homeland of some thirty unacculturated and often uncontacted Indian tribes representing six distinct language groups and sharing a profound knowledge of forest plants that had never been studied by modern science. Oklahoma and Mexico had been but a prelude. In Colombia Schultes embraced an entire continent, unknown peoples, and a forest that stretched to the Atlantic. It was, as he would write nearly fifty years later, a land where the gods reigned.

"Don Guillermo!" I turned to see Pedro gesturing frantically by the side of the bus. I hurried off the hillside and leaped onto the *flota* as it pulled away from the checkpoint and began the long coast down to the lowlands.

Mocoa turned out to be one of those forgettable jungle towns that are not even sleazy enough to warrant a traveler's advisory note in the guidebooks. A book Tim and I had with us devoted 360 pages to Colombia and two lines to the capital of the Putumayo, one of which read, "There is no point in stopping here." Pedro and I did, but just long enough to persuade a mechanic to drop what he was doing and proceed the next day up the *camino a la muerte* to repair the Red Hotel. By late afternoon we had hitched a ride back up the road in the rear of a small pickup. Cold and wet, we arrived long after dark and rolled immediately into bed.

The next morning Armando Saa of the Taller Central arrived early, crawled under the truck, and emerged ten minutes later with the dam-

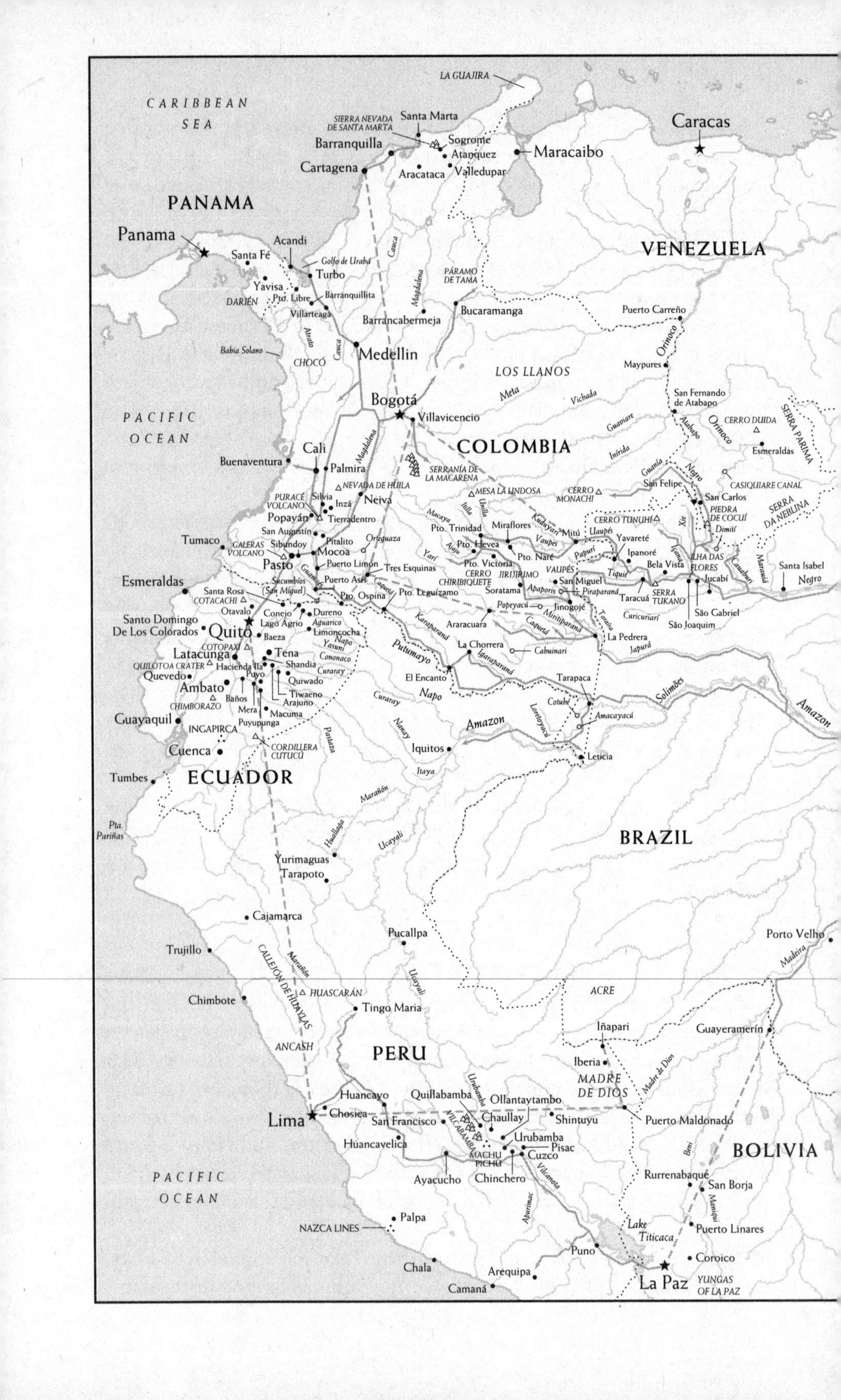
CARIBBEAN SEA
PANAMA
Panama
VENEZUELA
Caracas
Maracaibo
LA GUAJIRA
SIERRA NEVADA DE SANTA MARTA
Santa Marta
Barranquilla
Cartagena
Sogrome
Atanquez
Aracataca
Valledupar
Santa Fé
Acandi
Golfo de Urabá
Turbo
Yavisa
Pto. Libre
Barranquillita
DARIÉN
Villarteaga
Barrancabermeja
Bucaramanga
PÁRAMO DE TAMA
Medellín
Bahia Solano
CHOCÓ
Puerto Carreño
Maypures
LOS LLANOS
Bogotá
Villavicencio
PACIFIC OCEAN
COLOMBIA
Cali
Buenaventura
Palmira
SERRANÍA DE LA MACARENA
San Fernando de Atabapo
CERRO DUIDA
Esmeraldas
SERRA PARIMA
CASIQUIARE CANAL
San Felipe
San Carlos
PIEDRA DE COCUÍ
SERRA DA NEBLINA
NEVADA DE HUILA
PURACÉ VOLCANO
Silvia
Inzá
Neiva
Popayán
Tierradentro
MESA LA LINDOSA
CERRO MONACHI
CERRO TUNUHÍ
San Augustín
Pitalito
Tumaco
GALERAS VOLCANO
Sibundoy
Mocoa
Pasto
Puerto Limón
Tres Esquinas
Pto. Trinidad
Miraflores
Mitú
Yavareté
Ipanoré
Pto. Hevea
Pto. Naré
Pto. Victoria
CERRO CHIRIBIQUETE
JIRIJIRIMO
VAUPÉS
San Miguel
Bela Vista
ILHA DAS FLORES
Santa Isabel
Jucabí
Esmeraldas
Santa Rosa
Sucumbíos (San Miguel)
Puerto Asís
COTACACHI
Pto. Ospina
Pto. Leguízamo
Soratama
Piraparaná
Taracuá
SERRA TUKANO
São Gabriel
São Joaquim
Otavalo
Conejo
Dureno
Lago Agrio
Aguarico
Santo Domingo De Los Colorados
Quito
Baeza
Limoncocha
Popeyacú
Jinogojé
Araracuara
Curicuriarí
La Pedrera
COTOPAXI
Latacunga
Tena
La Chorrera
Cabuinari
QUILOTOA CRATER
Hacienda Ila
Shandia
Puyo
Quevedo
Quiwado
Ambato
Tiwaeno
Arajuno
Baños
CHIMBORAZO
Mera
Macuma
El Encanto
Tarapaca
Cotubé
Amacayacú
Guayaquil
INGAPIRCA
Puyupunga
CORDILLERA CUTUCÚ
Cuenca
Amazon
Iquitos
Leticia
Tumbes
ECUADOR
Pta. Pariñas
BRAZIL
Yurimaguas
Tarapoto
Cajamarca
Pucallpa
Porto Velho
Trujillo
CALLEJÓN DE HUAYLAS
HUASCARÁN
Chimbote
Tingo María
ACRE
Iñapari
Guayeramerín
ANCASH
PERU
Iberia
MADRE DE DIOS
Huancayo
Quillabamba
Ollantaytambo
Lima
Chosica
San Francisco
Chaullay
Shintuyu
Puerto Maldonado
VILCABAMBA
Urubamba
Pisac
Huancavelica
MACHU PICHU
Cuzco
BOLIVIA
Ayacucho
Chinchero
Rurrenabaque
San Borja
PACIFIC OCEAN
Palpa
NAZCA LINES
Lake Titicaca
Puerto Linares
Puno
Coroico
Chala
Arequipa
La Paz
YUNGAS OF LA PAZ
Camaná

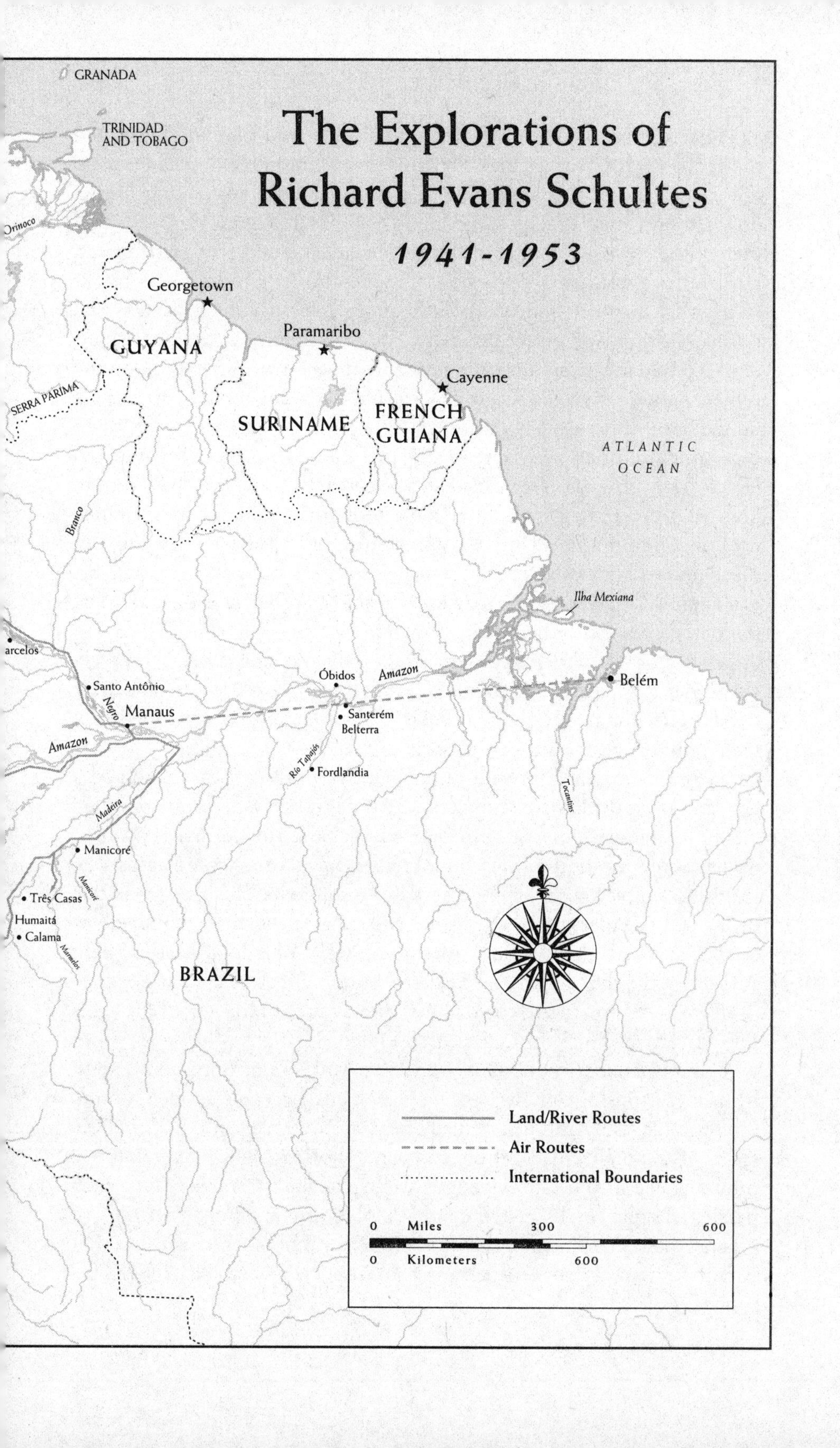
The Explorations of
Richard Evans Schultes
1941-1953
GRANADA
TRINIDAD
AND TOBAGO
Orinoco
Georgetown
GUYANA
Paramaribo
Cayenne
SERRA PARIMA
SURINAME
FRENCH
GUIANA
ATLANTIC
OCEAN
Branco
Ilha Mexiana
arcelos
Óbidos
Amazon
Belém
Santo Antônio
Negro
Manaus
Santerém
Belterra
Amazon
Río Tapajós
Fordlandia
Tocantins
Madeira
Manicoré
Manicoré
Três Casas
Humaitá
Calama
Marmelos
BRAZIL
Land/River Routes
Air Routes
International Boundaries
0
Miles
300
600
0
Kilometers
600

aged parts in hand. Pedro remained with Pogo, and Tim and I returned to Mocoa with Armando and our drive shaft. Colombian mechanics are notoriously poor at maintaining vehicles, but once something breaks down, their genius emerges. Armando Saa was a wizard. He was a thin man with long black hair, a bushy mustache, and deep-set brown eyes. His leather apron, dark with grease, made him look more like a blacksmith than a mechanic. With little more than an anvil, a sledge, a primitive forge, and a few wrenches, he hammered out the bent shaft, replaced the universal joint, and when dissatisfied with his work on the twisted yoke, fashioned a new one from a discarded hunk of steel lying on the floor of his shop.

By midafternoon we were ready to return once more to the Red Hotel. This time we sent Armando ahead in a jeep and followed an hour later in a freight truck piled high with green bananas and destined for Pasto. Ten minutes later, to our astonishment, the jeep carrying our mechanic and our drive shaft flew by on its way back down to Mocoa. We soon discovered why. Yet another landslide blocked the road just the other side of the checkpoint at Mirador.

Twenty or thirty men and women scrambled over the mud and rock, clearing out a path with shovels, picks, digging bars, poles, and bare hands. The riprap was ten feet deep, but with everyone pitching in, it took only an hour or so to scratch a track over the hump. The first vehicle to cross was a private jeep. Then a slightly larger pickup had a go. A bus followed, the passengers watching from a safe distance. When finally a large gasoline carrier bounced over the rubble, the remaining drivers threw caution to the wind. From then on they roared across the landslide as if it had been engineered into the road. Our truck was one of the last to cross. Belching smoke and lunging violently from side to side, it lurched ahead, the driver completely oblivious to the silent Andean gorge that lay exposed several hundred feet below his elbow.

Not a mile down the road, the traffic again ground to a halt. This time a cement truck from Pasto straddled the right-of-way, its front wheels suspended in the air beyond the bank. Two more hours went by as eight tons of cement bags were unloaded, a cable slipped around the rear axle, and the truck hauled to safety. We finally reached the Red Hotel at midnight, just in time to spend another night perched above the precipice in the rain, listening as the roadbed trembled with each passing freight truck, watching as their indifferent headlights swept around the corner in the mist. It was not until noon the next day that Armando had our truck back on the road and we were able to head for Mocoa.

• • •

Schultes remembers Mocoa as a "funny little town," mostly Indians of several tribes, Inganos and the less acculturated Sionas and Koreguajés from the Caquetá, Kofán, and the odd Bora and Witoto from the lower reaches of the Putumayo. The Mocoa that we knew was all dirt streets and sleepy vultures, mud walls splashed with urine, and melancholy Indians trapped in sad processions amid the bangs and cries of Catholic ritual.

It was *El Día de los Difuntos,* the second of November, the Day of the Dead, and in the barrio of San Agustín where we stayed with Pedro's sister-in-law Filomena, the holiday dawned in mourning. Filomena, her husband, Mauricio, and their seven children walked to the cemetery, lit candles, and prayed before an empty black hole in the ground. By midday the plangent sounds of the processions had faded, and there was a final gasp of lucidity before alcohol washed over the afternoon. For the rest of the day almost everyone we encountered was drunk. On the banks of the Río Mocoa we sought out a *curandero* Pedro knew, only to find him sprawled across the back of a mule, his face blanched and his mouth ajar. There were flies gathered at his eyes like animals at a salt lick. Pedro slapped him on the leg. He came to for a moment, let loose an incoherent howl, and then fell to the ground with a dull thud. We then went looking for Don Santiago Mutumbajoy, a venerable healer with whom Pedro and Tim had taken *yagé* eight years before, only to find that he, too, was in town "making himself drunk," as his wife put it. When we finally returned that evening to Mocoa, it was to a dismal house that reeked of *aguardiente* and fermented chicha.

"Es custumbre. Es custumbre," Mauricio said. It's the custom. He tried to ply us with huge calabashes of the stuff. Into the scene walked Pedro's sister-in-law. Her eye was black and her face disfigured with a nasty bruise.

"She fell by the river," Mauricio lied. Pedro said nothing. Later, long after we had gone to bed, I heard him arguing violently with his brother-in-law. I got up and wandered out back to take a leak. There was a full moon and a strange silence. Suddenly on a flutter of air came sounds from the forest. They faded with the wind and then returned, only to fade once more. Pedro came out and for a silly moment we stood shoulder to shoulder pissing in the moonlight.

"Mañana," he said, "we will go to see Don Jorge Fuerbringer."

"The German¿"

"Sí, el mismo." I had no idea that Schultes's old contact, the man who

had helped William Burroughs obtain *yagé,* was still alive and living in Mocoa.

"While we are here I must drink *yagé,"* Pedro said quietly. "Maybe Don Jorge can find us a *brujo* who is not drunk."

"There were two brothers," Tim explained as we drove out of town the next morning to look for Fuerbringer. "They left Germany together just after the First War. One went to New York and later became an editor for *Time* magazine. Jorge wound up in Mocoa."

"Why here?"

"Who knows. But he made enough money panning gold to carve a dairy farm out of the forest. That's where he was in 1941 when Schultes met him. He had already settled in with an Ingano woman. I don't know if they ever married. They had a son and two daughters. One of the girls had a scholarship to study medicine in Moscow. She became a doctor. The son became a cop."

"Were Schultes and he good friends?"

"Yeah, I suppose. I'm not sure friendship is the word. They were good to each other. Fuerbringer helped Schultes understand the Indians. Schultes brought Fuerbringer something of the world he had left behind. Just think of where he was living, three days on horse to get to Sibundoy. Going east, nothing but forest and rivers. Only the priests and the Indians to talk to. With Schultes he could speak his own language again. After twenty years in the jungle, that was probably more important than friendship."

In the settlement of Pepino, not far from Mocoa, Tim pulled up in front of a small nondescript house.

"Fuerbringer's?"

"No. Wait here for a second. I have to get something for Schultes."

Tim rapped on the door until an old woman appeared at a window. Tim said something, and both of them glanced toward a short bushy shrub planted in the front garden. Pedro laughed. Tim handed the woman some money and then clipped a few branches from the plant.

"What did you pay?" Pedro asked as Tim tossed the specimens onto my lap.

"She says it never flowers."

"How much?"

"Ten pesos."

"¡Ha! ¡Que engaño!"

"What are you talking about, Pedro?" I asked.

"Don Timoteo thinks he'll get a better price without you and me."

"It's *guayusa,"* Tim said as he backed the truck out of the woman's yard. "The leaves are full of caffeine. For a long time this was the only known locality for the plant in Colombia."

Tim explained that three species of hollies yield caffeine. The best known is *yerba maté, Ilex paraguariensis,* the national drink of Argentina. A second is the powerful emetic known in the Carolinas as *yaupon,* or the Black Drink, *Ilex vomitoria,* the only caffeine plant native to North America. The third and by far the most mysterious is *Ilex guayusa.* A tall tree native to the eastern *montaña* of Ecuador and Peru, yet sporadic in distribution, it has been collected only rarely. According to Tim it had never been found in flower.

Of the antiquity of its use as a tonic and stimulant there was no doubt. Schultes had analyzed a 1,500-year-old bundle of *guayusa* leaves found in a medicine man's tomb high in the Bolivian Andes, far beyond the natural range of the plant. In the lowlands of Ecuador, Jivaro warriors traditionally used infusions of *guayusa* to purify themselves and their families before shrinking the heads of their slain enemies. To this day they employ *guayusa* as a ritual mouthwash before making curare or taking *yagé.* When the Jesuits first contacted the tribe, they declared *guayusa* the "quintessence of evil." Half a century later they were growing it in plantations, having persuaded all of Europe that it was a proven cure for venereal disease, which it was not. Following the expulsion of the Jesuits in 1766, the forests reclaimed the plantations and commercial production of the leaves ceased. Trade continued, but on a much smaller scale.

"Schultes looked for *guayusa* when he came here in the forties, but he never found it," Tim said. "The Capuchins had heard nothing about it, though they knew a great deal about plants. Twenty years later one of his grad students, Mel Bristol, happened to buy some dried leaves from an herbalist up in Sibundoy. Turned out to be *guayusa.* Eventually Schultes traced the leaves to this very bush. The old lady, like most healers down in the lowlands, sold herbs and preparations to the Inga, who in turn traded them throughout the country. By then Schultes knew from early Church documents that there had been a Jesuit plantation somewhere around Mocoa in the eighteenth century. So he asked the *curandera* where she had obtained the original material for her shrub. She told him about some old trees growing near a little place called Pueblo Viejo. He took off immediately on foot, four hours on a muddy track. That's how he found the plantation, two hundred years after it had been abandoned."

"He was almost sixty when he did that," Pedro said knowingly.

Five minutes up the road, Tim pulled alongside a picket fence and a modest two-story house with a tin roof and a yard neatly surfaced with round river stones. Three Ingano men stood drinking beer at the base of a stairway that went up to a low wooden verandah that ran the length of the house. Asleep on the balcony in a cane chair was an old man, dressed in white cotton and wearing rubber boots. A straw hat covered much of his face, and from below what stood out were his thick and exaggerated hands that lay on his belly like gnarled roots. Beside his chair there was a glass of milk on a small table and an old radio that faded in and out with the music of a very distant symphony.

"Don Jorge!" Pedro called. The old man woke with a start and slapped a fly away from his face.

"¡Carajo!" he growled. He turned immediately to the radio and fiddled with the dial. The reception got worse. He carefully placed his milk on the floor, lifted the radio with both hands, shook it firmly, and then deliberately dropped it on the table. Still no improvement. Only then did he acknowledge Pedro's greeting.

"Pedrito," he said, slowly rising from his chair. He was a short man with a square face and a comfortable paunch. One look at us and the truck, and he knew the purpose of our visit.

"So, is my old friend Ricardo still collecting flowers¿" he asked after Pedro had made the introductions. From the way he spun out the phrase in Spanish—*"cojiendo flores"*—it was obvious that Fuerbringer never quite understood what Schultes had been doing.

"When he can," Tim replied.

"The way he worked!" Fuerbringer said. "The man was crazy for work. A demon for work. What a farmer he might have been! Still, I am glad for him. No one has ever made a happy dollar off the Putumayo."

Fuerbringer invited us into his parlor where a young Indian woman fed us milk, cheese, fresh bread, and mangos taken from the trees that brushed up against the house. Don Jorge spoke to her kindly. He seemed a simple and decent man. He had an air of detachment, especially when speaking of the past, and none of the impatience one normally expects from a German.

Mocoa had remained a sleepy town, he told us, even after the war with Peru and the building of the Capuchins' road. The changes really began in the late forties, long after Schultes's first visit, when the civil war in the highlands scattered *campesinos* like leaves into the jungle. Then Texaco discovered oil in the fifties, and towns like Puerto Asís tripled in size in three years, and before you knew it shopkeepers and engineers, soldiers and women in high-heeled shoes, outnumbered the Indians.

"Suddenly there was nothing here except their faces and fat asses and all their empty whorehouse music."

"Don Jorge," Pedro said, after a respectful silence.

"Sí."

"We have come because I thought you might help us find *el remedio."* Pedro referred to *yagé* simply as "the medicine."

"Sí, sí. I know. I can help you, of course. Lucho!"

A young boy of six or seven came running into the room. He had a pencil in one hand and an exercise book in the other.

"¡Qué bueno!" Fuerbringer said. *"Trabajando. Estudiando.* This is how you'll progress. Show the doctors what you can do."

The boy placed his primer on the table and spent the next five minutes working on his signature, slowly plowing his pencil through the fibers of the cheap lined paper. When he had finished, Fuerbringer patted him affectionately on the head.

"Qué bueno, Luchito. Now lend me your little book." Fuerbringer tore a page from the primer and scribbled a note.

"Lucho, *por favor.* Take this to Don José." Fuerbringer then turned to Pedro and told him where and when we should appear at the house of José Maria Jamioy, a *curandero* who apparently lived along the road halfway between Mocoa and the Río Rumiyacu. As effortlessly as a physician writing a prescription, Jorge Fuerbringer had arranged for us to take *yagé.*

The *curandero* held his sessions in a cinder block house that had the ambiance of a spartan rural clinic: tin roof and cement floors, low benches along the walls of a large consulting room, and a wooden table and chair positioned at the front of the room beneath a dangling light bulb. We arrived around nine, well after dark, and Pedro led us into the dim light where an indifferent woman instructed us to take our places on one of the benches. There was one other person in the room, a tall and thin *mestizo* named Mario who knew Pedro from Sibundoy. Originally from Mercaderes in southern Cauca, Mario came to Mocoa on a regular basis to take a purge of *yagé.* On this occasion he had been living in the *curandero's* house for the better part of a month. Pedro explained that he was hunting for magic.

"Los blancos son más inteligentes," he said. "The whites are more intelligent. That's why they come to the Indians to get healed."

"Don José understands this," Mario agreed. "He will make you hard so that nothing can touch you." Pedro smiled and nodded in the direction of the door.

"Buenos noches, caballeros."

The *curandero* was a large man with a smooth face and a deliberate manner. He greeted each of us formally and then took his place behind the table as he arranged and then rearranged his ritual paraphernalia: a fan of leaves, a smooth quartz crystal, red and blue macaw feathers, and several necklaces of seeds and jaguar teeth. He wore trousers, rubber sandals, and a cotton shirt that hung over his ample belly. He was perspiring freely, and as he leaned over the table, small drops of sweat ran off his face and splashed onto the feathers.

"Did you bring the *aguardiente¿"* he asked. Pedro handed him two bottles, which he placed alongside a glass jug containing perhaps two quarts of a dark liquid.

"Yagé," Pedro whispered furtively. "The necklace is the sound of the forest. The feathers paint the visions. The glass stones are forest fruits that come from the sky. They are the *curandero*'s lens. They allow him to see inside your body. The fan is the spirit, *waira sacha,* the wind's brush. With it he will sweep you clean."

Don José cracked the *aguardiente,* took a long drink, and then offered each of us the bottle. He bawled an order to his wife, who snapped off the light. A match was struck, and the glow illuminated the *curandero*'s face as he lit a kerosene lamp. A melancholy light infused the room.

"Now he must arrange the spirits," Pedro said softly.

The *curandero* poured *yagé* into a wooden bowl, which he set on a short tripod of sticks beside the table. He then sat on a small stool so that his legs flanked the stand and his entire body enveloped the potion. For five minutes he sat perfectly still. No one spoke. Gradually out of his hunched-up body came a low, guttural chant that ebbed and flowed and then faded like an echo. The rustling of a fan and leaves scraping air; the sound of water in a distant forest and the chant escalating in pitch; the tone of a language far too melodic and sweet to have been born in the jungle. He coughed, breaking the spell as he cleared his throat. Straightforward words of prayer followed, Catholic in tone, and then suddenly he drifted once more into this other velvety zone of sound. Pedro touched my arm.

"The songs release the wildness, stirring everything up so that with his fan he may sweep away the evil. Now he is asking that the paintings, the visions be strong."

With a small calabash the size of a cup the *curandero* dipped the *yagé* from the bowl and then poured it back, releasing a fecund scent that mingled with and then overwhelmed the sweet smell of the resin that was burning in an iron brazier by the door. He filled the calabash once more and drank the contents, gagging, spitting, groaning, and coughing.

"See how he barks," Pedro said. "Like a jaguar. He is born of the jaguar, and when he dies, he will become one again. All the living and dead jaguars come to us from their homes in the sky."

Beside me Tim shuddered.

"Are you all right?" I asked.

"I remember the taste," he said with a smile.

The *curandero* looked up. He had pounds of colored beads around his neck and long necklaces of palm fruits and jaguar teeth crisscrossing his chest and hanging to the waist.

"When did he put those on?" I asked.

"I didn't see," Tim replied.

The chant had turned into a soft whistling sound. The rattling and swish of the fan accompanied the language of prayer, words broken into syllables and sounds that built one upon the other into a melodious harmony that gave the *curandero* total command of the moment. Everything was still. He sang on and on, oblivious to the passing of time. When finally he stopped, he leaned over the *yagé* and blew a single breath of air across the surface. He made the sign of the cross and then spoke one last prayer, addressing the plant repeatedly by its Quechua name, *ayahuasca,* the vine of the soul. Then he nodded in the direction of Mario.

Mario dutifully took his place in front of the *curandero* and accepted a shot of *aguardiente.* Don José then passed him a cup of *yagé,* which he accepted with both hands. He drank it slowly, his face wincing with every sip.

"¡Aaaarh! Qué fuerte." Mario's body quivered as he tried to shake off the taste. Don José handed him another shot of *aguardiente.*

"To sweeten his mouth," Pedro confided earnestly.

"With *trago?"* I couldn't believe it. *Aguardiente* is arguably the most foul-tasting liquor in the world. Don José poured shots all around.

"Guillermo, go take your turn. Drink it quickly," Pedro said. The *curandero* cleansed the *yagé* with his fan and handed me a cup, which I took in a single swallow. The smell and acrid taste was that of the entire jungle ground up and mixed with bile. I welcomed the shot of *trago.*

Tim and Pedro drank in turn, and then we all entered a strange quiescence as we waited for the drug to take effect. Outwardly, of course, nothing had changed. Mario and Pedro chatted about farming and the recent floods, Don José busied himself at his table, and Tim leaned up against the wall, smoking and taking notes. I sat apart from the others on the floor on a reed mat. Pogo lay beside me, and he had the peculiar look on his face that he always affected when Tim and I were up to something like this. It was late, and we were showing no interest in sleep.

It was thirty minutes before I felt the first sensation, a numbness on the lips, and a warmth in my stomach that spread to my chest and shoulders even as a distinct chill moved down my waist and lower limbs. It was a surge of energy, part expectation, part enchantment. I heard a distant humming, which I took for cicadas or tree frogs, until I realized that the sound was vibrating beneath the surface of my skin. I glanced at Tim, who was holding his head with his hands. Pedro lay flat on one of the benches. I shut my eyes, and the world inside my head began to spin and pulsate with warmth and a sensual glow that ran over a series of euphoric thoughts, words that stretched like shadows across my mind, paused, and then took form as diamonds and stars, colors rising from the periphery of consciousness and falling like demons and angels in a chaotic mix of dream and paranoia. I opened my eyes to a flash of light, a passing headlight on the road, harsh and intrusive. I retreated again and felt myself fade into an uncomfortable physical body, prostrate on the mat, and tormented by vertigo and a mounting nausea.

I heard a whisper inviting me to move, but it was extraordinarily hard to do so. Pedro had me by the arm and was guiding me outside. The vomiting came suddenly, a short spasm like a child's stomach in rebellion, followed almost immediately by a violent retching convulsion. A great stream turned into a serpentine river rushing over black plants and flowing beneath stars and cold winds and colors that turned one into another. The soil was cool and wet on my face. I looked up from the ground and saw the night sky, beautiful and clear. Suddenly I regained perspective, a clarity of well-being in the midst of all the visual chaos. It was as if my stomach, acting as a conscious entity, had sought out and purged every negative thought and fear trapped within the maze of my mind. Free to engage or ignore the flood of stimuli that washed over my senses, my eyes roved at will across an utterly new and astonishing landscape.

"Don Guillermo." The voice seemed terribly far away. It was Pedro. His breath dissolved into a dozen textures; crushed roots and resins, smoke and ash, and the taste of moss in forest glens where light teases the orchids and carries small birds higher and higher into the canopy. Spiders and butterflies, bats taking wing from cavities in trees, leaves falling onto the backs of ants that disappear into holes in the earth, where in the dampness they grow mushrooms that glow in the dark.

"Don Guillermo. Pay attention. Take hold of one image. Ignore everything else." It was an absurd demand. The images changed at the speed of thought. I opened my eyes and saw an iguana lying impassively on a limb, insects feasting in a great gash that ran along its backbone. I blinked and saw Tim stumbling out the door. Pedro was gone. I heard a dismal chorus of retching and gagging. I crawled toward the sound and saw the moon illuminate a tremendous emptiness, heaving with wind. Tim was facedown on the earth. Pedro was on all fours. I held his head back by his hair as he vomited. His back trembled beneath my other hand. When he was finished, I half-carried him back to the house. Tim staggered into the room after us and fell into a fetal rest on the floor.

The night dragged on, as if each minute were something heavy and tangible that had to be pushed away to make way for the next. Gradually the tone changed, the visions softened, and I fell away into a not entirely unpleasant delirium. Minutes and hours went by. The air grew cold and wet with the coming of dawn. I went in and out of consciousness, only vaguely aware of events beyond my immediate physical needs. I was cold, dirty, and tired of the scent of vomit. Sometime during the night I came around as a spectator in a trance, to see a weeping Indian woman sitting on the chair as the *curandero* created a halo of light and sound around her body and that of a sick child who lay across her lap. When I woke next, she was gone. The others were asleep, and I felt for an instant a tremendous emptiness, all the fear and uncertainty that haunts the early morning hours.

The *curandero* instructed me to come forward to his chair. Once I was seated before him, he began a prayer that flowed into a hypnotic chant. Taking his fan in his hand, he swept my body in a series of carefully orchestrated gestures. The fan was made of dried grass, and as he brushed my back I sensed for a moment the possibility of transformation, as if the kinetic movement and the touch of the grass might cause feathers to come from the skin. As he completed each slow movement with the fan, he snapped his wrist as if to discharge energy. Then, taking my head between his hands and placing his lips above my eyes, he drew in three short gasping breaths, holding the last until he could rush to the window of the room and blow it away into the forest.

From that point on the sequence is vague. There was more chanting, and I recall his mouth close to my ear, the guttural sound of the word *"yagé"* repeated over and over until it caused sensations up and down my back. Moaning and chanting, rustling his fan in rhythm, he sent a ritual mist of saliva and *aguardiente* over my face, chest, arms, and the small of my back. He sucked at the base of my throat, above the heart, and once again at the top of my head. Finally he took the smooth quartz crystals and rubbed my back and stomach, before placing them in the palms of my hands. I held them and felt a warm surge as he said a final prayer and dispatched a last spray of saliva into my hair. When he said that my head was that of a jaguar, I realized that he, too, remained in the throes of visions.

After the cleansing was finished, I wandered outside. Dawn was coming. I felt exhaustion mingle with a deeper sensation, an intuition that what I had experienced, a confusion of random visual and auditory hallucinations without form or substance, was only a crude approximation of something indescribably rich and mysterious. No doubt, as Schultes had written, a power lay within this plant. To understand and perhaps experience even an inkling of its true potential would demand a journey far beyond the roadside of Mocoa. Where that might lead was the thought that carried me back into the *curandero*'s house, where I found a place on the floor beside Tim, Pogo, and Pedro, and fell into a fitful sleep.

Don José woke us around nine for a final treatment. We stood in a circle as he passed around a plate of burning resin, instructing each of us to inhale the smoke. At the moment I felt less desirous of further purification than desperately in need of a bath. A final brushing with the fan, a short prayer, and then we each paid him fifty pesos and went on our way. Too tired and dazed to compare notes, we drove back toward Mocoa without speaking, pausing only to wash off the night in the cold water of the Río Rumiyacu.

It was not just *yagé* that had drawn us to the lowland forests of the Putumayo. On December 6, 1941, while working in the vicinity of Mocoa, Schultes had collected a specimen of coca that he never identified to species. His field notes record only that the plant was used by the Inga, who chewed the leaves "for strength." Like the Paez, the Inga consumed the leaves whole, adding to the quid a powdered lime made from the ashes of a burned white stone. Five months later and nearly two hundred miles away at the confluence of the Río Orteguaza and the Caquetá, Schultes had come upon a completely different way of using the plant. There the Koriguaje Indians dried the leaves over a fire in enormous ceramic vessels and then pounded them in a large wooden mortar and pestle. Their source of alkali was the ash of certain jungle leaves, which they added directly to the powdered coca in the mortar. The mixture, once completely pulverized, was sifted through palm fiber to yield a fine gray-green powder the consistency of talc. This was carefully placed in the mouth, formed into a paste, and gradually consumed in its entirety.

Here was the problem. The Koriguaje method had been reported throughout the Northwest Amazon, and Tim suspected that the species being used was *Erythroxylum coca,* the coca of Peru. If so, how and when had it made its way out of the southern Andes and north to the Colombian Amazon? And what was the plant that Schultes had collected at Mocoa? Was it the lowland form of Peruvian coca, which had come upriver and was being consumed by a technique borrowed from the highlands? Or was it the coca of Colombia, *Erythroxylum novogranatense,* which had been carried to the lowlands together with the established highland tradition of using powdered limestone to potentiate the cocaine? Where was the boundary between the methods of preparation?

It took several days before we finally accepted that these questions might never be answered. We were a generation too late. Not only had the Ingano ceased the use of coca, but the forests of the upper Putumayo no longer existed. Along the foothills that fall away to the vast lowland plain at Villagarzon, then south through a number of small and insignificant settlements that litter the road toward Puerto Asís, we saw nothing but cattle ranches and fields of yuca and rice, seedy little towns with open sewers, cement churches, fat merchants, and rows of shops selling televisions, cassettes, and useless ointments imported from France.

Late one evening, after yet another long and suffocating day searching for remnants of undisturbed forest along the dusty roads south of Mocoa, we found ourselves once more at Pepino with Jorge Fuerbringer, sitting on his balcony and listening to the despondent tunes that crackled on his radio. We had returned to Don Jorge because after a week in the upper Putumayo it was clear that what we had come looking for would be found only in the memory of men and women like him. With Pedro asleep in a hammock and both Tim and I struggling to remain awake, the conversation dipped in and out of focus.

"There was a trade in visions," Don Jorge told us at one point, "and the Siona were the masters. They carried their knowledge into the mountains." He glanced toward Pedro. "Has he told you about *wixa?*"

"No," I said as Tim shook his head.

"It's his ritual language. You hardly hear it anymore. Pedro knows it, and so does Chindoy when he's sober. It's closely related to Siona, who call their own ritual language by the same name."

"But what about the man we were with last week?" I asked.

"He is good, but he's Ingano."

"What's that mean?"

"The Ingano were nothing. They didn't follow the taboos that protect the plant. They used *trago,* cheap liquor, which the others scorned. It was their weakness and fear that led them to the bottle. In the old days a Siona would never consider marrying an Ingano. How could a people who knew how to travel the heavens have faith in the visionary knowledge of those who drank the white man's liquor? It could not be."

"But the Ingano supply all the *yagé* in the highlands," Tim said.

"Only because the others are dead—not the people, the shamans." Don Jorge reached across the table and snapped off the radio. "The Siona once ruled the upper Putumayo. Now there are only perhaps two hundred and fifty alive in this region, mostly below Puerto Asís. But without their shamans they are lost. Their world had many layers, all of them inhabited by people, animals, spirits. The shaman alone mediated between these realms. *Yagé* allowed him to do it. It took him there. It gave him the clothing of the jaguar, and once in flight, he and the jaguar were the same—or he put on the skin of the anaconda or the hair of the wild pig."

"Did you know shamans like that?" I asked.

"Of course. So did your professor. He took *yagé* many times with them."

"How did one become a shaman?" Tim asked.

"It took months, years of hard work. First one had to master the basic visions. You had to be able to bring forth specific visions when you took the medicine. You had to learn to bend the visions with the songs. Then there was the terror. The master shaman conjured up snakes wrapped in fire, thousands of angry claws tearing at the sky. The apprentice had to face them, upright, no hesitation, with power. Then and only then could he suck at the breast of the jaguar woman. Just when he was getting comfortable, she would fling him away into a pit of vipers. One of the snakes would carry him away to heaven, where the *yagé* people introduced him to the spirits of the dead. Only after many such terrible journeys did the initiate meet God. He stood before a solitary tree and a door that opened into nothingness. The initiate had to walk into that emptiness. Only when he realizes what lies beyond the door can he receive his staff and the summons from God to be the protector of his people."

"And the Ingano have nothing of this?"

"They make a business. Of course, they have their own power. A man like Don José is a good friend to have. But most of them just sell colors."

"Pedro said that the Ingano still recognize seven kinds of *yagé.*"

"And what did he say they do with them?" Fuerbringer asked.

"I don't know."

"Nothing!" he snorted.

"He said each one sings in a different key."

"That's true. The visions determine the classification. But there is something more. The Siona have at least fifteen classes. When a plant is passed on in trade, so is its specific vision. A Siona cannot classify a plant without knowing its trading history. Every plant thus has a lineage that links it through all time to every other. It's like the family tree of an English prince."

"Where can we go to find people like these today?" I asked. Fuerbringer hesitated for a moment.

"The Sionas once ruled the spirits. When your professor was here, the shaman was the central authority. Through the plant he influenced every aspect of life. But since 1950 only a handful of Siona have tried to become shamans, and not one has achieved full mastery. In the thirties and forties there were still many. The Siona blame their demise on the sorcery they worked on one another. I blame Texaco."

"So it is completely dead."

"No, not at all. What do you mean by death? The people live. Their children face a new destiny, that's all."

"But surely somewhere there is a people or a man who still knows these things."

"I wouldn't be so sure," Fuerbringer said. "Still, the Siona always turned to the Kofán to learn about medicinal plants. The Kofán came here to learn about *yagé* and its power. Perhaps the roles were reversed. Maybe there is a Kofán who still knows. But it won't be in Colombia. If you want to find these things, I would try Ecuador. The Río Aguarico. Perhaps there. I don't know anymore."

"Perhaps we will try," Tim said as he leaned over and nudged Pedro in the hammock. "But for now, we'd best get some sleep. Don Jorge, *con su permiso.*"

"*¿Cómo no? Buenos noches y que le vaya bien.*"

We shook hands and said good night.

That was the last we saw of Schultes's old friend Fuerbringer, who passed away a few years after our visit. Tim had had enough of the Putumayo, and in the morning, just as the servants were beginning to stir, we left to return to the highlands. There was only one more mishap awaiting us on this leg of our journey. The road to Sibundoy was clear, and we reached the valley without incident, but as we climbed out of the valley en route to Pasto, Tim was distracted by a remarkable orchid growing on the cutbank. As he barreled up the road, gazing out the window and delighting Pedro with a salacious joke about pollination, the truck pounded into a hole on the open side of the road, bounced into another, and came to rest. I looked out the passenger window and saw nothing but air. The truck was straddling the edge of a cliff, with both right wheels completely suspended. I told Pedro and Tim to get out slowly. They did.

It took an anxious hour of digging before I had a large jack in place beneath the front of the truck. Pedro flagged down a jeep and a dump truck. By hooking cables to the front and rear axles, and carefully orchestrating the other two vehicles, we were able to pull the Red Hotel back onto the roadbed. By then Tim was nowhere to be found. Twenty minutes passed before I finally saw him, hundreds of feet above us on the mountain, sitting on a rocky outcrop, staring into the sun. Pedro shook his head.

"El botánico loco."

"What the hell is he doing, Pedro¿"

"Cojiendo flores," Pedro replied with a smile. "Collecting flowers and making plans for your journey to the Kofánes."

"Ecuador."

"Sí, claro."

CHAPTER SEVEN

The Sky Is Green and the Forest Blue,

1941–42

ON THE MORNING of Monday, December 8, 1941, Richard Evans Schultes awoke suddenly, his sleep interrupted by a terrible noise rising from the nave of the church next door to the Capuchin convent where he was staying in Mocoa. In the middle of Catholic mass a three-piece Indian band, all cracked trombones and brass tubas bent out of tune, was playing "Roll Out the Barrel." He hadn't heard or seen anything like it since the highlands near Bogotá. There in a little village a farmer had witnessed an oil painting of the Madonna appear as a vision in a mud puddle. Thereafter the townspeople commemorated the event each year by parading the painting around the plaza while their band played "Yes, Sir, That's My Baby."

"Where the hell do they get this music?" Schultes said, laughing, as he rolled out of his hammock and reached for the brown calabash balanced on the metal trunk that doubled as a night table. Inside the bowl was a thin, light brown liquid. Schultes lifted the gourd toward his mouth and then hesitated.

"Pedro!" he called softly. Young Pedro Juajibioy appeared from beyond the doorway that opened onto the plaza.

"Buenos días, Doctor."

"Muy buenos días, Pedro. One small question."

"¿Señor?"

"Yesterday it was white. Why is it now dark?"

"Because that was *yoco blanco.* This is *yoco colorado."*

"Oh," Schultes said. He took a small swallow. The taste was the same, and like the preparations he had been drinking each of the past four mornings, it was an infusion of the bark of a liana. Schultes drained the gourd and reached for another. He knew that within ten minutes his fingertips would begin to tingle.

The botanical identity of the plant remained a mystery. Schultes had first read of *yoco* in *Northwest Amazons,* the journal of Captain Thomas Whiffen, a British soldier who had spent the year 1908 in the lower Putumayo. Seventeen years later a Belgian botanist, Florent Claes, accompanied the Capuchin priest Gaspar de Pinell on one of his evangelical excursions into the lowlands and reported that the stimulant was a large forest liana in the genus *Paullinia,* an observation confirmed by subsequent collections in 1931 by Guillermo Klug, a plant explorer based in Iquitos, Peru. In 1940, Schultes's new friend and colleague José Cuatrecasas, working out of the Instituto de Ciencias Naturales in Bogotá, found *yoco* at Puerto Piñuna Negra on the Putumayo. Unfortunately, the plant had neither fruits nor flowers, and thus the species could not be described.

When Schultes had passed through Sibundoy the previous week, one of the padres, Marcelino de Castellví, had encouraged him to search for flowering material that would finally allow a complete botanical determination. A day after arriving in Mocoa, he met Jorge Fuerbringer, who knew *yoco* and had Indians in his employ who were able to guide Schultes to a living specimen. Unfortunately, the plant was sterile. From the Indians Schultes learned that virtually every household maintained a ready supply of *yoco* stems. Cut from the wild and stored in three-foot lengths away from the sun, the plant retained its stimulating properties for at least a month. The beverage was prepared by rasping the outer bark and expressing the milky sap into cold water. Taken at dawn, *yoco* allayed hunger for four to five hours, allowing men and women to march twenty miles or more through the forest without fatigue.

From chemical analysis undertaken by Claes, Schultes knew that the bark contained caffeine, roughly three times as much per weight as coffee beans. To make a single dose, the Indians of the Putumayo used roughly one hundred grams of bark, as compared to the eleven grams of ground beans in the average cup of strong coffee. In other words, even allowing for the relative inefficiency of cold water extraction, in drinking their morning calabash of *yoco* the Indians knock back the equivalent of more than twenty cups of coffee. They are not, as Schultes would later recall, a people to do things in half measures.

Suspended between the first flush of well-being and the hyperkinetic aftermath of caffeine intoxication, Schultes made plans for the morning. Across the room his companion, a young American botanist named C. Earle Smith, still slept. A college sophomore, Smith had joined Schultes in Bogotá in late September, and together they had accompanied José Cuatrecasas to the Páramo of Tamá, a remote and windswept massif that straddles the Colombian-Venezuelan border. Returning after a month to Bogotá, they reached Sibundoy at the end of November. Three days later, guided by Pedro Juajibioy, they traveled by mule to the lowlands, where Schultes expected to remain for several months.

Mocoa was no more than a collection of huts and small houses, perhaps a hundred in all, clustered around the church and convent. From the edge of the small plaza it was but a short walk to the forest, and although fields had been laid out for cattle and cane, the scent and sounds of the wild still dominated the town. Apart from the Capuchin priests and the nuns who ran the school, the only outsiders were the few merchants whose shops dominated the plaza, a pair of frustrated American missionaries, and a corpulent official who simultaneously represented all branches of the Colombian government.

As Schultes stood by the door of the convent, he felt the morning light warm his face. Behind him, Smith groaned in his hammock.

"Smithy, you awake?" Schultes said.

"Yeah, I suppose." Beneath his blanket Smith was a short man with a round face and a fuzzy growth of beard that made him appear even younger than he was.

"Good. There's a bowl on the floor. Drink it down, and let's get started on the plants."

"What about breakfast?"

"Lord, I forgot." Schultes looked around the room for a solution. "Maybe Pedro can get you something while I heat up the dryer."

"Right."

While Smithy found Pedro, and Pedro sent a young Indian girl to pester the priests for eggs, milk, and bread, Schultes went about organizing the stacks of specimens that covered the floor of the room. There were a hundred different plants in all, most collected in sets of six. A few he had managed to pick up along the trail coming down from Sibundoy; the majority had been found close to Mocoa.

Looking over the collection, Schultes marveled at his good fortune. On his first complete day of botanizing in the Amazon, he had discovered a new species of wild cacao, later named *Herrania breviligulata,* together with a new species of fish poison, *Serjania piscetorum.* Known to the Ingano as *sacha barbasco,* it was one of four fish poisons collected that first morning. The other three, all members of the Spurge family, remain unidentified to this day. Placed in slow-moving bodies of water, these poisons interfere with respiration in the gills of the fish. The fish float to the surface and are readily gathered. Many of the Amazonian fish poisons contain enormous concentrations of rotenone. Thus, in uncovering their identity, Schultes stumbled on the source of the most commonly used biodegradable insecticide available to the modern world.

This was just the beginning of a remarkable collection. Naturally his eye had been drawn to *achiote,* the yellow-orange pigment used as face paint throughout the Amazon and employed in Europe and America as a dye to color butter and margarine. There were two other dye plants, three rare and unknown fruits, and one hallucinogen, a sterile specimen of *borrachero* that came complete with a recipe: "two leaves for a drunk lasting all morning, four or five for all day." More significant was his collection of a nondescript tree in the genus *Croton,* one of twenty-five medicinal plants gathered in one day. Schultes noted that the red latex was "used by the Ingas to calm pain in aching molars." He recognized it as *sangre de drago,* dragon's blood, a plant he had first encountered the week before in Sibundoy. Now he learned of its extraordinary curative properties. Placed on an open wound, the resin dries into an antiseptic seal, a protective cover known to the Indians as a liquid bandage. In ways that modern science has yet to understand, the compounds in the resin accelerate healing in a remarkable manner. Wounds and lacerations that in the tropics would normally fester instead heal within days, without infection and without leaving a scar. Finally, his field notes record that on December 6, 1941, he made his first collection of coca. The Inga chewed the leaves "for strength," adding to the quid a powdered lime made from the ashes of a burned white stone.

Schultes set up the dryer in an adjacent room and moved in most of the plants. When there was still no sign of Smithy, he lit a cigar and sat down to wait on the cement floor. Glancing up at the thatch roof, he suddenly smiled. Every botanist drying plants in the field dreads the thought of burning down some poor *campesino*'s house. So far Schultes had met only one who had actually done it. José Cuatrecasas was a wild Spaniard, a royalist veteran of the civil war who had at least two abiding passions in his life besides botany. He hated priests and hated spending money, in that order. While collecting in Mitú, a small town located on the Vaupés River five hundred miles southeast of Bogotá, he had spurned an invitation to live with the Capuchins and instead moved into a beautiful thatch house just built as a wedding present for a headman's daughter. Too cheap to hire an Indian lad to stand guard, Cuatrecasas left the dryer on and went collecting. He returned the day before the wedding to find the house burned to the ground. Standing in the ashes was the Cubeo headman.

"Mi casa, mi casa." My house, my house.

"Your house?" Cuatrecasas wailed as he stood beside the elder. *"¡Mis plantas! ¡Mis plantas!"* My plants! My plants!

The memory of this incident caused Schultes to drag his dryer farther away from the wall of the room. He checked the skirting and made sure the canvas was not getting too hot. He had just decided to go ahead and begin loading the presses when Pedro came rushing into the room.

"Doctor! Doctor!"

"What is it?" Behind Pedro were Smithy and Padre Ildefonso de Tulcán, the head of the Capuchin mission. The padre wore a dark habit bound at the waist with a cord. His face was deeply lined, his dark features leathery and grave. There were small bits of bread and other signs of breakfast in the thick white beard that hung a foot below his chin. He placed his hand on Schultes's shoulder.

"I am afraid you will not be with us for as long as we had hoped."

"What do you mean?"

"There has been an attack—"

"The Japs bombed Hawaii," Smithy said.

"When?"

"Sometime yesterday. It just came through on the radio. It looks like an invasion."

"It is war for you," said Padre Ildefonso, "but it won't last long."

"I would not be so sure," said Schultes. He stood quietly for a moment, looking over the room. For four years he had known that war would come. He had raced through his thesis and left directly for Colombia from Mexico without returning home, all with the hope of fulfilling his dream of living in the Amazonian forests before the outbreak of hostilities. Now, barely a week after arriving in the lowlands, he had no choice but to abandon his expedition. He knew immediately what had to be done.

"We'll dry these plants," he said, "and then complete our collections in Sibundoy. In ten days we can be in Bogotá."

By the time Schultes and Smith reached the highlands, Germany and America were at war and the Japanese had launched their amphibious assaults on the Philippines and Malaya, the first moves in a strategic offensive destined to secure the wealth of the Dutch East Indies by the end of February 1942. Within a week of Pearl Harbor, on the day that Schultes stumbled on a new species of *Espletia* on the *páramo* above Sibundoy, British defenses in northern Malaya were overrun, and the rubber plantations on which the outcome of the war might depend were plucked from Allied control. It was a historical event that would transform Schultes's life in ways that he could never have imagined.

He and Smith left Sibundoy on December 17 and, traveling overland through Popayán, Cali, and Ibagué, arrived in Bogotá five days before Christmas. Young Smith embarked immediately for the States to enlist, but Schultes, soon to be twenty-seven, was too old for the initial call-up. On the morning of December 21, 1941, he reported to the American embassy, where he encountered only confusion. Those familiar with his work implied vaguely that he ought to remain in the city, that the government had something in mind for him. But no one in authority seemed willing to divulge his fate. No one knew what the war would bring, least of all the American diplomats isolated in the landlocked Colombian capital ten thousand miles from the nearest front.

He left the embassy that morning around eleven, in time to meet his colleague and friend from the Institute, Hernando García Barriga, at a small café in La Candalaria. While a dozen men in dark suits and felt hats drank *tintos* and discussed the problems of the world, the two botanists compared notes. García Barriga, who would one day write the definitive three-volume work on the medicinal plants of Colombia, was a superb field man, the only other botanist to have worked in Sibundoy. If Schultes connected with Indians through decency and good faith, García Barriga did so with humor. Beyond his dashing appearance, the slicked-back hair and dark eyes, the classic Latin mustache and aquiline nose, he had the look of the consummate trickster. In old age the lines in his face would betray a lifetime of laughter.

Leaving the café, Schultes invited Hernando to join him for lunch. Together they headed for his *pensión,* located on Carrera 7, between Calle 18 and 19. Crossing the boulevard at the corner of Avenida Jiménez and Septima, and heading toward the Parque Santander, Schultes suddenly heard a violent scream just behind his shoulder. On the steps of a colonial church stood a wiry old woman, dressed head to toe in red. She was shaking her fist and bellowing.

"¡Viva el partido liberal, Carajo!"

"Who the hell is that¿" Schultes asked.

"Oh," replied Hernando, "that's *La Vieja Margarita.* It's her mission to save the Liberal Party. She's been here for years. Don't worry, she's not all there."

"No kidding," Schultes said. They walked on, stopped in at the Librería Mundial where Schultes bought some writing supplies, and then began to cross Septima to head into the park. As Schultes stepped off the curb, García Barriga placed a hand on his arm.

"Careful," he said. There was a bright blue trolley bearing down the boulevard. Running before it and frantically gesturing was a young man dressed in bits of uniforms, evidently scavenged from every branch of the Colombian armed forces: a military cadet's overcoat, tattered blue trousers from the navy, and a mariner's cap beneath a Prussian helmet pinched from the Presidential Guard. As he came closer Schultes could see the blond wig, white face paint, and bright red lipstick. With a wild glare the figure forced Schultes and García Barriga back onto the sidewalk.

"El Bobo de los Tranvías," Hernando explained over the clatter of the passing tram.

"The fool of the trolleys¿" Schultes said.

"Sí. He thinks that they can't run without him clearing the way. It's his civic duty," Hernando confided. He touched his finger to his temple. "One has mercy," he said.

Hernando led Schultes into the park where they stopped en route to the *pensión* to have their shoes shined by an old man who had set up his station beneath the shade of an enormous ficus tree. Just behind them the steeple of the San Francisco church rose above the red-tiled roofs of the city. Four stories high, it was as tall as any building in Bogotá.

"Orange peels¿" the bootblack asked as Schultes placed his oxfords on the small wooden step.

"Excuse me¿"

"Do you want me to use orange peels, Doctor¿"

"For the leather," Hernando explained.

"Sure, why not?" Schultes said. "But how did you know I was a doctor?"

"In this country," the man said casually as he rubbed the fruit along the instep of the shoe, "every son of a bitch with a tie and a pair of bifocals is a doctor."

"I see." He turned to Hernando. "Ten years of training, and all I really needed was a pair of glasses and a proper suit coat."

"¿Periódico?" the man asked.

"Gracias." Schultes accepted a well-read copy of the morning edition of *El Tiempo,* the city's leading paper. He scanned the first page. The lead story in the left-hand column heralded the German assault on Russia: nine Soviet armies enveloped, twelve cities captured, thousands of airplanes destroyed, hundreds of thousands of prisoners taken. The right-hand column, highlighted by a headline of equal size and sensation, announced the outcome of a major battle closer to home. In the Putumayo the Peruvians had advanced on the town of Puerto Zancudo—Port Mosquito—capturing from the Ecuadorians one house, one saber, and a flag. Buried in the florid prose was the admission that there had been only one casualty, a Peruvian cook who sliced off a finger in the line of duty.

"What a country," Schultes said quietly.

"¿Qué dice?" asked Hernando.

"Nothing. Sometimes I wonder—What the hell is that?" Schultes looked up at a strange figure dressed like Louis XV: buckle shoes and silk stockings, satin waistcoat and lacy ruffles fluttering beneath a tricorner hat.

"Oh, him," Hernando said. "That's *El Conde de Kuchicuti.* He's a real count. He went to Spain and purchased the title. He's worth a fortune—*haciendas* all over the savannah run by *mayordomos.* He never leaves the capital. He holds court on the corner in that café. There's a large table by the window that he rents on a permanent basis." García Barriga glanced at his watch. "They just adjourned, I suppose. Every day from eleven until noon. He pays for the drinks, you know. Everyone loves him."

So it appeared. With the insouciant air of an eighteenth-century gentleman, the Count strolled along the sidewalk, greeting each pedestrian as if he or she was a lost and needy wayfarer—a slight bow for the secretaries, the tip of his preposterous hat for the men, a gentle wave of his silver-tipped cane for a passing priest. The effect was subtle but undeniable. The swagger of society ladies softened, and harried businessmen in their bowlers and suits of heavy tweed appeared visibly to relax, if only slightly.

"The *bogotanos* need him," Hernando said. "He reminds them that they, too, are Latins."

"Listen," Schultes said with a laugh, "we'd better get going. The old lady hates it if you're late for lunch."

By this time Schultes was beginning to understand that beneath the pampered comforts of Bogotá anything was possible. As he and García Barriga walked along Septima toward the Pensión Inglesa, they passed the German legation, a severe building with a flat façade of smooth granite and a large swastika hanging on the rooftop flagpole. Right next door, in a modest colonial structure, was the headquarters of the Colombian Anti-Nazi Committee. Schultes didn't even bother to comment. The war seemed impossibly distant.

The Pensión Inglesa was run by Mrs. Katherine Gaul, an elderly woman from Devon who, with her high lace collars and pillbox hats decorated in pink flowers, had the air of a slightly bewildered Queen Mary cast adrift beyond the frontier of the Empire. The bride of an Englishman who owned a farm in the hot country of the Magdalena Valley, she had first arrived in Colombia in 1911. Two years later, after her husband had died from malaria, she moved to Bogotá and set up a boardinghouse catering to British bachelors working in the capital. Her "boys"—in more than fifty years she never admitted a single woman—came largely from the English banking and insurance concerns that at the time dominated commerce in Colombia.

Schultes had stumbled on her establishment quite by chance on his fourth day in Bogotá back in September. He needed a permanent base in the capital, and when he inquired at the embassy, a staff member recalled receiving a letter recently from Mrs. Gaul. Having spent two miserable nights in a French *pensión,* suffering through greasy food and "interminable yapping" of the guests, Schultes was delighted at the prospects. On Saturday, he went calling and was met at the door by a short woman in a white apron.

"I can't shake hands," she said. "I'm baking bread."

English cooking and fresh bread, a fussy old lady and the musty atmosphere of an old boardinghouse, the place had Boston written all over it. Schultes was delighted. The building itself was rather modest, a three-story structure built in the Spanish style with wooden bay windows overhanging the street. A shoe shop occupied the street level. Mrs. Gaul rented the upper two floors, where a dozen rooms faced onto a damp and somewhat dingy patio. The rooms were small, clean, and simply appointed, with solid oak beds, thick Scottish blankets, a table and chair, and an armoire that served as a closet. Offered a corner room on the third floor, Schultes accepted casually, unaware that the Pensión Inglesa would be his home for the next twelve years.

Schultes and García Barriga never did get lunch. They had climbed the creaky stairs that rose to the second-floor landing and were just about to enter the dining room when they learned that Mrs. Gaul had been arrested. Apparently while taking a walk she had been accosted by a policeman—a routine security check, quite common during the war. But she had no identity papers and had lived in Colombia since before passports were issued. Asked if she was a foreigner, she had replied, "No, I'm British." The poor policeman couldn't make out a word she said. After thirty years in the country, only two Colombians could understand her Spanish: her cook and her maid. From what Schultes could gather, seventy-three-year-old Mrs. Gaul had been taken downtown. He turned to García Barriga.

"I had better call the ambassador."

The British ambassador, Sir James Joint, unable to attend to this delicate matter himself, dispatched his cultural attaché, Julio Tobon de Páramo—"Julio the Big Tub of the Páramo"—who apparently looked the part. After two hours at the station this force of nature finally managed to set Mrs. Gaul free on the condition that Schultes and García Barriga guarantee that she would obtain proper documents. The arrangement went down with several handshakes. By then Schultes and García Barriga barely had time to jump on a trolley and make their way out to the university for the dedication ceremony celebrating the opening of the new home of the Instituto de Ciencias Naturales.

More trouble lay ahead. The keynote speaker and honored guest at the dedication was the president of the Republic, Eduardo Santos, who also happened to be the owner of the newspaper *El Tiempo.* Schultes had just walked in the door when he was cornered by Armando Dugan, the director of the Institute, who introduced him to the president. Schultes naturally knew nothing of the man's business background.

"So, Dr. Schultes," Santos said, "what do you think of the press in Colombia?"

"Actually," Schultes replied, "I rarely have time for the papers in my own country. Here I never read them at all."

There was an awkward silence until Schultes added, "I do see *El Tiempo* quite often."

Santos seemed pleased. So did Dugan, who had been frantically trying to get Schultes's attention. It was President Santos, after all, who had authorized the funds for the new building.

"And why do you choose *El Tiempo?*" Santos continued.
"Because it's the most *absorbente.*"
"Absorbing? Well, I'm glad of that. I'll have to tell my editorial staff."
"Yes. I press my plants with it. I buy it by the kilo."

Christmas passed, and Schultes grew tired of hearing nothing from the American embassy. By mid-January he had processed his Putumayo collections. A short excursion to the Páramo de Guasca with Roberto Jaramillo, a Colombian expert on the Andean flora, gave him a respite from the capital but only whetted his desire to return to the field. When he returned to Bogotá on January 23 and still had no instructions from the embassy, he decided to go on his own to the Putumayo. He did so in the first week of February, arriving once again in Sibundoy on the eleventh. From what he had been told by contacts at the embassy, he had three months to make his mark in the field of tropical botany. After that, they implied, he would be swept up by the war.

Schultes's fellowship from the National Research Council called for a general survey of the medicinal and toxic plants employed by the tribes of the upper Putumayo, and in the course of this work he hoped to determine once and for all the identity of *yoco.* His precise mandate, however, was to join the hunt for the botanical sources of the arrow and dart poisons known throughout the Amazon as curare. Finding and identifying these plants were of paramount medical importance. In 1935 a British chemist, Harold King, had first isolated an active principle, d-tubocurarine. Experiments had long indicated that this paralytic poison was an extremely effective muscle relaxant. If it could be used in conjunction with anesthetics in surgery, researchers argued, it might be possible to save thousands of lives. Until the botanical identity was determined and a steady supply of the poison secured, research could not proceed.

Schultes's itinerary would take him from Sibundoy to Mocoa, down the Putumayo to the confluence of Río Sucumbíos, and then upriver to the homeland of the Kofán, by reputation the best poison makers in the Amazon. From there he would make his way one thousand miles downriver to the Colombian-Brazilian frontier where, if all went well, he hoped to get a flight back to Bogotá from the military outpost at Tarapacá.

Of all the mysteries of the New World, few struck such fear into the hearts of early explorers as did the legend and reality of the flying death.

In an era when firearms were in their infancy and battles still fought with lance and sword, the existence of vegetable poisons that could kill silently and discreetly posed a formidable psychological and physical threat. Christopher Columbus off the shores of Trinidad in 1498, Vicente Pinzón at the mouth of the Amazon in 1500, Sir Walter Raleigh as he sought El Dorado on the banks of the Orinoco—virtually every European explorer returned with tales of plants laden with "evyll frutes" and a black venom that caused wounds to fester and rot. If a man even slept beneath a poison tree, the Spaniard Gonzalo Fernández de Oviedo wrote to Charles V, "he hath his head and eyes so swollen when he riseth, that the eye liddes are joined with the chekes. And if it chance a droppe of dew of the saide tree fall into the eye, it utterly destroyeth the sight."

During five expeditions up the Orinoco, Sir Walter Raleigh saw the effects of the plant firsthand. It was a fearsome potion known to the Indians as *ourari,* killer of birds. In the chronicle of his voyage, Raleigh reported that those exposed to the poison suffered "an ugly and lamentable death, sometimes dying starke mad, sometimes their bowels breaking out of their bellies: which are presently discoloured as black as pitch, and so unsavoury as no man can endure to cure." There was an antidote, he wrote, but he added that no amount of torture could pry the secret from the "soothsayers and priests . . . who conceal it and only teach it . . . from father to son."

Knowledge of curare did not increase appreciably until the eighteenth century when the French mathematician Charles-Marie de La Condamine, dispatched to the Americas to determine the size and shape of the earth, became the first scientist to travel the length of the Amazon. The expedition's primary task, initiated in highland Ecuador in 1736, was to take measurements at the equator to calculate the precise length of one degree of latitude, a determination essential not only for improved navigation but also to ascertain whether the earth was flat at the poles or the equator. The work took the better part of a decade, and it was not until March 1743 that La Condamine made his final astronomical observation. Ready to return at last to France, this intrepid explorer headed east into the Amazon. Though his mission in the highlands was successful, it is for the subsequent four-month river journey, not the seven years of arduous surveys, that La Condamine is today remembered.

Traversing the Andes and reaching the Jesuit mission at Borja on the upper Río Marañon in what is modern-day Peru, he continued down to the Huallaga and on to the Ucayali, all the while becoming more and more enchanted by this "freshwater sea . . . penetrating in every direction the gloom of an immense forest." In addition to major geographical discoveries, including the prophetic suggestion that the Amazon and Orinoco rivers were linked by a waterway, La Condamine observed and collected medicinal plants, including quinine, and he was the first European to note the value of rubber, which he formed into bags to protect his scientific instruments. He also described the use of curare, a "poison so active," he wrote, "that, when it is fresh, it will kill in less than a minute any animal whose blood it has entered." He noted that certain tribes concocted arrow poisons from roots and leaves of thirty different plants; others employed only three or four. Many possessed a variety of poisons, each with different constituents, and his subsequent experiments in Europe revealed that the different curares had different strengths. Here were the first indications of the complexity of the preparations, a variability that would confound ethnobotanists for almost two hundred years.

The botanical identity of the source plants became an issue of intense speculation. Exploring the Guianas in 1769 the physician Edward Bancroft became the first scholar to see the poison made. He described a complex recipe that included numerous plants, few of which had been mentioned by La Condamine. Thirty years later Hipólito Ruiz, returning from his explorations with José Antonio Pavón in Peru, identified a liana used to make curare as *Chondrodendron tomentosum.* Alexander von Humboldt knew of this determination, for he consulted Ruiz in Spain in 1799 on the eve of his own epic voyage to the Americas. A year later, however, on the banks of the Orinoco near the Venezuelan outpost of Esmeraldas, Humboldt and Aimé Bonpland saw curare prepared; they reported that the source was not *Chondrodendron* but a liana of the genus *Strychnos,* a plant related to *nux vomica,* the East Indian source of strychnine.

Attempting to solve the mystery once and for all, Charles Waterton, an eccentric English planter, set out alone in 1812 to cross the heart of the Guianas. Traveling up the Demerara River, he portaged to the Essequibo and then traversed the Kanuku Mountains to reach a tributary of the Río Branco that flowed via the Río Negro into the Amazon. En route, this fearless and somewhat comical animal lover captured a ten-foot boa by punching it in the nose, and later secured a large caiman by jumping onto the creature's back, rolling it over, and dulling its senses by massaging the belly. His work with curare was more serious. Among the Macusi Indians he took note of the ritual prohibitions that circumscribed the preparation of the poison, and he documented the ingredients in great detail: red pepper and various roots, venomous ants, the crushed fangs of deadly bushmasters and fer-de-lance, and the rasped bark of *wourali,* a forest liana, "thick as a man's body." The final product, he wrote, was a glutinous pitch, the result of a slow concentration over a low fire.

While still in the forest, Waterton made several critical observations. He noted that the quantity of poison required to kill was proportional to the size of the prey. A wild boar shot in the face died after running 170 paces. A dog wounded in the thigh survived fifteen minutes. A thousand-pound ox, wounded three times, died within thirty minutes. As for botanical antidotes, Waterton correctly surmised that none existed. He did note, however, that "wind introduced into the lungs by means of a bellows would revive the poisoned patient, provided that the operation be continued for a sufficient length of time." No one knows how Waterton came up with this prescient idea. Even as he wrote it down in his journals in the forests of Guiana, experiments in England by Benjamin Brodie and Edward Bancroft were indeed showing that the only effective antidote was artificial respiration, continued until the effects of the poison subsided. Waterton could not have been aware of these experiments, and one can only assume that he learned of the treatment from his Indian hosts.

Months later, upon his return to England, Waterton himself undertook a series of extraordinary curare studies, administering the poison to dozens of animals ranging in size from chickens to oxen and mules. In every case the poison brought back from the Guianas proved fatal, with death occurring within twenty-five minutes. By this time Waterton was familiar with the work of Brodie and Bancroft, and one morning he decided to experiment with their technique. He began by injecting the poison into the shoulder of a female donkey. In ten minutes the creature appeared to be dead. Waterton, being rather accomplished with a blade, having bled himself on at least 136 occasions, made a small incision in the animal's windpipe and began to inflate its lungs with a bellows. The donkey revived. When Waterton stopped the flow of air, the creature once again succumbed. Resuming artificial respiration, he nursed the animal until the effects of the poison wore off. After two hours the donkey stood up and walked away. This treatment marked a turning point in the history of medicine. In demonstrating that curare caused death by asphyxiation and that the victim could be kept alive by artificial respiration, Waterton, like Brodie and Bancroft before him, revealed how the remarkable properties of this muscle relaxant might be used therapeutically in modern medicine.

From this point research took two distinct avenues. On the one hand, understanding of the physical effects of the crude poison progressed at a consistent pace. By the 1850s the French physiologist Claude Bernard had revealed precisely how the poison killed. In the body, impulses from the brain activate the nerve endings to produce a chemical called acetylcholine, which in turn stimulates the muscles. In a classic series of experiments Bernard showed that curare acts to block the transmission of the nervous impulse to the muscle, thus precipitating paralysis. This discovery led to immediate clinical applications. Throughout the nineteenth century curare was used to treat diseases that cause severe muscular pain and contractions. The results were mixed, in large part because the preparations were inconsistent. Even as Victorian physicians were administering crude Amazonian arrow poisons to victims of rabies and lockjaw, no botanist had managed to identify the raw constituents. In contrast to the medical advances, the pace of botanical discovery had been slack.

Since the time of Karl Friedrich Philipp von Martius, who traveled in the Northwest Amazon in 1820, plant explorers had recognized that curare had two major botanical sources. In the Guianas and south across the Brazilian frontier to the lower Amazon, the main plants appeared to be species of *Strychnos,* just as Humboldt had reported from the Orinoco in 1800. Twenty years after Waterton walked across the Guianas, the botanist Robert Schomburgk confirmed that the curare of the Macusi Indians was derived from *Strychnos toxifera.* In the western Amazon, however, virtually every report suggested that arrow poisons were made from species of *Chondrodendron* and related genera in the moonseed family.

By the last decades of the nineteenth century the situation had become absurd. Physicians in increasing numbers required the muscle relaxant for clinical use, and they wanted a standardized supply. Into the vacuum created by botanical ignorance entered a German, Rudolph Boehm, who classified curare not on the basis of the botanical sources, which remained largely unknown, but on the type of container that the poison arrived in. There was, to be fair, some basis for this. Tube curare, as it was called, came packed in bamboo and generally was made from *Chondrodendron tomentosum.* Calabash curare usually consisted of *Strychnos toxifera* as the main ingredient. Pot curare, which came in clay vessels, often was *Strychnos castelneana,* the basis of the poisons first studied by La Condamine. In general, however, this classificatory scheme was about as reasonable as determining the quality of wine not on the basis of vintage but by the shape of the bottle.

By the time Schultes came on the scene, the biological action of curare had been understood for a century. By contrast, botanical knowledge was still embryonic, and the quality and inconsistency of supplies remained a major impediment to research. In 1935, when Harold King first isolated d-tubocurarine, the main paralytic alkaloid, he did so from a twenty-year-old sample of crude tube curare found in the basement of the British Museum. He had no idea of the identity of the source plant. Taxonomic studies that might finally delineate the basic species used for curare began at the New York Botanical Garden in 1936 but would not be published until 1942, and not until 1943 would Harvard researchers produce d-tubocurarine from a properly determined botanical specimen.

Into this void walked Schultes, together with a distant comrade, a man with a colorful past. Richard Gill, son of a Washington doctor, realized in the early 1920s that he had no interest in medicine. He went to sea on a tramp steamer, worked for a year at a whaling station in the South Georgia Sea, and eventually took a job for a rubber company in South America. Enchanted by the exotic beauty of the mountains and forests of Ecuador, he left his work and set out with his wife, Ruth, on an eight-month search for the perfect place to establish a ranch. They found it in 1929 on the eastern slopes of the Andes on the upper Pastaza, not far from the town of Baños. For three years they lived in the forest, becoming close to the Canelos Quichua Indians and studying the extraordinary native pharmacopoeia. Through sentiment and interest, Gill became an ethnobotanist.

In 1932, just prior to returning to the United States for a holiday, Gill fell from a horse, an accident he later blamed for the onset of neurological symptoms that left him, by October 1934, almost completely paralyzed. Physicians diagnosed the disease as multiple sclerosis, but Gill never accepted their verdict. During his protracted recovery, one of his doctors mentioned that his painful muscle spasms might be relieved if consistent supplies of curare were available. In that moment Richard Gill, with a supreme act of determination and will, vowed to return to his jungle home and secure the drug that scientists had been seeking for so long.

After months of self-imposed physiotherapy, Gill could feel his fingers. After two years he could walk. After four years, with the aid of a stout stick, he walked into the Ecuadorian jungle at the head of a massive ethnobotanical expedition. Twenty days on foot from his ranch, he established a base camp where he remained for four months. At the end of 1938, just as Schultes was winding up the first phase of his work in Oaxaca, Richard Gill returned to the United States with proper voucher specimens as well as twenty-five pounds of crude curare. The collections included three species of moonseeds, *Chondrodendron iquitanum, C. tomentosum,* and *Sciadotenia toxifera,* as well as *Strychnos toxifera,* which the Quichua Indians considered a minor ingredient. Here at last was good material. Extracts from Gill's collections were standardized at the University of Nebraska in 1939 and introduced into medical practice for a host of muscular and neurological ailments. In January 1942, Harold Griffith in Montreal used a curare extract to induce muscular relaxation in patients undergoing general anesthesia. Six months later he and Enid Johnson published the paper that would revolutionize modern surgery. They revealed that by administering d-tubocurarine it was possible to use a far lower level of anesthetic, thus reducing the risk of general anesthesia and the postoperative nausea and vomiting. Over the next fifty years d-tubocurarine would save far more human lives than curare had ever taken.

All of this remained unknown to Schultes, who began his own search for arrow poisons on the Putumayo in January 1942. As Schultes began his explorations, a journey that would yield no fewer than fourteen sources of curare, he did so charged with the excitement and anticipation of a scientist poised on the edge of discovery.

Within two weeks of leaving Bogotá, Schultes made one of the most important discoveries of his career, though at the time he had only a vague, intuitive sense of its significance. What's more, it had little to do with curare. Leaving Sibundoy on February 18, Schultes and young Pedro Juajibioy crossed the Cordillera Portachuella in the company of a Swedish gold miner by the name of Hansen who had a camp south of Mocoa on the Río Uchupayaco. Schultes made the camp his base, and leaving the bulk of his supplies and equipment in Hansen's care, marched two days north, overland to Puerto Limón, an Ingano village on the Río Caquetá. There he spent a week describing the trees used to make weapons and canoes, and the varieties of palms used for food and thatch. He made a second sterile collection of *yoco,* observed aroids crushed to treat stingray bites, roots cooked to relieve hysteria, the latex of fig trees drunk as milk by children. For the first time he found the cold fever tree, *chiricaspi,* the plant that thirty years later would nearly kill Tim Plowman and Pedro on the Río Guamués. It was, Schultes noted with characteristic understatement, a "serious intoxicant."

He also collected *yagé,* and on the evening of February 28, 1942, made the following notes:

> Yagé is taken often by some, infrequently by others. It is a most violent purge and often acts as a vomitive. Extremely bitter. Some say the after effects are an exhilaration and feeling of ease and well-being; others that it is a day of discomfort and headache. The bark of yagé

> is scraped off and small pieces are heated in water. The water is drunk. People take it alone or in small groups in houses, often with a sick person who is to be cured. The curandero takes yagé to *see* the proper herb or herbs the sick man needs. Usually taken alone, but in Puerto Limón it is taken sometimes together with the bark of another vine—the *chagropanga.* It is said to be almost the same leaf, but a harder and stouter vine.

Schultes was not sure what to make of this, but two themes intrigued him. First was the realization that the healer embraced *yagé* both as visionary medium and as teacher. The plant made the diagnosis. It was a living being, and the Ingano acknowledged its magical resonance as reflexively as he accepted the axioms of his own science. Second, at the same time there was evidence here of pure empirical experimentation of a specificity he had never before encountered. In Oklahoma and Mexico, and more recently among the Kamsá of Sibundoy, he had always seen psychoactive plants taken alone, not in any sort of combination. Now his Ingano informants, including an old man named Jeremiah Zambrano, insisted that by manipulating the ingredients of the preparations—in this case by adding a plant known as *chagropanga*—it was possible to change the nature of the experience.

Schultes did not question the word of his informants; instead he elected to test their preparations on himself. At Puerto Limón he drank an infusion derived solely from the bark of the liana *Banisteriopsis caapi.* The visions that came were blue and purple, slow undulating waves of color. Then a few days later he tried the mixture with *chagropanga.* The effect was electric; reds and golds dazzling in diamonds that turned like dancers on the tips of distant highways. If *yagé* alone felt like the slow turning of the sky, the addition of *chagropanga* caused explosions of passion and dreams that collapsed one into another until finally, in the empty morning, only the birds remained, scarlet and crimson against the rising sun.

What Schultes had stumbled on was a bit of shamanic alchemy that in its complexity and sophistication had no equal in the Amazon. The psychoactive ingredients in the bark of *yagé* are the beta-carbolines harmine and harmaline. Long ago, however, the shamans of the Northwest Amazon discovered that the effects could be dramatically enhanced by the addition of a number of subsidiary plants. This is an important feature of many traditional preparations, and it is due, in part, to the fact that different chemical compounds in relatively small concentration may effectively potentiate one another.

In the case of *yagé,* some twenty-one admixtures have been identified to date. Two of these are of particular interest. *Psychotria viridis* is a shrub in the coffee family. *Chagropanga* is *Diplopterys cabrerana,* a forest liana closely related to *yagé.* Unlike *yagé,* both of these plants contain tryptamines, powerful psychoactive compounds that when smoked or snuffed induce a very rapid, intense intoxication of short duration marked by astonishing visual imagery. The sensation is rather like being shot out of a rifle barrel lined with baroque paintings and landing on a sea of electricity. Taken orally, however, these potent compounds have no effect because they are denatured by an enzyme, monoamine oxidase (MAO), found in the human gut. Tryptamines can be taken orally only if combined with a MAO inhibitor. Amazingly, the beta-carbolines found in *yagé* are inhibitors of precisely this sort. Thus when *yagé* is combined with either one of these admixture plants, the result is a powerful synergistic effect, a biochemical version of the whole that is greater than the sum of the parts. The visions, as the Indians promised Schultes, become brighter, and the blue and purple hues are augmented by the full spectrum of the rainbow.

What astonished Schultes was less the raw effect of the drugs—by this time, after all, he was becoming accustomed to having his consciousness awash in color—than the underlying intellectual question that the elaboration of these complex preparations posed. The Amazonian flora contains literally tens of thousands of species. How had the Indians learned to identify and combine in this sophisticated manner these morphologically dissimilar plants that possessed such unique and complementary chemical properties? The standard scientific explanation was trial and error—a reasonable term that may well account for certain innovations—but at another level, as Schultes came to realize on spending more time in the forest, it is a euphemism which disguises the fact that ethnobotanists have very little idea how Indians originally made their discoveries.

The problem with trial and error is that the elaboration of the preparations often involves procedures that are either exceedingly complex or yield products of little or no obvious value. *Yagé* is an inedible, nondescript liana that seldom flowers. True, its bark is bitter, often a clue to medicinal properties, but it is no more so than a hundred other forest vines. An infusion of the bark causes vomiting and severe diarrhea, conditions that would discourage further experimentation. Yet not only did the Indians persist but they became so deft at manipulating the various ingredients that individual shamans developed dozens of recipes, each yielding potions of various strengths and nuances to be used for specific ceremonial and ritual purposes.

In the case of curare, Schultes learned, the bark is rasped and placed in a funnel-shaped leaf suspended between two spears. Cold water is percolated through, and the drippings collect in a ceramic pot. The dark fluid is slowly heated and brought to a frothy boil, then cooled and later reheated until a thick viscous scum gradually forms on the surface. This scum is removed and applied to the tips of darts or arrows, which are then carefully dried over the fire. The procedure itself is mundane. What is unusual is that one can drink the poison without being harmed. To be effective it must enter the blood. The realization on the part of the Indians that this orally inactive substance, derived from a small number of forest plants, could kill when administered into the muscle was profound and, like so many of their discoveries, difficult to explain by the concept of trial and error alone.

The Indians naturally had their own explanations, rich cosmological accounts that from their perspective were perfectly logical: sacred plants that had journeyed up the Milk River in the belly of anacondas, potions prepared by jaguars, the drifting souls of shamans dead from the beginning of time. As a scientist Schultes did not take these myths literally, but they did suggest to him a certain delicate balance. "These were the ideas," he would write half a century later, "of a people who did not distinguish the supernatural from the pragmatic." The Indians, Schultes realized, believed in the power of plants, accepted the existence of magic, and acknowledged the potency of the spirit. Magical and mystical ideas entered the very texture of their thinking. Their botanical knowledge could not be separated from their metaphysics. Even the way they ordered and labeled their world was fundamentally different.

It was in Sibundoy, en route to the Putumayo, that Schultes first learned the Indians did not distinguish green from blue. "For them", Padre Marcelino had explained, "the sky is green and the forest is blue. The canopy shelters their lives, and the sky beyond, so rarely seen, is but another layer, protecting the living from the darkness beyond." This strange concept lingered in his imagination as he entered the lowlands. It surfaced when he confronted yet another botanical enigma: the manner in which the Indians classified their plants. The Ingano at Puerto Limón, for example, recognized seven kinds of *yagé.* The Siona had eighteen, which they distinguished on the basis of the strength and color of the visions, the trading history of the plant, and the authority and lineage of the shaman. None of these criteria made sense botanically, and as far as Schultes could tell, all the plants were referable to one species, *Banisteriopsis caapi.* Yet the Indians could readily differentiate their varieties on sight, even from a considerable distance in the forest. What's more, individuals from different tribes, separated by large expanses of forest, identified these same varieties with amazing consistency. It was a similar story with *yoco,* the caffeine-containing stimulant. In addition to *yoco blanco* and *colorado,* Schultes collected black *yoco,* jaguar *yoco, yagé-yoco, yoco* of the witches—fourteen categories in all, not one of which could be determined based on the rules of his own science.

Though trained at the finest botanical institution in America, after a month in the Amazon Schultes felt increasingly like a novice. The Indians knew so much more. He had gone to South America because he had wanted to find the gifts of the rain forest: leaves that heal, fruits and seeds that supply the foods we eat, plants that could transport the individual to realms beyond his imaginings. Yet within a month he had learned that in unveiling the indigenous knowledge, his task was not merely to identify new sources of wealth but rather to understand a new vision of life itself, a profoundly different way of living in a forest.

In the first week of March, Schultes carried these thoughts away from the Caquetá and headed south by mule along the Capuchins' road, reaching Puerto Asís, the head of navigation on the Putumayo, on March 8. Though heralded in the Bogotá press as "an emporium of wealth and a bulwark of Colombian integrity," Puerto Asís was actually a sleepy Capuchin mission animated only slightly by the movement of military supplies slipping over the border to sustain the Ecuadorian forces in their border war with Peru.

Schultes remained with the padres at Puerto Asís for two weeks and then made his way one hundred miles down the Putumayo to Puerto Ospina, a military post located at the mouth of the Río Sucumbíos. There he was fortunate to run into Colonel Gomez-Pereira, the Colombian army officer responsible for security in the borderlands. When Schultes mentioned his desire to reach the upper Sucumbíos, Gomez-Pereira offered to mount an expedition. The Sucumbíos—or the San Miguel, as the Capuchins insisted on calling it—rises on the eastern flank of the Andes and for much of its length forms the boundary between Ecuador and Colombia. A frontier patrol, the Colonel suggested, was long overdue. Thus on March 27, 1942, escorted by the Colombian army, Schultes headed upriver into Kofán territory.

The military launch was the *Mercedes,* an old Amazonian riverboat, narrow at the beam, with a flat tin roof and a shallow hull just rotten enough to keep the crew attentive. There were six on board besides Schultes and the Colonel, including one young boy of twelve, a runaway from Pasto. By night they slept in hammocks strung up one above the other; by day the soldiers huddled around the wheel or kept themselves busy working the bilge and servicing the engine that coughed and sputtered and every so often spewed oil about the galley. They fished, ate rice and plantains cooked in river water, and shot what they could, butchering the meat on the rear deck and tossing the entrails into the wake of the boat.

Schultes spent most of the journey on the roof in the hot sun, working with his plants and, with the aid of the Colonel, learning the rudiments of the Kofán language. Gomez-Pereira was an educated man, a proficient linguist well versed in local history and remarkably sympathetic toward the Indians. From him Schultes learned how little was known about the Kofán. Isolated at the foot of the mountains, they spoke a language that, with its strange pattern of glottal stops, was related to no other living tongue. The word "Kofán" had no meaning in the language; it was simply the term traders and the early Spanish missionaries had used to describe the tribe. Some believed it was derived from *ofanda,* a term in the Sucumbíos dialect for the wood of the blowgun, a reference perhaps to their reputation as poison makers. Their word for themselves was *a'i,* meaning simply "the people."

Before the arrival of the Spaniards, the Kofán had been a powerful tribe—a small nation, in effect, strong enough to incur the wrath of Inca Huayna Capac, who made war against them as he attempted to expand his empire to the north. By the time the Jesuits appeared in 1602, the Conquistadors had stripped the gold-bearing sands of the upper Aguarico and Sucumbíos, and enslavement and disease had reduced the population to twenty thousand. By the mid-nineteenth century, when Colombia and Ecuador first mapped their eastern lowlands, geographers described the Kofán as a warlike tribe of perhaps two thousand, a figure that remained more or less constant until 1899 when, after a hiatus of more than two centuries, missionaries returned. The latest blow had come in 1923 when a measles epidemic, introduced by the Capuchins, killed half the tribe. Now, the Colonel informed Schultes, the total Kofán population was less than five hundred.

Those who survived were a riverine people, living in scattered households and small communities, and entering the forest only to hunt and seek medicinal plants. They oriented themselves in space by the flow of the rivers, in time by the fruiting cycle of certain trees, which formed the basis of their calendar year. There were four villages in Ecuador on the Río Aguarico, a like number along the Sucumbíos, and two on the Río Guamués, the next drainage to the north in Colombia. Their territory ran only seventy-five miles east to west along the rivers and had a width of no more than fifty miles. It was as if the entire tradition had come down to a string of small settlements clinging to the banks of three forgotten rivers.

The Colonel's idea of a frontier patrol was a casual affair. With the Colombian flag hanging limp around the flagpole, the *Mercedes* inched its way upriver, pausing at every clearing just long enough for the Colonel to shake hands with the local elders, dispense a few machetes and aluminum pots, and perhaps grace the wife of the chief with a bolt of cloth. As they went along Schultes collected what he could and between stops worked on his list of local plant names. During the first half of the journey the names that appear in his notebooks are mostly Ingano and Siona, with a smattering of Kofán. Then, after March 29, all the entries are Kofán. It was on this day that he arrived at Quebrada Conejo, roughly 150 river miles above Puerto Ospina.

The *Mercedes* pulled into shore just downriver from a landing where several dugout canoes were tied up at the foot of a muddy bank overgrown with moon flowers. The Colonel instructed his crew to wait on board while he and Schultes climbed the slippery path and made their way past plantings of manioc and peanuts to a clearing that spread a hundred yards away from the river. Peach palms and planted fruit trees softened the edge of the forest, and on all sides was evidence of the wondrous disorder of Indian gardens: clumps of cane and plantains, cashew trees, *achiote* and papaya, sweet potatoes and bottle gourds twining around the foot of tall daturas. The houses in the middle of the clearing were raised bamboo platforms with thatch roofs and walls of split bamboo sheets. Some were open to the air, and Schultes could see the shadowy figures of women huddled over the cooking fires.

It was several minutes before anyone came to greet them. The man who did walked alone. In an instant Schultes realized where the Ingano and Kamsá healers had learned to dress. The Kofán shaman was an old man, and he wore a *cushma* of bright blue trading cloth that fell as a mantle well below his knees. Mounds of colored beads hung around his neck, and necklaces of seeds and shells, jaguar teeth and boar tusks draped in concentric rings down to the middle of his chest. His face was broad, his eyebrows plucked and painted, his lips dyed indigo, his wide and flattened nose decorated with a blue macaw feather that ran through the septum. Across his forehead and on both cheeks were intricate patterns of lines and dots, painted in orange and blue, motifs like visions.

There were bamboo tubes and feathers in his ear lobes. His hair had been shaved short and allowed to grow to a uniform length that tickled the tips of his ears. On his head was a magnificent feathered headdress with an iridescent band of turquoise hummingbird feathers, a circle of red and white toucan breasts, and a corona of green parrot feathers that created a strange halo effect as he moved. Emerging from the top of the headdress was a fan of five scarlet macaw feathers. Two long bandoliers of jungle seeds crossed his chest, iguana skins and leaves circled his wrists, and great manes of scented palm fiber hung from his upper arms. When he turned to lead the Colonel into the center of the village, Schultes saw the long train of parrot feathers that fell over his back.

"Imagine what they look like when they dress up," the Colonel said as they stepped up the log ladder into the old man's house.

The shaman, whose Spanish name was Miguel, invited the Colonel and Schultes to sit beside him on the floor. A woman approached with a calabash of *chicha.* She knelt before the Colonel, dipped her fingers into the liquid to remove some coarse fibers, and then graciously offered him the bowl. The Colonel nodded and drank the preparation with a single swallow, straining the fibers with his teeth. When he was finished, he leaned over a crack in the floor and spat.

"Anduche kiiki," he said. The woman smiled. Like the men she was beautiful. She wore necklaces and feathers, and patterns of color on her face.

"This is a banana drink," the Colonel said to Schultes. "Take it." Schultes and the shaman drank in turn. After a few rounds of the sweet, mildly alcoholic *chicha,* the stiff posturing of the old man softened, and he spoke a few words in Spanish. He and the Colonel exchanged gifts, and after a suitable interlude Schultes brought up the purpose of his journey. Not only was the shaman not suspicious, he acted as if a visit by a student of plants was the most reasonable idea he had ever heard from a white man. Then as the afternoon turned toward evening, one of the young men brought a smaller gourd, which the shaman accepted casually. He drank and handed back the empty vessel. When the calabash came to Schultes, he took note of the ritual decorations around the lip and the thin greenish brown liquid within. He was about to drink when he felt the Colonel's hand on his wrist.

"I believe that is *va'u,"* the Colonel said quietly. He leaned over and whispered something in Schultes's ear. The shaman strained to hear. Schultes turned away from the Colonel and as politely as he could passed the gourd back to the youth. When he had recovered his composure, he realized that not only had he almost drunk an infusion of *borrachero,* or tree datura, the most dangerous of all hallucinogenic plants, but he was now living with a people who evidently took the drug as casually as Englishmen take tea.

• • •

The following morning the *Mercedes* departed early as planned, leaving Schultes to complete his collections and make his own way back downriver to Puerto Ospina. Schultes saw the Colonel off at the landing and then left immediately for the forest with the shaman Miguel. To understand what happened next, one must have some sense of the historical moment. The Kofán, living alone on the upper reaches of the Sucumbíos, had had little contact with the outside world. The few whites they had met, mostly missionaries, soldiers, rubber tappers, and the odd explorer, had generally viewed their society and their forest with fear and contempt. Only a decade earlier the Capuchin priest Gaspar de Pinell had with perfect sanctimony described a sojourn in a land where "tall trees covered with growths and funereal mosses create a crypt so saddening that to the traveler it appears like walking through a tunnel of ghosts and witches. There, far from civilization and surrounded by Indians who could at any moment kill and serve us up as tender morsels in one of their macabre feasts, we spent spiritually blissful days."

The British army officer and explorer Thomas Whiffen, whose book Schultes had read, wrote that the forest was "innately malevolent, a horrible, most evil-disposed enemy. The air is heavy with the fumes of fallen vegetation slowly steaming to decay. The gentle Indian, peaceful and loving, is a fiction of perfervid imaginations only. The Indians are innately cruel." Living for a year among them, he noted, was to become "nauseated by their bestiality." In offering advice to future travelers, he suggested that exploratory parties be limited to no more than twenty-five individuals. "On this principle," he wrote, "it will be seen that the smaller the quantity of baggage carried, the greater will be the number of rifles available for the security of the expedition."

Whiffen, who traveled the Putumayo only a generation before Schultes, claimed to have come upon cannibal feasts, "prisoners eaten to the last bit, a mad festival of savagery . . . men whose eyes glare, nostrils quiver . . . an all pervading delirium." Other academic explorers of the era, if somewhat more restrained, nevertheless subscribed to what Michael Taussig has charitably called the "penis school of physical anthropology." The French anthropologist Eugene Robuchon, who descended the Putumayo during the rubber boom, reflected that "in general the Huitotos have thin and nervous members." Another chapter in his book begins: "The Huitotos have gray-copper skin whose tones correspond to numbers 29 and 30 of the chromatic scale of the Anthropological Society of Paris." A footnote in Whiffen's book reads, "Robuchon states that the women's mammae are pyriform, and the photographs show distinctly pyriform breasts with digitiform nipples. I found them resembling rather the segment of a sphere, the areola not prominent, and the nipples hemispherical."

Needless to say, Schultes had no interest in measuring the penises, breasts, or skull size of the Kofán. Nor did he want to take their land, profit from their labor, or transform their souls. He was alone and unarmed. He was a botanist who respected their knowledge of plants and who revered their forest. He described the Kofán leaders as "friendly, helpful, intelligent, trustworthy, and dedicated." In a prose that is today archaic in tone but thoroughly modern in sentiment, he noted that "the naturalist, interested in plants and animals, both close to the Indian's preoccupations, usually is immediately accepted with excessive collaborative attention. These leaders are gentlemen, and all that is required to bring out their gentle manliness is reciprocal gentle manliness. Until the unsavory veneer of western culture surreptitiously introduces the greed, deception and exploitation that so often accompanies the good of ways foreign to these men of the forests, they preserve characteristics that must only be looked upon with envy by modern civilized societies."

These convictions, once translated into gesture and repartee, immediately distinguished Schultes from any other outsider the Kofán had known. They understood his passion for plants, appreciated his patience, and responded enthusiastically to his curiosity. In ethnobotany he had the perfect conduit to culture.

For the first three days at Conejo, Schultes and Miguel followed the same quiet rhythm; slow mornings harvesting plants, afternoons on riverbanks concocting preparations, evenings in the shelter of the shaman's home preparing the specimens and recording on paper knowledge that the Kofán had passed on by word of mouth for generations. Within a week they had worked the forests of the flood plain, wading through inundated palm swamps and exploring the streams by canoe. In two days they walked overland to the Río Tetuye, following the traditional route to the Río Aguarico in Ecuador. Never had Schultes made collections of such significance.

Over the course of two hundred years of research, plant explorers in South America, beginning with Humboldt and Bonpland, have identified only twenty-one species of the genus *Strychnos* used as arrow poisons. In one week on the Sucumbíos, Schultes found eight of these, each believed by the Kofán to have unique chemical and magical powers. He also collected *Chondrodendron iquitanum,* one of the curare plants brought

• Mrs. Gaul's Pensión Inglesa, Carr. 7, Calle 18–78, Richard Evans Schultes's first home in Bogotá. Photographed here on January 9, 1942, it was destroyed during the Bogotá riots in April 1948.

• Pedro Juajibioy, photographed in Sibundoy, December 15, 1941.

• Kofán preparing curare at Quebrada Conejo, Río Sucumbíos, April 5, 1942.

• Schultes and Nazzareno Postarino leaving La Chorrera on June 12, 1942, heading down the Río Igaraparaná. Schultes stands in the rear, wearing a pith helmet.

• *(Above)* Schultes on the summit of Cerro de la Campana, Río Ajaju, June 1–6, 1943.

• Schultes and crew hauling a boat over the Naré Trocha, Río Apaporis, September 3–17, 1943.

• *Left to right:* Loren Polhamus, E. W. Brandes, and Robert Rands, the scientists who directed the USDA Cooperative Rubber Research Program, Washington, D.C., 1943.

• The falls of Jirijirimo, Río Apaporis, photographed in October 1943.

• Richard Spruce at age seventy-two, four years before his death in 1893 (from the frontispiece of *Notes of a Botanist on the Amazon and the Andes,* published in London in 1908 by Macmillan & Co.).

• Rubber tappers bringing their daily harvest of latex, Río Loretoyacu, October 1944.

• Richard Evans Schultes heading up the Río Negro, September 1947.

•Schultes sorting some of the six hundred thousand rubber seeds he collected at Leticia, March 1945.

•Schultes with Miguel Dumit to his right, Captain Liebermann, and an unknown associate meeting in Bogotá in March 1951 to plan Soratama, the rubber station on the Río Apaporis.

•Schultes with Pacho López on the Río Negro, 1947–48.

•The wilderness base at Soratama, Río Apaporis, established June 1951.

•The Catalina arriving at Soratama to resupply the rubber station.

•Near the headwaters of the Río Miritiparaná, in a village on the Río Guacayá, Schultes poses with Yukuna and Tanimuka dancers during the celebration of the Kai-ya-ree, April 1952.

•Schultes taking tobacco snuff, May 1952.

•Makuna boys dancing during nocturnal ceremony, Río Piraparaná, September 1952.

•A Makuna shaman and a young boy collecting *yagé*, Río Popeyacá, June 1952.

•*Yagé* images painted on the walls of a Makuna maloca, Río Piraparaná.

•Schultes and Makuna boys taking shelter at the falls of Yayacopi, Río Apaporis.

•The descent of the Río Vaupés, April 1953.

•Tim Plowman with Pogo, Ollantaytambo, Peru, January 1975.

back from the Canelos Quichua by Richard Gill, as well as two species of *Abuta,* a related genus in the moonseed family. He also described for the first time the use of *Schoenobiblus peruvianus,* a plant known to the Kofán as *shira"chu sehe"pa,* a poison used exclusively for hunting birds.

It was not just curare that drew his attention. He collected dozens of folk remedies, stimulants, hallucinogens, fish poisons, and wild fruits. More than the collections, it was the opportunity to be with a people who manipulated plants with such dexterity that changed forever his perceptions about ethnobotany. Living with the Kofán he slipped into a world of phytochemical wizardry unlike anything he had ever known. Psychoactive and toxic plants touched every aspect of their lives. At dawn men and women went to the river to bathe, rinsing their mouths with water, preparing their palate for the morning dose of *yoco.* There followed an hour of ritual grooming as each adult made ready for the day—plucking facial hair, painting intricate designs on the skin, donning the elaborate costume that mimicked the dress of the spirit beings encountered during *yagé* intoxications. En route to the fields or forest, the men carried with them a paste of fermented banana, which, when mixed with water, could be whipped into a frothy alcoholic beverage. They pounded the leaves of shrubs, the bark of lianas, the roots of small trees, and placed the pulp in streams, gathering the stunned fish in great baskets woven from the fiber of palms. To make curare they manipulated a dozen recipes, producing poisons of various strengths. Every adult man possessed his own repertoire, and there were specific poisons for each animal and bird of the forest. Returning to the settlement in the late afternoon, men joined the women and children in the evening meal, sharing the events of the day as they imbibed more *chicha* and smoked cigars the size of a child's arm.

In Kofán the word for medicine and poison is the same, for just as curare-tipped darts bring death to an animal, so the medicinal plants bring death to evil spirits that cause misfortune and disease. Thus, in a manner that made perfect sense to the shaman, Schultes moved from the study of curare to an examination of the art of healing, which led him inevitably to *yagé.* For the Kofán, he learned, disease is caused by magic arrows cast into the victim's flesh by the avenging souls of malevolent sorcerers. The duty of the shaman was to free his own soul to wander so that he might find and remove these forces of darkness. It is *yagé* that allows him to soar away. The image of flight is invoked in the colors of the corona and the train of feathers winglike on the shaman's back. The feathers, the shaman explained, are the memories of birds that can only be seen on *yagé.* They are the masters, the patrons of ecstatic intoxication.

For the Kofán, Schultes discovered, *yagé* is far more than a shamanic tool; it is the source of wisdom itself, the ultimate medium of knowledge for the entire society. To drink *yagé* is to learn. It is the vehicle by which each person acquires power and direct experience of the divine. The teachers are the *yagé* people, the elegant beings of the spirit realm, the dwelling place of the shaman grandfathers. Expressing themselves only in song, the *yagé* people give each and every Kofán an image, a song, and a vision that become the inspiration for the designs painted on the skin. No Kofán shares the same motif or the same song. There are as many sacred melodies as there are people, and with the death of a person the song disappears.

When Schultes asked the shaman how often the people drank *yagé,* his response suggested that the question had no meaning: during illness, of course, and in the wake of a death; in times of need or hardship; at certain passages in life; when a young boy of six has his initial haircut or when he kills for the first time. And naturally, the shaman suggested, a youth will drink *yagé* at puberty, when his nose and ears are pierced and he obtains the right to wear the tail feathers of the macaw. As a young man he may drink it at his leisure to improve his hunting technique or simply to flaunt his physical prowess. The message that Schultes received was that the Kofán took *yagé* whenever they felt like it—at least once a week and no doubt on any occasion that warranted it, such as the eve of his own departure from Conejo. With the people dancing, the men facing the women in long lines moving forward and then back, turning from side to side, the dancers stamping the ground lightly in time with the drums, this solitary student of plants took *yagé* with, as he would recall fifty years later, "the whole damn village."

The next morning, his face still painted and his mind swirling with the sounds of chants—*ya-jé, ya-jé, ya-ya-ya, ya-jé, ya-jé, ya-ya-ya*—Schultes paddled up the Sucumbíos in a dugout canoe. His head throbbed. Unbeknownst to him, the preparation had contained bark of the *tsontinb"k'o,* the cold fever tree, together with *su-tim-ba-che,* a root said by the Kofán to cause a "drunk worse than *yagé."* Indeed it had. Despite his discomfort, Schultes made a note to investigate the admixtures at length. Thirty years later he did, by dispatching his best graduate student to the Putumayo. After fifteen months in the field, Tim Plowman emerged with the definitive study, which included descriptions of several new species and varieties. The plant Schultes drank at Conejo that gave him such a headache is now known botanically as *Brunfelsia grandiflora* ssp. *schultesii.*

Schultes remained on the Sucumbíos for another two weeks. He collected around the headwaters at Santa Rosa, returned to Conejo, and poled up the Quebrada Hormiga, reaching the height of land and crossing on foot to the drainage of the Guamués, which he entered on April 13. Returning after a few days to Conejo, he traveled downstream and arrived a week later in Puerto Ospina, where he remained for only twenty-four hours before heading up the Putumayo once again in the *Mercedes* to explore the lower Guamués.

Ten days later he was back in Ospina, making ready for his next destination, the lower Putumayo, when his plans were interrupted by the unscheduled arrival of a military plane. It was a tri-motor Fokker destined for Tres Esquinas, an army post and Indian settlement one hundred miles to the northeast at the confluence of the Río Orteguaza and the Caquetá. Military flights left Tres Esquinas periodically for Bogotá. Schultes leaped at the opportunity to get his curare and *yoco* specimens, particularly the living material, back to the capital before he began the long journey down the Putumayo. Thus, quite unexpectedly, he spent the first days of May on the banks of the Caquetá, living in the long houses of the Koreguaje at Nuevo Mundo, waiting for the Bogotá plane, and enjoying for the first time the smoky taste of Amazonian coca.

In mid-May, after only eight days in the capital, he returned to the Putumayo, flying directly to the naval base at Puerto Leguízamo at the mouth of the Río Caucaya, one hundred miles downstream from Ospina. There he rendezvoused with a young Italian, Nazzareno Postarino, one of Fuerbringer's men from Mocoa whom Schultes had hired as a field assistant for the expedition. A native of the Italian Alps, blond and blue-eyed, Nazzareno was a good man who worked hard, ate what was put in front of him, and paid no attention to where he was asked to sleep.

Schultes and Nazzareno had no plans, only a rough sense of itinerary. In the 700 miles of river that lay between Caucaya and the Brazilian frontier, there are only two major tributaries, the Karaparaná and the Igaraparaná. Flowing from the northwest and running roughly parallel to each other, they reach the north bank of the Putumayo approximately 450 and 300 miles above its confluence with the Amazon proper. The Karaparaná, the first of the two heading downstream, is navigable for much of its length. The Igaraparaná, by contrast, is dissected by a dozen major rapids. The traditional homeland of the Witoto, Bora, and Andoke, these rivers were largely unexplored and the peoples unknown until 1886 when rubber gatherers first penetrated the region.

Schultes's goal was to travel by riverboat to the Karaparaná and then head overland northeast to the mission at La Chorrera, located halfway up the Igaraparaná. There he would purchase a dugout and with the help of Nazzareno and Witoto guides make his way down the Igaraparaná through the rapids to its confluence with the Putumayo. From there they would paddle downriver to the military camp at Tarapacá. The first destination, located a day by canoe up the Karaparaná, was an old rubber depot with the charming name of El Encanto, the Enchantment. On May 19, after four days of collecting at Caucaya, Schultes and Nazzareno embarked on a wood-burning, three-decked paddle wheeler known as the *Ciudad de Neiva.*

Schultes did not notice the holes in the gangplank at the landing at El Encanto. Nor did he see the shadows in the wood, all that remained after thirty years of the dark stains that once colored the stocks. The iron bolts that had held the chains were flaked with rust. Schultes did not know the history of the two-story house with the galvanized roof that stood high on the hill above him overlooking the turbid waters of the Karaparaná. He could not have known that from its open balcony, with the cedar wood floor and iron bath, the overseer Miguel Loayza had soaked in scented waters as he contemplated the usefulness of punishment. The stone foundation that Schultes passed as he made his way up the bank had been a house packed with Indians on the night that Loayza's men had deliberately torched the thatch so that their master could practice his marksmanship as the burning bodies ran from the flames.

Nor could Schultes imagine the dying and diseased lying about the house and adjacent fields—women and children weak with hunger, waiting for the sun to release them from their suffering. In the courtyard between the wash houses and storage buildings, dirt now covered the pits that had held the cages where men were confined in chains for months, nursing madness and a hunger that made them impatient for maggots to mature in their wounds. The instruments of torture had been removed from the patio. Gone were the six-tailed whips that had once flayed men to the bone, the stocks where women were raped, the poles where naked men were staked in the sun, the ropes that suspended children of ten high above the ground to be flogged.

The outline of the "convent" had long since been erased from the ground. It had been located close to the house, a palm-slatted hut with a plank ceiling that sheltered fifteen girls aged nine to thirteen. These were Loayza's concubines, Indian children who grew into adolescence physically deformed, their hips weak and permanently dislocated from intercourse. By day they were locked in the shack. At night one or two would be led to his room. Only four times a year, when the riverboat arrived from Iquitos and Loayza shared them with the crew, did they all see the sun.

Schultes knew little of this history. Climbing past the broken-down ruins, he and Nazzareno walked toward a row of Indian houses where they were met by the *capitán* of the surviving Witotos. A taciturn man perhaps thirty years old, he wore dirty khaki pants, a canvas shirt, and a belt stuffed with weapons. His face had a livid scar over one eye. Hesitantly, he invited Schultes into a small hut. The ground was strewn with fish bones and armadillo shells, tin cans eaten by rust. The air reeked of fermentation. In a dirty hammock slung by the fire lay an old man. The father of the *capitán,* he rose slowly to greet them. His hand was missing three fingers, and when he turned into the light, Schultes saw that an ear had been sliced off at the scalp.

That night, and the next morning as he stood above the river and watched a dozen Witoto in fresh white clothes arrive to have their blood sampled by the priests, and the following night when he took *yagé* prepared by an old man who refused to drink it with him, Schultes peered through a shadow world that left him anxious for the forest. He and Nazzareno left El Encanto after only a few days and, guided by the *capitán,* whom they now knew as Raphael, marched overland for La Chorrera. It took three days along an overgrown track that ran through a vault of immense trees. There were moments of beauty—epiphytic orchids translucent in a glimpse of sunshine, the flight of a blue morpho, bird songs that began each day. But all along the path, spaced at random intervals, were the trees that had been the reason for the suffering. When Schultes paused to ask Raphael about the crude markings on the bark, the incisions thick with scar tissue, the Witoto responded in a low voice with few words.

They arrived on the bank of the Igaraparaná late in the afternoon, just as the shadows began to form and the sun softened the gentle pastures that spread across the river beyond the mission at La Chorrera. Just upriver, bamboo and small trees swayed in the wind that rose from the water as it broke over a broad shelf of rock. Crossing by canoe, with all sounds muffled by the falls, Schultes focused intently on the buildings ahead, large two-story structures of stone, far more substantial than the ruins at El Encanto, with freshly painted balconies spreading beneath overhanging tin roofs and gardens of planted palms and neatly cropped grass. As they neared the far riverbank, dozens of small children appeared, all dressed in white uniforms and chattering away, darting in and about the legs of a priest who was standing at the top of the steps that rose from the river. Moving back and forth across the bank, gathering the children, were two sisters, cloaked head to foot in white cloth. As the canoe came close, the priest flung his arms out and dropped them onto his belly. Schultes could hear the laughter as he stepped onto the shore.

"¡Gracias a Dios!" Padre Xavier said as he greeted Schultes and Nazzareno. "For a moment I thought he had come back."

"Excuse me¿" Schultes said.

"The Bishop. That son of a bitch. I thought he'd never leave." Padre Xavier smiled, and Schultes could see that his teeth were stained green. As the sisters led Schultes and Nazzareno along the path toward the mission, the garrulous priest startled them with the details. It was one thing for the Bishop of Sibundoy to drop in once a year and start ordering him around. It was quite another to have to go a week without coca. The final straw had been the Bishop's insistence that sermons be delivered in Spanish, not Witoto.

"The fool. What's the point if they can't understand¿" Padre Xavier glanced back at the sisters, who shuffled along, heads bowed, pretending to ignore what they were hearing. The padre took a small jar from beneath his habit, snapped off the cap, and dipped the tip of his finger into the black gooey syrup.

"People from the city, they know nothing," he said with a sigh, slipping his finger into his mouth. "He didn't dare stop me taking tobacco. Here, try some. The Witoto make the best."

Schultes accepted the jar and daubed his gums with the syrup. By the time they had walked the hundred yards to the courtyard of the mission and Padre Xavier had found his keys to the shed where they were to stay, Schultes's brow was damp with sweat, his head spun, and his stomach was about to give way.

Once he had recovered from his first dose of Amazonian tobacco, Schultes enjoyed his time at La Chorrera. Traveling alone on the upper Putumayo, he had never been able to study the Indian gardens, for they are the domain of women and the place where couples make love. Among the Kofán and Ingano, his queries about sweet manioc cultivation had invariably provoked fits of laughter. Now with the enthusiastic support of Padre Xavier and chaperoned by the nuns, Schultes was able to spend time in the fields with Witoto women. He collected avocados, sweet potatoes, cashews, maize, beans, chocolate, unusual squashes, and a dozen local varieties of pineapple, a wondrous plant born of the Amazon.

In the houses he watched food being prepared, *chicha* fermented from peach palm fruits, a hot paste condensed from chilies. He made a special study of bitter manioc, the poisonous tuber that was the mainstay of the diet. In order to extract the poison, the women peeled the roots and soaked them overnight in warm water. In the morning they grated the roots with a graceful rhythmic motion, on beautiful wooden boards with a surface of small slivers of quartz set in a thick layer of resin. After kneading the yellow mash through tightly woven sieves, they placed it on a mat of rough fiber. Rolled up over the manioc, the mat was hung from a house rafter and twisted with a long pole, round and round, tighter and tighter until all the poisonous fluid was squeezed out. Dried in the sun, the manioc was filtered once more through a sieve to produce a fine white flour. When Schultes asked about the origins of the plant, he learned that it had been a gift of the Sky God.

At night he gathered around the men's circle in the long house, taking coca and tobacco, watching and listening as the voices became more and more animated. The talk was less conversation than ritual discourse. Reviewing the important events of the day or anticipating future problems, the *capitán* would begin a long rambling monologue that was soon echoed by three or four other speakers who would rework the same ideas over and over until the sounds blended into one another. Finally, as the thoughts and opinions achieved a certain harmony, the men would nod and one by one dip their fingers into a large pot of tobacco placed prominently in the center of the ring.

The men often stayed awake all night, preparing coca or making tobacco syrup. They took the plants together, the strong pungent flavor of the tobacco burning the tongue, stimulating saliva, invoking the dark serpentine world of the nightshades. Coca, far gentler, had a smoky flavor of powdered leaves and ash, and gave a mild sense of well-being. The only trick with coca, Schultes discovered, was learning to form the fine powder into a moist paste without sneezing. Tobacco was another story. Neither sold nor traded, the preparation was derived only from the large green leaves of *Nicotiana tabacum,* which the Witoto cultivated with great care. The leaves were boiled in clay pots for ten hours, and the concentrate was mixed with the ashes of certain palms and salt prepared from the roots of forest trees. The Witoto sweetened the syrup by storing it in the pods of a wild cacao. They called the drug *ye-rrás,* or honey, and recognized it as the mediator that had first brought the people together with the jaguar.

Sometimes at dawn the Witoto played the *manguaré,* two enormous log drums hanging side by side, suspended from the rafters of the long house and bound to the earth by rope ties. As large as a man and hollowed out with burning stones, each had a narrow opening that ran along the log, slightly off center. Striking the drum on either side of the hole resulted in different sounds. The drum with the thicker and more dense wood had deeper tones, whereas the smaller log had a higher pitch. Thus the drummer, standing between the *manguaré,* had four notes to choose from. Combining these in an ingenious manner and using prearranged codes, the Witoto were able to send complicated messages as far as ten miles across the forest.

What was odd about these drums, Schultes noted one morning as the sun broke over the river and he found himself once again without sleep, was that within the long house the sound was quite tolerable. Yet just a few feet away, beyond the thatch, it was enough to drive one to cover. He walked away from the men's circle and placed his hands on their polished surface. He examined one of the drumsticks. They were wood with the heads bound in thick layers of wild rubber. How strange it seemed. Such a reasonable use of the latex.

He thought of Padre Xavier. In the ten days that he and Nazzareno had stayed as guests at the mission, the priest had told them in some detail about its history. Schultes liked the man, his wiry grin, his affection for coca, his unexpected irreverence in this island of redemption that the Capuchins had built on top of the past. It was indeed the past that made the presence of the mission, with its simple church and schoolhouse, the healthy fields and kitchens, seem reasonable, perhaps even necessary. The young Witoto girls and boys, separated by sex, dressed in white smocks and school uniforms, parading to church twice and sometimes three times a day, would never know the forest world of their grandparents, but neither would they experience the horrors that marked the lives of their parents.

The walls of the mission were built of cut stone two feet thick. Carved into the stone, facing on the cobbled courtyard beneath the stairway that rose to the second story, was an arched doorway with a heavy wooden door bolted shut. The door had once been painted white like the walls, but years of use had worn bare the wood. One evening Padre Xavier decided to show Schultes and Nazzareno what was inside the room.

"We will always leave this as it was," the priest said as he opened the lock. Schultes and Nazzareno stepped through the doorway, their eyes adjusting to the dark. Padre Xavier opened a small shutter, and faint light spread into the room, illuminating a row of stocks. Chains and iron bars protruded from the walls; a line of hooks hung from a rafter.

"Some men were kept in chains for a year. The color of the skin changed. The Witoto say that it became yellow like the death of the sun."

Schultes and Nazzareno stood apart, stunned by the instruments of torture.

"Cruelty invaded their souls," the padre said. "In those years the best that could be said of a white man in the Putumayo was that he didn't kill out of boredom."

The Indians called it *caoutchouc,* the weeping tree, and for generations they had slashed its bark, letting the white milk drip onto leaves where it could be molded by hand into vessels and sheets, impermeable to the rain. Columbus encountered Arawakans playing games with strange balls that bounced and flew. La Condamine wrapped his precious instruments in sheets of cloth painted with the latex and hardened by the smoke of fires and the heat of the sun. Joseph Priestley, the English clergyman who discovered oxygen, found that small cubes of *caoutchouc* were ideal for erasing his lead pencil notations. He told his friends Thomas Jefferson and Benjamin Franklin, and since they believed the plant hailed from the East Indies, the substance became known as India rubber. In fact, it came from the Amazon, and there the king of Portugal had already established a flourishing industry that made rubber shoes, capes, and bags. In 1823 a Scotsman, Charles Macintosh, dissolved rubber in naphtha and made a pliable coating for fabric, which led to the invention of the mackintosh, the world's first raincoat.

All of these products, however, had a major flaw. In cold weather the rubber became so brittle that it cracked like porcelain. In summer heat a rubber cape was reduced to a sticky shroud. Then in 1839, quite by accident, an inventor from Boston, Charles Goodyear, dropped a mixture of rubber and sulfur onto a hot stove. It charred like leather and became plastic and elastic. This was the birth of vulcanization, the process that made rubber impervious to the elements, transforming it from a curiosity to an essential product of the industrial age. In the next thirty years annual production of wild rubber in Brazil soared from 31 to 2,600 tons.

In 1888 the son of Irish veterinarian John Dunlop won a tricycle race in Belfast with inflated tires that his father had invented to challenge the solid metal wheels of the competition. Seven years later in France the Michelin brothers stunned critics by successfully introducing removable pneumatic tires in the Paris-Bordeaux car rally. By 1898 there were more than fifty American automobile companies. Oldsmobile, the first to be commercially successful, sold 425 cars in 1901. Less than a decade later the first of 15 million Model T's rolled off Henry Ford's assembly line. Each vehicle needed rubber, and the only source was the Amazon. By 1909 merchants were shipping 500 tons of rubber downriver every ten days. In one month alone, March 1908, three foreign ships sailed from Pará, each bound for a different port and each carrying a rubber cargo valued in excess of $5 million. By 1910 rubber accounted for 40 percent of all Brazilian exports. A year later production peaked at 44,296 tons. It was worth, at a conservative estimate, more than $200 million.

The flash of wealth was mesmerizing. In Pittsburgh, steel tycoon Andrew Carnegie lamented, "I should have chosen rubber." In London and New York, men and women flipped coins to decide whether to go after gold in the Klondike or black gold in Brazil. At the height of the rush, five thousand men a week headed up the Amazon. Almost overnight a forgotten land of jungle and rivers became the destination of an army of dreamers and thieves, merchants and their barbarous lackeys who in time would be known to the Indians simply as "the people eaters."

Manaus, situated at the heart of the trade in Brazil, grew in a few years from a seedy riverside village into a thriving city where opulence reached bizarre heights. The governor, Eduardo Ribeiro, a man of unlimited ambitions, laid out boulevards paved with cobbles imported from Portugal and lined with ornamental trees from Australia and China. He installed the first phone system in Brazil, built a racetrack, a bullring, and dozens of schools as well as hospitals, churches, banks, and a Palace of Justice that alone cost over $2.5 million. The city's filtration system was capable of providing 2 million gallons of pure water each day. At a time when New York and Boston still had horse-drawn trolleys, Manaus had sixteen miles of streetcar tracks and an electric grid built for a city of a million, though the population had yet to reach forty thousand.

Such public extravagance, made possible by a 20 percent export tax on rubber and a wild optimism that seduced bankers and merchants alike, paled in comparison to the excesses of individuals. In a city cut off from the world by an enormous expanse of forest, flaunting wealth became sport. Rubber barons lit cigars with $100 bank notes and slaked the thirst of their horses with silver buckets of chilled French champagne. Their wives, disdainful of the muddy water of the Amazon, sent linens to Portugal to be laundered. At banquets, guests sat at tables of Carrara marble, in chairs of cinnamon and cedar wood shipped from London. They ate food imported from Europe: Russian caviar, Danish butter, English meats, German potatoes, and pickled vegetables from Belgium. In the wake of opulent dinners, some costing as much as $100,000, men retired to any one of a dozen elegant bordellos. Prostitutes flocked to Manaus from Moscow and Tangier, Cairo, Paris, Budapest, Baghdad, and New York. Prices were fixed. Four hundred dollars for a thirteen-year-old Polish virgin. As much as $8,000 a night for the most desirable women, those who took shower baths in iced champagne while their clients knelt and lapped it up. Men often paid with jewels, the tiaras and necklaces that in 1907 made the citizens of Manaus the highest per capita consumers of diamonds in the world.

In a city with the motto *Vale Quem Tem,* "You're worth what you've got," the great symbol of excess was the Opera House, a monumental Beaux Arts extravaganza designed by a Portuguese architect and built over a seventeen-year period ending in 1896. The builders, rejecting construction materials of local origin, imported ironwork from Glasgow, marble and gold leaf from Florence, crystal chandeliers from Venice, and sixty-six thousand ceramic tiles from Alsace-Lorraine. Even the massive murals of local jungle scenes were painted in Europe and shipped to Manaus. By opening night, January 6, 1897, when 1,600 gathered to hear the Grand Italian Opera Company in Ponchielli's *La Gioconda,* the project had exhausted over $2 million. Operating expenses included subsidies of more than $100,000 per performance, the cost of luring established performers across an ocean and a thousand miles up the Amazon to a lavish venue built in the midst of a malarial swamp.

All of this wealth derived from the latex of three closely related species of wild *Hevea* that grew scattered across 2 million square miles of tropical rain forest. In this vast expanse, an area the size of the continental United States, there were perhaps 300 million individual trees worth exploiting. Finding them was the challenge. In nature, rubber trees grow widely dispersed in the forest, an adaptation that insulates the species from their greatest predator, the South American leaf blight. This disease, found only in the American tropics, invariably proves lethal once trees are concentrated in plantations. It was this accident of biology that forged the fundamental structure of the wild rubber industry.

To make a profit the merchants had to establish exclusive control over enormous territories, lands that in the Amazon generally corresponded to particular river drainages. They accomplished this with private armies, gunboats commissioned in Liverpool, and enough loose capital to buy off any government official who stood in the way. Once their lands were secure, they needed workers, thousands of them, in order to gather the latex from the wild. Labor supply was a constant problem. On the Río Madre de Dios one rubber baron created a stud farm, enslaving six hundred Indian women whom he bred like cattle. Others dispatched agents to the impoverished northeast, the dust bowl of Brazil, to secure by contract hollow-faced peasants who were desperate and hungry. In time, as the industry expanded, it soaked up entire tribes of Indians into an atrocious network of debt peonage from which there was no escape.

The system was quite simple. Workers under contract were legally bound to their employers for a two-year period. After that they were free to change jobs or return home, provided they were debt free. This was the catch. Arriving upriver at some remote camp, the tapper already owed $150 for his passage. The overseer then advanced him three months' worth of supplies—food and clothing, buckets and knives to collect the latex, perhaps a Winchester rifle and ammunition. These goods, bought by the merchant for a fraction of what the recipient was charged, created a debt that could be liquidated only by delivering rubber.

Everything, including the rubber, was priced to ensure that before any worker could pay off his debt, he would be obliged to take out another loan, more food, and more supplies merely to stay alive. With each cycle the debt mounted. Rifles sold for the equivalent of two years' labor. Threadbare blankets were priced at five dollars. Latex paid out at five centavos a kilo. After six months of work a man still owed for the manioc flour consumed during the first weeks. Debts accumulated that would take a century to pay off. Even death itself could not free the rubber tapper, for his obligations passed by law to his children. Sons inherited the debts of fathers; daughters were sold into prostitution by desperate mothers. It was, in effect, a new form of bondage in which the overseer enslaved not only the man but also all of his offspring, including those unborn.

In servitude, the workers endured boredom and the endless rounds of the tapping circuits, which carried them into the forest each dawn. They lived for the most part in makeshift camps, often in the middle of unhealthy swamps. Their trails ran a mile or two into the forest, winding past two hundred and sometimes three hundred rubber trees, each rising high into the canopy. In the morning they visited each one, making an incision in the silver bark, watching as the latex started to bleed into a small tin cup. They finished the circuit at noon, ate a sparse meal of manioc and dried meat, and then returned with buckets to gather the latex in the early afternoon. Arriving once more at their base, they began the slow process of curing the latex, pouring it little by little over a ball being gradually rotated in smoke rising from a slow fire. It was hard, miserable work. It took three hours to process a day's harvest, and the ball, weighing as much as two hundred pounds as it grew, had to be rotated more than 1,500 times. In the end the best tapper, working twelve hours, could produce only twenty-five pounds a day. For some rubber barons this was not good enough.

Julio Cesar Arana began life in the eastern foothills of the Peruvian Andes in the small town of Rioja. The son of a hatmaker, he left school at fourteen, entered the family business, and became a peddler, selling straw hats in sleepy Andean cities. Restless and ambitious, he soon tired of the stagnant trade. Before he was eighteen he had traveled the length of the Amazon. At twenty-four, in the same year that Dunlop invented the pneumatic tire, he opened a trading post at Tarapoto on the rubber-rich Huallaga River. Selling everything from canned meats and beads to cheap cologne and bullets, he devised a system of barter that was commercially unbeatable. He bought rubber on credit, latex that had yet to be harvested, always at the price of the date of the initial transaction. Essentially he was trading futures on a commodity that only increased in value. His average profit was 400 percent.

In 1890, Arana bought a rubber circuit in the forests near Yurimaguas, also on the Huallaga, and for the first time as a rubber producer confronted the inherent problems of the industry. Each of his laborers, brought in from Ceará and Piauí and equipped with a basic kit, cost him roughly $400, a steady drain of capital that only compounded his own debts with the trading houses of Manaus. Men in the forest, no matter how productive, had to be maintained, even during the six months of the rainy season when the waters flooded the forest and the latex would not congeal. The answer was to find men to whom you owed nothing, men whose disappearance would be of neither interest nor concern.

Arana found such a place in 1899 when his launch first steamed into the mouth of the Río Putumayo, some six hundred miles west of Manaus. Navigable for most of its length and densely populated with Indians, the Putumayo was claimed by both Colombia and Peru, yet was occupied by neither. The only settlers were a handful of Colombian rubber traders who maintained tenuous supply routes over the mountains to Sibundoy and Pasto. These Colombians—Benjamin Larrañaga and his brother Rafael at La Chorrera, the Calderón brothers at El Encanto, and Gabriel Martínez at Remolino—initially welcomed Arana's offer to establish a downriver trade directly to Iquitos and Manaus. They knew nothing of his character or intentions, nor did they recognize that by becoming dependent on the riverboats they would inevitably lose control of the trade.

Almost immediately Arana began buying rubber holdings. By 1905, after he had purchased the operations at both El Encanto and La Chorrera, he owned over twelve thousand square miles of the Putumayo. Only four independent Colombian rubber traders remained. These were dealt with in a style for which Arana would become famous. In December 1907 he dispatched his overseer at El Encanto, Miguel Loayza, up the Karaparaná to encourage David Serrano, one of the Colombians, to abandon his post at La Reserva. The Peruvians attacked in overwhelming force, bound Serrano to a tree, raped his wife before his eyes, and then left the Colombian to die as they returned downriver with his son, who was later forced to work as a servant at El Encanto. Serrano's wife, also kidnapped, was held by Loayza as a concubine.

Though cruelty and sadism came easily to the men dispatched by Arana throughout his growing domain, the horror that he unleashed was, at least initially, the result of cold calculation. For all its advantages, the rubber trade in the Putumayo had three drawbacks: The distance to market, over one thousand miles from certain depots, increased the cost of supplies and the risk of loss. The second problem was the quality of the rubber. The highest prices in Manaus went for rubber derived from the latex of *Hevea brasiliensis,* a species that grows primarily in the forests south of the Amazon River. The source of the Putumayo rubber was *Hevea guianensis* and its variety *lutea,* trees found over the entire range of the genus, from eastern Brazil west to the foothills of the Andes, but that produce a distinctly inferior grade of rubber.

Arana overcame the first impediment with a fleet of twenty-three armed trading launches that both safeguarded his goods and provided absolute control of the river. The second problem took care of itself. In the frenzy of the era, as prices soared by the day and more than $70 million of rubber sailed down from Manaus each year, any cured latex found a market. The third challenge was more difficult. Some means had to be devised to secure the labor of thousands of Indian men and women who, when confronted with ordinary conflict, could flee into the forests they knew so well. Arana decided on terror.

He began by importing overseers: criminals and deviants who arrived in his debt. In 1904 he placed two hundred wardens from Barbados under contract and charged them with the task of hunting down those who attempted to escape. From the ranks of the Indians he drafted four hundred boys who grew up nursed on violence and rewarded for barbaric deeds from which they could never retreat. Behind a veil of isolation, Arana swept aside all ethical and moral constraints, made financial gain strictly dependent on commissions, and let loose these savage men onto the land.

Rubber traders, legally permitted to "civilize" the Indians, attacked at dawn, trapping their victims within their long houses and then offering the gifts that were the excuse for enslavement. Once caught in debts they could not comprehend, with the lives of their families at stake, the Witotos labored to produce a substance for which they had no use. Those who failed to deliver their quota, who watched as the needle on the scale stopped short of the ten-kilo mark, fell prostrate on the ground to await punishment. Some were flogged and beaten, others lost hands or fingers. They submitted because if they resisted, their children and wives would be the ones to suffer.

With each incident the terror grew. Armed thugs dispatched to the forest returned with severed heads wrapped in heliconia leaves. Overseers cut off ears for sport. For diversion they bound Indians to trees, spread their legs, and lit fires beneath their bodies. Children were tortured to make them reveal the whereabouts of their parents; young girls were sold as whores; infants were cut up to feed guard dogs; boys and young men were bound and blindfolded so that the rubber traders could win bets as they shot off the genitals. One overseer hung an Indian from a tree, toyed with him as he swung in agony, and then, in order to amuse a colleague, bit off one of the victim's toes, spitting it to the ground. A rubber agent named Aquileo Torres placed a rifle barrel into the mouth of a man and told him, as a joke, to blow. When he did, Torres blew his head off. Another time he blindfolded a young girl, told her to walk away and then shot her in the back. This sadist once sliced the ears off a man, bound him to a tree and forced him to watch as he took a torch to his wife. José Fonseca, the head agent at Ultimo Retiro, celebrated Easter of 1906 by personally shooting 150 Indians with his carbine. The wounded were heaped into a grotesque pile, doused with kerosene and burned alive.

On women, there was open season. Rafael Calderón, a twenty-two-year-old bandit who tethered Indians for target practice and once gave a Witoto child fifty lashes for stealing a loaf of bread, lived by the motto "Kill the fathers first, enjoy the virgins afterward." When a woman refused to sleep with one of his men, Armando Norman wrapped her in a kerosene-soaked Peruvian flag and set it on fire. Other women were kept in stocks for weeks, available to anyone. When the station manager at Atenas discovered that a young Indian girl he had raped had venereal disease, he had her tied to the ground and flogged while a burning firebrand was inserted in her vagina.

As the Indians died, rubber production soared. In 1903 the Putumayo yielded 500,000 pounds. Two years later the figure was over a million. In 1906, at a time when even the scrapings of the tin cups attached to the trees to collect latex fetched a price, the output rose to 1.4 million pounds. In the twelve years that Arana operated on the Putumayo he exported over 4,000 tons of rubber, earning more than $7.5 million in the London market. During that time the native population on the Putumayo fell from over fifty thousand to less than eight thousand. For each ton of rubber produced, ten Indians were slaughtered and hundreds left scarred for life with the welts and wounds that became known throughout the Northwest Amazon as *la marca arana,* the mark of Arana.

• • •

"I have not slept in two nights and have had a touch of dysentery," Schultes wrote on the eve of his journey down the Igaraparaná from La Chorrera to the Putumayo. In fact, he had come down with malaria, and as they embarked from the mission, Nazzareno encouraged him to rest beneath the canvas shelter they had rigged over the middle of the canoe. Schultes declined. He lifted his pith helmet, waved toward the Witoto on the riverbank, and then looked ahead at the broad river.

He stood behind the shelter, just in front of his plant press, which was wrapped in a rubber tarp. Nazzareno sat up ahead. A young Witoto perched on the bow, reading the river. His brother paddled in the stern. Both were strong, content to make their own way back upriver from the mouth. Schultes admired them for the ease with which they moved about the country. It was almost two hundred miles to the Putumayo. In the margin of his collecting book, Schultes had scratched a few notes concerning the trip, the location of Bora and Andoke villages, the distances between the major rapids, the anticipated length of the journey: "large canoe 26 ft., 4 *bogas* (paddlers), 11 days . . . small canoe 8 days." Though the canoe he had bought was over thirty feet long, he would reach the Putumayo in less than a week.

As they passed out of sight of the mission, the Indians steering the canoe into the current, a flush of fever and nausea took Schultes out of the sun and into the shelter, where he lay lengthwise, his head propped on a bundle of bedding. He had no plans to collect on this portion of the trip. With limited supplies and a great distance to travel, there would be little opportunity to set up the dryers. What's more, the canoe was already filled with gear and valuable specimens. There comes a time in every expedition when the initial momentum wanes, and the only challenge is to get home safely with the collections intact. That time had come, and Schultes knew it as he drifted down the river, listening as the silt polished the bottom of the canoe.

He had only one nagging regret: Having inquired in every locality, he still had not managed to find flowering material of *yoco,* the stimulant that had so impressed him on his arrival in Mocoa six months before. From his many sterile collections he knew that the plant had a large natural range. Curiously, not all Indians exploited it. The use of *yoco* appeared to be restricted to a few tribes living along the base of the Andes, the Inga and Siona, the Kofán on the Guamués and Sucumbíos rivers, and the Koriguajés of the Río Caquetá. Though *yoco* grows in Witoto territory, the people at La Chorrera and El Encanto were unaware of its value. Even those Witoto who had moved into the territory of the other tribes showed no interest in the plant. They preferred coca, which they consumed in copious amounts. The Koreguajé enjoyed both plants, but in general the use of the two stimulants was exclusive. For Schultes this observation reinforced his growing conviction that the use of any drug is firmly rooted in culture.

There were four major rapids on the river and dozens of patches of white water, any one of which could topple the overloaded canoe. Focusing on the river, anticipating the currents, watching as the boiling waves broke over the sides, climbing over slippery rocks to steer the canoe across a rocky shelf, portaging the gear around the worst of the cataracts, Schultes abandoned himself to the journey. It was only in the evenings, with the return of his fever, that memories of La Chorrera came back to him.

Two images in particular lingered in his mind. He recalled the shadow of an iron chain growing across the wall of the mission as the sun dropped in the sky. And he remembered the procession of young children, all dressed in white, passing out of the church to celebrate June 6, the Feast of Corpus Christi. It was insane. He thought of the books he had read before going to the Putumayo, Whiffen and Robuchon, the German anthropologist Konrad Preuss. Although all had traveled through the region at the height of the rubber boom, not one had mentioned the atrocities.

They wrote instead of Witoto cannibalism, of lurid accounts of orgiastic Indian feasts, of warriors tearing into the bodies of enemies, consuming hearts, kidneys, livers, and bone marrow. In the twelve days that Schultes spent with the Witoto at La Chorrera, he heard no tales of cannibalism, nor did he see empty skulls hanging from the rafters of houses. Whiffen described Indian rituals as a "mad festival of savagery." It was a phrase, Schultes now realized, that applied not to the Witoto but to the white traders who had, among other things, cooperated with Whiffen and made his journey possible. Now, a generation later, Witoto children parade on the Feast of Corpus Christi, the Catholic holiday that celebrates the mystical transformation of bread and wine into the actual body and blood of Christ, the spiritual sleight of hand that by definition turns holy communion into a cannibalistic rite.

On June 18, six days after leaving La Chorrera, Schultes and Nazzareno reached the Putumayo. Bothered throughout the night by a plague of blackflies and having parted ways with their Witoto guides, they left immediately the next morning for Tarapacá. Floating along, watching islands of vegetation break off from the shore and great thunderclouds gather in the afternoon sky, Schultes waited for the days to pass. The scale of the Putumayo amazed him. A third of a mile wide and only one of a thousand known tributaries of the Amazon, it was nevertheless the largest river he had ever seen.

A week after setting off, Schultes and Nazzareno recognized the Isla Amelia and realized they were within a day or two of the Brazilian frontier. Neither knew what to expect of Tarapacá. Schultes assumed it would be a small village; he was therefore surprised when, on the evening of June 27, they rounded a bend and saw a lone sentinel standing at the mouth of the Río Cotuhé. Only when they saw the Colombian flag drifting in the breeze on a flagpole high up on the riverbank did they paddle toward shore. Had they not seen the flag they would have continued downstream, with the next major settlement Manaus, six hundred miles away.

The soldier was an enormous man of African descent whose accent betrayed his origins on the Pacific Coast. It turned out the entire garrison came from Tumaco, with the exception of the two officers, a lieutenant and a major, both from Bogotá. The major was waiting for Schultes at the top of the steps.

"Buenas tardes," Schultes said. "I am—"

"Yo sé. I know who you are. It is my business to know." The officer spoke with such authority one might have thought he had followed Schultes's progress by military radio, as opposed to the sporadic riverboats that were his only source of intelligence. Without introducing himself, the major spun on his heels and marched toward the compound, a number of shacks perched in a clearing hacked from the forest and connected by wooden walkways.

"Mi capitán," Schultes said. The major froze. "I was wondering if we might catch a plane for Bogotá."

"I am Major Gustavo Rojas Pinilla."

"I beg your pardon, Major," Schultes said. "About the plane, I understand that—"

"There is service twice a month. The next plane is due in a week." As discreetly as possible Schultes asked if there might be a place to stay. The Major walked toward Schultes, looking him over as if he was a fresh recruit. Schultes glanced down and noticed for the first time his bare feet, the tattered legs of his trousers, the khaki shirt quite literally rotting on his back. He apologized for his appearance.

"¿Puede jugar ajedrez?" the Major asked.

"Excuse me?" Schultes said.

"Carrajo, hombre. Can you play chess?"

"Sure," Schultes replied, remembering how much he hated the game.

"Why didn't you say so? They stick me in this hellhole of a jungle with a bunch of *animales* and one white man, and the son of a bitch can't play chess. Evangelista!" A soldier came stumbling out of one of the shacks.

"Mi mayor."

"Take these men to the *bodega* and get them in uniform. Dr. Schultes, come along with me. We have time for one match before dinner."

It was the first of many. For the next seven days, dressed in an oversized uniform of a private in the Colombian army, Schultes played chess from morning until night. He lost every game.

The Catalina flew low over Tarapacá, banked to the west, and fell into the wind. Schultes watched its lumbering flight and then turned in haste to gather his specimens and gear. Until this moment he had been more or less at ease with his circumstances. Now with the plane in sight, he suddenly felt that one more day at the Major's chess table would be his undoing. It was not the longest wait he would experience; that would come a few years later when, stranded on the Apaporis, he would linger sixty-two days for a plane. Still, enough was enough.

The Major had hoped to squeeze a few more days out of a weather front that was rolling in from the north. Nevertheless, he took the Catalina's arrival in good spirits. He and Schultes said good-bye in his office. A few minutes later, as the amphibious plane taxied into position, Schultes peered out one of the small round windows and saw him standing alone at the top of the bank. As the plane accelerated into the current, the Major came to attention with a perfect salute. It would be eleven years before they met again, in Bogotá at the corner of Septima and Avenida Jiménez. Schultes would be buying a newspaper. The Major, by then a general, would be at the controls of a Sherman tank leading four others as part of the military coup that sent the conservative president Laureano Gómez off to Spain and installed himself, Rojas Pinilla, in the presidential palace. Approaching the center of the city and ever respectful of the law, the revolutionaries in their tanks had stopped for a red light.

The Catalina took Schultes only as far as Puerto Ospina. There, waiting for another plane, he found himself once again the guest of Colonel Gomez-Pereira, who used his influence to find him a cabin on the Cañonero Cartagena, a naval gunboat tied up alongside the army post. After six months of almost continuous fieldwork, Schultes welcomed the comfort of the ship. He was thin and exhausted, and the ulcers on his legs, which had been bothering him since the Igaraparaná, had begun to fester. When he bathed he picked off a dozen ticks that dirt had concealed. All he really wanted to do was rest. Instead, clean-shaven and outfitted in army fatigues, he emerged from his cabin and joined the Colonel, his men, and several bottles of good scotch for a raucous celebration, in his honor, of the Fourth of July.

The next morning there was a knock at his door just after dawn. It was one of the marines. Someone was waiting to speak with him. Schultes walked onto the narrow deck, looked both ways, and was about to return to bed when he heard a quiet voice coming from below the side of the ship. Leaning over the railing he saw a small dugout canoe, and in the bow was one of the Kofán he had met on the Sucumbíos. In the man's hand was the foliage of a plant. It was *yoco,* and apparently the liana was in full flower.

"Where did you find it?" Schultes asked.

"Aquicita, no más," the Kofán answered, pointing upstream. Just a little ways.

Schultes smiled. *Aquicita.* That could mean anything. He thought of his infected legs and how little he relished going back into the forest. What if the Kofán had seen the flowers of some other liana intertwined with the foliage of *yoco?* And if it was *yoco* in bloom, would the flowers still be out by the time they found the plant? He hesitated for a moment. Then, his mind having run through all the reasons not to go, he walked into his cabin, grabbed his kit, and climbed over the side of the ship.

After twenty hours of hard paddling and a stroll through a swamp, they reached a place in the forest a mile from the river where the ground was covered with thousands of tiny flowers. Working together they cut down four large trees until finally the enormous liana fell to the ground. It was the plant that he had been hunting since his first days in the Putumayo. A new species, he called it *Paullinia yoco,* using the common name as a way of honoring the people who first discovered its remarkable properties.

Three days later, still elated by the discovery, Schultes flew out of Puerto Ospina for Villavicencio, the first leg of the trip back to Bogotá. Still dressed as a Colombian private, he arrived in the capital in the early afternoon, dropped in at Mrs. Gaul's *pensión* for a change of clothes, and went immediately to the American embassy. This time he did not encounter confusion or indecision. Most of the faces had changed; in its own way this diplomatic outpost reflected the resolve with which America, in the wake of Pearl Harbor, had gone to war. The receptionist at the front desk greeted Schultes as if he had been expected. Within minutes he was escorted into the office of the military attaché. There he was given his assignment. American armies needed rubber, and with Malaysia and the East Indies occupied by the enemy, there was only one place to get it. Schultes's task would be to return to the Amazon, organize the Indians, and gather as much wild rubber as possible.

CHAPTER EIGHT

The Sad Lowlands

THE FIELDS OF wheat and barley lie on a restless land, dissected by snowmelt rivers and streams that cut deep gullies across the flanks of the mountains. At eleven thousand feet the equatorial air is clear and the sky a refined blue, thin and transparent. The houses on the scattered farms are made of dirt and straw, sunk into the ground. There is a great silence on the bare earth, a stillness broken only by herds of wild horses and the winds that fall away from the glaciers that cover the summits of immense volcanoes: Cayambe, Cotopaxi, Chimborazo. There is no place on earth where the sun comes closer to the surface of the planet.

For several days Tim and I had been wandering over this ancient landscape, across the slopes of Cotacachi and along the rim of a crater lake called Cuicocha where five species of yellow calceolarias grow side by side in lush meadows on rich volcanic soil. The mornings were always bright and clear. You could see far down the mountain, past fields of daisies being harvested by women in felt hats, and beyond to the outskirts of Otavalo, an Indian town with phoenix palms in the plaza and red-roofed houses piled up around a small church. On the mountain the shepherds in their mud-stained ponchos did their best to warn us about *la neblina,* the mist that invariably rolled over the summit at noon, condensing into a thick fog that by early evening softened each sound and obscured every landmark. More than once we became lost and found ourselves foolishly following animal paths that went everywhere and nowhere, disappearing into caves and rocky outcrops, stunted groves of polylepus and dense thickets of gleichenia ferns and ephedra. Wet and miserable, we would wait in the dark for a truck to pass on the Apuela road and then chase its shadow as it sputtered and backfired down the grade. It was at moments like these that Pogo came close to going his own way. But always there was another morning and a new sun to wash over the mountain.

We were staying in a shallow draw beside a stream, and late one evening, with the dew turning to frost, I returned to camp with a bundle of dry stems to kindle the fire. Tim was sitting on a flat stone, whittling a cane from a twisted tree branch. He was wearing a red poncho with a high collar, and the light from the fire illuminated the mist that swirled visibly past his head. Pogo was curled up at his feet, nose to the flames.

"Anything happen?" he asked.

"Nothing," I said.

"Me, neither. You probably have to grind them up."

"I just swallowed them whole."

Tim nodded. That afternoon we had collected *Coriaria thymifolia,* an alpine shrub with fernlike leaves and long racemes of purplish black fruits. Farmers in Colombia believe that the plant is poisonous, but according to Schultes the seeds induce an illusion of flight, the sensation of soaring over open ground. We had each eaten a handful to no avail.

"How did he take it?" I asked.

"I don't think he ever did."

"You're kidding."

"He just heard about it somewhere. You know how he is."

"You should have told me."

"Then how would we have known?" Tim laughed.

"You could have said something."

"I knew we'd be okay."

I shook my head and knelt down to tend the fire.

"So where to next?" I asked.

"I'm not sure," Tim said. "Ecuador's a mystery." He lay down his knife, drew his poncho tight to his chest, and turned toward the fire. "There should be coca, but it's never been found. It was definitely here in the time of the Inca. This was one of their richest provinces. Roads ran everywhere. From where we are, they could send messages to Cuzco in eight days. Twelve hundred miles by courier in eight days."

I lifted a pot of tea to the edge of the coals and poured two cups. "So what happened?"

"A cataclysm. Everything was destroyed." Tim teased an ember from the fire and lit a cigarette. "The first great Inca conqueror was Pachacuti. His son Topa Inca Yupanqui took Ecuador and left his heir at Tumibamba. That boy was crowned Inca in 1493, a year after Columbus landed in America."

"Here. Take this."

Tim reached for the cup. The son of Yupanqui, he explained, was Huayna Capac, the Inca responsible for expanding the empire into what is today southern Colombia. Like his father before him, Huayna Capac retired to Tumibamba, now the Ecuadorian city of Cuenca, where he built palaces of the finest stonework, masonry inset with jasper, and temples draped in sheets of gold. The climate was ideal and the land bountiful. As Inca, Huayna Capac lived at Tumibamba for a decade, and it was from there that he first heard of a strange ship approaching the coast from the west. This was Pizarro on his second voyage, in 1527. Only months later smallpox swept into the mountains, coming not from the coast but from the north, from Colombia and the Caribbean, sweeping far ahead of the Spanish, ravaging tribe after tribe so that by the time the Conquistadors touched land, 200,000 had died, including the Inca, whose shrouded body was carried south to Cuzco and the Temple of the Sun.

A battle for succession then convulsed the empire. The rightful heir was Huascar; the usurper was his half-brother Atahualpa. The civil war lasted six years, finally ending in 1532 with the death of Huascar on a battlefield close to the modern Ecuadorian city of Ambato. That same year saw the return of Pizarro with a band of hardened men who would capture Atahualpa at Cajamarca, lay siege to Cuzco, and destroy the empire. With the death of Atahualpa, the presence of the Inca in Ecuador vanished. The royal road that once traversed the highlands, passing beneath the majestic volcanoes and dazzling all who saw it, was broken up for cobbles. Gone were the storehouses and nunneries, the rest houses and palaces, the scores of ceremonial and stately complexes that still give life to the landscape of Peru and Bolivia.

"There is only one surviving ruin," Tim continued, "in the south, near Cuenca." With a stick he cleared away the ashes at the edge of the fire and sketched a map of the country in the dirt. We were in the north, an hour or so from Quito. The Andes formed the spine. To the west was the coastal plain of Guayaquil, to the east the Amazonian lowlands. Three hundred miles would take a bird clear across the country, provided it could soar above twenty thousand feet.

"It's called Ingapirca. There are several ruins, but the main one is shaped like this." He drew a figure in the dirt with straight sides and rounded ends.

"To the Spaniards it was just a mound of stones. They didn't notice that the diameter of each end was exactly one-third the length of the structure. The circumference of the building contains three circles of equal size, aligned precisely on an east-west axis."

Ingapirca, Tim explained, may well have been a solar observatory, with each of the circles representing positions of the sun: *Anti* for dawn and the east, *Inti* for the overhead sun at noon, and *Cunti* for the evening and the west. By observing the shadows cast at dawn and then again at midday by a piece of rope stretched the length of the platform, the priests could measure the movement of the sun through the heavens and thus ascertain the timing of the equinox and the moment of the solstice. The ritual cycle of the Inca revolved around these astronomical events, which in turn profoundly affected daily life, determining the onset of agricultural seasons, the proper time to wed or give birth, a propitious day on which to die. Reverence for the sun and the moon, together with a complex understanding of the esoteric significance of the four directions, provided the foundations of the Inca world. They called the empire Tawantinsuyu. *Tawa* meant four; *suyu* was region or quarter. Fusing the two concepts is *ntin,* the principle of unity embodied by the Inca himself, the mediator between the living and dead, man and the gods, human society and the land. Coca was integral to the cosmology. It was the sacred leaf, the most revered plant in the empire.

"The earliest evidence of coca in all of South America is here in Ecuador on the coast at Valdivia: lime pots and figures of coca chewers dating to at least 2000 B.C. And the same pottery, or something very much like it, turns up in the highlands outside of Otavalo at San Pedro, so there was definitely trade. Coca thrived on the coast, and it almost certainly was brought into the mountains, and very early."

Tim got up slowly and made his way toward the back of the truck. He returned with a handful of papers.

"Listen to this." He crouched beside the fire and turned toward the light. " 'Coca leaves,' " he read, " 'are extremely strengthening and provide a food of incredible virtues, since the Indians, with no other provisions beyond these leaves, make journeys lasting for weeks, and appear to grow stronger and more vigorous every day. There is a trade in this plant in nearly all parts of the country.'

"That was written by a Jesuit in Quito in 1789. So here's the curious thing. We know that coca was everywhere at the time of the conquest, and it was like that for two hundred years. Then somehow and for some reason the plant disappeared, but only from Ecuador. No one knows why." He reached for his knife. "In all the literature there are just a handful of reports of Indians still using it—the Colorados at Santo Domingo and here among the Kofán." He pointed on the ground to the west and then to the east side of the Andes.

"So where are we going first?"

"To the Kofán," Tim said. I smiled and threw a bit of wood onto the fire. Pogo snapped awake and watched the sparks spin away into the night.

Leaving Quito by road, Tim and I traveled east, climbing through driving rains and violent winds, past beautiful elfin groves of polylepus trees hung with fuchsia, to a 13,400-foot-high pass that looked out over sparse fields dusted with snow. Across the divide the road fell away and began a long descent into the cloud forests of the Río Papallacta and beyond toward Baeza, a small mission outpost perched on a rise in the Quijos pass, the traditional route to the Amazonian lowlands.

The rains were the heaviest the mountains had seen in a decade, and the road, built only four years before to service the oil fields to the east, gave way at a dozen points. Tim naturally took advantage of each delay, scrambling onto the pipeline that ran alongside the road and following it into the cloud forest. In vegetation so dense it was sometimes impossible to find the ground, we collected wild begonias and salvias, delicate ferns and oxalis, boehmerias and at least three unknown plants, a new species of *Dalbergia,* a new *Centropogon* with a brilliant red corolla, and an exquisite aroid later named for Tim, *Caladium plowmanii.*

Until the late 1930s there were no roads across the Andes in Ecuador. Then explorers for Royal Dutch Shell found asphalt oozing out of cliff formations and discovered pools of oil floating just six hundred feet beneath the surface of the eastern foothills. A notoriously corrupt government granted the company exclusive mineral rights to the entire Oriente. In 1938, Shell blasted a route along the cliffs of the Pastaza canyon to a staging area in the *montaña* that became known as Shell-Mera. From there crews advanced into the forest to drill test wells and carve out airstrips.

Meanwhile, farther south, engineers working for Standard Oil of New Jersey were exploring the frontier regions of Peru. War broke out in 1941 between Ecuador and Peru, with each side backed by its respective oil partner. A year later the United States imposed a peace treaty that obliged Ecuador to sign away almost half of its national territory, lands that today contain the bulk of Peru's oil reserves, including one field currently pumping 100,000 barrels a day. The concession owned by Shell proved less productive. In 1949, after investing over $40 million and losing fourteen of its oil workers to spearing raids by Indians, Shell abandoned its Ecuador holdings without pumping a gallon of commercial oil.

Less than twenty years later Texaco hit pay dirt on the banks of the Río Aguarico, only 120 miles north of Shell-Mera. Beginning in 1970, from a base established at Santa Cecilia, oil drills spread throughout the forest and construction began on the 315-mile pipeline that now traverses the Andes, reaching the Pacific at Esmeraldas. In time the Napo field would pump 275,000 barrels a day, elevating Ecuador into the ranks of OPEC and accounting for over 60 percent of the country's export earnings. The center of the field was Lago Agrio, a town established in 1970 and named for Sour Lake, Texas, the site of Texaco's first gusher back in 1902.

Tim and I rolled into Lago Agrio late in the afternoon, just as the eastern sky began to glow with the orange light of flaming natural gas. Although the town has now become the fastest-growing center in the Oriente, then it was little more than a row of clapboard cantinas and brothels stuck up against the fresh paint of the Hotel Oro Negro. We took a room across the road at the Hotel Utopia. Taped to the wall behind the desk was a notice from the Ecuadorian Ministry of Health announcing the weekly inspection schedule for the women. Right beside it was a list of names—Maria, Suzie, Beatrice—and a price list in U.S. dollars: five dollars for twenty minutes, fifteen dollars for the night.

The young boy who led us to our second-floor room flung open the shutters, and all the clutter and noise of the street tumbled into the room.

"What do you think?" Tim asked.

"It has a good view." I looked out on shopkeepers sleeping by open sewers, women in hot pants and platform shoes struggling through the tar and gravel, and highland Indians wrapped in wool in the stifling heat. The traffic of heavy diesel trucks was constant. Storefronts blasted salsa, and red pickups lined up in front of country-western bars. Down the road you could see Texaco's fenced compound and a row of trailers, shiny and clean. Across the way, beyond the Hotel Oro Negro, shimmered the silhouette of the distant rain forest.

"I guess it isn't what I expected," said Tim.

"What do you mean?"

"Utopia."

"No."

"Look down there." He pointed to a pair of money traders and a barefoot Indian bent double beneath a load of cinder blocks.

"He's Kofán," Tim said.

"He's wearing bellbottoms."

"Ten years ago this was all forest. Texaco dropped its operation right in the middle of their land."

We left Lago Agrio in the morning, headed for the Río Aguarico, and found that the bridge had been torn and twisted by the recent floodwaters. In some haste we transferred our gear to a barge that ferried traffic across the river. As we pulled away from the shore I turned suddenly to Tim.

"Where's Pogo?"

Tim looked aghast. Standing on the riverbank was a frantic dog who clearly had no interest in being abandoned in Lago Agrio. We watched helplessly as Pogo leaped into the river and was promptly swept away by the current. I was certain he would drown. As soon as the barge touched the far shore Tim raced down a sandbar, yelling his name. After an agonizing few moments a pair of brown ears popped up in the white water. Pogo struggled to shore just a few feet above a stretch of rapids that no doubt would have killed him.

The crisis resolved, we returned to the landing and hired a dugout to take us two hours downriver to the small Kofán village of Dureno. There we unloaded our gear, climbed the bank to the high ground, and made contact with Basilio, whose name had been given to Tim by Homer Pinkley, a student of Schultes who had spent a year among the Kofán in 1965. Basilio, a gentle and retiring man, found us a place to stay and offered to accompany us in the forest. For the next three days we played a lot of soccer, collected in a desultory fashion, and generally waited for a diplomatic moment to leave the community.

The problem was neither the forest nor the people but rather the overwhelmingly depressing prospects of their situation. To visit Dureno was to bump into the history of South America condensed in a single generation. As recently as 1953 the Kofán of the Aguarico, though living only 120 miles by air from Quito, remained completely isolated. The first sustained contact did not occur until the following year when two Protestant missionaries arrived by river and built an airstrip. Their names were Bub and Bobbie Borman, and their goal was to translate the New Testament into Kofán. They came in the rainy season, found the entire tribe awash in *chicha,* and discovered to their dismay that the only time the Kofán did not drink *chicha* was on the not infrequent occasions when they took *datura* or *ayahuasca,* as *yagé* is known in Ecuador and Peru. The Bormans did everything they could to discourage such practices. Still, a decade later, when Homer Pinkley came to the village, the old men gathered as they always had: in a small hut across the river where, dressed as visions, they ingested their sacred plants and conversed with spirits known as the "heavenly people."

In his unpublished thesis Pinkley describes a remarkable day in the life of the Kofán he knew. A chief had died, strangled at an oil exploration camp. For several weeks the people had taken *ayahuasca* as they tried to decide whether to move their village away beyond the reach of the foreigners. On the morning of May 28, 1966, a jaguar appeared at the edge of the forest. For the rest of the day, in a heavy rain, the men tracked the animal, finally killing it at dusk, a deed that evoked both admiration and fear in the village. That night twenty men and boys, adorned in feathers and palm fronds, came together in the ceremonial house to take *ayahuasca.* Every moment of the hunt, each gesture of the jaguar, was analyzed and discussed. Incense glowed in the darkness. The conversations faded in and out until finally, just before dawn, the shaman's song rose above the fire. The people chose to stay where they were, where they had been born.

In the morning Pinkley examined the *ayahuasca* cauldron and found three seeds of *oprito,* an admixture he later identified as *Psychotria viridis,* a plant in the coffee family containing small amounts of several tryptamines. Though a specimen of the plant had been collected a decade before by William Burroughs, of all people, this was the first documented account of its use in sacred ritual. It was one of the most significant ethnopharmacological discoveries since Schultes himself, while living among the Ingano on the Caquetá in the spring of 1942, had stumbled upon *chagropanga,* the tryptamine-containing liana also used by the shamans to enhance the brilliance of the *yagé* visions.

Homer Pinkley was the first botanist to reach the Río Aguarico. In his thesis he notes rather casually that while he was among the Kofán anthropologists came upon a completely unknown tribe living nearby. By the time of our visit, less than a decade later in 1974, the shaman whom Pinkley had worked with was dead, and his son had taken a job with Texaco. Roads had replaced hunting trails. References to *ayahuasca* provoked giggles and the admonition that it was a plant of the devil. The Kofán knew coca and had a name for it, *itifasi sehe,* but there was no evidence that they had used it in years.

The night before leaving Dureno, Tim and I found ourselves on the riverbank, lying in the middle of a spreading meadow of moon flowers, a morning glory with large white blossoms that open in the evening with a speed readily seen with the naked eye. Pollinated by moths, the flowers have a long tubular corolla and a sweet aroma that hovers low over the ground throughout the night. Just before dawn the blossoms wilt. By midday they have turned brown and have begun to rot. In the moment the plant's ephemeral beauty seemed symbolic of so much of what we had seen in South America. At one time there were thousands of cultures around the world and probably as many as fifteen thousand languages, each like a flash of the human spirit. Today perhaps six thousand are still spoken. I mentioned this to Tim.

"More than half are gone," I said. "In a century only a few hundred will remain, a few hundred out of thousands."

There was a chill in the air, a cool night wind blowing from the north. We sat still and said nothing. You could hear the flow of the river, the distant calls of night birds mingling with the sound of frogs and cicadas, a canoe rubbing up against the shore, dogs barking. Tim stood up and moved to the edge of the meadow.

"Kofán," he said, "will not be one of them."

From the summit of Chimborazo, the highest point in Ecuador, to the tropical plantations that blanket the western lowlands, the distance is only thirty miles. The vertical drop is more than twenty thousand feet. The handful of roads that tumble to the coast have gradients steeper than any in all of South America. We chose a route that began in the clouds on a rolling *páramo* cradled by the flanks of two extinct volcanoes and then fell precipitously into the narrow gorge of the Río Pilatón, eventually bringing us to Santo Domingo, a bustling commercial center and road hub founded in the territory of the Colorado Indians. The only evidence of Indian life was in the town plaza. There, beside a box camera and a placard noting the price of a photo, stood an old man, face painted with black stripes and hair sculpted and encrusted with brilliant red *achiote* dye. Tim passed through town without stopping. Heading south toward Quevedo, we drove past wooden signs advertising the services of Colorado *curanderos,* again with price lists, and endless plantations of bananas, oil palm, and papaya. In a hundred miles even Tim could not find a plant worth collecting.

From Quevedo we turned east toward the mountains, climbing in a morning from the lowlands to a high pass that fell away to a spectacular valley of snow-capped peaks and rugged cliffs. It was cold country, and the people lived in round houses, low to the ground and covered in thatch. We stopped often, and with each passing mile and each passing field the people appeared poorer, the lilt of the language softened, and for the first time we found ourselves addressed as "your grace" by men and women who still refer to their landlords as *mi patroncito,* my dear master.

After a windy night on the *páramo* we made our way to Quilotoa Crater, a dormant volcano with a pure emerald lake filling the bottom and a view beyond of Cotopaxi, Ecuador's second highest mountain. Late in the afternoon, having made camp on the lip of the crater, Tim and I followed Pogo over the edge, down a narrow path that zigzagged across a steep slope and dropped toward the lake below. It was already cold, and both of us were wearing new woolen ponchos. Tim had brightly colored ribbons tied to his waist and a red bandana around his hair. With a collar of dangling beads, even Pogo looked flamboyant.

The land inside the crater was dry and rocky, and seemed impossible to till. Yet every inch of it was divided into fields, and where no crops grew, goats and sheep grazed on roots and stubble. Tending the herds were the people of the crater. As we approached, a dozen of them rose to meet us. All were the size of dwarfs, stunted and inbred. Several clasped stones in their hands. After a confused conversation, we gave them a little money and made an awkward retreat. As we turned to head back up the side of the crater, I heard one of them say, *"El perro también es gordo."* Even the dog is fat.

The next morning had us once again on the road, heading for the central valley and south beyond the cities of Latacunga and Ambato to Baños, the gateway to the Pastaza gorge and the road back across the Andes to the Amazon. Following the route pioneered in the thirties by the engineers of Shell, we reached Shell-Mera, the expediting center long abandoned by the company.

Having crisscrossed the country for a month without success, Tim no longer expected to find coca in Ecuador. Our journey to the Kofán had proved only that the world that had inspired Schultes and led Jorge Fuerbringer to direct us to the Río Aguarico no longer existed. Faced with disappointment, Tim retreated into plants. For a week we stayed at the end of a dirt road in the shacks of a derelict farm and explored an astonishingly rich lowland forest. Little was in flower, but that did not deter Tim. He was on a ginger kick, and we collected several fascinating species, mostly *Renealmias* and *Costus,* as well as another new species of *Calathea* and a number of beautiful *Heliconias.* We also found a new species of tree, a melastom in the genus *Blakea* with a lovely pale green calyx. Only once did Tim become disoriented. I came upon him standing at the foot of a tree, staring in all directions, his hand clinging to a specimen, a liana in the potato family, a plant so rare that he initially thought it not only a new species but a new genus. Tim was still enthralled by his discovery when, tired and hungry, we found the trail back to the shelter, made our way into town, and met in a restaurant the drifter who first told us of the fate of the five missionaries.

Our new friend was a Canadian named Alberto who lived with his family up the road in the girl's dormitory of an abandoned school in Tena. Tall and lanky, he had long black hair and a mouth full of gold teeth, which left open the possibility that he had actually been a dentist as he claimed. Apparently he still pulled the odd tooth, particularly for the families of the mayor and local officials, but mostly he survived on the meager revenue of a pair of gold claims. He also was working on a bridge project. Asked what his job entailed, Alberto replied, "Hassling, bribing, conning, and flattering all the parties involved. That's the only way anything gets built. The mayor raked in ten million sucres before they finally put the son of a bitch in jail."

Clearly an educated man, with a wry sense of humor, Alberto had given a lot of thought to the situation of the missionaries. In fact, when he first came into the valley six years before, he had tried to become one.

"It's the Oriente's only growth industry besides oil, and that's not going to last forever."

Needless to say, the missionary organization rejected his application.

"It was a long shot," he said with a shrug. "But you have to understand they have a good scene. And if they're right, if God is a Christian God and Jesus is the only route to heaven and all who do not know his word are condemned to burn in hell and all who follow any other way are doomed, then everything they do is pretty incredible. I mean, what generosity! What sacrifice!"

Alberto reached across our table and helped himself to some beer.

"But," he said, "if they are wrong, if religion is a metaphor and the nature of God is unknowable, if every sincere seeker is deserving of revelation and every religion is by definition legitimate, then what those five did was an act of sheer folly."

I glanced at Tim, who was listening carefully.

"What I'd like to know," Alberto said, "is what was going through their heads when they saw the first of them die."

Alberto proceeded to tell us what he knew of the history of the tragedy. It was a thin sketch but one that stayed with me until I had a chance to complete the story myself.

In the early 1950s the isolation that enveloped the Kofán also insulated the other indigenous peoples of the Oriente. To the north of the Río Napo lived the Siona-Secoya. In the south of the country, beyond the Pastaza and on both sides of the Cordillera de la Cutucú, were the Shuar and Atshuar, two related but antagonistic Jivaroan peoples notorious for shrinking the heads of their enemies. Along the flank of the Andes were two Quichua-speaking cultures, the Quijos Quichua and, farther south, the Canelos Quichua. With a population of more than thirty thousand, the Quichua were and are by far the most numerous Indians of the Oriente. They call themselves Runa, "the people," and address each other as Alaj, meaning "mythic brother." Living at the foot of the mountains, with a long history of sporadic contact with the Spanish, the Quichua have historically been a conduit for the flow of trade goods and information from the highlands to the more remote tribes living in the forests farther to the east. For the most part the interactions have been peaceful, with one notable exception.

Only sixty miles to the northeast of Shell-Mera lived an unknown people whom the Quichua called Auca, a pejorative word meaning "savage" or "barbarian." No one knew how many they were or exactly where they lived, but the Auca controlled eight thousand square miles of forest, some 7 percent of the national territory, and were the most feared Indian group in Ecuador. Traders and adventurers who drifted into their land often did not return. In 1947 a disastrous attempt by a Swedish explorer, Rolf Blomberg, to contact the tribe ended in a bloody ambush. Five years earlier Auca men had speared to death three oil workers at Arajuno, one of Shell's exploration camps. A year after that eight other employees were killed. The death of three more workers, again by spearing, and the ensuing panic among the local *mestizo* and Quichua population provided one of the incentives for the company to pull out of Ecuador in 1949. As late as 1957 there had never been a peaceful contact between the Auca and the outside world.

Even before Shell abandoned the Oriente, American missionaries—many of them former servicemen searching for new adventure—had set up shop in the company's facilities. By 1954 there were twenty-five evangelists working out of nine stations, all interconnected by short-wave radio and supplied by air by the pilots of the Missionary Aviation Fellowship. Founded by Nate Saint, an ex–Army Air Corps mechanic, and based at the old company airstrip along the Río Pastaza at Shell-Mera, the M.A.F. had achieved a certain notoriety in Ecuador by refusing to allow Roman Catholic clergy the use of its planes and services. The rivalry between the Italian and American missionaries in particular had led one official to question whether any of them realized that the war in Europe had ended.

For the Americans a beacon of light in the Oriente was an old Englishman, Wilfred Tidmarsh, who with his wife had been "in the fields of the Lord" in Ecuador for twelve years. In September 1951, with his wife seriously ill, Tidmarsh received a letter from Pete Fleming, a twenty-three-year-old aspiring evangelist associated with the Plymouth Brethren. Fleming, and his friend Jim Elliot, a recent graduate of Wheaton College, a Bible school outside of Chicago, proposed to re-open the mission station among the Quichua that Tidmarsh had been forced to abandon because of his wife's condition. The motives of the young missionaries were simple. "I dare not," Elliot confided in a letter to his parents, "stay home while Quichuas perish." Elliot had grown frustrated by his "sterile days" on the evangelical circuit in the United States: youth rallies in southern Illinois that provoked at best a tepid response. "I do not understand," he wrote, "why I have never seen in America what missionaries write of—that sense of swords being drawn, the smell of war with demon powers." Tidmarsh accepted their offer, and late in the summer of 1952, after several months of language training in Quito, Elliot and Fleming found themselves living among the Quichua at Shandia on the banks of the Río Napo, thirty miles to the north of Shell-Mera.

The decision by Elliot and Fleming to devote themselves to the Quichua was a small part of a wave of evangelical fervor that swept over the Oriente. Nate Saint and his wife, Marj, had arrived at Shell-Mera to establish the Missionary Aviation Fellowship in 1948. The Summer Institute of Linguistics, the field arm of the Wycliffe Bible Translators, came in 1952. In 1953 the Gospel Missionary Union penetrated the territory of the Shuar. A year later one of its members, Roger Youderian, a twenty-nine-year-old former paratrooper, successfully contacted the Atshuar and established a station at the abandoned oil camp at Wambini. At Dos Rios there was a group called the Christian and Missionary Alliance. Other evangelists represented the Moody Bible Institute, the American Bible Society, Christian Missions in Many Lands, and HCJB, a Christian radio station based in Quito. In 1954 the contingent of the Plymouth Brethren was reinforced by the arrival at Shandia of Ed McCully, a college friend of Jim Elliot's from Wheaton.

The most potent of the new evangelical forces was the Summer Institute of Linguistics. Founded in 1936 by former Bible salesman William Cameron Townsend, with the goal of translating the New Testament into every language, the S.I.L. was on its way to becoming the most extensive linguistic organization in the world. By the 1970s it had 4,300 missionaries at work in more than seven hundred cultures, with a new language being embraced every eight days.

In 1952 the S.I.L.'s first representative in the Oriente was Rachel Saint, sister of the pilot Nate Saint and arguably the most formidable missionary the forests of eastern Ecuador would ever see. As a young woman of eighteen, Rachel had a vision of living with a tribe in a green jungle. She later spent twelve years drying out drunks in a mission in New Jersey before joining Wycliffe in 1949 at age thirty-five. They sent her to Peru, where she lived for two and a half years among the Piro and later the Shapra Indians. Periodically she would travel north to Ecuador to visit her brother, and it was on one of these trips that she first heard of the Auca. The very idea of an uncontacted people so fierce that even her brother admitted flying around and not over their territory sparked not only evangelical zeal but a quiescent sibling rivalry that in the end left one of them dead. Even before Wycliffe was established in Ecuador, Rachel Saint had decided that the Auca would be her tribe.

The others were not so sure. After many months in the field even the most ardent of the missionaries were becoming disillusioned by the resistance of the Indians to the word of God. In two and a half years among the Shuar and Atshuar, Roger Youderian had made so little headway that he had fallen into a spiritual crisis and was ready to go home. At Shandia and the other Quichua mission at Puyupunga, Jim Elliot and the others had come to regard the Quichua as children, constantly playing off the Protestants against the Catholics. After two years of concerted evangelical effort, the Quichua still considered their dreams a source of inspiration, still maintained that learning involved both knowledge of the outer world and understanding of inner realms revealed in visions by sacred plants. Their shamans were still strong. Even those who did convert, who volunteered their labor to the church, dressed in western clothing, recited Christian prayers, and sang on Sunday, reverted to their former ways as soon as they returned to the forest.

Increasingly frustrated by their efforts among the Quichua, the missionaries were drawn to the pure challenge presented by the Auca, "an unreachable people," as Pete Fleming put it in his journal, "who murder and kill with extreme hatred." The desire to lead such a savage group out of the darkness rekindled their faith in the simplicity of the world. By September 1955, Ed McCully and his wife, Marilou, had moved to Arajuno, an abandoned oil camp located on the edge of Auca territory twenty minutes by air from Shell-Mera. Jim and Betty Elliot were at Shandia, Pete and Olive Fleming at Puyupunga. The Youderians, Roger and Barbara, were still in Shuar territory at Macuma. Nate and Marj Saint maintained the communications base at Shell-Mera.

Life for the McCullys was especially tense. It was at Arajuno that the Auca had speared three Shell employees in 1942. The mission Quichua were so afraid of another attack that they never spent a night by the airstrip on the Auca side of the river. When Nate Saint flew in with supplies, Ed McCully and the Indians gathered to unload the plane with guns drawn. To relieve the pressure, the missionary families often gathered at Nate and Marj Saint's house at Shell-Mera. There they chatted and prayed, and on evenings when the wind blew the clouds off the mountains, they watched awestruck from the deck as streamers of red light exploded like fireworks from the summit of Sangay, an active volcano forty miles away. On one such evening, in October 1955, as they drank hot chocolate and studied maps spread on the living room floor, Nate Saint, Ed McCully, and Jim Elliot decided to initiate what they called Operation Auca, the first peaceful attempt to contact the unknown tribe.

The idea had been on their minds for some time. On September 19, Ed McCully and Nate Saint had flown over the Nushiño river and located an Auca clearing fifty miles to the east of Arajuno. Ten days later Saint, flying with Elliot and Fleming, came upon several large Auca houses fifteen minutes by air from the McCullys' home. They swore each other to secrecy and code-named the site Terminal City.

Their plan was to initiate a gift exchange by using a clever drop technique pioneered by Nate Saint. He had found that if you let out 1,500 feet of rope with a canvas bucket attached to the end and then put the plane into a tight turn, the drag of the rope would overcome the centrifugal force tending to throw the bucket outward. As the Piper Cruiser described a slow circle in the air, the bucket moved to the center, eventually hanging motionless at the bottom of the vortex.

While circling far above the forest canopy, Saint could drop a bucket at the foot of a man standing more than 1,000 feet below. By substituting a telephone line for rope and concealing a speaker in the bucket, they could have a conversation. On October 6, 1955, Saint placed a small aluminum kettle, a bag of salt, and twenty brightly colored buttons on the ground at Terminal City.

Between October and December, Nate Saint made fourteen additional drops at the clearing. On October 14 the missionaries saw their first Auca: half a dozen agitated and naked men gathered around a gift machete. On the fourth flight they rigged a battery-powered loudspeaker in the basket, and Saint flew low to the ground as Jim Elliot shouted the few words of Auca he had picked up from the Quichua: "I like you! I am your friend! I like you!" On the sixth drop the Auca responded, tying to the line a basket containing a feathered headdress. After that there was an exchange of goods on every flight, the missionaries giving clothes, machetes, pots, ax heads, and the Auca returning cooked fish, live parrots, and monkey meat. At each exchange the Auca, as Elliot put it, acted like "women at a bargain counter."

By this time events were moving quickly. At Terminal City the Auca had cleared land, built platforms to direct the plane, and placed a three-foot model of it on top of one of the shelters. Jim Elliot returned from one of the flights certain that they had waved at him, beckoning him to land. He was ready for ground contact. "God, send me soon to the Aucas!" he exclaimed that night in his journal. The pressure was indeed on. The Quichua at Arajuno were no fools. On November 26 they confronted the missionaries, demanding to know why the Americans were giving to the Auca what they had to work for. On the next flight Nate Saint reported seeing Quichua strip off their clothes and prance around with long sticks, hoping that the missionaries would drop some gifts on them. Clearly it would not be long before word of the attempted contact spread throughout the Oriente. It was not just interference from the Ecuadorian military that the missionaries dreaded. Their bigger concern was Rachel Saint.

Convinced that it was her destiny to contact the Auca, Rachel Saint had been living since February 1955 at Hacienda Ila, a farm carved out of the forest within a day's walk of the mission at Shandia. The land belonged to Don Carlos Sevilla, a legendary figure in the Oriente and a man with more personal experience of the Auca than any living Ecuadorian. In twenty-six years he had been speared six times. In 1914 the Auca killed eight of his workers on the Río Curaray. Five years later fifteen of his rubber tappers were speared on the Tzapino. In 1925 Sevilla himself was attacked twice in four months. The worst encounter occurred at an ambush on the Río Nushiño. Five Quichua died immediately. Sevilla and one other fought their way through a hail of spears, killing two attackers before being severely wounded. Eight days later he crawled onto his hacienda at the Ansuc River. Yet another attack in 1934 finally drove him out of Auca territory.

What brought Rachel Saint to Hacienda Ila was the opportunity to learn the Auca language from a young captive woman named Dayuma who had fled her homeland in the summer of 1947. Christened Catherine by a Catholic missionary, she had worked for eight years as a field hand in exchange for her food. Pregnant by one of Sevilla's sons, she was hooked up with a Quichua named Padilla and sent off to another farm. When that husband died, she returned to Hacienda Ila, where she continued to work as a slave for her patron, the grandfather of her son. The boy's name was Ignacio Padilla, but not for long. Ignacio was far too Catholic a name for a boy Rachel Saint intended to convert. She insisted it be changed to Sam and in due course it was.

Rachel Saint knew nothing about Operation Auca. Had she, reports Betty Elliot in her memoirs, "she would have put in so many obstacles that it would have been impossible." She was, Elliot continues, "very possessive." In the fall of 1955 while Rachel was away on leave, Jim Elliot went to Hacienda Ila and obtained a copy of Rachel's Auca word list from Dayuma. Meanwhile, the flights continued. On December 10, Nate Saint identified "Palm Beach," a stretch of sand four and a half miles from Terminal City where he could land his plane. That same day the missionaries dropped from the air four large photographs of themselves, each bearing the insignia of the operation and a drawing of a small yellow airplane. The gestures and smiles appeared, by our standards, friendly and benign. In his journal Nate Saint wrote, "What wouldn't we have given to see those boys studying our pictures and see their reactions." Had he been able to do so, he might have been less enthusiastic. Living in the forest, the Auca had never seen anything two-dimensional in their lives. They held the photographs and looked behind them to try to find the form of the image. Seeing nothing, they concluded that the portraits were calling cards from the devil.

By mid-December the missionaries knew that it was now or never. The Quichua were on to them. The Auca, doing a little scouting of their own, had turned up at the mission at Arajuno. One violent confrontation could ruin the entire operation. Within the month floodwaters of the rainy season would wash over Palm Beach and eliminate access by plane. The ideal time for "the establishment of their beachhead in Auca territory" would be early January during the full moon. Nate Saint spent a melancholy Christmas saddened by "these two hundred silent generations who have gone to their pagan graves without a knowledge of the Lord Jesus Christ. . . . These have no Christmas! . . . May we who know Christ hear the cry of the damned as they hurtle headlong into the Christless night without ever a chance!" Writing home to his parents, he spoke of their desire to "charge the enemy with all our energies in the name of Christ."

Every aspect of the operation, including the rhetoric, was conceived in military terms. In recommending Roger Youderian to the group, Nate Saint described him as a true soldier of Christ. "He knows the importance of unswerving conformity to the will of his Captain. Obedience is not a momentary option; it is a diecast decision made beforehand. He was a disciplined paratrooper. He gave Uncle Sam his best in that battle and now he is determined that the Lord Jesus Christ shall not get less than his best. Everything that made him a good soldier has been consecrated to Christ, his new Captain!" Once recruited, Youderian did indeed organize the team with military precision, assigning specific responsibilities to each member, breaking down the operation into hourly and daily objectives, determining code words for each phase of the process.

On January 3—D-day, as they called it—each man checked his equipment on the airstrip at Arajuno. They pulled straws to see who would be left alone at Palm Beach following the initial flight. The families then ate breakfast and prayed. At 8:02 A.M., Nate Saint and Ed McCully were airborne. Fifteen minutes later they landed on the sand by the Curaray. There was no sign of the Auca. Five flights later all the gear was in place, and work was under way on the prefabricated tree house where they would sleep. On the last flight Nate Saint passed over Terminal City and with a loudspeaker invited the Auca to "come tomorrow to the Curaray."

The next day was anticlimatic. Alone on the sand, the men fished and made lunch, cleared a few trees at the end of the runway, drank lemonade, and ate hamburgers. Wearing shorts and a pith helmet, Nate Saint spent much of the afternoon in the river reading *Time* magazine. Jim Elliot wandered up and down the beach preaching to the forest. On Thursday the men reported by radio, "All's quiet at Palm Beach." It was not until the next morning, Friday, January 6, that the missionaries heard a booming voice in the forest. Jim Elliot waded across the river and met an Auca man, a young girl, and an older woman. They immediately nicknamed the man George, the young girl Delilah.

George was naked, save for a G-string that tied his penis up against his belly. His hair, cut across the forehead in bangs, hung past his shoulders. He had no eyebrows, and in the lobes of his ears were large bright white disks of balsa wood. The hair of the women was cut in the same style. All three jabbered away in Auca, quite unaware that none of the missionaries could speak the language. Gifts were given and photographs taken. The Indians were introduced to balloons and yo-yos. Delilah looked at the pictures in *Time.* George ate hamburgers and had his back doused with insect repellent. He was given a shirt and late in the day, after Delilah had rubbed her body up against the fabric of the plane and imitated its movements through the air, he was invited for a flight. While Nate Saint banked the plane over Terminal City, George leaned out the door, fearlessly imitating the gestures of the missionaries he had studied so carefully from the ground.

Once back at Palm Beach, the missionaries knelt in prayer, arms uplifted to the sky. George was not impressed. He wanted another ride. Delilah followed Jim Elliot to the tree house and returned disappointed. She appeared petulant and stalked off down the beach. George followed her into the forest. The older woman remained by the cooking fire, talking nonstop at Youderian. When he finally retired to the tree house, she spoke to the stars.

That night Nate Saint and Pete Fleming flew back to a quiet celebration at Arajuno. Rachel Saint was by then at Shandia, still unaware of the operation. The next day there were no Auca visitors to Palm Beach, but on an afternoon flight Nate Saint spotted George back at Terminal City. From the air he appeared agitated. On Sunday morning, January 8, Saint and Fleming left Arajuno with a special treat of blueberry muffins and ice cream for the men. Passing over Terminal City they saw no sign of the Auca. Certain that the Indians were on their way to Palm Beach, Nate Saint glanced at his watch and radioed Shell-Mera. It was 12:30 P.M. "Looks like they'll be here for the early afternoon service. Pray for us. This is the day! Will contact at four-thirty." At 3:12 P.M. the watch was smashed by a stone and the hands stopped moving as the body that wore it slipped beneath the muddy water of the Río Curaray.

The first corpse was seen from the air three days later, a quarter-mile downriver from the beach. A second was spotted in the sand about two hundred feet from the camp. The fabric covering the plane was torn off, the wings reduced to a metal frame. When the search party reached the site late on Thursday, they found another victim caught under the branches of a fallen tree, with only a gray foot protruding above the surface of the water. The remains of four of the missionaries were identified. Each had been speared. The body of the fifth, Ed McCully, had been found the day before by the Quichua.

Word of the killings flashed around the world. *Life* magazine had Cornell Capa on the scene by Friday, just in time for him to photograph the discovery of the bodies. Wrapped around one of the spears taken from Nate Saint's body was a Gospel tract that he had dropped from the air on Terminal City. His camera was found at the bottom of the river and the film was developed; it revealed that the last photograph taken was a portrait of Delilah. In one hand she clasps a number of gifts, and in the other a large paper cup, no doubt containing lemonade.

When word of her brother's death reached Rachel Saint, her first reaction, by all accounts, was not grief but anger at the thought that he had dared to contact the Auca without her. Her eight-year-old nephew Stevie had a different response. In a radio broadcast he explained, "I know why Daddy got to heaven before we did—because he loved the Lord more than we did."

Seven years after Tim and I met Alberto in Mera, I traveled alone to the land of the Auca to see what two decades of contact had brought. By then Tim had moved from Harvard to the Field Museum in Chicago, I had become a graduate student under Schultes, and the Oriente had become transformed in ways that even Tim and I could not have anticipated in 1974. Lago Agrio had become a major commercial center, with three flights a day to the capital and buses leaving every three hours for Conejo, the Kofán village on the Sucumbíos that had so impressed Schultes in 1942. Throughout the lowlands, land that belonged to Indians was being given away to colonists for the price of a survey.

Within an hour of arriving at the sprawling missionary base of Limoncocha I had transferred my gear to a small Cessna and was once again airborne, heading southwest, across the oil-stained Napo River and over the traditional territory of the Auca, a tribe I now knew by its proper name, Waorani, meaning "the people." As the plane rose above the heat, falling through rolling banks of clouds, dipping and darting like a dragonfly, the pilot, a garrulous fellow from Kansas, called out the names of the rivers—Indillana, Tiputini, Tivacuno, Cononaco. North of the Napo the forest is relatively flat. To the south the palm swamps give way to a series of broken ridges and hills that run up against the Andes and fall away to the east, running parallel to one another and rising six hundred feet or more above the floodplains of the rivers. From the air, the land and forest appear formidable.

"Nushiño!" the pilot shouted above the noise of the engine. "Tzapino coming up. Means fish river."

The names were becoming familiar. I glanced at the map spread across my knees. The Curaray was next. I looked to the west and saw the flank of the Andes.

"Can't make out anything now," the pilot said. "River's too high. But that's where they died. Just around that turn."

The bend in the river looked like any other, a brown meandering stream overhung with vegetation.

"Where's Tiwaeno?" I asked. The pilot pointed over his shoulder to the east. Tiwaeno was both a river and a Waorani settlement, the place where the first peaceful contact was established in 1958. It was made by Betty Elliot, widow of one of the martyrs, and Rachel Saint—two women who in character and temperament could not have been more different.

In the wake of the massacre Rachel Saint never forgave her brother Nate and his companions, but she did come to view their deaths as part of God's plan. The unmarked graves, she wrote to her parents, "were five grains of wheat planted way down the Curaray River in Auca soil." The harvest would be the Bible translated into their language, a task to which she set herself with all her considerable energy and ambition.

The key was Dayuma, the only "real Auca in captivity." When Carlos Sevilla realized that he had an international celebrity in his midst, he elevated her from field hand to domestic servant. Dressed in calico, she was taken to Quito, where the missionaries worked with her on Waorani language and grammar. Then, in the spring of 1957, Cameron Townsend instructed Rachel Saint to bring Dayuma to the United States, ostensibly to continue the work at the Summer Institute of Linguistics' language school at the University of Oklahoma. Townsend's real purpose was to drag the pair of them through a bizarre series of public appearances designed to promote the national image of the Wycliffe Bible Translators. On June 5, 1957, Dayuma appeared before 30 million television viewers on *This Is Your Life.* Later that summer Billy Graham introduced her at Madison Square Garden in New York. Before appearing on the podium, a nervous Dayuma was advised by Rachel, "Think of all those people out there as the turkeys you saw on a farm recently, and you'll feel better."

In the spring of 1958, in a highly publicized baptism at Wheaton Evangelical Free Church, Dayuma was proclaimed the first Auca Christian. No one noted that her confession of faith had come in the midst of a near fatal bout of Asian flu. Nor did they know that in quiet moments she still spoke of jaguars wandering through her dreams. Above all, the participants at the ceremony happily ignored the fact that Dayuma had already been baptized years before in Ecuador by Padre Cesar Ricci, a Catholic priest and arch foe of the Protestant evangelists.

While Rachel Saint and Dayuma were promoting the missionary

cause in the United States, events in Ecuador were moving ahead without them. Following the death of her husband, Jim, Betty Elliot had remained at Shandia, witnessing to the Quichua and hoping that one day she might be able to bring the Word to the Auca. Her opportunity came in November 1957 when two Waorani women unexpectedly wandered out of the forest. Their names were Mankamu and Mintaka, and they were looking for their niece Dayuma. Elliot rushed to meet them at Arajuno and soon discovered that one of them, Mintaka, was the older woman who, along with George and Delilah, had first met her husband on Palm Beach. Elliot perceived the hand of the Lord at work.

After several weeks of language study, Betty Elliot and the two women moved cautiously into Waorani territory with the hope of building a mission on the banks of the Curaray. The expedition collapsed on the first night when the body of one of the Quichua porters was found stuck with eighteen Waorani spears. Bound around one of them were pages ripped out of the New Testament. Unfazed, Elliot and her party returned to Arajuno and continued to work on the language. By May 1958, when the S.I.L. finally brought Dayuma back to Ecuador, Betty Elliot had learned in six months as much Waorani as Saint had mastered in three and a half years.

Once Dayuma met Mankamu and Mintaka, she decided to go home. On September 8 the three women headed into the forest, became lost, and then eventually made their way to the Curaray and beyond to the Tiwaeno. As they walked, Dayuma grew frightened. Never had she forgotten the spearing raid eleven years before that had killed her family and forced her to flee to the outside. From her youth she remembered young children being buried alive to accompany their dead fathers. From her aunts she had learned of a recent incident: A solitary white explorer, besieged by the Waorani, went mad, and shot his gun randomly into the trees until finally, on the fifth day, he took his own life. The Waorani speared him anyway and took the teeth from his jaw.

Dayuma dealt with her fear by sharing her new faith with her aunts. By now she knew that there was no snake guarding the approaches to heaven, that spirits did not rush through the air in the wind, that those who failed to enter heaven did not fall into oblivion, become termites, and die. No. They fall into great fires that rage beneath the earth, and they live, burning but never dying, in mortal agony forever.

On September 28, after just two weeks in the forest, Dayuma returned to Arajuno with seven Waorani women, all of them singing "Jesus Loves Me" in English. The Indians invited Rachel Saint and Betty Elliot to visit their relatives. Saint wanted to go alone. The S.I.L. insisted they enter as a team. On October 6 they set off, accompanied by ten Waorani, dozens of Quichua, and three-year-old Valerie Elliot. Their route carried them overland to the Río Oglán, down the Curaray, up the Añangua, and then overland once more across a high ridge to the Río Tiwaeno. Betty Elliot carried fifteen pounds of gear for her and the child. She had no idea how long they would be in the forest, or what fate awaited them. Late in the day on October 8, 1958, she entered an encampment of some fifty-six Waorani, including many of those who had speared her husband.

"It's coming up pretty soon," the pilot said. "Quiwa's river." The plane dropped over a ridge and came in low over a broad stream, scattering a pair of macaws from the crown of an immense tree. The pilot pointed to his left, and suddenly we were roaring across a small clearing.

"Quiwado!" he shouted as he pulled the aircraft into a steep climb. Behind us on the ground, a dozen or more children were running along the riverbank and through a cluster of thatch houses toward a grass runway carved into the forest just beyond the edge of the settlement. By the time we had turned back into the wind, most of the village appeared to be waiting on the grass. Among them was an American anthropologist, Jim Yost, a linguist with the S.I.L. who had been living with his family among the Waorani for seven years. One of a handful of outsiders to speak the language fluently, Jim was eager to learn more about their use of plants, just as I wanted to know more about their life in the forest. We had corresponded for some months and met once before in Quito where we had planned our collaboration. As the plane landed and taxied toward the village, the thought crossed my mind that I would probably never again have the opportunity to study the ethnobotany of a people who but a generation ago had lived in almost complete isolation.

I stepped out of the plane and was immediately accosted by an old Waorani man. Unlike the others who were clothed in an odd assortment of dresses, shirts, and gym shorts, he was naked except for a cotton G-string. His hair was cut in the traditional style: bangs in front, long down the back, and shaved above the ears. A scarlet feather pierced his septum, and there were large balsa disks in his earlobes. Dangling from a cord strung around his neck was the jaw of a piranha. He held a blowgun in one arm, and a dart quiver hung over the opposite shoulder. He seemed incredibly fit for someone who revealed only a pair of teeth as he spoke. I couldn't understand a word he was saying.

"He just gave you the name of his dead brother." I turned and shook hands with Jim. A slight man with fair skin and a thin red beard, he was wearing cut-off shorts, sandals, and a broad straw hat. "Now he wants an ax," he said with a grin. "You'll have to give him one. He's a little concerned. He can't believe I've invited you to stay in my house with my kids and we're not related. He thinks you may try to kill us."

I glanced at the old man and attempted a smile, which appeared only to confirm his suspicions. Jim spoke briefly with the pilot while several of the younger Waorani gathered my gear and the supplies for the mission. Then we all watched as the plane taxied and took off. The Waorani stared as if witnessing the spectacle for the first time. I had seen that same look on other Indian faces. Pedro Juajibioy once told me that when an airplane first flew over Sibundoy, the Kamsá thought it was a great crucifix with a priest floating in the sky. When the Waorani saw Nate Saint's plane flying over Terminal City, they discussed endlessly what it might be. Initially, because of the drone of the engine, they thought it was a giant wood bee. But since bees don't have people inside, they decided it was a devil with devil cubs within, a conclusion that provided part of the reason for the massacre.

After the plane disappeared over the forest, we walked single file toward the village. The old Waorani, whom I later knew as Kowe, stayed close by my side, talking incessantly.

"He came out of the forest only a year ago," Jim explained. "He still thinks that everyone speaks Wao. It's all he's ever known. Their word for hearing is the same as for understanding. He calls outsiders the earless ones. When he met me, he was surprised, as he put it, that I had 'holes in my ears' and could understand him."

Jim spoke directly to Kowe, calming him down. I was curious about what he had said.

"I told him that you would give him an ax and that you had taken the name of his older brother. I asked him what more he could possibly want."

"What's the name?"

Jim waited until the others were a few steps away. "Eweme," he said quietly. "Normally it's not a good idea to ask someone's name. For them it's an act of aggression. Here one is known or one is not known. If the person is a stranger, it's best to find out his name before he knows yours. The name reveals the kin ties. If he turns out not to be related, you may have to spear him."

"That's why the old man—"

"Right. That's why he named you. Otherwise you'd be an enemy. Plus, since you now are related and since you are white and have more than he does, you'll bring him goods."

"The ax."

"Perhaps more."

"What happened to his brother?"

"He was even older. We don't know how much. They have no measure of age in our sense. But I figure he was at least in his early eighties when he died."

"Speared?"

"Actually there's not much of that anymore—at least not here. No, he was gathering fruit. He fell out of the top of a palm tree."

The village was not unlike most lowland Indian settlements: a scattering of thatch huts, mostly elevated on stilts, surrounded by informal, even chaotic gardens. Even before contact the Waorani had begun to build their houses like the Quichua. Their traditional dwellings, of which there was only one in the village, were large A-frame structures roughly fifty feet long, twenty feet wide, and twice the height of a man. With bamboo ridge poles, rafters of *caña brava* bound by lianas, and a roof of palm fronds split down the middle and lashed to the reed rafters, they could be built by a man in a day. With fire smoke leaving a thick coating of soot and tar on the interior thatch, protecting it from insects, the shelters might stand for a year, about as long as the Waorani ever expected to remain in one place.

Jim's small hut was at the edge of the village, close to the river. By the time we reached it, the crowd from the airstrip had melted away. There was little of the frenzy that usually marks one's entry into an isolated village. Once I arrived, and after the flurry of activity at the airstrip, there was dead calm. It was as if I had always been there. Jim's wife, Kathy, whom I had met previously in Quito, was standing at the head of their ladder. After saying hello, I mentioned how unusual the arrival had seemed. She was a lovely woman from Iowa, thin and, like her three young children, blond and fair.

"It used to unnerve me," she said, "so little sense of personal space. People always around yet in a way strangely invisible. Everyone arriving and leaving without ever saying hello or good-bye. The words don't even exist for them. Then I finally understood."

"Just what you went through coming in," Jim added. "A person is either part of the social group or a threat, of the people or a *cowode,* a cannibal. Once Kowe gave the name and you offered the gift, you became part of this place."

I glanced at Kathy, who smiled.

"Jim," she said, "show him where he can string his hammock."

CHAPTER NINE

Among the Waorani

THROUGHOUT THE REST of the afternoon, as we arranged equipment and made plans for the collecting, I continued to struggle with the symbols of a culture that had been so misunderstood by the outside world. Walking around Quiwado at dusk, as clouds gathered and cool winds heralded a coming storm, I saw a woman milking her breast to feed a baby monkey, children with jaguar teeth hanging from their necks, and young boys shooting darts at a captured bird. By the river I came upon Jim's eldest girl, Natasha, and several of her Waorani friends frolicking like seaside children in shallows that no doubt concealed several species of piranha. As I wandered about the village enjoying a strange and unfamiliar anonymity, I tried to reconcile the tranquillity of the scene with the certain knowledge that there was not a woman over twenty-five who had not lost a parent or close relative to a spearing raid, or a man who had not been responsible for such a deed.

It was not until that evening after the storm had passed and the last of the visitors had returned to their homes that I learned from Jim something of the history of the Waorani. Sitting around the fire, chewing on the stringy flesh of a monkey killed that day, I felt as if transported through a looking glass that finally revealed the world as seen from the other side.

No one, including the Waorani, knows how long they have lived in the forest. For the Waorani experience and time cannot be divided into discrete elements. There are no abrupt transitions between night and day, between sleep and wakefulness. When the sun rises, mist envelops the morning, and the children move toward the warmth of the fire. Songs that begin and end with an explosive burst of air, two or three notes to a tune, the same words sustained over hundreds of repetitions, lead the darkness into the day. By night there is always movement, feet shifting the logs of a fire, voices discussing dreams or the memory of a raid, a man and woman coupling in a hammock.

There is a sense of the moon, its position in the sky, the ebb and flow of its shadow. Yet the year itself has no mathematical associations. All reference points are to the natural landscape, and the seasons, such as they are, are measures of the productivity of the forest. The year begins in our month of April when the fruiting cycle of the *chonta,* or peach palm, ends. May and June correspond roughly to the three phases of the fat season, the time when spider monkeys and howlers gorge on fruit, start to get fat, are fat, and begin to get lean. The summer is marked by the Atta ant season, a one-week period when the ants emerge to establish new colonies and the people gather to roast them as a delicacy. There follows a long unnamed gap in the Waorani year until the *chonta* season, which lasts six months, comes again in November.

For the Waorani the universe is a disk, an undulating surface surrounded by great waters and covered by a dome that lies just beyond the clouds. Below the earth is the underworld, a replica of this life, complete with trees, rivers, and hills but inhabited by *babitade,* mouthless creatures that cannot speak or eat. Above are the heavens, the destiny of the dead. Each Waorani has a body and two souls. At death the flesh rots or is transformed into jungle animals. The soul that lives in the heart becomes a jaguar, the one lodged in the brain ascends to the sky where it meets a sacred boa at the base of the clouds. If and only if its nostrils have been pierced and decorated by the finest of feathers can the soul enter heaven. If turned away, it falls back to earth and is consumed by worms. Once accepted into heaven, the Waorani live as they always did, hunting, fishing, and spearing. The animals of heaven are themselves the souls of creatures that once inhabited the earth, and thus by killing in this life, a Waorani hunter ensures a good supply of food in the life to come. It is to accompany deceased elders on this perilous journey that the Waorani occasionally buried children alive.

Both in this life and the next, the Waorani perceive themselves as a people of the forest. In heaven as on earth, the rivers are the domain of the *cowode,* the outsiders, all of whom are cannibals. Traditionally, the Waorani lived on the ridge tops between the rivers, deliberately avoiding the valley floodplains. Fishing only in the small feeder streams of the upland forest, they tabooed catfish, most waterfowl, turtle eggs, and many sources of food commonly eaten by other Amazonian peoples. They did not swim, and they did not use paddles, rafts, or dugouts. So rudimentary was their knowledge of canoe building that Jim once met a Waorani elder who had cut down a hollow tree, hoping to avoid having to burn out the wood. It had not occurred to him that there might be a problem sealing both ends of the vessel.

Linguistic evidence suggests that at the time of contact the Waorani had been isolated in the forest for many generations. Once their language was understood, only two words were found that had been borrowed from surrounding tribes. Trade with the *cowode* was nonexistent. As late as 1957 the Waorani had yet to adopt metal tools.

If the Waorani were uncertain about the world beyond their borders, the people on the outside viewed the Waorani as savages, symbols of the demon heart of the wild. Although the tribe before contact never numbered more than five hundred, Ecuadorian officials regularly estimated a population of several thousand, partly because of the vastness of the Waorani lands and partly due to sheer hysteria. Waorani raiding parties regularly covered as much as forty miles a day through the forest. Within a week the same Indians might be responsible for incidents occurring two hundred miles apart, killings that the government attributed to different groups of raiders. What's more, each Waorani carried several spears and thrust more than one into a victim. No corpse was left with fewer than eight. In 1972 the body of a cook for an oil company was discovered stuck with eight spears; the authorities suggested that twenty Waorani had been involved but, in fact, the killing was the work of only three.

In many ways the vast extent of the Waorani territory, roughly fifteen square miles for every man, woman, and child, proved to be a curse for the people. At any one time they occupied but a small part of their land. Typically a settlement consisted of two or three houses built in a clearing cut from the forest. Thirty minutes away would be another house site, settled by closely related kin. A series of such settlements formed the community of extended kin with whom one interacted throughout life. At the time of contact there were four major neighborhood clusters —the Guiquetaidi, Baiwaidi, Ñiwaidi, and Wepeida—all of them mutually hostile, none of them certain where the others lived. They referred to each other simply as upriver people, downriver people, overland people, and people of the ridge. Cut off from the world outside by generations of conflict and buffered from one another by great stretches of forest, the Waorani, by their own accounts, lived in almost constant fear and suspicion, certain that at any moment an enemy might bear down on them.

The reasons for killing were many: the death of a child at the hands of a distant shaman, the birth of a deformed baby, frustration at the loss of a woman, the simple need to avenge earlier killings. On the eve of a raid the men prepared the spears, decorating each with a specific pattern, ensuring that the victims would know the identity of the killers. Then, in the morning of a night with no moon, in the wake of a thunderstorm with lightning to carry the souls of the intended victims to heaven, the raiders departed, traveling for days or even weeks through the forest. Upon reaching the enemy settlement, they watched for known relatives. If none were present, the attackers would remain hidden until dark, when their silhouettes could not be seen against the night sky. Approaching with stealth in the early hours of the morning, the raiding party would slip into the shelters and kill indiscriminately. Once the village was destroyed, the victorious raiders returned home, whereupon they beat their own young sons so that the boys would be certain to mature into powerful warriors.

This cycle of war and vendetta determined the settlement pattern of the tribe. For months following a raid attackers lived in fear of retaliation. Abandoning their old fields, they relocated to another clearing where they dwelt within barricaded walls, protected by harpy eagles and caracaras tethered as guard animals at the approaches to the settlement. Thus each extended family required several living sites, and the structure of the household units was constantly shifting as individual couples and their children moved about. This fluidity left little room for hierarchy. There were no chiefs. A Wao could be a leader for a specific act, but each man remained intensely independent, and the society as a whole was completely egalitarian.

Balancing the authority of the individual was the obligation of kinship. Marriage for the Waorani was a relatively simple affair, an arrangement between parents formalized at a public celebration. The two people to be wed were often the last to know. Without notice a mother or aunt would lead a young girl to a hammock. A boy would be eased to the front of a line of dancers and then taken to be seated next to the girl. A marriage song would seal the arrangement.

The rules of kinship, however, were far more complex. A Waorani child addressed his father and his father's brothers by the same term. Similarly, his mother and his mother's sisters fell into an identical kinship category; therefore, the brothers of one's father became the husbands of one's mother, and after a long journey it was common for a man to share his wife with a brother. The children of one's father's brothers and mother's sisters—what we call first cousins—were considered siblings by the Waorani. But the children of a mother's brothers and father's sisters—again to us first cousins—were not viewed as siblings but indeed as cousins, or *qui.* For the Waorani to marry any sibling was to commit incest. Yet to marry someone not *qui* was to have a "wild" union. All marriages were expected to be between what anthropologists call cross-cousins.

The result was an astonishingly tight network of relationships. Among the current population of 630, Jim had found that only 20 individuals could not trace their ancestry to some common source. Hence the importance of names. Knowing a person's lineage determined how one was expected to relate. Thus the first thing the Waorani discuss when they gather is genealogy, with the challenge being to discover the other's bloodline before he or she uncovers yours. It was by tracing kinship through time that Jim first understood the extent to which warfare had dominated the lives of the people. To his astonishment he learned that over the last five generations, no less than 54 percent of all Waorani men and 40 percent of the women died as the result of spearing raids. One in five was shot or kidnapped by outsiders. Over 5 percent of the mortality was due to individuals fleeing by their own volition to the lands of the *cowode.* Presumably they felt that life even among cannibals was preferable to the world they knew. In his seven years with the tribe Jim heard of only three instances of what we might consider natural death. For months the Waorani implied that the individuals in question had grown old and passed away. Then one day a young Wao inadvertently let slip that one of the men had indeed grown old, so old that the people decided to spear him anyway and throw his body into the river. The man, the youth explained, had "died becoming old."

• • •

There was a half-moon, and I awoke to the sounds of the forest: cicadas and tree frogs, the piercing notes of a screech owl, the caw-caw-caw of bamboo rats. At one point I thought I heard a jaguar but wasn't sure. I looked about, saw smoke seeping out of the thatch of Kowe's house, heard soft voices and the swoosh of a feather fan bringing a fire to life. Someone was singing on the other side of the village, a far-off nasal chant, difficult to distinguish from the other sounds.

It was good to wake up in a house without walls. I glanced at Jim and Kathy's children, squirreled away in their hammocks, sleeping peacefully, and saw in the distance a woman in a calico dress walking to the forest, a burning ember in one hand, a pot in the other, an empty basket on her back. The forest on the other side of the clearing was still in shadow, but there was a thin streak of violet above the kapok trees and colors moving slowly in the east. The grasses and sedges by the river's edge had been flattened by the rain, the red clay glistened, and the canoes at the landing moved up and down slowly with each surge of the river. A hunter with a blowgun passed by just below my hammock.

Another Waorani was standing at the foot of the ladder. I assumed he was Wepe, who was going to guide us into the forest. Our plans were straightforward. In the seven years that Jim had been living among the Wao he had recorded the name of every plant he had seen used. We would begin by working our way through his list, knowing that the process itself would elicit from the people additional information, other plant names and uses. My task—collecting voucher specimens and seeing that they were properly identified—promised to be relatively simple. The only challenge was to cross-reference the names and collections with Waorani men and women from each of the major dialects. For that Quiwado was an ideal locality, for living at the clearing were representatives from all regions of Waorani territory.

Wepe, I would learn, had killed at least fifteen enemies. His people had, within memory, lived north of the settlement at Tiwaeno, close to the site of the Palm Beach massacre. Long before that tragedy his group had been pushed out of the region by a raid from the north. Fleeing downriver toward the Tiputini and Yasuní rivers, they had met Kowe's people. The bands lived together peacefully until the early 1960s when Wepe's group raided one of Kowe's house sites. Wepe and his people then retreated deeper into the forest. Becoming known as the Ridge Waorani, they lived undisturbed until the early 1970s when Texaco thrust exploration roads across the Río Napo into the heart of their land. There was a killing, and the government pressured the S.I.L. to pacify the Ridge Waorani and relocate them to Tiwaeno.

At the time, Wepe's half-brother Toño, whom he had never met, was living at the Tiwaeno mission. To draw Wepe and his people out of the forest, the S.I.L. equipped Toño with a radio and dispatched him to make contact. After a year of garbled radio messages, word reached Tiwaeno that Toño was dead. His mistake had been to arrive at Wepe's clearing wearing clothes. His ears were not pierced, further proof that he was *cowode,* an outsider. His fluency in the language had made no difference. As far as Wepe was concerned, all human beings spoke Waorani. Within hours of Toño's arrival, the Ridge Waorani killed him with an ax. The voice on the radio throughout the following months was that of a nephew, Kiwa, imitating Toño. When the S.I.L. arrived by military helicopter, Wepe said that the nephew was Toño, a lie soon exposed by Toño's widow. Wepe was then shown a map indicating the network of roads projected for his territory. Once he understood the implications, he fainted. When he came to, Wepe agreed to move.

None of this history showed on his face that morning as he led us into the forest. From time to time the narrow trails broke out onto an open beach, but for the most part they followed the ridges, dropping away only to cross streams and climb again into the forest. Unlike the cautious pace of most other lowland hunters, Wepe moved quickly along a trail, becoming more and more animated. Every sign of game—a broken twig, the pungent odor of a deer, scratchings of wild pigs in the mud—provoked memories that erupted into loud frantic monologues, certain to scatter any wildlife. Wepe's behavior, Jim noted, was typical. So abundant was game that the Waorani hunted less by stealth than movement, finding their prey purely by sound and smell. In the midst of the rain forest they could detect the scent of animal urine at forty paces and correctly identify the species from which it came.

Wepe, like all the Waorani I met, turned out to be not only a keen observer but an exceptionally skilled naturalist. He recognized such conceptually complex phenomena as pollination and fruit dispersal, and he understood and could accurately predict animal behavior. He could anticipate the flowering and fruiting cycles of all edible forest plants, list the preferred foods of most forest animals, and identify with precision the places where they slept. It was not just the sophistication of his interpretations of biological relationships that impressed me; it was the way he classified the natural world. He often could not give you the name of a plant, for every part—roots, fruit, leaves, bark—had its own name. Nor could he simply label a fruit tree without listing all the animals and birds that depended on it. His understanding of the forest precluded the narrow confines of nomenclature. Every useful plant had not only an identity but a story: a pungent leaf used for fever, a poison capable of killing fish in half a mile of river, a solanum first planted by the jaguar, another employed as a treatment for scorpion bites.

Late in the day, just after we had feasted on the fruits of a wild cacao, we came upon a three-toed sloth climbing slowly through the upper branches of a cecropia tree. The two creatures, animal and plant, are in many ways the perfect symbols of the Amazon. Every species of cecropia has living within the hollow nodes of its trunk a distinct species of fire ant. The ants live on protein nodules secreted by the plant, and in exchange they protect the tree from its major predator, the Atta leaf-cutting ants.

The three-toed sloth is a gentle herbivore. Its slow movements, together with its cryptic coloration, protect it from its major predator, the harpy eagle. Viewed up close, the sloth appears as a hallucination, an ecosystem unto itself that softly vibrates with hundreds of exoparasites. The animal's mottled appearance is due in part to a blue-green alga that lives symbiotically within its hollow hairs. A dozen varieties of arthropods burrow beneath its fur; a single sloth weighing a mere ten pounds may be home to over a thousand beetles. The life cycles of these insects are completely tied to the daily round of the sloth. With its excruciatingly slow metabolism, the sloth defecates only once a week. The animal climbs down from the canopy, excavates a small depression at the foot of the tree, voids its feces, and then returns back up. Mites, beetles, and even a species of moth leap off the sloth, deposit an egg in the dung, and climb back onto their host for a ride up the tree. The eggs germinate, and in one way or another, the young insects find another sloth to call home.

Why would this animal go down to the base of the tree, exposing itself to all forms of terrestrial predation, when it could just as easily defecate from the treetops? The answer provides an important clue to the immense complexity and subtlety of the Amazonian ecosystem. Biologists have suggested that in depositing the feces at the base, the sloth enhances the nutrient regime of the host tree. That such a small amount of nitrogenous material might actually make a difference suggests that this cornucopia of life is far more fragile than it appears. The tropical rain forest, though home to tens of thousands of species, is in a sense a counterfeit paradise, a castle of immense biological sophistication built quite literally on a foundation of sand.

Faced with the wonder of these creatures, I watched to see how Wepe would react. Without hesitation he blew a dart into the sloth. As soon as the poison took effect and the animal fell to the forest floor, he borrowed a machete and cut down the tree. He called the cecropia fruit *mangimeowe* and noted that it is eaten and dispersed by toucans and piping guans. He stripped the fruit from its stem and passed me a moist handful. It had the luscious taste of ripe figs.

In the evening we bathed with the children and then lay in the sand watching the lazy flight of herons against the soft light of the forest. Across the river and downstream, Kowe—the old Waorani I'd met on my first day in the village—and an old woman appeared, walking cautiously along the river's edge, stirring the water before them with a stick.

"Stingrays," Jim said. "It's about the only thing they're afraid of. You can't see them even in the clearest water. They lie just beneath the sand."

Something splashed along the bank—perhaps a turtle or caiman, maybe just a falling branch. Kowe glanced toward the sound and then continued to walk, poking the water as he went.

"But the current's too fast," I said. Jim nodded.

"I don't know why he's worried here. Maybe just an old habit."

"Do they get hit often?" I asked.

"Just look at their legs," Jim said. "It's a very serious wound, extremely painful. Even morphine won't do any good. The tail has a barbed spike like a saw tooth that drives right into your leg. The wound always gets infected. Even with antibiotics it can take weeks to heal. Without drugs, the wounds fester for months, leaving a hole a couple of inches across. It's about the only serious infection the Waorani ever get."

"How do they treat it?"

"They have a few plants. One's a big tree with enormous leaves. They call it *boyomo.* It means stingray leaf."

"Do they make a poultice?" I asked.

"No, they suck the fruits."

The next morning when we went looking for the plant, I found that it was the shape and size of the leaf, not the properties of the fruit, that mattered to the Waorani. Their logic was similar to that underlying the famous Doctrine of Signatures elaborated to such a high degree in medieval Europe. People believed that God in leaving his mark on creation had provided clues as to the medicinal properties of plants. Thus European monks treated liver ailments with hepatica because its leaf was shaped vaguely like the human liver. Although discredited by

Western medicine, this intuition appears often in folk traditions around the world.

As we moved through the forest, other Waorani notions of health and healing came into focus. With us was Geke, a youth who just three weeks before had been mauled by a wild boar. Evacuated by air to the hospital at Limoncocha, his spirit had collapsed, and physicians feared for his life. Once back with his family, however, his morale had soared. Younger and more enthusiastic than Wepe, he led us through the jungle, pointing out dozens of plants as he scrambled up trees with an agility and grace born of the wild. At one point he reached into a tangle of lianas and was almost struck by a poisonous viper. He leaped to my side and stood laughing as he pointed to a small scar on the back of his hand.

"He's been bitten before," Jim said, "just like everybody else."

There were, I discovered, no fewer than nine venomous snakes native to the Waorani lands. The most toxic but the least dangerous is the coral. Its fangs are at the back of the throat, and to inject the venom it must take hold of the prey and slowly gnaw the flesh. If struck, one always has time to tear the snake away.

"You have to fight to get bitten by a coral snake," Jim said, "and as for a bushmaster, it can only hurt a blind man. It's ten feet long and as fat as a fire hose."

The fer-de-lance was something else indeed. Just twenty inches long, quick and aggressive, it had struck over 95 percent of adult Waorani. Nearly half the men had been bitten twice. It was the highest rate of snakebite ever recorded for a human population. One death in twenty-five among the Waorani was caused by a serpent.

This talk of poison led back to plants and the various antidotes. Geke showed us four different species used to treat snakebite. Two were renealmias, tall, lush, and aromatic herbs in the ginger family. In each case the stems were pounded, mixed with water, and drunk each day until the victim was well. The stem and root of a Philodendron, crushed in hot water and administered thrice daily, was employed specifically to treat the bites of a snake called *cayatamo.* The fourth antidote was a stinging nettle. The botanical origins of these plants, together with the way in which they were used, suggested that their value was not based on pharmacology but on their magical resonance as perceived by the Waorani. As Geke explained, the strong scent of these plants, their inherent spirit power, repelled the symptoms, forcing them to leave the body.

The full significance of his comment did not become clear until later in the day. By then we had collected a number of other medicinal plants. The Waorani used the sap of a tree fern as an anesthetic to soothe toothache. They dealt with botflies, a noxious parasite that burrows beneath the skin, by suffocating the larvae with a topical application of latex obtained from a forest tree. The bark of a tree in the bean family served as both a fish poison and a medicine to treat fungal infections. We also found *oonta, Curarea tecunarum,* the dart poison that provided the basis of the hunting technology. These collections revealed another side of Waorani knowledge: the ability both to identify pharmacologically active plants and to treat certain ailments symptomatically in a manner more or less consistent with the practices of modern scientific medicine.

Waorani medicine, in other words, operates on two quite different levels: the material and the immaterial. At the root of their system is a non-Western notion of the origin and nature of disease. For the Waorani, as for many indigenous peoples, good or bad health results not from the presence or absence of pathogens alone but from the proper or improper balance of the individual. Health is harmony, a coherent state of equilibrium between the physical and spiritual components of the individual. Sickness is disruption, imbalance, and the manifestation of malevolent forces in the flesh.

In general, physical ailments that can be treated with herbal remedies are considered less serious than the troubles that arise when the spiritual harmony of an individual is disturbed. In such cases it is the source of the disorder, not its particular manifestation, that must be challenged. For the Waorani the origin of all such evil and misfortune is the *ido,* dark shamans who invoke from the heavens the *wenae,* the malevolent spirit helpers who dispense the magic arrows of disease and death. The Waorani solution, and in a sense their ultimate medical act, is to seek out these rival shamans and spear them.

The means by which an *ido* elevates his spirit and enters the realm of death is a plant known as *mii.* With some reluctance and trepidation young Geke took us to a place at the edge of the forest where it grew. To my surprise it was *Banisteriopsis muricata,* a relative of *ayahuasca* never before reported as a hallucinogen in the Amazon. Geke explained that when he was a boy, his grandfather had taken the windpipe of a toucan and blown a small piece of wadded *mii* into his lungs, thus ensuring that he would grow up to become a great hunter. His grandfather, he added, had been a Snake Shaman, a healer capable of drawing a viper out of the forest to find out if it had left poison in a wound.

• • •

Tomo was one of the Waorani hunters who could smell prey and know from the sound of a rustling leaf whether an animal was worth killing. As an infant he had impaled spiders with tiny pieces of broom straw. He had snared bumblebees, tethered them with a thin piece of fiber, and wandered about his village flying them like model airplanes. By the time he was five he could hit with a blowgun targets of hanging fruit at thirty paces. At ten he could shoot a small bird out of the air. Before he reached puberty he knew how to imitate the call of almost every bird in the forest. He understood nesting habits and breeding behaviors, anticipated feeding cycles, and could list the trees where each bird preferred to dwell. Before he married he had killed a wild pig with a twenty-foot palm lance, skewering the creature from a distance of 4 feet, pinning it to the ground with one arm while the other thrust a second and third spear into its flank. With a blowgun he could drive a dart clear through a squirrel at 40 feet, knock a hummingbird out of the air, and hit a monkey in the canopy 120 feet above the forest floor.

The day before the hunt, in Kowe's house, we had watched Tomo prepare curare, scraping the bark, placing the shavings into a funnel of palm leaves suspended between two spears, slowly percolating the water through, and collecting the drippings in a small clay vessel. He had then heated the dark liquid over a fire, slowly bringing it to a frothy boil, letting it cool, and then firing it again until a thin viscous scum formed on the surface.

All this time Kowe sat quietly on a wooden stool, his feet resting on a steel ax, as he twisted the stems of a vine into a large gathering basket. The ground around him was littered with fish bones, wads of cotton, dry leaves, and small bits of paper. Spears decorated with brilliant feathers hung in the black and smoky rafters.

Kowe had ignored Tomo. Then, just as the curare had congealed and the darts lay bare on the dirt, he moved to Tomo's side, displacing him by the fire as he reached for a dart and spun it in the liquid poison. He was, Jim had explained, a Jaguar shaman, the one best prepared to empower the darts. He was the one who could sing the forest into being, sliding his voice up and down, falling away into a chant that only the Jaguar mother had the power to translate for the world. One by one Kowe placed the poison darts in the earth by the fire. Slowly the tar hardened into a jet-black lacquer, ready to be used.

The next morning Tomo had these darts in his bamboo quiver. Hanging from his neck was the piranha jaw he would use to notch the tip of each one. This ensured that the poisoned tip would remain embedded in the flesh even if the rest of the dart was swatted away by the prey. Squatting by the fire, he drank a calabash of *tepae,* the thick and mildly fermented beverage prepared by the women from the masticated roots of manioc. Like most adult Waorani, Tomo would drink almost two gallons a day. Although *tepae* is the major source of carbohydrate in the diet, no Waorani considers it food. No matter how they are consumed, all fruits, roots, and seeds are said to be drunk. Similarly, a garden is not harvested, it is drunk. Only meat is eaten, for it is the only true food in the forest.

From the rafters, Tomo selected a short blowgun, just over six feet long. Without a word or gesture to his family he led us out of the shelter and onto the narrow path that led away from the airstrip and into the forest. The sun was high but the air cool and still beneath the canopy. For an hour or so we walked slowly along a trail, effortlessly gathering useful plants. There were any number of fruits and edible seeds, several dye plants, a wild grass employed as a paintbrush, leaves that served as dishes and serving bowls. A number of palms provided food, thatch, and wood for blowguns and spears. Two different species of pipers were used to blacken teeth and prevent tooth decay; two other plants yielded latex that treated tooth decay once it had occurred. Finally, growing alongside the trail was a short bambazoid grass whose inner shoot yielded a razor-sharp knife. It was this plant that women sought when they were about to give birth.

When a woman was ready, Jim explained, she carried her wedding hammock into the forest and hung it between trees. Straddling a hole cut into its tough fibers, her feet firmly on the ground, she leaned forward and waited for the infant to drop on a mat of heliconia leaves spread beneath the hammock. After it was born, she sliced the umbilical cord with the bamboo shoot and then carefully examined the child. If deformed, unwanted, or a twin, she would bury the newborn alive in a shallow hole in the soil. If the baby was to be kept, she held the child to her breast, carried it to the river to wash, and placed it in the bark cloth sling that served both as blanket and diaper. Once the baby was accepted, it was never put down. Loved and cared for, it received all the mother's affection.

The trail opened into the harsh light of a clearing. The plantings rose over an undulating hill to a ridge of dead snags and peach palms and the distant wall of the forest. Tomo hesitated and then skirted the field to one side, avoiding the sun. Later we would return here with the schoolteacher Nange and his wife, Oncaye, and find among the chilies, yams, peanuts, plantains, and bananas no fewer than twenty varieties of manioc, many named for forest animals. The field, Tomo explained as we climbed through a tangle of deadfall, was almost exhausted. The family that owned it would already have begun to farm another site—

the men working in the shade of the trees, clearing the underbrush, planting cuttings, felling the largest trees, leaving the leaf litter on the ground to protect the soil from the rain and sun. It was slash and rot, as opposed to slash and burn. Rarely did the Waorani torch a field. Manioc grown in these soils matured in nine months, and the roots would remain in the ground without rotting for a year. Harvested once, the land would be abandoned and never used again.

We had barely entered the forest again when Tomo froze, dropped into an attack crouch, and slipped away from us, moving silently and steadily through a thicket of heliconia until stopping at the base of an enormous tree sixty feet from the trail. In a single gesture he had withdrawn a dart, notched its tip, deftly spun the kapok fiber around the base, and placed it in the mouth of the blowgun that now hovered motionless above his head. His cheeks suddenly puffed out with tremendous pressure, which was released in an instant. A moment later he was lunging through the vegetation, laughing and shouting. By the time we caught up, he held a rufous mot mot in his hand. The bird was still alive. Tomo had managed to reach it before the poison took effect. He dropped the frightened creature into his basket and placed the dart conspicuously in the notch of a tree so that all would know an animal had been taken.

As we continued along the trail, still amazed by the accuracy of the shot, Tomo snapped off a small branch, broke it open, and rubbed the inner cavity on the inside of his cheek. The plant was *Duroia hirsuta,* a common tree in the coffee family. Ants live within the stems, and the Waorani, it turned out, apply their concentrated pheromones topically to relieve the pain that sometimes results from using the blowgun too much.

"It's a macho thing." Jim smiled. "Often they don't really have to blow as hard as they do." He then explained that the volume of air in a typical blowgun is less than a tenth the capacity of the lungs. Thus it is not force but control that counts, judging the distance to the prey, the angle of ascent, the proper trajectory.

"The longer the blowgun, the greater the velocity of the dart. Up to a point. Then resistance takes over. Finding that perfect balance, the right length is something they're always looking for."

Though a gifted hunter with a dart, Tomo confessed that he, like most Waorani, preferred shotguns. It was something that had initially confused and concerned Jim. When he first arrived, the entire tribe possessed only three shotguns. Six years later one hunter in three owned one. For the most part they were miserable weapons: single-shot breechloaders cursed with weak firing springs that rarely lasted a year.

A small box of shells cost the equivalent of three blowguns, as much cash as a Waorani could earn in a week if there was work. To make the purchase required a journey of four days. It just didn't make any sense. At close range a shotgun was useful for large terrestrial animals, provided that it worked, but for birds and monkeys and anything that lived in the canopy, the blowgun was by far the superior weapon. One day it finally dawned on Jim that the Waorani affection for shotguns had little to do with efficiency. It was the intrinsic attraction of the object itself, the clicking mechanisms, the polished stock, the power of the explosion. As one Waorani hunter explained, "It makes such a beautiful noise."

The trail reached a small creek. As Tomo demonstrated the proper way of poisoning fish, smashing the liana on a log, placing the bark in a small pool of water, he noticed a footprint in the mud on the far bank. Moving to examine it closely, he named the person who had left it. Sure enough, a mile or two up the trail we heard shouting and came upon the man in question and his wife, standing over a collared peccary that he had run down and speared. He acted as if we had been expected. Swatting away the flies and sweat bees, the hunter launched into a full account of the kill, evidently sparing no detail, for it was fifteen minutes or more before we finally pushed on. By then Tomo was wired. The trail climbed high onto a ridge of boulders and moss and twisted lianas. Suddenly he began to shout.

"It was here. It was here. It was here that I speared it!" He then explained to Jim that he and two friends had once tracked a wild pig to these rocks. They had heard something just below them and in their excitement had jumped off the bluff, fully expecting to land just behind the pig, ready for the kill. Instead, they came down face-to-face with an immense jaguar. Each of them had three wooden spears. They had no choice but to try to kill it, which they did.

"Have you ever seen a jaguar up close?" Jim asked me.

"Once, but just for a second. Then it ran away."

"The teeth are sunk into an incredible jaw, perfectly designed for tearing flesh." We both looked at Tomo, who was still laughing at the memory.

Only later in the day did I realize that Tomo was as intensely interested in my world as I was in his. We had just collected *Jessenia bataua,* a lovely palm known to the Waorani as *petowe.* Each part of the tree had a name and a separate use. One of Schultes's students, Mike Balick, had studied the palm and discovered that the oil in the seed was, in terms of both taste and chemistry, indistinguishable from olive oil.

Single-handedly he had offered the Brazilian government the opportunity to reduce its annual trade deficit by hundreds of millions of dollars.

It was hot, and we were resting. I was thinking of Mike when I realized Tomo was asking me a question.

"What was that?" I said.

"He wants to know how many brothers and sisters you have," Jim said.

"Two."

"So few." Tomo was laughing again. He spoke and Jim listened.

"He wants to know if you have to buy your wives."

"No, not that I know of."

"It's because the Quichua do. There's a bride price, and they tell the Waorani it's the civilized thing to do," Jim said, turning toward me. "You have to understand that every contact he's had with the outside has left him utterly perplexed."

Tomo's first exposure, Jim explained, had occurred in 1975 during a shamans' war between two rival groups of lowland Quichua. Knowing that the Waorani were eager to establish trading relations, one side asked three Waorani men, including Tomo, to kill a rival shaman, assuring them that such a murder was acceptable among the Quichua. The only stipulation was that the deed be done with spears. Tomo agreed and was promised a radio, a pair of boots, a shotgun, and ongoing access to trade goods of all sorts. As soon as the shaman was dead, those who had contracted the murder betrayed the Waorani to the authorities. The military flew in by helicopter to the Quichua village where Tomo was staying. The soldiers landed, weapons in hand, firing indiscriminately into the treetops. Tomo was on the ground watching, but he soon fled and remained on the run for more than a year.

Undaunted by the experience, Tomo retained a genuine hunger for the outside world. Invariably his encounters were bitter, especially once he began, like so many of the Waorani, to work for the oil companies. Hired by an agent of Texaco to cut trail, he worked for three months, living on a diet of white bread; in the end he received no pay. It was only then that he denounced the *cowode* and returned to the forest.

Tomo's father had died at the hand of his own brother, Dabo. Now Dabo is Tomo's father, and Tomo addresses the killer of his natural father as father. Tomo's mother's sister is Meñemo. Four times in her life spearing raids had forced her to flee to the forest. When she was five, she survived three months alone in the forest. The insects were so terrible that she buried herself in the swamps to sleep. For food she ate clay. She lost all of her hair, but she lived. When she married, she once again found herself on the run, this time with two children. One kept crying and had to be choked to death so that the other might live. It was a harsh world but one that Tomo at least understood.

Now Tomo himself was a father. He had six children, including a stepson whose real father, Néngkiwi, had been one of the Waorani who killed the missionaries in 1956. Néngkiwi was a boastful character, always threatening to spear people. Finally one day Giketa, a respected elder, asked, "Who is going to take care of this guy?" A Wao called Dyowe killed Néngkiwi with a spear, and Tomo married his widow. Thus Tomo's word had weight when he spoke to us that afternoon of the history of the massacre of the missionaries. It was all a terrible mistake, he explained. George, whose real name was Gimari, was angry and jealous that Delilah, or Naenkiwi, was flirting with the white strangers. He told the Waorani at Terminal City that the missionaries were evil. The women disagreed, but Gimari's opinion held. Thus, the entire time the five Americans waited on Palm Beach, there was never a doubt that they would be killed.

On Easter Sunday I awoke to soft voices and hymns sung in a gentle manner, innocent and pure. I looked up and saw Kathy and her children huddled by the fire. The morning was cold. Thick ground fog hung over the village. I fell back into my hammock and lifted the thin blanket over my shoulders. Drifting back to sleep, I heard other voices from the night before, the slow back-and-forth rhythm, the subtle tone shifts of the Waorani chants, songs so old the people no longer understand their meaning—like singing hymns in Middle English. For several hours we had danced at the house of a Waorani named Kento, forward and back, sweaty hands on one another's shoulders, feet flat on the ground lifting dust that mingled with the scent of the aromatic plants the Wao had crushed onto their skin. Then the songs had seemed to emerge from a trance. Now in the uncertainty of dawn, hymns did not seem out of place.

Breakfast brought me back to the forest: smoked meat, a calabash of thick banana drink, and beetle grubs roasted over the fire. The day before, Tomo had killed four monkeys and Kento bagged two. Jim figured they had each covered sixty miles of ground. They would have shot more, but they had broken taboo by drinking *chicha* the morning they made the poison.

"It's sort of like eating your cousin," Jim said, smiling, as I worked my way through the forearm of a howler monkey. By this time I had eaten them all—capuchin, howler, woolly, spider—and I still couldn't tell the meat apart.

"There's a squirrel that lives on one kind of palm nut," Kathy said. "That's what the Wao really like. They roast the stomach in ashes and eat the whole thing."

"It's pretty good," Natasha added. I laughed. She was a sweet girl, strong and very brave. Two nights before she had been struck by a scorpion, and she didn't cry, not even when all the Wao gathered around her and the old lady had rubbed the solanum fruit on the wound.

After breakfast we wandered over to the small church the Waorani had built by the airport. Many of them were already inside—Tomo and Kento, Kowe, Geke, Wepe, and their families—all seated stiffly on the wooden benches, talking casually, apparently oblivious to the bossy young woman who paced back and forth in front, barking orders and dismissing them as savages. It was disconcerting to see the numb faces of the Waorani men enduring the taunts of this woman, dressed stiffly in a fresh cotton smock. She, too, was Wao, but her manner betrayed her. She was Wepe's daughter, and for the last two years she had lived outside, working as a maid for a *mestizo* family in Puyo. Throughout the Oriente, owning a Waorani is something of a status symbol. They speak no Spanish, are subject to rape, and after several months are dismissed, with perhaps a dress as payment. Wepe's daughter was wearing hers.

Jim and Kathy and their kids sat quietly on a bench at the back. I took a place beside young Natasha. It all seemed strange. I remembered Betty Elliot's accounts of the first Waorani church services, with Dayuma very much in the role of Wepe's daughter. Before dawn she would shout to the clearing, "Everyone come out. I am going to speak about God." The people would then gather. For them the idea of sitting and listening was ridiculous. Waorani speak when they have something to say. So Dayuma kept telling them all to shut up. Then, since there was no word for prayer in the language, she would announce that it was time for them all to sleep. "Close your eyes and sleep!" she would yell. For the Waorani, a church service meant having this odd woman wake you before dawn just to tell you to go back to sleep.

The humor of it was not lost on Betty Elliot, Jim Elliot's widow, who emerges from the pages of her memoirs as an amazing woman, possessed of a simple faith, tempered by tragedy. Rachel Saint, by contrast, was a zealot, informed by fear. Welcomed at the Waorani settlement at Tiwaeno in part because they were women and thus no threat to the Wao, Saint and Elliot—one a spinster, the other a young mother—had completely different experiences. Surrounded by men singing love songs, young girls slipping into hammocks, guiltless lovemaking all around, Rachel Saint was soon writing home of the "scum of heathens" and the need of Christian marriage. Elliot saw the positive. Intoxication was unknown, men never beat their wives, criticism was public, and relations between husbands and wives were equitable. "Many of our civilized sins," she wrote, "were conspicuous by their absence. There is no gossip, vanity, personal pride, stinginess, or avarice. I was faced with the fact that socially I had nothing whatever to offer the Aucas."

Inevitably, Elliot and Saint would clash, and there was never a doubt who would come out on top. "Rachel," Elliot later wrote, "is 100 percent dogmatic. I couldn't discuss anything rationally with her. She simply would not give way on anything." The final split occurred over a trivial matter: whether or not the Waorani ought to be exposed to communion. Elliot maintained you could use grape juice, though the Bible specified wine. The very idea of introducing a Catholic ceremony deeply offended Saint. For Betty Elliot it was the final straw. Having lost her husband to the Waorani and raised her daughter among them, Betty Elliot left Tiwaeno in 1961. Perhaps more than anyone she understood the people. "The Auca," she would later write, "has no form of religion. He knows nothing of prayer, sacrifice, worship, placating evil spirits, or adoring the good. He is not consciously seeking after anything."

With Elliot out of the way, Saint installed herself and Dayuma as power brokers, mediating not only the delivery of new religious ideas to the Waorani but, more important, the flow of goods from the outside. For the next decade, as more groups of Waorani were drawn to Tiwaeno, swelling its population to three hundred by 1969, virtually all contact with the Waorani filtered through Rachel Saint. In 1964, when the Ecuadorian government established a reserve for the tribe, an area of less than half of 1 percent of their former territory, it was officially known as "Dayuma's Auca Protectorate." Journalists began to write of the Auca queens, and the myth of the Auca matriarchy was born. By the early 1970s, Saint had become almost psychotically attached to her tribe. She built a little empire in the forest, controlled by a select number of women whose power derived strictly from the Waorani hunger for goods and information about the world outside.

Saint's downfall began when Jim Yost, then a young anthropologist, issued a report criticizing Dayuma's monopoly and urging that the flow of trade goods be more carefully monitored. He also pointed out the danger of concentrating the Waorani in a single settlement, citing the polio epidemic that had swept through Tiwaeno in 1968, killing sixteen and paralyzing six. Saint was livid. She had herself been stricken by the disease and knew it to be the judgment of God. Without her presence, she maintained, the Waorani would kill each other. Without her protection, they would kill Yost and his family. Jim called her bluff. The S.I.L. leadership backed him, and Saint was discreetly retired.

Wepe's daughter finally sat down, and in her place stood the schoolteacher Nange. I tried to imagine what he was saying. Jim had explained how difficult the translation work had been. How do you give English words to a language so rich in onomatopoeia and punctuated with the sounds of the forest? How do you translate the Bible when nothing in it—places, names, objects, let alone themes—makes any sense? The Waorani had no words or reference points for rich people and poor, for specialized labor, hierarchy, prayer, buying and selling, towns, kingdoms, or nations. So coins become "fish scales." Paper and bread become "wasp's nest." But even if one can understand the sounds of Waorani and forge them into words, how do you translate the words into meaningful phrases? Take a basic Waorani sentence: *Bitö maomomi hemoi.* Literally it says, "You carrying in hand come give you take I you follow." What it means is simply, "Bring it to me." Add in the challenge of a tonal language, and you have, in Jim's own words, "Fellini in the forest."

I looked around at the congregation standing stiffly as they mouthed a melancholy hymn. Clearly, I thought, it had not been the promise of the afterlife that drew the Waorani to Christianity. They already believed they would go to paradise when they died, and their heaven had animals in it. As Jim had written, the magic of Christianity was its potential to break the vendetta and end the cycle of violence. And it had. Looking around this small gathering provided ample evidence. A generation ago Tomo would have had to avenge Toño's death by killing Wepe. Wepe's group had been raided by Kento's father, Ñíiwa. Wepe in turn had raided Kowe. None of them could be sitting together. Whatever else it had wrought, Christianity had stopped the spearing raids, the killing of innocent women, infanticide, and the live burial of children.

Christianity also provided a model for interacting with the world. Rachel Saint may have thought she was in control, but in many ways it was the Waorani. Even before contact they had been fascinated by the possessions of the *cowode.* Once they knew they could trade in peace, their hunger for goods and their insatiable curiosity drove the entire dynamic. Just the summer before while in the capital Jim had received a phone call from the Presidential Guard at the National Palace. Three Waorani whom no one understood were sleeping on the sidewalk waiting, it turned out, to ask the whereabouts of a minor government official they had met in the forest.

When Jim had tried to encourage the Waorani to disperse, both to establish their inherent rights to the land and to stem the flood of goods that he thought threatened their way of life, the Waorani became furious. In 1975, when the S.I.L. did attempt to control the flow of radios, T-shirts, sunglasses, and baseball hats, the Waorani responded by seeking other trading contacts in the oil camps south of the Río Napo or among the tourists Dayuma was bringing in each year to see the "wild Auca savages." At Tzapino the Waorani cleared their own airstrip. They invented rituals, imitated the activities of an oil camp, and sang songs to the helicopters, with the hope that they would unleash a rain of gifts.

Finally, Jim realized that he could no more control the flow of goods than he could reverse the process begun so long before on Palm Beach. "As romantics," he told me, "we idealize a past we never experienced and deny those who knew that past from changing. We forget perhaps the most disturbing lesson of anthropology. As Lévi-Strauss said, 'The people for whom the term cultural relativism was invented, have rejected it.' "

It was at the end of my third week in the village that Jim and I noticed a peculiar pattern in our work. By then we had managed to collect at least 80 percent of the plants he had recorded over the years as well as many others. Some of these, including a new species of basidiolichen reputedly employed as a hallucinogen by shamans four generations ago, were exceedingly rare and had never been seen by Jim. But it was a very common plant that tipped us off—that plus a serendipitous visit from an old lady. The plant was *Brunfelsia grandiflora* ssp. *schultesii,* the admixture that had given Schultes such a headache at Conejo when he took *ayahuasca* with the Kofán and that Tim later named for him. Used by dozens of tribes to treat fever, it is one of the most prized medicinal plants of the Amazon. The Waorani agreed that it was useful but only as wood. It seemed incredible to me that a people who had such a profound knowledge of the forest would have failed to recognize the medicinal properties of the plant. When I later reviewed my notes, I discovered to my surprise that, out of all our collections, there were only thirty-five medicinal plants.

It was early evening, and Jim had been playing a tape of a group of

Waorani discussing a spearing raid. Several of the Quiwado people had gathered around and were arguing with the machine. One of them was a thin, elderly woman. Suddenly she exploded in rage, flung her arms in the air, and then lifted her blouse for all to see. There were two neat scars where the razor-sharp triangular spear had perforated her belly and emerged from her back. It had happened years before, Jim explained, when the woman was young. The raiders had left her for dead, but when her kin returned, they treated her. Because of the reversed barbs on the spear, they could not risk removing the weapon. Instead, they cut it off back and front and, leaving the remainder of the spear in her, plastered the wound with the standard treatment: mud from the watering hole of a peccary. They then carried the woman to her hammock, where she remained for a couple of weeks. As the tissue around the wound became necrotic, she felt well enough to return to work in the gardens. One day as she bent over to harvest manioc, the long spear fragment slipped out.

I asked the obvious question: "What about infection?"

"Wasn't any," Jim said. "She was lucky. No internal organs were hit. But I've heard similar stories. And I've seen lots of hunters accidentally skewer their feet on punji sticks on the trail."

"Again no infection?"

"No. They were a healthy people."

"How healthy?" I asked.

"Very. It's one of the few societies that we really know about."

Jim explained that not only had the Waorani access to Western medicine since their first sustained exposure to Western disease but a team of medical experts had compiled a complete profile of their health at the time of contact.

"I have copies of their papers here," he added. "I can dig them out for you."

Later that night I had a chance to read several of these reports, the most important of which had been written by James Larrick of Duke University Medical Center. Jon Kaplan of the University of New Mexico and Jim were coauthors. The results were astonishing. Among the Waorani the medical team had found no evidence of hypertension, heart disease, or cancer. There was no anemia. Hemoglobin values were equal to or better than North American standards. The common cold was absent. The Waorani ranked as one of the few populations in the world where blood pressure does not increase with age. The people had practically no internal parasites and virtually no secondary bacterial infections. They had never been exposed to polio or pneumonia, nor was there any evidence that smallpox, chicken pox, typhus, or typhoid fever affected the tribe. There was no syphilis, tuberculosis, malaria, or serum hepatitis.

Although on the whole the Waorani were remarkably healthy, the medical researchers found a number of chronic ailments. Yellow fever was endemic, as was hepatitis A and herpes simplex. The Waorani also suffered from fungal infections and external parasites such as lice and scabies. They had terrible teeth. Naturally they experienced burns, wounds, and various bites, ranging in severity from that of a congo ant, which could paralyze an arm for a day, to a snakebite, which could kill.

With this information in hand, Jim and I had a closer look at our botanical collections. Of the thirty-five medicinal plants, thirty were used to treat one of six conditions: fungal infections, snake bite, dental problems, fevers, insect stings, pains and traumatic injuries such as animal bites, spear wounds, and broken bones. The remainder were valued for treating some idiosyncratic ailment. Two species of hot peppers, for example, were used by a shaman's wife to bring him down from a hallucinogenic intoxication. In each of the six conditions the Waorani had sought a set of medicinal plants representing many distinct botanical families. In other words, they had very carefully sampled their forest to find treatments for the ailments they suffered at the time of contact. In the end they had needed and come up with very few preparations.

This limited and highly selective use of medicinal plants stood in marked contrast to that of neighboring tribes such as the Canelos Quichua, a people who have been repeatedly exposed and ravaged by Western diseases for hundreds of years. The Quichua have hundreds of plants used for dozens of conditions, including such vague and possibly European concepts as "bad air," "kidneys," and "liver." The difference between the two tribes forced us to consider a basic question: Was the Waorani use of plants an anomaly, or might it represent the pre-contact state of affairs throughout the Amazon? I knew, and told Jim, that ethnobotanists working among the Yanomamo, another recently contacted group, had reported low numbers of medicinal plants. If the Waorani and other isolated groups are in fact indicative of the pre-contact era in regard to their use of medicinal plants, it would seem that the vast pharmacopoeias reported for the more acculturated tribes reflect, at least in part, the chaos of contact and the accelerated experimentation that took place in response to the arrival of Western diseases. This idea, while challenging the notion that indigenous knowledge of medicinal plants necessarily developed slowly, over hundreds of years, in no way denigrates native healing practices. On the contrary, it revealed native healers, including those of the Waorani, for what they are: active scientific experimenters whose work reflects social needs and whose laboratory happens to be the rain forest.

On the day before I left Quiwado, Jim and I poled upriver with young Geke and Kento to collect bark cloth, a specimen of yam bean, and a number of other plants I had yet to find. Kento pointed out the places on the bank where animals had drunk that morning, and a clearing where an old man had died and been buried, accompanied by his three-year-old son, who was buried alive. We passed a youth fishing on the shore and watched as he lunged his spear into the water and then whipped it out, a bright fish spinning on the sharp tip. Around the next bend in the river we came upon a red caracara perched on a branch. Kento shot it with a dart.

"I thought you said they were taboo," I said to Jim.

"You're not allowed to eat them. That doesn't mean you can't kill them."

We passed a quiet eddy, and Geke wanted to fish. He reached into his fiber bag and pulled out a stick of dynamite. Jim told him to put it away. They argued. Jim prevailed, and Geke sulked the rest of the morning.

"They get it from the local officials," Jim explained, "a fuse connected to a blasting cap. Jam it into a stick of dynamite, tie it to a rock, light it, and let her rip. The concussion kills all the fish. The rivers around Tena are all wiped out. If it's not dynamite, it's DDT."

"What do you mean?"

"All the villages get sprayed. The World Health Organization gives it to the Ecuadorian government. Health officials bring it in for malaria, and then they sell it by the hundred-pound sack to the Waorani, who use it as fish poison."

"Don't the Indians know?"

"Know what?"

"Can't they see what it does to the rivers?"

"Sure. And I've asked them. I once asked Kowe if he had noticed anything about yields. 'Yes,' he said. 'When we first started using it, we got lots of fish. Now not so much.' What about your children? I asked. Without the slightest sign of remorse he calmly said, 'Oh, they won't have any fish, but we will.' "

"What does that tell you?" I asked.

"Do you remember when we got the honey?"

"Sure." I could still see the excitement on Tomo's face, his head and sweaty body covered by thousands of stingless bees, his hand uncovering the red cone of cecropia seeds made solid by saliva and wax, the outer casing of the hive, the thick black wax prized by the Waorani as resin and glue. Then the massive comb itself, great cavities of thick honey, poured onto maranta leaves and carried home to the children.

"Tomo must have eaten a quart of it," I said.

"When he was in the middle of cutting down that tree, do you remember what he said?"

"I remember you saying something."

"Suddenly, in all his excitement, he turned to us and said, 'One time I killed a howler monkey, and it was no farther than from me to you.' Then he went right back to cutting the tree. There was this wild association between hunting and cutting the tree. It's what they most like to do."

"It wasn't just the honey."

"No, I don't think so. I've seen it so often. Nothing thrills the Waorani more than killing game and cutting down big trees. It's what so many people don't understand who haven't lived in the forest. You don't have to conserve what you don't have the power to destroy. Harming the forest is an impossible concept for them. The fact that they use every part of an animal has nothing to do with a conservation ethic, and everything to do with hunger."

"They don't know what it means to destroy."

"They have no capacity to understand. In a world of such abundance, the word 'scarcity' has no meaning. It's what makes them most vulnerable. It's the same with their culture. When you've lived in complete isolation, how can you understand what it means to lose a culture? It's not until it is almost gone and when people become educated that they realize what's being lost. By then the attractions of the new way are overpowering, and the only people who want the old ways are the ones who never lived it."

Late in the afternoon, just after we had returned to Quiwado, there was a Waorani visitor waiting for us. He had come from downriver, five hours away, not far from the military outpost on the Curaray. He wore wraparound shades, canary yellow trousers, a white polyester shirt, bright sneakers, and a hat emblazoned with the logo of a German company. The immaculate condition of his clothes revealed that he had changed at the edge of the clearing, just before entering Quiwado. He had a huge silver watch without hands and an Instamatic camera that didn't work. His name was Nénkiwi. He couldn't read and so had no way of knowing that scrawled across his chest was the lettering AMOR-ECUADOR MILITARY POST # 5.

Nénkiwi had come because he had something to sell: a stone ax that Jim identified as Late Napo phase, A.D. 800–1200.

"Might as well buy it," Jim said. "Otherwise someone else will."

I bought it for a few dollars, and Nénkiwi was delighted.

That night I sat alone in my hammock and held the ax in my hands. It seemed incredible to me that just a generation before, during the years when Schultes lived and traveled in the Amazon, these Indians were still using such an implement to clear their land. I imagined a Waorani man standing in the shade of the forest and pulverizing the cambium of some tree. They did not even know how to carve stone. They found the ax heads in the jungle, gifts of their creator, *Waengongi.*

I thought of what Jim had told me and remembered that out beyond the mauritia swamps there were still Waorani who had yet to be contacted—a small band splintered from the people at Yasuni, running scared in the forest. Just three years before they had killed six oil workers, having given the intruders ample warning to get away from their land. By now they had speared or been speared by every other Waorani group. No one knows how many there are—perhaps just eight or ten, brothers and sisters, probably without children. In the moment, as I listened to the wind brush against the trees, I hoped only that somehow they would stay away.

CHAPTER TEN

White Blood of the Forest, 1943

ON NOVEMBER 20, 1942, a tall, thin, strikingly handsome man entered the Washington, D.C., headquarters of the Rubber Investigations Division of the Bureau of Plant Industry, U.S. Department of Agriculture. Dressed in a dark suit, he cut an impressive figure, quite unlike the rough-and-tumble rubber explorers the secretaries had grown accustomed to. Handing his card to Miss Price, the receptionist, he asked to see Robert Rands, senior pathologist and head of the *Hevea* rubber project. The card read simply: *RICHARD EVANS SCHULTES, BOTANIST, HARVARD UNIVERSITY.* Miss Price passed the visitor along to Mrs. Bedard, Rands's personal secretary, who ushered Schultes into a small waiting area. Returning once more to her desk, Mrs. Bedard telephoned Rands.

"Sir," she said, "there is a young botanist to see you. He's from Harvard, but I think he's English."

"Is the name Schultes?"

"Yes."

"Show him in right away."

Mrs. Bedard found Schultes where she had left him. "Dr. Rands will see you now."

"Thank you."

The office was located at the end of a long, narrow corridor hung with old photographs of plantations and coolies and colonial officials in pith helmets and white linen suits. Rands, who was waiting at the door, greeted Schultes affably and led him into the corner office. The room was basic government issue: wooden desk and chairs, metal filing cabinets, a large dark fan, and windows shaded by dusty venetian blinds. A map of the world covered most of one wall. From another hung a Javanese shadow puppet, the only memento of Rands's many years working in the rubber industry of the Dutch East Indies.

"Please, do sit down," Rands said. He was a middle-aged man, fit and of average height, with a bald head and thin fringe of gray hair that made him appear older than he was. He wore rimless bifocals, which he adjusted several times as he leafed through a file on his desk.

"I understand from our people in Bogotá that you're a good field man. That's what we need."

"Sir, I'd rather not work for the government. At least not as a civil servant."

Rands looked up from his notes.

"Son, you're already hired. That happened before you even walked out of the jungle in Colombia. And now that you're here, I'm going to let you in on something those boys back at Harvard know nothing about. Fact is, no one out there does, or we'd have a serious problem."

Rands stood up, took off his jacket, and hung it on a corner coat rack. Then, returning to his desk, he reached into a drawer and flung a handful of photographs across the desk.

"Take a look," he said. Schultes thumbed through the images: mobile artillery, army trucks, a tank battalion in action, barrage balloons over London, a pair of GIs manning a sentry post in a rainstorm.

"What do you make of them?" Rands looked directly at Schultes.

"Rubber," Schultes said.

"Right. Everything in this war depends on it. Those Sherman tanks have twenty tons of steel and half a ton of rubber. Five hundred pounds

The Indigenous Peoples of the Colombian Amazon

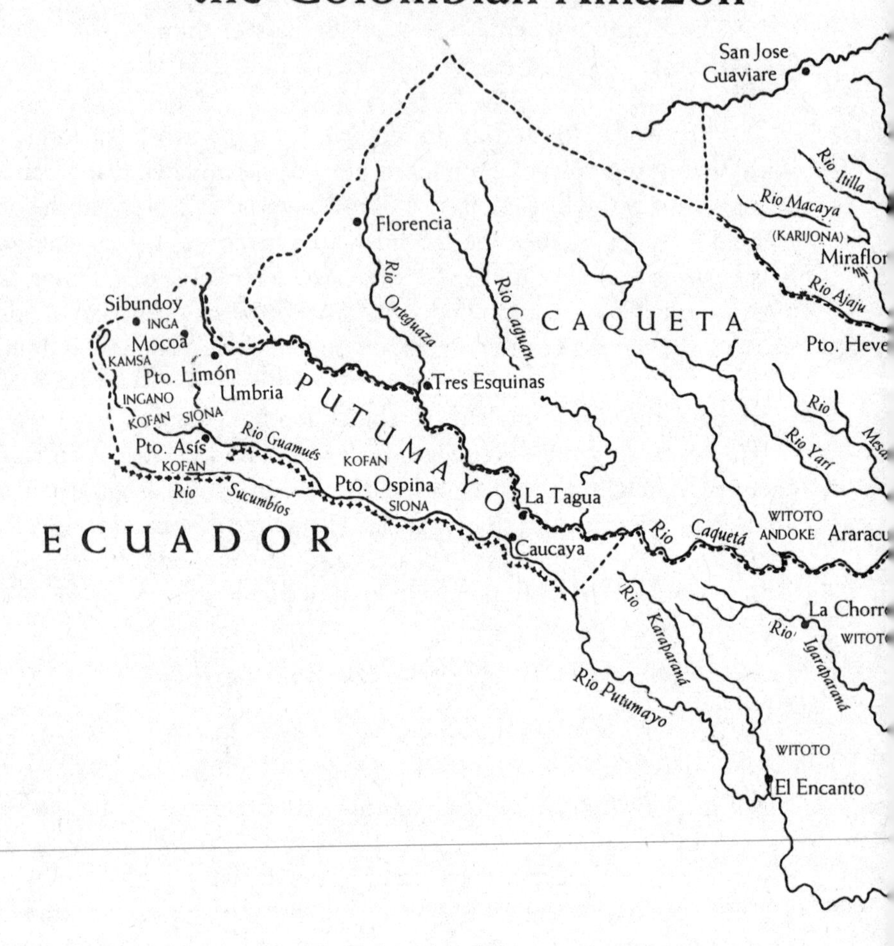

Rio Inírida
KURIPAKO
KURIPAKO
Rio Guainía
San Felipe
VAUPES
Rio Querari
Rio Kuduyarí
Naré
KARIJONA
Rio Kubiyú
Vaupés
Mitú
KUBEO
Victoria
DESANO
GWANANO
Rio Papurí
Apaporis
Rio Kananarí
TUKANO
BARASANA
MACÚ
Rio Piraparaná
KARAPANA
MAKUNA
PUINAVE
TAIWANO
BRAZIL
NIMUKA
Soratama
Jirijirimo
San Miguel
Caño Guacayá
Rio Popeyacá
Yayacopi
Jinogojé
MAKUNA
Rio Tarafra
YUKUNA
Rio Miritiparaná
MATAPIE
Rio Caquetá
MIRAÑA
La Pedrera
MAZONAS
Arica
MUINANE
Tarapacá
Rio Cotuhé
Rio Loretoyaco
TIKUNA
Pto. Nariño
WITOTO
Rio
Amazonas
Leticia

are in that Dodge truck. Almost a ton in a heavy bomber. Each one of the battleships sunk at Pearl had more than twenty thousand rubber parts, more than one hundred sixty thousand pounds all told. Every hose on every ship, every valve and seal, every tire on every truck and plane we ship to the Russians. It's wrapped around every inch of wiring in every factory, home, and office in America. Conveyor belts, hydraulics, inflatable boats, gas masks, rain gear, it's all rubber."

Rands stood up once more from his desk, grabbed a pointer, and moved across the room.

"The year before Hitler started this mess we imported over six hundred thousand tons. Cost us $300 million plus. We were making seventy percent of the world's cars, something like eighty million over the last forty years, and each one with four tires and a spare, and each tire with thirteen pounds of rubber in it. We were buying over half the international production. That added up to about an eighth of the value of all our imports."

Rands stopped in front of the map and pointed to the Far East.

"Here's the dirty little secret. In the last year of peace ninety-nine percent of the world's rubber came from the plantations of Southeast Asia—mostly British Malaya and the Dutch East Indies, almost all of it concentrated within fifteen degrees longitude and latitude of Singapore."

"First place the Japanese planned to hit."

"That's correct. Our entire economy is dependent on a product that has to travel halfway around the world, seven weeks in slow-moving freighters, from lands and across sea-lanes now controlled by the enemy. We only got through the first months of the war because of our stockpile. Technically, we had half a year's supply in reserve, but some of it was in transit. A lot was merely contracted for, and a great deal was stored in warehouses in shipping ports thirteen thousand miles away. At any one time we never had more than half the reserve on American soil."

"But we've been at war for nearly a year."

"For once the government saw it coming, or at least someone did."

By the end of 1939, Rands explained, rubber inventories were down to 125,800 tons. On May 1, 1940, on the eve of Hitler's assault on France and with America threatened in the Pacific by Japan, President Franklin D. Roosevelt was informed that the nation had, at best, a three-month supply of a vital raw material without which its military and civilian economies could not function.

"First thing the president did was set up the Rubber Reserve Company. Congress authorized it to spend $140 million, and with the help of the big rubber firms—Goodyear, Firestone, Goodrich, U.S. Rubber—we secured three hundred thirty-three thousand tons on the international market. That bought us the breathing space."

"What about synthetics?" Schultes asked. Rands shook his head.

"We'll see what they can do. Right now the quality is not there—or the quantity. In 1939 the entire industry produced only twenty-five hundred tons."

"So how can I help you?"

"We need to get hold of every bit of rubber—find it in the forests and squeeze it out of plants even you've never heard of. We're checking out every latex-bearing plant in the Americas. I have Russian dandelions growing in forty-one states!"

Rands swung away from the map and tossed the wooden pointer across his desk.

"I want you and young men like you to go to the source. I want you in the Amazon. I want to know where the best hevea rubber is, how much you can get out, and how soon. Let us worry about the cost. You find the rubber."

"When do I go?"

"Within a week," Rands answered. "You'll be assigned as a field technician to the Rubber Reserve Company. It's Commerce, but you'll be working closely with us. And one more thing: This time we're going to make our own plantations."

"But what about the blight?"

"That's where we come in. Within a decade you're going to see healthy rubber plantations in this hemisphere. Never again are we going to be sitting ducks dependent on British and Dutch ships to bring us an indispensable product from lands they can't manage to control."

There was a knock at the door, and Mrs. Bedard looked in.

"Excuse me, Dr. Rands. Your next appointment—"

"I'll be right along."

"Yes, sir."

"And Mrs. Bedard—would you please take Dr. Schultes upstairs to fill out the various forms? He's working for the government now." Rands looked at Schultes, who grimaced at the thought. They stood up, shook hands, and Rands led him to the door.

"Dr. Schultes—Dick, if I may. Whatever you do, let's not be too vocal about all this. The newspapers are making enough of a disturbance, and they don't know the half of it. It wouldn't help for people to know how close we came."

"I understand."

• • •

With only a week to prepare for the field, Schultes returned immediately to Boston. He had more than logistics on his mind. Before heading overseas there was someone he had to see. The truth was that Dick Schultes had been spoken for, although he didn't quite know it. Two months before, on September 25, 1942, Schultes had met at his sister's wedding a young singer, a Scottish soprano named Dorothy Crawford McNeil. For Dorothy, who already had her own radio show and was as liberal as he was conservative, it was love at first sight. Right from the start, though, she sensed that things might be moving a little slowly, especially when Dick invited his sister Clara along as a chaperone on their first date. Never could she have imagined that it would take seventeen years for the courtship to lead to marriage.

For much of that time he would be in the forests of the Amazon. She spent the war with the U.S.O. in the Aleutians, flying from Attu to Point Barrow, singing her cords dry in over 250 concerts in less than eight months. There were five young women in her concert division, all with the rank of captain, all happy to be guarded with machine guns as they danced through the nights with hundreds of lonely GIs. The end of the war found Dorothy in Italy, living outside of Naples in the King's Palace. On New Year's Day, 1946, she sailed home on a Victory ship, one of five women among five thousand happy and pleasantly drunk soldiers. Through the late forties she continued to sing in Italy, Austria, at the Edinburgh Festival, and most frequently with the New York City Opera. Whenever she was in the city, she moonlighted as a vocalist with Phil Spitalny's all-girl band. It helped pay the bills while she waited for Dick to return. She never considered another man. They were finally wed on March 26, 1959.

The chain of events that brought Richard Evans Schultes to the center of the wartime rubber emergency began nearly a century before in London, in the Whitehall chambers of Clements Robert Markham, an ambitious official of Britain's India Office. Markham was no botanist, but he had a prescient understanding of the economic potential of certain tropical plants. In 1854 he had been instrumental in securing from Peru and Ecuador the cinchona seeds that became the basis of the quinine plantations of Ceylon and India, an industry so successful that the price of the antimalarial drug had dropped by a factor of sixteen in less than a decade. In 1870 Markham turned his attention to rubber.

It was already well known that rubber could be derived from as many as one hundred different species of plants. For sixty years the British had depended on *Ficus elastica,* a native of Asia that flourished along the floodplain of the Brahmaputra River. Ruthless exploitation of the wild populations and failure to successfully establish the tree in plantations forced them to seek alternatives. In the Congo the Belgians were harvesting the thick latex of *Landolphia,* a forest liana. The natives of Mexico and the West Indies extracted good rubber from *Castilla elastica,* a tree in the fig family. Northeast Brazil was the home of Ceara rubber, a wild relative of tapioca. At the India Office, Clements Markham's first challenge was to choose which species to go after, a critical decision that he delegated to one of his consultants, James Collins. The task did not prove difficult. By far the best rubber on the market and the most expensive was called Pará Hard Fine. Derived from a species of *Hevea* of uncertain origin and named for the Brazilian city at the mouth of the Amazon, it was already coming into Britain at a rate of three thousand tons a year, at a cost to British manufacturers of some £720,000 sterling.

Having decided to break the Brazilian monopoly, Markham turned to Sir Joseph Hooker, director of the Royal Botanic Gardens at Kew. For some months Hooker had been in touch with an eccentric English planter, Henry Wickham, who was living at Santarém, a small Amazonian town located four hundred miles upstream from Belém. An aspiring artist and the son of a London milliner, Wickham had fled England in his early twenties and wandered for several years through the wilds of Nicaragua and Venezuela, eventually drifting up the Orinoco and down the Amazon to Santarém, where he established a modest farm. In 1872 an account of his travels appeared in London. It was this rather mediocre book, replete with many enticing though imprecise references to hevea rubber, that drew the attention of the botanists at Kew. In July 1874, Hooker and Markham agreed to pay Wickham ten pounds for every thousand rubber seeds he managed to ship out of the Amazon.

The first attempts failed. The seeds, rich in oil and latex, which readily ferment and turn rancid, could not survive the long Atlantic voyage in sailing ships. Of ten thousand seeds gathered in April 1875, not one germinated in the greenhouses at Kew. Wickham's luck did not change until February 1876, when he received in Santarém a last-minute invitation to dine aboard the *S.S. Amazonas,* a one-thousand-ton ocean liner then inaugurating the Liverpool to Manaus run. It was the first modern steamship to sail the Amazon. Wickham enjoyed a pleasant evening with Captain Murray but thought no more about the ship until a month later when word reached Santarém that corrupt business associates in Manaus had left Murray without an ounce of cargo for his return voyage to Britain.

Wickham moved quickly. Though penniless, he sent a message upstream offering to charter the vessel in the name of the government of India. Then, leaving Santarém at dawn, he traveled by canoe sixty-five miles up the Tapajós to a trading post where he hired Tapuyo Indians to scour the forest for ripe seeds. By chance the ship had arrived at the perfect time of year, and the silence of the forest at noon was broken everywhere by the sounds of exploding capsules of hevea fruits scattering seeds one hundred feet from the bases of the tall, silvery trees. Within a fortnight Wickham had secured over seventy thousand seeds, all carefully packed in moist sawdust and banana leaves in baskets woven on the spot by Tapuyo women.

Just how he and Captain Murray managed to dispatch these seeds to England continues to be a source of controversy. Brazilians, conveniently forgetting that their entire agricultural economy is based on six imported plants—African oil palm, coffee from Ethiopia, rice from India, cacao from Colombia and Ecuador, soybeans from China, and sugarcane from Southeast Asia—still speak of the "rubber theft" as a moment of infamy. Wickham himself, in his memoirs, lent a note of mystery to the deed, no doubt intending to elevate his own profile in the eyes of his peers. In fact, all evidence suggests that the exportation was a straightforward affair conducted in the open and actively facilitated by the Brazilian authorities in Belém.

There was at the time no law prohibiting the export of hevea rubber seeds. Wickham's customs declaration described the cargo as "exceedingly delicate botanical specimens specially designated for delivery to her Britannic Majesty's own Royal Botanic Garden of Kew." Fully aware of the fragility of the shipment, the Brazilian official in charge went out of his way to expedite the paperwork. Nor is there any evidence that the Brazilians were concerned even when the precise nature of the cargo was known. Indeed, it would be eight years before the country imposed so much as a modest duty on the export of hevea seeds, and four decades before the trade was finally prohibited. At the time the entire idea of establishing viable rubber plantations was dismissed by Brazilians as a fantasy. A representative of the Manaus Chamber of Commerce declared, "If it was not our duty to keep abreast of scientific developments, we could completely ignore these foreign plantations."

All initial indications suggested that the Brazilians had little to worry about. Following a rapid ocean voyage, Captain Murray and the *Amazonas* docked at Le Havre on June 9, 1876. Wickham raced across the Channel, woke Hooker in the middle of the night, and urged the immediate dispatch of a royal train to Liverpool, where the ship was expected the following morning. For the next week the gardeners at Kew worked around the clock, clearing the greenhouses, preparing the beds, and planting seed. By June 15 all seventy thousand seeds had been sown. The first seedlings appeared on the 19th. By July 7 hundreds had sprouted. In all, some 2,800 plants germinated, a remarkable survival rate of better than 4 percent. Protected by specially designed portable greenhouses and escorted by the Royal Navy, the first shipment of seedlings left England on August 12, bound for the Peradeniya Botanical Gardens in Colombo, Ceylon. The total cost of the operation, including the efforts of Wickham, was just over £1,500.

Although the entire rubber industry of the Far East would eventually be built on the progeny of these original Wickham seeds, the initial introduction proved a disappointment. Convinced by the Colonial Office that the Amazon was one vast swamp, the Ceylonese planted the first generation of Asian seedlings along the Kaluganga River in a region that receives over 170 inches of rain a year. The seasonal floods wiped out more than thirty thousand young plants. Even in areas suitable for rubber, the British had difficulty persuading farmers to try the new crop. Tea in Ceylon and coffee in Malaya were established and profitable, and no one knew quite what to make of rubber. One planter in northern Borneo grew one hundred hevea trees to maturity and then sent workers scrambling up the trunks. When no rubber balls were found dangling from the branches, he ordered that the plantation be cut down. Others insisted that there was no point growing rubber in the East, when in the Americas it could be mined from the earth. It would be almost twenty years before the potential of the plant was realized.

The pivotal events occurred in the last decade of the century. In 1888, Henry N. Ridley, a young protégé of Sir Joseph Hooker, took over as head of Singapore's botanical garden. Among the garden's inventory were nine rubber trees and one thousand young plants, all descendants of twenty-two seeds introduced to the Malay Straits in 1877. With an annual budget of just one hundred pounds, Ridley set out to build an industry. Importing seeds from Ceylon, he immediately raised eight thousand additional plants. He then turned his attention to the problems of production.

At the time planters maintained that only mature, twenty-year-old trees could be tapped, and then only once every two years. To do so they cut out segments of the bark or used an ax to make deep incisions up and down the trunk. Both methods resulted in the formation of large burrs that eventually made the trees untappable. Ridley showed that by slicing very thin layers of the bark, it was possible to cut across the latex-bearing vessels without damaging the cambium, the layer of cells that provides the base for the growth of the wood. With care, trees as young as four years could be tapped on a daily basis and almost indefinitely. Yields were highest, he proved, if the wound was reopened in the morning, in a spiral fashion with the sloping, latex-bearing channel cut into the bark from left to right. By 1897 all rubber tapping in Asia was based on Ridley's method. Yet another of his innovations was the use of acid to coagulate the rubber, a procedure that allowed the latex to be processed on an industrial scale.

It was at this time, just as Ridley was pushing for the establishment of plantations, that three critical factors converged: The price of tea collapsed, a devastating fungal disease struck the Malaysian coffee crop, and Americans embraced the automobile. In November 1895, Ridley finally persuaded two young coffee growers to plant two acres in rubber. By 1907, just twelve years after this initial experiment, plantations containing 10 million rubber trees grew on three hundred thousand acres of land in Ceylon and Malaya. In that year alone a wave of 387,000 Chinese and Tamil immigrants flowed through Singapore to find work in rubber. Production doubled every twelve months. By 1909 Malaya had planted over 40 million rubber trees, each spaced just twenty feet apart in neat rows that allowed a single worker to tap over four hundred trees a day. Each tree produced eighteen pounds of rubber a year, roughly five times the yield of even the most prolific wild hevea in the Amazon. In 1910 fine amber-colored rubber from four-year-old plantations in Malaya fetched twelve shillings tenpence a pound on the London exchange. The cost of producing this rubber was approximately ninepence a pound, a fifth of what the Brazilian rubber barons were spending. In a move symbolic of the shifting fortunes of the era, the United States Rubber Company, the company that built and paid for the celebrated streetcar system in Manaus, abandoned Brazil in 1910 and invested in a ninety-thousand-acre rubber plantation in Sumatra.

It was in Sumatra and in the Dutch colonies beyond the Java Sea that the experiments took place which guaranteed the final demise of the Brazilian rubber industry. Working again with material derived from the original Wickham stock, a Dutch botanist named Pieter Cramer revealed for the first time the remarkable variability that characterizes the principal species, *Hevea brasiliensis.* In plantations grown from seed there was no way to predict latex production in individual trees. The key, Cramer determined, was to select high-yielding clones and propagate them vegetatively. Instead of dispensing seeds, the Dutch patented a method of marketing budwood from the very best trees. The results were astonishing. An acre of hevea planted from seed produced 350 pounds of rubber a year. Yields from the first selected clones doubled that figure, and with each successive generation, production doubled once again. In less than a century some plantations would experience a sevenfold increase in yield, with the best sites ultimately producing as much as 3,000 pounds of rubber an acre.

With the success of the plantations in the Far East, the Amazonian rubber boom imploded. In 1910 Brazil still produced roughly half the world's supply. Of the 94,000 tons that came onto the international market that year, the Eastern plantations contributed only 10,916 tons. Within two years, however, the output of the Far East equaled that of Brazil. By 1914 the yield of the plantations was more than twice that of all South America. Four years later, at the end of World War I, the plantations produced over 80 percent of the world supply. In 1934, largely in response to the ever increasing demand for automobile tires, international production for the first time surpassed a million tons. In that year plantations covering 8 million acres of the Far East produced 1,006,000 tons of rubber. In South America the industry was dead. Total production for the entire continent did not exceed 10,000 tons. By 1940, as the Japanese finalized their plans for conquest, Brazil produced only 1.3 percent of the international rubber supply, and the nation had become a net importer of the product she had given to the world.

Seven days after meeting with Rands in Washington, Schultes passed through Barranquilla en route to Bogotá, where he was immediately engaged in the search for wild rubber. His first assignment took him south to Cali and Palmira to chase down a rumor that feral populations of *Cryptostegia,* rubber-bearing vines originally native to Madagascar, were growing throughout the valley of the Cauca River. Meeting his old friend José Cuatrecasas, he drove north, and then south, finding in four days only a few individuals of *Cryptostegia grandiflora* planted as ornamentals. His report, filed in Bogotá on December 10, 1942, suggested that the only way to derive rubber from the species would be to propagate the vine from seeds obtained from the cultivated plants. As for the extraction of latex, he noted, neither he nor anyone else had the slightest idea how to go about doing it in a manner that would prove efficient and economical.

If this initial foray proved somewhat frustrating, Schultes's next commission was both quixotic and dangerous. In the fall of 1942, Alfredo Londoño, a Colombian lawyer representing three wealthy landowners, approached the Rubber Reserve Company with a proposition. Londoño's clients owned three extensive concessions, covering in total some 130,000 acres along three small affluents of the upper Caquetá River. Title to these holdings had been granted in 1891, and all were rumored to be rich in rubber. According to Londoño, the lands also contained gold and tin, extensive salt deposits, groves of quinine, and entire valleys where the air was sour with the scent of petroleum. The Colombians were willing to finance a proper survey, provided the United States Embassy could supply the scientific personnel.

On December 19, Jules de Wael Mayer, the Dutchman in charge of the rubber effort in Colombia, sent Schultes a memorandum that read simply, "The object of your trip will be the investigation of the varieties of rubber trees that exist within the limits of these three concessions and the sampling of various areas in order to establish an estimation of the number of trees which these concessions may contain." On paper it was a straightforward task. The problem, as Schultes would soon discover, was the location of the concessions. All three lay south and east of the last range of the Andes, in a remote and wild mountainous region that was uninhabited. According to the maps available at the embassy, the westernmost holding, containing perhaps twenty thousand acres, was a rectangular plot on either side of the upper Villalobos River, a tributary that drains into the Caquetá thirty miles above Puerto Limón. The concession farthest to the east was a diamond-shaped piece of land somewhere along the headwaters of the Pescado, a small river that flows into the Orteguaza. The largest of the three lay in between, spread across ninety thousand acres of the Fragua drainage, a Caquetá affluent located fifty miles downstream of Puerto Limón.

With the exception of the odd hunting party of Indians, nobody had entered the region in thirty years. One who had tried—a surveyor named Alfredo Perry, hired by the Colombians the previous year—had given up after a week. The owners themselves had never been in the forest and had only a vague sense of the boundaries of their lands. That did not stop them, however, from insisting that the best approach was from the north, from the small highland town of Pitalito, across the mountains and down the slope to the headwaters of the Villalobos. Schultes, who had been on the upper Caquetá a year before during his Putumayo expedition, wanted to enter the concessions from the south, traveling upriver. To his later regret he relied on the judgment of his sponsors and thus ended up cutting one hundred miles of trail through one of the wettest and most inhospitable regions of South America.

The expedition began the day after Christmas. Leaving Bogotá by train, Schultes reached the city of Neiva, where he stopped long enough to secure a permit to carry arms and then proceeded by local transport along the one-lane dirt road that headed south to Pitalito. His bus, like all those making the run for Transporte Huila y Caquetá, had four tires instead of the usual six, and the spares were stuffed with sawdust. The rubber shortage, Schultes noted, had reached the hinterland. In Pitalito there was bad news. Workers hired by the local agent of the owners to slash a trail from the mountains through to the Caquetá had run out of food and were forced to return. One had succumbed to malaria; another had been critically injured in an accident.

After five days spent meeting with local officials, hiring new men, and organizing supplies, Schultes left Pitalito on Sunday, January 3, accompanied by two representatives of the owners, a botanical assistant named Villareal from the Universidad Nacional in Bogotá and a crew of fifteen *campesinos* employed to carry baggage and cut trail. For the next two weeks, over terrain too rough for mules, Schultes led the party south across a series of high ridges and through the forested headwaters of a dozen small rivers and streams that drained into the Villalobos. It was in these wet and cold upland forests that he found *caucho blanco,* white rubber, a species in the genus *Sapium* that yielded a quality of rubber equal to that of the finest Amazonian product.

Unfortunately, the trees were neither sufficiently abundant nor large enough to warrant commercial exploitation. On even the richest sites along the steep slopes at the headwaters of the narrow *quebradas,* in a cloud forest dappled with the silvery white foliage of *yarumo blanco,* there were never more than two or three individual rubber trees per acre. Few of these were more than six inches across; fewer still had been growing for more than thirty years. *Sapium* may be tapped, but in the early days of the rubber boom it was common for the trees to be felled and the latex gathered on leaves wrapped around the trunk. After just a week in the forest Schultes realized that the wild stands of *Sapium* had been ravaged by the *caucheros* who moved through the country at the turn of the century. The few trees that had escaped the ax bore the characteristic tapping scars: horizontal bands roughly eight inches apart, climbing like tracks up their trunks.

Still eager to reach the Caquetá, Schultes dispatched several of his men to cut a trail along the Villalobos to the Indian settlement at Yunguillo. By this time he knew that the only way to reach either the Pescado or Fragua concessions would be by river, as he had initially suggested. To go by land was a journey of a month and a half, over ground that had nearly broken the spirit of his men in a fortnight. Instead of waiting for the trail to be opened along the Villalobos, Schultes turned north, extending his survey in a series of broad sweeping movements through the mountains, eventually returning once more to Pitalito in the third week of January. There he met several old rubber workers, including a crusty fellow named Daniel Sanchez who had once worked for Arana at El Encanto. Sanchez confirmed that nowhere in the Villalobos were there merchantable stands of *caucho blanco.* After a month in the forest Schultes left his botanical assistant Villareal to wrap things up while he returned to Bogotá. His crew, hired to cut a trail down to the Caquetá, never reached its destination. Rains flooded the forest, the rivers became torrents, and finally, after seventeen days, the men returned sick and exhausted to Pitalito. Arriving in Bogotá on January 27, Schultes had little to report, save that the forests of the upper Caquetá were of no value to the war effort. In his wanderings he had made one important discovery, a new source of rubber known as *caucho colorado,* a tree that turned out to be a new species in a genus then known as *Piratinera.*

Needless to say, the officials safely sequestered in Washington did not fully appreciate the logistical difficulties faced by the rubber explorers in the field. Poring over brightly colored maps that bore no resemblance to reality, setting hopelessly unrealistic production quotas, the officials of the Rubber Reserve Company spun out reams of paperwork, endless memos that shifted men around like chess pieces. For the most part these notices could be ignored. The rivers of the Amazon, Schultes would claim years later, were clogged with all the government forms he had dumped over the side of his canoe.

Caught between the bureaucrats in Washington and the small band of explorers was the old Dutch man known always to the men as Mr. Mayer. A portly figure well past retirement age, Mayer had worked for years in the plantations of the Far East. Posted to Bogotá as senior field technician, he had taken on the task not only of maximizing Colombian rubber production and exploration but of insulating his men from the whims and idiocies of his superiors back in Washington. For that he had earned the affection of all those working under him.

In January 1943, while Schultes was in the forests beyond Pitalito, an exchange of letters occurred between Mayer and the head office in Washington that determined the young botanist's fate for the coming year. For some time Mayer had been under pressure to begin rubber production in the Vaupés, a vast roadless track of the Colombian Amazon that falls away to the eastern frontier with Brazil. In a letter dated January 16, 1943, Mayer reminded his boss, Earle Blair, that a basic survey of the region had yet to be completed. In fact, the technician responsible for the work, Everett Vinton, was currently in Gorgas Hospital in Panama and would not be available for some time. With characteristic understatement Mayer wrote: "It is much to be regretted that Mr. Vinton's work has been interrupted by a combination of Dengue and malaria, the more so as Dr. Schultes is making an inspection of the upper Caquetá. This has left us without a field man to continue Mr. Vinton's work."

Blair's response on January 26 was typical of the head office. As long as Schultes was working in the area, he wrote, perhaps he might have a look at a rubber operation in the vicinity of the military post and Catholic mission at La Pedrera. Blair had no idea what this casual suggestion entailed. For Schultes to get to La Pedrera from where he was meant not only finding a route overland to the Caquetá but also securing a dugout canoe for a six-hundred-mile paddle downstream through country inhabited only by Indians.

Such a journey was, in fact, exactly what Mayer had in mind for Schultes—only the river was the Apaporis, the distance to be covered was 1,350 miles, and the banks of this black water river were inhabited by no one. Unlike Blair, however, Mr. Mayer was fully aware of the inherent dangers of such an expedition. He wrote Blair on January 16, "I share your anxiety to penetrate the Apaporis basin; the local legend has it that this is the real mother lode of Colombian Syringa [rubber]. . . . I hope that during this year we will be able to make a thorough exploration of this river area, so that its development can be undertaken on a far greater basis of security than has been the case with the Vaupés, where we are still guessing." Dismissing aerial surveys as undependable, Mayer continued, "The only way to make safe estimates is the hard way, and for lack of time and manpower, under present circumstances, it is just as necessary to take risks in the hope of success, as it is in warlike operations. I strongly believe in this so long as the risks are carefully studied before they are embarked upon."

The river in question was the least known and most isolated of all the major waterways of Colombia. Formed by the confluence of the Ajaju and the Macaya, rivers themselves born not on the slopes of the Andes but farther to the east in the midst of the trackless savannahs of Caquetá and Meta, the Apaporis flows through a series of flattopped sandstone mountains, remnants of the ancient landmass that once rose above the forests of the Guianas, Venezuela, and Colombia. Eroded by time and transformed by rain, these isolated massifs stand as lonely sentinels, serene and otherworldly, above a river that winds like a serpent.

There are treacherous rapids in the Macaya and in the first thirty miles of the Apaporis as it flows past the base of a long ridge known as Chiribiquete. Then for three hundred miles the river moves gently through undisturbed forest until suddenly the somnolence is broken by the great falls and cataracts of Jirijirimo. Here the river, perhaps half a mile wide, narrows to one hundred feet, drops seventy feet over a ledge, and enters a chasm nine miles long and fifty feet wide with towering rock walls on both sides. Disappearing into a tunnel of stone, it flows, deep and silent, through a mysterious fault until emerging on a beautiful plain of low forest and savannah spread across dazzling white sands. Below Jirijirimo there are another four hundred miles of rapids until the Apaporis finally meets the Caquetá. The most dramatic of the rapids is Yayacopi, a wall of rock forty feet high over which flows the entire river.

According to the Makuna Indians who live up several of the tributaries of the lower Apaporis, the falls at Yayacopi, like all the rapids of the river, were breathed into being by a primordial shaman who struck a deal with the gods. For generations the Makuna had been attacked and once nearly annihilated by a ruthless tribe of cannibals who dwelt at the headwaters of the river. Seeking an end to the terror, the shaman drank *yagé* for seven days, and in the course of his visions, the spirits agreed to transform the land, creating mountains and impassable cataracts that would forever separate the Makuna from their pitiless enemies. Since that time the Makuna have never traveled above Yayacopi, and the banks of the upper Apaporis have remained uninhabited.

The whereabouts of this legendary tribe of cannibals was of some concern to Schultes as he arrived by plane on March 3, 1943, at Miraflores, the rubber station recently carved out of the forest on the north bank of the upper Vaupés River. His assignment was to survey the upper Apaporis, identify potential rubber-gathering sites, and then proceed downstream, mapping the river and measuring the concentration of hevea over its entire length. Each task required men, and Schultes always preferred to work with Indians. From what he could gather the tribe in question was the Karijonas, a Carib-speaking people who had originally migrated to the headwaters of the Vaupés and Apaporis from the upper Orinoco. A fierce nomadic people, numbering perhaps twenty-five thousand at the turn of the century, their population had collapsed as a result of bloody intratribal warfare and the spread of smallpox. No one in Bogotá had the slightest idea whether any of them remained at the headwaters of the Apaporis, on the Ajaju and Macaya.

"Doctor!"

Schultes turned to see none other than Nazzareno Postarino, the blond Italian who had traveled with him down the Putumayo, standing alone and waving at the edge of the grass airstrip. It was a good sign. Postarino worked hard, and Schultes would need men like him on the river. After the Putumayo, Schultes had recommended him to the embassy.

"How was your flight¿"

"Wonderful," Schultes said, "except for him." He glanced toward the plane where an extremely unhappy mule was being unloaded.

"By God," Nazzareno suddenly exclaimed. "There's a hole in the side of the airplane."

"Yes." The aircraft was a German trimotor Fokker, a frame of steel covered with corrugated sheets of aluminum. Apparently just after take-off at Villavicencio the mule had broken loose and kicked one of the panels off the side of the plane.

"That pilot has a wonderful way with animals. He just swore at it, and somehow he settled the thing down."

"Who flew the plane?" Nazzareno asked.

"I did, naturally."

"I didn't know you—"

"Someone had to," Schultes said as they made their way toward the row of shacks that made up the settlement. "Now, tell me, Nazzareno. What's the word on the Indians?"

"You can ask for yourself. They're at Puerto Naré, just a few miles downriver. About twenty families. That's all that's left. They moved over to the Vaupés five years ago. Their chief's named Mora. He's blind, but he tells me there's not a Karijona anywhere on the Apaporis."

"I see."

"He says the place is deserted. And he says you should know about the rapids."

Schultes hesitated for a moment. Behind them, they could hear the clang of hammers on metal as the crew tried to repair the plane.

"Nazzareno," he said, "you have shaved that godawful beard of yours."

"Yes. Everyone thought I looked like a priest."

"Everyone?"

"The girls, mostly."

After two days at Miraflores, Schultes knew he was in trouble. The supplies for the expedition, though promised, had not arrived. Among the workers hired by the Colombians, Schultes had expected to find a few familiar with the region, perhaps the sons or grandsons of the men who had combed the Apaporis for balata rubber during the boom. A nonelastic rubber much in demand before the advent of plastics, balata had been used primarily for covering underwater cables. The trade had died on the eve of World War I, but not before the *balateros* had a chance to discover and report vast concentrations of hevea in the upper Apaporis. According to rumor there were still *balateros* in the area, and it was these men Schultes hoped to meet at Miraflores. Instead he found that all the workers were *llaneros,* ranch hands from the eastern savannahs who knew little of the forest.

Schultes visited Mora, the old Karijona chief at Puerto Naré, who spoke of terrible rapids, but there was no way of locating them with precision. Maps meant nothing to the Indians, and those that Schultes had brought from Bogotá were worthless; each placed the river and its affluents in different alignments. One, he would later discover, had the Apaporis going over the top of a mountain range. The only map he could trust had been prepared by the American Geographical Society. At least it was honest. It depicted the upper three-quarters of the river with a broken line, indicating that its course was unknown.

Returning to Bogotá on the first available plane, Schultes approached the U.S. military attaché Colonel Manuel Ascencio and requested a reconnaissance flight over the Apaporis and Macaya. By Tuesday, March 9, he was back at Miraflores, flying southeast toward the Apaporis in a small plane piloted by a Colonel Greenbank. Once over the river, they turned east and followed the valley to within sight of the Brazilian frontier where the Caquetá and the Apaporis flow side by side, separated only by a narrow neck of land. There in the distance loomed a large and conspicuous cone-shaped mountain, which Schultes correctly identified as Mount Kupatí, the sandstone massif first explored in 1828 by the German botanist Von Martius. At its base, just below a series of violent rapids on the Caquetá, lay the military post of La Pedrera.

Turning upriver, Colonel Greenbank throttled back and flew low to the ground so that Schultes could take the river's measure. There were dozens of rapids, at least fourteen that were clearly impassable. The gorge at Jirijirimo stretched for miles, and the Indian settlements rumored to exist at the mouth of the Río Kananarí, just above the falls, consisted of two huts on an elevated clearing. The thatch was bleached white by the sun, the gardens overgrown, the land deserted. There was no sign of life, there or farther up the Kananarí, a black water river itself marred by rapids. Beyond Jirijirimo the Apaporis widened to a broad silt-laden river contorted into immense loops that significantly increased the length of a journey by canoe. The distance to be traveled by water, Greenbank cautioned Schultes, was over eighteen hundred miles, a third again as far as his superiors in Bogotá had indicated.

Schultes quickly formulated a plan of action, the details of which he expressed in a memo sent to Mr. Mayer on March 15. Logistics promised to be a serious problem. Simplicity would be essential. Thus, instead of two boats and crews, Schultes would make do with one, a thirty-foot dugout with an outboard, a machinist, and a limited supply of gas. To navigate shallow rapids they would secure two balsa logs alongside the canoe. As crew he would take Nazzareno, two Karijonas, and possibly one white worker. Materials would be taken upriver from Miraflores along the Río Vaupés, up the Río Itilla, and then carried overland to the Macaya across a forested trail once used by the *balateros.* The canoe would be built on the Macaya, but the bulk of the expedition's gear would come overland by a second route, a two-day portage from Miraflores southeast to the Apaporis. There a camp would be built and a resupply system established by means of runners moving constantly between the drainages. It was essential, Schultes noted in closing, that he be able to depend on the Colombians at Miraflores after the expedition was under way. Once on the river he and the men would be completely cut off from the world.

On April 3, after two weeks in Bogotá, Schultes left Miraflores by river and began the 180-mile journey up the Río Vaupés to the confluence of the Unilla and Itilla rivers, and beyond to a landing on the south bank of the Itilla known to the *balateros* as Puerto Trinidad. There he left Nazzareno and one of the workers to construct a thatch shelter and camp while he and the rest of the crew moved on. Finding no sign of a trail, Schultes walked south-southwest, following a compass bearing that took him through dense forest and across the high and dry *planada* that separated the Itilla from the Macaya. His goal was to scout a route to the river, establish a base, and determine whether it might be possible, as one of the Colombians at Miraflores had suggested, to requisition canoes on the spot. From what he had seen from the air, Schultes thought it unlikely that anyone was living on the Macaya. Still, he retained some hope that they might be spared the arduous task of hollowing out logs and building boats in the middle of the forest.

There were three men with him: a Karijona guide known as Barrera and two porters, *llaneros* named Julio and Franco. Both were young and strong. The traverse was expected to take six hours. After more than eight, Julio insisted on pushing ahead.

"You'll find your way¿" asked Schultes.

"Of course," Julio replied. "The river must be just a little ahead."

"Barrera¿" Schultes looked toward the Karijona.

"Es más allá," the Indian said.

"Lo que es más allá, es más allá," said Franco. What's farther ahead is farther ahead.

"What will I say to your boss if you get lost¿" Schultes smiled.

"Tell him that I was taken by the Karijona and that I'll see him on the savannah. I'll be the sweat of his horse that spits in his eye."

"Franco, what about you?" Schultes said with a laugh.

"I'm staying with him," he said, pointing to Barrera.

"Fine," Schultes said. "All right, Julio. Go ahead, but be careful. And keep the sun on your right."

"In this jungle? If there is a sun, who could find it? But I'll see you in camp."

"¡Ojalá!"

"Señor Julio," Barrera said quietly. "When you see *seringa* growing alongside *juansoco,* the river will be close."

"Bueno. Entonces nos veremos más tarde." We'll see each other later.

"Sure."

Three hours later Julio did see the crew again, but it was not on the banks of the Macaya. Schultes had been walking slowly, collecting the odd specimen of hevea and informally surveying the latex-bearing trees along the route. So far he had found at least five species of hevea that Barrera distinguished as *seringa blanca* and *seringa amarilla,* based on the white or yellow color of the latex. There was also *caucho negro,* named for the black bark of two species of *Castilla,* considerable amounts of balata, and several *juansoco* trees, a source of chewing gum. Schultes had paused in a shallow draw and was preparing a specimen of *Hevea guianensis* when suddenly Julio appeared, eyes to the ground, walking with great confident strides from the opposite direction. He was almost upon them before he looked up.

"What are you doing here?" he asked.

"Collecting plants," Schultes said. "What are you doing here?"

Unfamiliar with the forest and burdened by a heavy load, Julio had walked in a great circle.

"Perhaps from now on," Schultes said, "we had better follow the compass."

Julio evidently agreed. As the party pushed on, he stayed right behind Schultes, whose steady faith in the compass was, for the *llanero,* a thing of mystery and intrigue. Finally, after more than fourteen hours, they reached the river, emerging from the forest just a few hundred yards above a violent cataract they would later name Cachivera del Diablo, the Devil's Cataract. That evening young Julio once again strayed from the group. Attempting to swim the river, he was taken by the current and flung into the boiling white water. His body was never recovered. The next morning Schultes returned to Miraflores to report the death.

• • •

On the day that Julio died, Mr. Mayer dispatched a letter from Bogotá to the Rubber Development Corporation, a subsidiary of the Rubber Reserve Company, indicating that the scale of the Apaporis expedition was about to increase significantly. "In view of the fact that this river is totally uninhabited and completely isolated from the rest of Colombia . . . I thought it advisable to send two men on the exploration party rather than to assume the risks that would be involved if only one man was sent on this mission." With Schultes would be Everett Vinton, an accomplished explorer now fully recovered from the various diseases that had nearly killed him two months before. Each man, Mayer added, would have his own crew and thus "the time necessary to make this preliminary survey will be cut in half." Left unsaid was Mayer's desire to assess the potential of the Apaporis as quickly as possible so that resources could either be concentrated there or diverted to the Vaupés, a more accessible region of lesser promise. With this in mind, Mayer summoned Schultes yet again to Bogotá on April 10 for a round of meetings with Vinton. It was not until April 18 that Schultes was back in Miraflores to lay the groundwork for the expedition.

In the week before Vinton's scheduled arrival on April 24, Schultes went downriver to the Karijona settlement at Puerto Naré, walked the beginning of the second overland route to the Apaporis, and began assembling supplies and a crew. By the time the expedition finally motored upstream, destined once again for the Macaya, it consisted of six carpenters, two sawyers, two cooks, two foremen, two boatmen, and a dozen miscellaneous workers, a total of twenty-eight men who had to be fed and supplied by means of a tenuous overland trail that ran twenty miles, as Vinton put it, from "nowhere to nowhere."

For the next month Schultes was completely preoccupied with the logistics of the expedition. Camps had to be established at Trinidad on the Río Itilla, at the Cachivera del Diablo on the Macaya, and at a point in the forest midway between the two. The trail between the rivers, little more than a vague route, needed to be slashed out to accommodate mule trains. River transport on the Itilla and construction of the boats on the Macaya had to be supervised. Men had to be kept alive. Maintaining the supply lines to Miraflores and beyond was a constant struggle. An order for screws and nails placed in Villavicencio in mid-April, for example, did not arrive for two months. This was typical. It would take until June 11 to accumulate sufficient food at the Macaya camp to begin the exploration of the Apaporis.

While waiting, Schultes and Vinton put the men to work clearing a tract of land and building yet another camp on the high point overlooking the confluence where the Macaya comes together with the Ajaju to form the Apaporis. Impressed by the concentration of rubber trees, they named the site Puerto Hevea and decided to make it the administrative center for the entire rubber effort in the upper Apaporis. Equipped only with axes and machetes, and working with the materials at hand, they built a large complex complete with kitchen, dining hall, storerooms, and sleeping quarters for thirty men. Just across from the camp, on the north bank of the Macaya, Schultes identified a long stretch of flat, level ground ideal for an airstrip.

While Vinton and his crew began clearing the dense secondary forest, Schultes finally had an opportunity to botanize. It had been a long time coming. Over the past four months he had been in the forest for much of the time, but he had collected only thirty plants. Such a hiatus was, for him, not just unusual but profoundly unsettling. He was the first botanist ever to enter this part of the Amazon. Every day he had walked past plants unknown to science. There were moments on the trail when he had buried his face in his hands just to avoid seeing yet another new species that he was unable to collect.

His release came on the morning of May 14 when he began a slow climb across the flank of Cerro Comejen, now known as Cerro Chiribiquete, one of the sandstone mountains that had teased him for so many weeks. These upland formations stretch in a great arc, dividing the Macaya from the Ajaju and continuing south one hundred miles as far as Araracuara on the Caquetá. Known collectively as the Sierra de Chiribiquete, they are, in fact, isolated relics that from the ground appear to emerge randomly from the midst of the forest. Some are immense flattopped ridges with one side sloping gently toward the forest floor, and the other a thousand-foot wall of yellow stone surmounted by jutting strata draped in vegetation. Others are dome shaped, with perpendicular cliffs on all sides dropping to a series of broad and flat sandstone shelves. Massive and remote, veiled in mist and often grotesquely eroded, the mountains seemed to Schultes to echo the dawn of time, rising like giant sculptures left over from God's first workshop. It was from these first tentative experiments, Schultes mused, that He had gone out and built a world.

As Schultes moved slowly across the sandstone ledges, working his way up the narrow draws, he observed in the flora a gradual transition as the lowland elements fell away and were replaced by rare and novel plants, many of them endemic and adapted specifically to the peculiar conditions of the uplands. There were, to be sure, isolated pockets where sand and soil had lodged and a low scrubby vegetation had taken hold. But for the most part on the exposed rock, especially across the summits, the conditions were those of a desert. In the dry season when no rain may fall for a month, the hot sun bakes the sandstone. When the rains do come, they pour off the surface in great sheets, forming waterfalls and torrents that rage for a few hours and then are gone. In the evening throughout much of the year, a curtain of mist drops over the hills, but by dawn it is dispersed and the plants are once again exposed to the intense heat and radiation of the equatorial sun.

What Schultes found on the summit was a grassland interspersed with dense brush of low gnarled shrubs, an island of savannah perched a thousand feet above a tropical rain forest. Adapted to the dry conditions, the plants were reduced in size, and many bore glossy leathery leaves, often coated with heavy waxes or dense pubescence. Their bark was either thick and corky, or thin and coated with wax. Epiphytes had exaggerated pseudobulbs for water storage, and many plants grew low to the ground and had dense rosettes of leaves. The roots were especially well developed, penetrating the cracks and fissures in the rock, reaching like veins across the face of cliffs. The growth forms were exceedingly strange, the overall aspect of the flora elfin and bizarre. Something of the magic of the place is revealed in the names chosen months later by Schultes for two new species of low trees and shrubs discovered that day: *Graffenrieda fantastica* and *Vellozia phantasmagoria.*

Many of the plants produced resins or latex. In fact, as Schultes slashed his way across the ledges of the mountain, he found himself completely covered with white sticky latex that mixed with the heat and sweat to make the climb thoroughly miserable—that is, until he realized that the stand he was cutting through, which had a density of some five thousand individual shrubs per acre, was made up of only two plants, both new to science. One species, which was closely related to balata, he later named after the mountain, *Senefelderopsis chiribiquetensis.* The other was a new variety of rubber, *Hevea nitida* var. *toxicondendroides.* Thus Schultes ended his first foray on the mountains perched above an open cliff, listening as the wind rustled the leaves of these two rubber plants that had never before been seen by a botanist.

The first of the two boats was finally finished at the beginning of June. A simple dugout with sides and decking of sawn timber, and a capacity of two and a half tons, it would have moved effortlessly through the water had the outboard engine arrived from Miraflores. Since it didn't, Schultes and his small crew had no choice but to make their way up the Ajaju the hard way: paddling against the current for more than 150 miles. For the most part the river was wide and meandering, with the only dangerous passage being the narrows of Macuje, a series of standing waves and swift-churning water located thirty miles upstream. It was the point, Schultes noted, that separated lands rich in rubber from those more impoverished. Below the narrows the banks were high and firm and had as many as twenty trees per acre. Farther upstream the land was flat and low, flooded for much of the year. It was the one region of the Apaporis drainage where rubber did not flourish.

Schultes continued up the Ajaju until he was certain there were no areas of commercial potential, and then he drifted back downstream, counting hevea trees as he went. On the way he explored the Yaya, the only notable affluent, and found it full of rapids. The other major tributaries, he discovered, though carefully drawn on the official map of Colombia, did not exist. What he did find was the Cerro de la Campana, the westernmost extension of the Chiribiquete uplands and by far the most beautiful of all the ancient formations. On a day that began in wonder, in hidden caves etched with images of animals and stories scratched into the rock by the ancestors of the Karijona, he found himself alone, peering like a raptor across an endless forest that rose in waves against the flank of the mountain. At the summit he came upon a thinly eroded slab of rock that when grazed by a stone gave birth to fierce thunderstorms and torrents. When struck by the blow of a shaman, it sent forth a bell-like tone that carried across the velvet surface of the canopy, touching all creatures with the breath of life. This, at least, was the story told to him by Barrera, who guided him that day.

The next challenge was the Apaporis itself. Returning to Puerto Hevea, he went downstream along the open face of the mountains and through a series of rapids known as the Cachivera de Chiribiquete. The first impediment was only a whirlpool, and for the next twenty miles the river ran deep and wide, dwarfed by the vista. Then quite suddenly the first of the rapids was upon them. The power of the water, the futility of the paddles, and the weight of the boat left them momentarily stunned. Then a boulder sent the canoe past a shelf of rock close to shore. Barrera dove into the water, fell away beneath the wash, and emerged on the bank, clinging to a liana. They threw a rope. He bound it to a tree and together they pulled themselves to the edge of the forest.

It was just the start of ten miles of rapids. There would be no river transport here. On June 16, Schultes and two of his crew returned through the jungle to Puerto Hevea on foot, escorting an injured man and seeking additional workmen for the portages that lay ahead. Six days later they pushed on, committing their boat to the river. That night cloudbursts soaked and damaged most of their food and equipment. Another man was injured. The rapids were more formidable than

Schultes could have imagined. When finally the rest of the party emerged unscathed fifty miles on, he elected to abandon the canoe and return to Puerto Hevea on foot along the north bank. Ostensibly it was to allow him a more precise estimate of the distribution of hevea rubber; in truth, his men cut through the forest because there was no other way of getting back. Still, it was this trek that yielded a true understanding of the immense potential of the land. As he moved along the shore he realized that the counts he had been making from the water had been consistently low. When he factored this error into his survey results, he discovered that the upper Apaporis together with the Ajaju and the short stretch of the Macaya that lay below the Devil's Cataract supported more than a quarter million rubber trees. Properly exploited, they would yield almost a million pounds of rubber a year. And this was just the beginning, the first short stretch of a river that ran for almost one thousand miles.

Schultes first noticed the faint coloration on his left arm as he made his way overland from Puerto Hevea to Trinidad en route back to Miraflores on the last day of July. The fevers came a day later. He knew it wasn't malaria, for in all the Apaporis they had not seen one anopheles mosquito. If it was yellow fever, he would have expected nausea with bloody vomiting. Something was seriously wrong with him, and he did not know what it was. By the time his launch approached Miraflores there were streaks of red running up his arm, and the pain was intense. His plan was to take the first plane out to Villavicencio and from there catch a bus to Bogotá. It was a fortunate choice of itinerary. A landslide had closed the Bogotá road, and he was forced to seek medical attention in Villavicencio. Had he flown all the way to Bogotá or had he been able to travel by land, there is a good chance that the blood infection running up his arm would have killed him.

As it was, a Colombian physician in Villavicencio nearly did. The doctor seemed to know what he was doing. Perched on the edge of a table in a perfectly respectable clinic, Schultes watched calmly as he received an enormous injection of some mysterious blue compound. He then promptly passed out. He came to in a bed, propped up in a tent lined with burning light bulbs; his arm was wrapped in hot towels. At his bedside was Marston Bates, a biologist who, like Schultes, had received his doctorate from Harvard. An expert in the epidemiology of insect-borne diseases and the world's authority on yellow fever, Bates had been living in Villivicencio for some time, completing a project for the Rockefeller Foundation. Since he and his wife, Nancy, were the only

Americans in town, the Colombian physician had naturally brought Schultes to their door after things had gone wrong at the clinic.

Recognizing immediately the severity of the infection, Bates administered sulfa drugs and with his wife remained by the patient day and night, keeping heat on the arm. Schultes's subsequent recovery was due in no small part to this prompt attention, and to this day he credits Bates with saving his life. As for the Colombian physician, Schultes later discovered that he was a veterinarian.

After a short convalescence in Bogotá, Schultes returned to Miraflores on August 25. This time he planned to explore the middle section of the Apaporis, that part of the river bounded by the Cachivera de Chiribiquete—the rapids that had frustrated him in June—and the great cataract of Jirijirimo, located three hundred miles downstream. In addition to mapping and assessing the navigability of the river, he was to survey the rubber growing along its banks and find a suitable location for a landing strip. In their last meeting in Bogotá, Mayer especially stressed the importance of safety. With Vinton and Nazzareno committed to the construction of the airstrip at Puerto Hevea, Schultes would now be on his own, working with only a reduced crew. He was to proceed as far as Jirijirimo. Under no circumstances was he to attempt the rapids of the lower river. To reach the Apaporis below the Cachivera de Chiribiquete, he would travel overland by a second route, the Naré Trocha, a trail that ran from the Karijona village of Puerto Naré on the Río Vaupés thirty-six miles through the forest to a point on the Apaporis that Schultes would name Puerto Victoria. The condition of this trail was of particular concern to Mayer, for it would be the only means of getting rubber out of the drainage should the program enter the development stage.

As it turned out, Schultes got to know the Naré Trocha well. Instead of building a canoe on the Apaporis side, the company elected to have a sixteen-meter, 1.5-ton boat hauled overland, yard by yard, through the jungle. Under the best of circumstances a loaded mule required three and a half days for the one-way trip. A unencumbered man could make it in two. Schultes hoped to complete the traverse in ten days.

It was a horrendous passage. From Naré to the first camp at Tacunema, a distance of eight miles, the path was broad and straight. But the soil was thick clay, and the rains that fell every afternoon and evening turned it into a muddy slough. What few bridges there were could not accommodate the weight or scale of the load. The men had to inch the boat down each embankment and drag it across the ground on rollers of wood cut from the forest. Between Naré and Victoria there were 120 creeks, and each had to be negotiated in this manner. Photographs taken at the time reveal exhausted men straining against a pole bound by rope across the bow of the boat. Their clothing hangs limp with sweat. Many are barefoot. All are filthy, including Schultes, who works beside them, distinguished only by his thick beard, pith helmet, and the cigar that appears to burn incessantly in his mouth.

Fortunately, the terrain improved once they passed from the Vaupés into the Apaporis drainage. The poorly drained and impervious clays gave way to porous soils of sandstone and lateritic gravel. The first sandstone outcrop lay just beyond their overnight camp at Guaduales, twenty miles inland from Naré. From there on the streams ran clear, the waters drained into the Apaporis, and the sounds of the forest were often subdued by the roar of rapids.

At night they bivouacked in the forest, sleeping beneath thatch *miriti-sabas,* small shelters woven on the spot from palm and heliconia leaves. With no time to hunt, they ate manioc flour, *farinha,* augmented with U.S. Army–issue rations that the men disdained. They drank water and thick coffee boiled with sugar on an open fire. In the evening during the fleeting moments of the tropical dusk, Schultes went alone into the forest, watching for the smooth white bark of hevea. The number of rubber trees was impressive, particularly as they got closer to the river. In well-drained ground it was not unusual for him to encounter as many as twenty mature trees in an acre of forest.

After fourteen days Schultes, his crew, and the boat finally reached the river, on the morning of September 17. His mood was subdued. The condition of the men weighed heavily on him. They were worn and weary, and had little of the spark of those embarking on a long journey. It was no way to begin an expedition. Schultes had argued against dragging a boat overland. His plan had been to walk swiftly over the trail and with the men move upriver until locating the boat that he and Vinton had abandoned in June. With a crew of twelve, he had told Mayer, he could manage the rapids of Chiribiquete. Mayer had remained adamant, insisting on a new boat airlifted to Miraflores from Villavicencio and taken over by land. Now, Schultes thought, he had to deal with the consequences. It left him frustrated and angry, at least until the next morning when, moving slowly up the Apaporis, he saw splintered boards spinning like flotsam in a whirlpool at the edge of the river. They were the remnants of the boat he had left behind. Broken loose from the shore, it had been crushed by the river. He laughed aloud. Mayer had been right.

Schultes had no precise idea where he was, nor did any of the men.

His assignment left him no choice but to head upstream from Victoria, hoping to reach the point where he and Vinton had been forced to abandon their survey in June. From there he could begin. For transport on the river he depended on a twenty-two-horsepower outboard hauled across the Naré Trocha on the back of a mule. As luck would have it, the motor arrived broken. Leaving most of the crew at Victoria to build a camp and repair the engine, Schultes and five others moved onto the river. The only landmark they recognized was a distant ridge of mountains, an extension of the Sierra Chiribiquete that hovered in the south, close to the horizon.

After five days paddling against the current, Schultes reached the base of the cataracts. Just beyond, three miles up the shore, were the remains of El Morichal, the camp that Barrera and the men had made in the forest two months before. This was the tie point for the survey, the farthest downstream position reached and recorded during the earlier work. From here the expedition could begin mapping the rest of the river. Schultes arrived in the late afternoon, and finding no sign of the old boat save for a frayed rope drifting in the current, he entered the forest to look for hevea. By evening he had counted seventy mature trees just in the vicinity of the camp. It was the highest concentration of rubber trees in all the Apaporis.

At dusk one of the men shot a tapir. There was no salt and, with the rains, no means of drying the meat. They ate all they could, butchered a hindquarter for the trip downstream, and packed the rest of the animal onto the boat. Two days later they delivered what remained of the carcass to the men waiting at Victoria. By then the motor was repaired. Carrying as much gasoline as possible, Schultes and five men left immediately for downriver. They expected to be away no more than ten days.

From below the last rapids of Chiribiquete—the Raudal de la Grulles—the Apaporis flows unimpeded for nearly three hundred miles. Deep and wide, with few sandbars and no breaks in the channel, it runs beneath high banks, some merging into great cliffs that tower above the river. Although these fall away as one proceeds downstream, there are few places where the banks are low enough to be flooded even at the height of the rainy season. Beyond the shore the land drops off, and a hundred yards inland there is an almost continuous stretch of low swampy ground. These *rebalses* drain into the river, and as a result the banks are interrupted at very short intervals by numerous creeks and rills, which make for broken ground. This very much complicated

Schultes's task of identifying a level area large enough for a good landing field.

His other work proceeded effortlessly, the monotony of the survey being more than offset by the wonder of drifting down an unknown river. He could not collect and for the first time in his career did not even carry plant-collecting equipment. There was no time and no room to spare on the boat. In a sense this freed him to experience the forest in a different way. His main responsibility was to make an actual count, mile by mile, of all mature and tappable rubber trees whose crowns emerged from the canopy or protruded along the edge of the shore. This figure would later be doubled to account for both banks. The terrain was ideal for rubber, and as he proceeded downstream Schultes had no difficulty distinguishing the foliage of several species. As in the upper Apaporis the dominant commercial trees were *Hevea guianensis* and its variety *lutea.* Abundant in the forest but of no value as sources of rubber were *Hevea nitida* and *Hevea pauciflora* var. *coriacea.*

The actual mapping of the river was a relatively simple affair. Several times a day Schultes would pace a kilometer along the shore and mark either end with white flags. By measuring how long it took for the boat to float that distance he could calculate the speed of the river. A hand compass gave him the bearing, and knowing the current, he was able to estimate the distance traveled between points as the river shifted direction. It was a crude technique and tedious to execute, but the results were surprisingly accurate. Schultes's map would hold up for fifty years, with only minor modifications made once aerial photographs were available.

For the most part Schultes remained on the main trunk of the river, recording the location and status of the various tributaries but making no effort to ascend them. On the evening of September 29, however, he noticed a distant ridge that stretched away from the Apaporis, almost at right angles to the river. As the boat drew near, the sandstone cliffs of Cerro Isibukuri hung above the forest, and the sun fell upon more than a hundred waterfalls plunging into the canopy. Schultes landed on the north bank, just above the mouth of a large black water affluent he knew to be the Río Kananarí, an important river that drained all the western flank of the Isibukuri massif. Standing at its mouth was a solitary fisherman, the first Indian encountered by Schultes in six months on the Apaporis.

The fisherman was Taiwano, one of two peoples of the river, and the next morning he led Schultes and his crew up the Kananarí, along the low and flat riverbanks, past an overgrown and long-abandoned rubber station, and through the first of three sets of rapids that blocked the river twenty miles upstream from its mouth. They spent the night at a small settlement of Kabuyarí Indians on the Caño Paco, in a longhouse, or *maloca,* decorated with elaborate abstract motifs painted in yellows, reds, browns, and blacks with pigments made from leaves, dirt, and roots. The images were reminiscent of the petroglyphs carved by the ancestors on boulders and into the riverbed of the Kananarí, engravings that were said to represent *yagé* visions experienced by the gods.

Dawn found the men asleep in hammocks beneath the rafters of the *maloca.* Schultes was out early, as always, enjoying the quiet and cool of the morning. "There is a feeling," he would write of the Apaporis fifty years later, "of living within a fleecy dome, a welkin of unreality which gradually fades away with the strengthening sun." He bathed at the landing and then watched as the mist spread to reveal the silhouette of the forest and the tall, narrow trunks of the *caranaí* palms he had so admired coming up the river. A new species, soon to be known as *Mauritiella cataractarum,* the graceful plants appeared to grow only beside rapids where their delicate foliage might shade the edge of the river. The headman had confirmed that they were found nowhere else in the jungle. They had been deliberately planted on the rocky shores before man came to earth from the Milky Way during the battles of the gods when opposing sides flung up waterfalls as ramparts in their struggle to dominate the world.

The headman also warned Schultes about the rapids of Jirijirimo. There were only six men and boys in the settlement, and Schultes had asked if there might be others available to work rubber. In all the Kananarí, he learned, there were no more than thirty adult men. Many more were rumored to be living below the great cataract, but the headman had never been there and had no intention of going. Jirijirimo was a place of danger. Boulders encased the souls of dead shamans, cliffs bore the faces of spirits and demons, whirlpools swallowed mountains and led to the dark center of the world. Schultes assured him that he had no intention of proceeding farther downriver. In fact, having reached and surveyed the Kananarí, he intended to return up the Apaporis to Puerto Victoria.

It is not clear from the archival records precisely when Mayer declared Schultes missing and presumed lost. Word that the party was overdue at Miraflores reached Bogotá in the first week of October. The last contact had been September 2. Mayer waited a week and then dispatched a military aircraft from Villavicencio, ordering it to fly the length of the Apaporis. It returned without having sighted the expedition. Mayer informed the ambassador and was on the verge of notifying Schultes's parents when a cryptic message arrived from the Colombian Ministry of War. An American, claiming to work for the rubber program, had turned up at La Pedrera, the frontier military post on the lower Caquetá. A radio transmission had requested that a plane be sent to pick him up. Apparently there was no hurry. He was being well cared for by the Capuchin priests who had put him to work painting the inside of their church. Mayer glanced at the map. La Pedrera was four hundred miles by air from where Schultes was supposed to be. It meant only one thing: Rather than returning to Victoria as instructed he had gone down the entire Apaporis, through the cataracts of Jirijirimo and the falls of Yayacopi. In other words, he had traveled the entire length of the river.

For Schultes there had been little choice. When he returned to the mouth of the Kananarí, the engine failed. Gas was running low. With luck, provided they could repair the outboard, the expedition had enough fuel to motor the short distance from the Apaporis confluence up the Caquetá to La Pedrera. They could never have reached Victoria. With food almost exhausted and no certainty of killing game, Schultes elected to continue downstream, although it meant a passage of several hundred miles.

The most dangerous of the rapids was Jirijirimo, located just below the mouth of the Kananarí. Schultes approached the falls at dawn. In the stillness of the morning, with the slow, steady movement of the river beneath the boat and the dark bands of vegetation on the distant shores, he witnessed for the first time the great plume of mist that rises from the water and hangs ominously above the cataracts as if to conceal the wild beauty of the river from the sun. As the river narrowed, funneling half a mile of water toward a chasm sixty feet across, he heard the dull thunder of the entry falls, a one-hundred-foot wall of water leading to the gorge.

The crew pulled to shore on the right bank and began the difficult task of dragging the boat over the first of three short portages. On the Naré Trocha they had been a fresh crew of twelve; now they were six, all tired and worn after a month on the river and a diet consisting mostly of farinha, sugar, and coffee. Fortunately, the ground was flat and dry, the river low, and the distance to cover half what it would have been in the rainy season. Schultes worked alongside the men and ordered frequent breaks. While the men rested he explored the edge of the cliffs that fell two hundred feet to the river.

Walking back to the first rapids, he climbed down over giant boulders and saw to his amazement the hardiest of plants, an algalike growth clinging to the rocks in the midst of seething cauldrons of white water. That a plant could establish a foothold in such a place! It was an astonishing adaptation. In time he would learn that the leaves of this river weed flush out at the very height of the rainy season when the river runs hardest. A member of the Podostemaceae, a family of strictly aquatic species, it flowers and sets fruit on an annual cycle designed by nature to take advantage of the full sweep of the floodwaters. Almost a decade later Schultes would return and collect the plant with the Makuna, who called it *moo-á* and reduced it to ash to make salt. For the time being, however, he could only admire its wild beauty and imagine what other wonders awaited him downstream.

From Jirijirimo the river widened and ran beneath high banks for nearly thirty miles before coming upon a treacherous bit of water known appropriately as El Engaño, the deceitful one. A short portage on the right bank brought them back to the river, which they entered for only a mile before coming upon the falls of Yayacopi, a shelf of rock spread in the shape of a horseshoe across the entire river. Another portage, another stretch of open water, and then another run of rapids six miles long known as the Cachivera Carao.

Through all this, Schultes continued his survey. Without fuel, however, it was impossible to explore even the most important tributaries—the Popeyacá on the south bank or, later, the Taraíra, the river that forms part of the boundary with Brazil. He did go up one north bank affluent, the Río Piraparaná, paddling as far as the rock of Nyi, a petroglyph carved into a boulder of granite sitting precisely on the equator. Long sacred to the Makuna, Barasana, Tatuyo, and Taiwano, the image commemorates the visit of Father Sun when he first gave *yagé* to the people. Of these Indians of the upper Piraparaná, Schultes was told, they were wild and useless for his work. Perhaps for rubber, he thought as he moved back down the river. But he knew that one day he would return and reach the headwaters of not only this river but all those that drained into the Apaporis and still had populations of Indians untouched by the modern world.

The character of the Apaporis changed somewhat below the mouth of the Piraparaná. Previously all the rock had been sandstone. Now, as they passed through more rapids and boulder fields, Schultes noticed that the formations were granitic, and the plants on the riverbanks revealed the shift. It was curious and rare, but there was little time for reflection. There were three more dangerous cataracts, all within the final hundred miles before the Brazilian frontier and the confluence of the Caquetá. From there they turned north and, using their last reserves of gas, motored upriver to La Pedrera. When finally, on the morning of

October 15, Schultes and his men came within sight of the white statue of the Virgin Mary overlooking the river and the gray huts of the military post, their boat was battered, their food gone, and their fuel reduced to a single gallon. At the landing their motor sputtered and died.

Two weeks later Schultes was lifted out of La Pedrera by yet another trimotor Fokker, this time one with floats. In Bogotá he worked furiously for two months, completing his map and correlating the data from his survey. The results were as extraordinary as the achievement. With the exception of eight miles of the Jirijirimo gorge, where he had more pressing concerns, Schultes had counted every visible rubber tree along one thousand miles of the Apaporis and its tributaries. With Vinton he had established the basic infrastructure—camps, depots, trails, landing fields, boats—required for commercial exploitation. He had found that the highest concentrations of hevea coincided with the two longest navigable stretches of the river. In seven months he had actually counted 16,713 individual trees, which meant that within one thousand yards of the river's edge there were more than 1.5 million rubber trees. Based on his understanding of the ecology and flora of the drainage, together with information he had gleaned from the old *balateros,* he concluded with some confidence that the Apaporis basin supported 16.7 million rubber trees. Extracting the latex from just the trees within the immediate vicinity of the river would employ ten thousand men and yield 6,582,160 pounds of rubber a year. Based on these estimates, Mr. Mayer sent a letter to Washington on December 6, 1943, that stated unequivocally, "There can be no doubt but what from a long-range point of view, the Apaporis region offers the prospect of the largest potential tonnage of rubber in Colombia."

A week before Christmas, Schultes returned to Boston for a well-deserved rest and some quiet moments with his family and friends in New England. Just before leaving Bogotá, however, he stopped by the Ministry of Agriculture and spoke at length with the senior officials, urging them to seek protection for the mountains of Chiribiquete, lands that none of them had ever visited. Indeed, most had no idea that such a place existed. It would be nearly fifty years before Schultes's plea was finally acknowledged with the creation in 1990 of the Parque Nacional de Chiribiquete, an enormous tract of protected land encompassing all the mountains that fired his imagination as he rode that river into the forest. One of them, a beautiful plateau due south of the headwaters, would bear his name, the Mesa Schultes.

CHAPTER ELEVEN

The Betrayal of the Dream, 1944–54

As THE YEAR 1943 came to an end, Mr. Mayer found himself torn between loyalty to his men and growing doubts about the program for which they were risking their lives. Of the quality of the work there was no doubt. Vinton's data from the Vaupés and Schultes's rubber survey of the Apaporis and the resulting map, all twelve feet of it, were first-rate. Mayer took justifiable pride in dispatching their reports to the head office in Washington, and he seldom missed an opportunity to remind his superiors of what his men had gone through. In the cover letter that accompanied Schultes's Apaporis report, a document Washington considered several months overdue, Mayer wrote, "As you will undoubtedly realize . . . the work involved in gathering the material contained is substantially different to that of a Visiting Agent walking across a thousand acres of nice, clean rubber paths on an oriental plantation."

Had the men in Washington actually read the accounts, such an admonition might not have been necessary. A two-page itinerary report covering the months of May and June and dispatched from Bogotá in July 1943 is typical of what passed across Mayer's desk. The correspondent was Paul Allen, identified as an associate field technician and thus, like Schultes, one of the rubber explorers. On June 1, Allen was in Villavicencio, "awaiting guide, who had disappeared." Two days later on the banks of the Río Negro he had "hired a new man to replace one lost in Villavicencio." Ten days after that he had "discharged man hired at Río Negro, as being generally useless." The first three weeks of May found him in Bogotá, "sick with amoebic dysentery, tertian and quartan malaria, contracted on the Río Guayuriba, in regular line of duty." Such were the day-to-day lives of the young men whom Mr. Mayer sent off in search of rubber.

What troubled Mayer was the knowledge that he had neither the time nor the manpower to take advantage of the information provided by the rubber explorers. Schultes, for example, had concluded that the remote Apaporis might produce 3,000 tons of rubber, enough to employ ten thousand workers. (These tons are long tons, a British unit of measurement, standard in the rubber trade and equivalent to 2,240 pounds.) But where was Mayer to find these men? And even if he did succeed, how significant would the contribution be? The numbers simply did not add up. The Amazon supported an estimated 300 million hevea trees, theoretically enough to yield 800,000 tons of rubber a year. But these trees were scattered over millions of square miles of forest, with a density of perhaps one per acre. Even at the height of the boom, when prices rose as high as three dollars a pound and tens of thousands of rubber tappers spread throughout the Amazon, the total annual harvest for all of South America never surpassed 50,000 tons. Industrial America alone consumed fourteen times that amount each year.

With three decades of service in the Dutch East Indies, Mayer understood what real production was all about: trees planted twenty feet apart in neat rows running across millions of acres, plantations serviced by armies of men working for almost nothing, yields measured not in dozens but thousands of pounds per acre. In comparison, the work in Colombia could not promise a great deal.

Mayer had gone to Bogotá as a representative of the Rubber Reserve Company. In January 1943 his job shifted to the Rubber Development

Corporation, a new subsidiary established specifically to oversee foreign purchases of natural rubber. Although he never voiced his concern to his men, Mayer realized that the Rubber Development Corporation was more than anything a sign of the desperation of the times. By the terms of agreements negotiated with fifteen Latin American nations, including Mexico, most of the Central American countries, Brazil, Colombia, Ecuador, Venezuela, Peru, and Bolivia, the United States promised to purchase at a fixed price all rubber production until 1946. In exchange these countries offered to restrict their own consumption, provided that the U.S. supply them with essential manufactured rubber goods for the duration of the agreements. The most optimistic projections anticipated rubber yields of 30,000 tons in 1942, 40,000 in 1943, 60,000 in 1944, and 100,000 in each of 1945 and 1946. Even had these goals been achieved—and they were not by a long shot—the total production from fifteen nations over a period of five years would equal less than half of the annual U.S. consumption. Clearly, wild rubber from Latin America might help, but it would not solve the wartime emergency.

The situation was indeed dire. Although 1940 had seen record imports of 811,564 tons, the military buildup had absorbed all but a third, leaving by year's end a strategic reserve of only 288,864 tons, less than a six-month supply. In June 1941 rubber became the first commodity to come under direct government control. The Rubber Reserve Company, by then the only legal importer in the country, purchased nearly half a million tons in the last months of the year and contracted for an additional 1.2 million to be delivered in 1942. By the late fall of 1941 the national stockpile had risen slightly, and the outlook appeared more promising. Then came Pearl Harbor, and within three months the fall of Singapore and the Dutch East Indies.

Virtually overnight the rubber trade ended, and the United States faced potential disaster. With every conceivable source taken into account—the strategic reserve, supplies held by the major companies, freighters in transit on the high seas, reclaimed rubber stored in warehouses, the immediate capacity of the fledgling synthetic industry—the country had, at normal rates of consumption, perhaps a year's supply. Out of this dwindling reserve it had to satisfy its domestic needs as well as those of most of Latin America. It had also to meet the demands of its wartime allies, including Canada, which required 50,000 tons, and Britain, which itself had a stockpile of 100,000 tons, again about a year's supply. Above all, America had to fulfill the requirements of the greatest and most important industrial expansion in history: the arming of the Allied cause.

Four days after Pearl Harbor the government prohibited the manufacture and sale of new tires and outlawed the use of rubber in any product not deemed essential for the war effort. Within three weeks the production of cars and trucks destined for the domestic market ceased. The speed limit dropped to thirty-five miles per hour, tire inspections became mandatory, and gasoline rationing was introduced, not to conserve gas but to reduce the wear and tear on the nation's limited supply of tires. A massive recycling effort began in the spring of 1942. On June 27, President Roosevelt broadcast to the nation, urging every citizen to scour the neighborhood for old tires and scrap rubber of any description that could be redeemed for a penny a pound at one of 400,000 designated filling stations. In a symbolic gesture the White House staff collected 400 pounds, including the rubber bones belonging to Fala, the president's dog. All told, the country came up with 454,000 tons of scrap that, once processed, accounted for nearly 65 percent of the nation's 1942 consumption.

Still the clock was running down. Despite all the efforts of the Rubber Reserve Company, and later the Rubber Development Corporation, the importation of natural rubber dropped from over a million tons in 1941 to a mere 55,329 in 1943. America's strategic planners confronted a critical dilemma. Together, the remains of the national stockpile, the meager new supplies coming in from Ceylon, Liberia, Mexico, and South America, and the rubber salvaged during the reclamation drive might keep the country going through the end of 1943, but after that the Allied effort would collapse unless some alternative source was found.

It was an astonishing state of affairs. Various chemical substitutes existed, but none had been perfected. In 1941 the total output of the entire synthetic industry was just over 8,000 tons, mostly specialty rubbers useless for tires. Thus the nation's survival depended on its ability to manufacture over 800,000 tons a year of a product that had barely reached the development stage. No blueprints existed for the factories that would process this immense tonnage. No facilities had been built to produce an equal volume of the various precursors, the feed stocks from which the rubber would be made. American industry had never been called upon to handle such a task, to accomplish so much in such a short period of time. The engineers had two years. If the synthetic rubber program did not succeed, the nation's capacity to wage war would end.

Mayer had known this for some time, just as he had known that no amount of rubber his program might produce would ultimately make a difference. (From April 1942 to June 1946 the total output for all of Latin

America would be only 127,662 tons, with 80 percent coming from Brazil, Mexico, and Bolivia. Colombia, in the end, provided a mere 2,403 tons. The average cost across Latin America was 63.9 cents a pound, compared to 28 cents a pound for Ceylon. Since the Rubber Reserve Company allocated the product to manufacturers at the fixed price of 22.5 cents, the Latin American imports of wild rubber represented a loss of over $50 million.) Mayer recognized, of course, that natural rubber remained of vital importance. In quality it was unsurpassed, and mixed in a ratio of one to ten it significantly increased the usefulness of rubber reclaimed from scrap. But Mayer had no illusions that rubber gathered from the Amazon would fundamentally determine the outcome of the war. Hence, in the summer and fall of 1943, he increasingly turned his attention to the one part of the rubber initiative that promised, no matter what the fate of the synthetic effort, to make the United States forever independent of the plantations of the Far East.

For Schultes the timing could not have been better. In early August 1943, while he was on the Apaporis, the Bogotá office received a copy of a confidential memorandum dated July 17 that may well have terminated his botanical career had not Mayer intervened. Schultes, the letter noted, was a young man of twenty-eight, single, with a highly specialized training in botany and no commercial experience in rubber production. Consequently, the Washington office could see no reason to request his further deferment as a rubber technician. As soon as the Apaporis work was completed, Mayer was expected to "advise this office so the local board may be notified of his release for induction into military service."

Mayer, of course, had no intention of letting the army take away his best explorer and the only trained economic botanist on his staff. A letter to Rands, whom Mayer had known in the Far East, informed the government that Schultes's destiny lay elsewhere, as the young botanist himself discovered at a meeting held that summer in Bogotá. Present were Mayer and two representatives of the U.S. Department of Agriculture, Carl Grassl and Hans Sorensen. Grassl was a neophyte, a herbarium botanist with little field experience who devoted most of his spare time to collecting art as a hedge against retirement. Hans Sorensen was an older man, short and well built, with a thick Danish accent and the ruddy features of one who had spent his life in the forest. An agronomist with a long and distinguished record in the Far East, Sorensen had arrived in Colombia the previous December to oversee the establishment of three experimental plantations, all located in the northwest of the country near the Gulf of Urabá.

For the past several months Sorensen and Grassl had been traveling around Colombia, examining wild populations of rubber and searching for material with which to create an industry. Of critical interest were the rare varieties and ecotypes of hevea that showed resistance to the South American leaf blight, the disease that had frustrated all previous attempts to establish plantations in the Americas. Told by Mayer of Schultes's work in the Apaporis, Sorensen was especially intrigued by the dwarf variety of hevea found growing on the mountaintops of Chiribiquete. At his request Schultes brought to the meeting a number of herbarium specimens of the plant, which he spread out on the long wooden table that ran the length of the wall of Mayer's office.

"You say it's new," Grassl said. He was a tall man, thin and dressed from head to toe in pressed khaki. His high leather boots were of the sort Schultes had come to disdain—good for nothing but ulcers and trench foot. Sorensen, by contrast, wore a rumpled brown suit and scuffed oxfords.

"A new variety, I suspect," replied Schultes.

"The leaves appear to be clean. All the other species, I assume, have it."

"Some of the worst I've seen," said Schultes.

"But these don't."

"No. You can see for yourself. It's a strange thing, growing on those mountains."

"So there could be resistance," Grassl said.

"The blight's up there. It just doesn't seem to do much damage."

Sorensen stood up from a corner chair and moved toward the table. Lifting a specimen from the newsprint, he ran a finger over the leaves and then carefully examined the surface.

"Everything depends on resistance," he said.

Schultes glanced at Mayer, who turned away and stepped toward the window. Outside, the winter light fell on the slopes of Monserrate. Mayer looked out at eucalyptus trees planted in neat rows, the huts of the poor, smoke rising from fires burning at the edge of the city.

"What's your opinion?" he asked.

"Well," Schultes replied, "I am a botanist, not a plant breeder."

"Go ahead," said Mayer.

"The species is remarkably elastic. Just consider the increases in yield. It's been in cultivation for only fifty years, and we've seen yields go up tenfold. So a proper breeding program might produce what you want."

"High yields and resistance."

"Yes."

"And this new variety?"

"It can't be tapped, of course, because of its size. But it has good rubber. It could be useful."

"Good," Mayer said. "When you go back in, I'm going to send Grassl with you. I want you to take him up that mountain and get some living material—budwood, seeds if possible. Do you understand?" Schultes glanced at Grassl and did his best to ignore the boots.

"Yes, sir," he said as he began to make his way toward the door. Mayer motioned for him to stay where he was. Sorensen leaned across the table and pushed the specimens to one side. He then slowly unrolled a map of South America, weighing down the corners with books lifted from Mayer's shelf.

"You ever see an outbreak of leaf blight?" he asked Schultes.

"No, not in a plantation."

"It's the damnedest thing. The trees are perfectly healthy. They drop their leaves as they're supposed to. Then out comes a beautiful flush of new growth. And just like that it runs like wildfire through the rows, and before you know it there's nothing left but branches."

"Can't you spray?" Schultes asked.

"You can use fungicides," replied Grassl, "but it costs a fortune. Nothing you could afford on a plantation scale."

"There is a larger issue here," said Sorensen, sweeping his hand across the map and landing on the lower Tapajós, close to the mouth of the Amazon. "Wickham got his material here. Just a couple of dozen trees, all one local strain of *Hevea brasiliensis.* Now, as you know, those few seeds became the basis of the entire plantation industry of the Far East. Incidentally, Dr. Schultes, do you have any idea what would happen if leaf blight broke out in Southeast Asia?" Sorensen looked up from the map.

"No. I suppose—"

"It would mean the end of the industry," Sorensen said. "Twelve million acres, and I tell you it would go quickly."

Schultes looked over toward Mayer.

"They could never control it," Mayer said.

"I had no idea," said Schultes.

"All that has saved us so far is the thickness of the spore wall of the fungus. The fastest ships still take several weeks to get across, and the spores can't survive the voyage. So the blight has never gotten over there."

"But eventually," Grassl said, "every disease gets everywhere."

"So we're not just talking about this war. We're looking at the fate of the industry."

"I see," said Schultes.

"Fundamentally, we don't know a damn thing about this genus *Hevea.* Not even how many species there are, let alone their tolerances and local variations. We know that the best rubber comes from *Hevea brasiliensis.* It grows for the most part south of the Amazon, reaching across the river only at three points: the delta below Belém, at Manaus in the center of the basin, and here at Leticia, where it crosses north into Colombia. Then there's *Hevea benthamiana.* It yields a good but second-rate product. It's only found north of the main river, along the banks of the Río Negro and beyond to the Orinoco in Venezuela. The only usable species that grows across the entire range of the genus, from eastern Brazil west to the foothills of the Andes, is *Hevea guianensis* and its variety *lutea.* This, by the way, is the rubber of the Putumayo."

"And what you've been seeing in the Apaporis," Grassl added. Schultes nodded and looked back at Sorensen.

"But the relationship between these species and the identity of all the others is a complete fog. Already we've been noticing resistance in some, but until we have a handle on the entire genus, we won't even know what we're looking at."

"That's where you come in," said Mayer. "We need a young botanist to take on the genus, travel everywhere it grows, look for resistance, and figure out what the plant is all about."

"You'll have plenty of help, but you'll have to design your own program," added Sorensen.

"It's a big job," Schultes said.

"Yes, it is."

"Dick," said Mayer, "I don't know how much rubber we can get out of this country. Five thousand tons? Maybe more. And I don't know how long this war is going to last. I do know that we could be growing it here, and you realize what that would mean—to give this country such a product."

"I understand."

"Plantations," Mayer continued. "There's no reason why we can't have plantations by '47, '48 at the latest. It may be too late for the war—God knows, let's hope it will be too late—but it can and will happen, and when it does, the history of Latin America will change."

"What do you think?" asked Sorensen.

"Naturally, I'll do whatever I can," Schultes said.

"Good."

• • •

With this decision Schultes's role dramatically shifted. Until this time all of his explorations had been directed toward securing raw supplies of rubber for the war effort. This work continued, but beginning with his successful trip with Grassl back to the Apaporis in the last two weeks of July 1943, he became part of a team committed to establishing rubber plantations in the Americas. It was an obsession, he would discover, of all those who had worked in the Far East.

It was not as if they had not known about the danger all these years. Every major rubber company had attempted to break the Asian monopoly. In 1926 Firestone had set up a 30,000-acre plantation in Liberia. Goodyear tried in Central America, establishing holdings on the shores of Gatun Lake in Panama and farther north in Costa Rica. In the last years of his life Thomas Edison spent his fortune seeking sources of rubber in North America. Over seventeen thousand latex-producing plants were sampled and analyzed. Edison found rubber in milkweeds, wild lettuce, osage orange, Indian hemp, rabbit bush, and the common goldenrod. Of these, the most promising was goldenrod, and by the eve of the war Edison had sown its seeds across thousands of acres of the American Southeast. In the Southwest, the Intercontinental Rubber Company had 32,000 acres planted in guayule, *Parthenium argentatum,* a plant that yields 20 percent of its gross weight as pure white latex. In Haiti thousands of acres were planted in *Cryptostegia.*

The most powerful of the rubber men—Henry Ford, Harvey Firestone, Paul Litchfield of Goodyear—focused on species of hevea. Ford, who by 1926 was making half the automobiles in the world, was particularly enraged by a British initiative known as the Stevenson Plan, which in 1922 had restricted production in their colonies and led to a fourfold increase in the price of rubber. Under pressure from Ford, Secretary of Commerce Herbert Hoover authorized a government search for alternative sources. In 1923 and again the following year scientists from the USDA traveled throughout the Amazon identifying potential growing sites and acquiring seeds for the establishment of plantations in Panama and the Philippines, both American colonies. Based in part on the results of these surveys, Henry Ford decided to try his luck in Brazil.

In 1927 the Brazilian government granted him a concession nearly four times the size of Rhode Island, some 2.5 million acres stretching seventy-five miles along the east bank of the Tapajós River. The site was deliberately chosen because of its close proximity to the locality where Wickham had secured the seeds in 1876. Within months a small town grew up, complete with sewer and water lines, thirty-three miles of roads and railways, a deep water port, three schools, a fully equipped hospital, several churches, two hundred houses, as well as barracks for nearly one thousand single men. Beyond a neat row of brick and stucco bungalows, on a street lined with mango trees, there were tennis courts and an eighteen-hole golf course. The club and swimming pools—one for Americans, the other for Brazilians—were on top of the hill, overlooking the third fairway. Beautification plans called for lawns around every house and ornamental lights to highlight the palm and eucalyptus trees planted in the parks where square dances were to be held. The town was named Fordlandia.

It was an enormous financial investment and a tremendous risk. The first plantings were mostly of local material, gathered along the Tapajós and propagated in the company nursery. Seeds from elsewhere in Brazil, from the Río Negro, Acre, and the lower Amazon made up a small percentage of the crop. In 1933 representatives of the company traveled throughout Sumatra and Malaya and secured over two thousand living samples of fifty-three of the best of the Far East clones. Carefully boxed in sterilized sawdust and shipped halfway around the world, these arrived at the mouth of the Amazon in early February 1933, only months before the British, Dutch, and French prohibited the export of live material. Two weeks later the seedlings were planted at Fordlandia. Over half survived, and the initial results were promising. By 1934 approximately 8,400 acres had been cleared and planted with nearly 1.5 million rubber trees.

Then disaster struck. As the upper branches of the young trees began to close in, creating a canopy over the fields, South American leaf blight, always present in the forest, became virulent in the plantations. Within a year it ravaged most of Fordlandia. Trees derived from seeds from the Tapajós were completely defoliated. By far the most sensitive and the first to die were the high-yielding strains developed in the Far East. Each one of these clones was the descendant of one of the original Wickham seeds, which had been derived from only twenty-six trees, an astonishingly small genetic pool. In selecting for high yield, the plant breeders in Asia had inadvertently produced strains that were especially susceptible to the blight.

Undaunted, Ford instructed his managers to find a different site and try again. In 1934 the Brazilian government agreed to trade 700,000 acres of Fordlandia for a new concession located near the mouth of the Tapajós just thirty miles south of the city of Santarém. Once again thousands of acres were cleared, some 5 million seeds were planted, and the infrastructure of a small city was created. This time the scale was even larger. In addition to the hospital, churches, schools, movie theaters, sawmills, machine shops, barracks, roads, power plant, and water works, there were eight hundred houses, three recreation halls, a golf course, and five regulation soccer fields.

At first the managers at Belterra remained optimistic. They knew, of course, that leaf blight had wiped out not only Fordlandia but earlier attempts to establish plantations in Trinidad and the Guianas. At the Goodyear plantations in Panama less than 1 percent of the stock introduced from the Philippines had shown any resistance. Despite this record, the agronomists confidently predicted that rubber grown on the well-drained plateau of Belterra would escape serious damage. Perhaps they were blinded by the sheer scale of the endeavor. By 1941 there were seven thousand people living at Belterra, and 3.6 million trees had been planted.

Initially all went well, and for the first five years the plantation escaped the blight. But it did so only because the trees had yet to reach the age when they lose their leaves. Once they did, and the rainy season brought forth a predominance of highly susceptible young foliage, a devastating epidemic broke out. At this point things became interesting. At both Belterra and Fordlandia, agronomists had noticed that certain trees, notably those from distant parts of the Amazon, and especially plantings of *Hevea guianensis* and *Hevea spruceana,* displayed remarkable resistance to the blight. This variability in resistance had first been noticed in wild populations by Mayer's boss, Robert Rands, one of the leaders of the 1924 rubber survey and the scientist who later supervised the hevea rubber project at the USDA. At Belterra and Fordlandia the contrast between diseased and resistant trees was dramatic.

With nothing to lose, one of the rubber biologists at Fordlandia, James Weir, decided in 1936 to experiment by grafting the stems with healthy foliage from resistant trees onto the trunks of high-yielding but susceptible clones. The procedure was time-consuming but technically not difficult. The trees were decapitated at a height of seven feet, and the new shoots carefully sprayed until large enough to accept a graft. Provided the emerging growth was pruned and cared for, the result was a new crown of healthy green leaves. The results were so successful that beginning in 1941 the managers at Belterra undertook the enormous task of top budding more than 2 million trees, a job that would engage six hundred workers for four years. It was an expensive proposition and just one part of Ford's $20 million investment. But the result was a technique that could be readily employed to counteract the single most significant impediment to establishing rubber plantations in the Americas.

This horticultural breakthrough inspired the confidence of the USDA rubber researchers. Certain that a self-sustaining rubber industry was now possible, they turned to Congress and sought approval for a broad cooperative plan involving fifteen Latin American countries. For three critical years nothing happened. Finally, on June 22, 1940, Henry Wallace, head of the USDA—and later vice president—managed to get through a bill that authorized annual spending of $500,000. With this funding in place, specialized research and propagation stations were established throughout the Americas. There was an important project in Haiti, an island nation free of the blight, and remote bases in Mexico, Guatemala, Colombia, Ecuador, Peru, and Bolivia. At the mouth of the Amazon at Belém, the Instituto Agronomico do Norte became the center for the ecological and botanical study of the genus, and the analysis of resistant strains found throughout Brazil and beyond. The most important base was the USDA Cooperative Rubber Plant Field Station at Turrialba, Costa Rica. It was here, in a beautiful valley with ideal conditions for the blight—cool nights, prolonged morning fog, lingering dew—that new, resistant clones would be most severely tested. Fifty miles away and one thousand feet below on a broad coastal plain, the Americans established Los Diamantes, an experimental farm where hybrids developed at Turrialba and rare specimens gathered from the wild could be grown and stored as living trees.

The first shipment of germ plasm came from the Philippines: 96,000 seeds stuffed into the bomb bays of B-18 bombers and delivered to Turrialba in November 1940. Three months later 200,000 seeds arrived from West Africa. In September 1941, Firestone's Liberian operations dispatched an additional 2 million seeds. Six million more seeds arrived in October, with half going to Costa Rica and half to Brazil. In the last months of peace, Goodyear dispatched from its Mindanao plantation living material of 120 of the most important Eastern clones. The last consignment, containing 5,500 plants, arrived in the United States only eleven days before the Japanese invaded the Philippines.

By the end of 1941 a million seedlings had been planted in the Americas. Within the next two years that number would rise to 25 million as the rubber workers laid the foundation of a New World plantation industry. By 1947, the agronomists predicted, America would not only be free of its dependence on the Far East but it would possess a far better crop. The Asian plantations, productive as they were, had been built exclusively on Wickham's collections from the Tapajós, which everyone recognized as belonging to a distinctly inferior ecotype. Not only would the American plantations be immune to the blight but their genetic stock would be derived from the healthiest and most productive specimens in all the Amazon. Finding these trees and understanding their biology was the mission that carried Schultes back to South

America in the early spring of 1944, after a brief rest and holiday in New England.

Hans Sorensen was the first to draw attention to the strange rubber trees of Leticia. Shortly after arriving in Colombia in December 1942, the Danish agronomist had flown to the small jungle town, isolated in the extreme southeastern corner of the country. A military outpost with a population of perhaps six hundred, Leticia anchored Colombia's claim to the Trapezium, a narrow slab of territory that reaches south from the Putumayo River and provides the nation with its only access to the main trunk of the Amazon. Upstream from the Brazilian frontier, along the eighty miles of riverbank controlled by Colombia, lies one of the very few regions in South America where *Hevea brasiliensis* grows north of the Amazon River. It was within the forests of this floodplain, on the channel islands and along the banks of the Loretoyacú, Amacayacú, and other Amazon tributaries that Sorensen made his discovery.

In the first week of March 1943, in the middle of the fruiting season, Sorensen noticed that hevea seedlings sprouting on the forest floor appeared to be remarkably healthy. Leaf blight was present, and indeed in the Leticia area the conditions for it are ideal—eight months a year of warm and damp weather. Yet the seedlings, Sorensen observed, were only rarely attacked. In many parts of the forest they appeared to be completely untouched. Moreover, the seeds were consistently a third larger than the norm for the species. None of this proved resistance, but it suggested that the rubber trees growing in this small area of Colombia might represent an unusual and distinct ecotype worthy of further investigation.

Although hampered by the seasonal rains, which inundated the forest and made travel on foot all but impossible, Sorensen managed to gather in a fortnight well over 100,000 rubber seeds. With this precious cargo he flew out of Leticia at the end of March. Seventeen days later, having passed through Bogotá and Medellín, he reached Villarteaga, one of the experimental rubber stations the Colombian government had established in Antioquia, just south of the Gulf of Urabá in a region believed to be free of the blight.

Most of the seeds were planted at Villarteaga. A small collection, representing samples from individual trees of particular interest, went into the ground at Apartado, thirty miles to the north. The germination rate was exceptional—an astonishing 80 percent at Villarteaga—and by April the seedlings were ready for the field nurseries. Transplanted onto new land cleared and burned from the forest, they grew well. A month later, quite unexpectedly, a severe infestation of the blight savaged the plantation. Over the course of the next year all the rubber seedlings derived from Asian sources suffered immensely, as did those obtained from Belém at the mouth of the Amazon. The young plants from Leticia, however, were barely affected.

In a letter postmarked "Urabá Jungle, March 20, 1944," Sorensen shared his remarkable findings with Dr. T. J. Grant, director of the USDA research center at Turrialba, Costa Rica. The seedlings, Sorensen wrote, "have now been under spore bombardment for almost 10 months and only a dozen or so have been attacked, but so slightly that it is hardly noticeable. None of the few attacked leaves on these 10–12 plants have produced spores. You will therefore understand that the resistance is very unusual." Even more significant, Sorensen reported, many of the resistant seedlings from Leticia also had the potential to produce high yields of latex. Of great interest were the progeny of one tree that Sorensen had found growing completely free of the blight twenty miles west of Leticia. In a standard test designed to anticipate yields, more than half of its seedlings had scored in the highest category. "That mother tree really has something," he wrote Grant. "I only hope it is still there." It was to find this tree and follow up on Sorensen's discovery that Schultes was dispatched to Leticia in July 1944.

Of all the places Schultes lived in the Amazon, Leticia is the town he speaks of with the greatest affection. It was an odd little settlement, cut off from the world, with a shallow history and no pretensions. Originally founded by Peruvians, it had passed into Colombian hands only in 1922. From the cement steps of the church on the Plaza Santander you could watch the sun setting across the Amazon over land that still belonged to Peru. The Brazilian border was a fifteen-minute walk to the south, past the Colombian military hospital, through the fields, and along the steep banks that fell away to the muddy river. There were no cars, and the dirt streets ran past simple whitewashed houses built of rough boards and thatch. Mango and avocado trees, planted between the houses and along the narrow paths, provided shade and made the heat bearable until the gentle afternoon breezes blew off the river and cooled the town. If the power plant worked, which it sometimes did, there was electricity each evening between six and ten, just enough to charge the string of lights that ran past the church and illuminated the flagpole in front of the governor's modest wooden office. But there were no telephones and few radios. The night, for the most part, still belonged to the forest.

The river was the focus of the community, its lifeline and reason to exist. Two thousand miles from the mouth it was already a mile wide. Deep and brooding, pushed to the sea by the rains falling out of the distant Andes, it rose and fell with the seasons. In June canoes loaded with fish, manioc, wild fruits, and plantains formed a floating market that at any moment might drift past the riverbank onto solid ground. A month later children sent to gather water or women with clothes to wash had to clamber one hundred feet down a slippery bank just to reach the edge of the river. Sometimes entire sections of the shoreline slipped away and formed islands that floated in the stream and then drifted off downriver. At the height of the rainy season, forest animals lived in treetops, mules grazed in water to their haunches, and cows had their udders nibbled by fish.

Catalinas flew in once a month, weather permitting. River traffic was intermittent. The ships of the Booth Line, which made the run between Manaus and Iquitos, sometimes stopped, but there was no way of knowing when they would come, and their arrival was often marked only by a whistle blowing in the darkness from across the river at the small Peruvian settlement of Ramón Castilla. The only boats based in Colombia, wood-burning paddle wheelers like the *Ciudad de Neiva,* were even less dependable. Sailing down the Putumayo and back up the Amazon, they arrived in Leticia every month, give or take ninety days.

Those who lived in Leticia had no choice but to adapt to the easy rhythm of the river. Messages passed by canoe and word of mouth. Supplies, when they arrived, meant little to the Indians, who made up most of the population—Witotos and Boras who had fled the Putumayo atrocities, Mirañas who drifted to town when the balata boom collapsed, Tikuna who had lived in the region for all their history. Enamel pots that went for six dollars each hung like ornaments in the shops by the plaza. Those few who could afford such prices, government employees for the most part, found their choices severely limited. Lumber and nails were available, but paint was scarce. Canned butter and milk, soap, and aspirin came upriver from Brazil. Sugar and salt arrived brown with dirt, insects crawled through the flour, and maize came in great sacks riddled with worms. There were few ready-made clothes for sale. Muslin went by the pound, people sewed their own trousers, and a Peruvian cobbler provided shoes and boots. Laundry was done by Indian women who charged three dollars a month, provided the client supply soap and starch, and charcoal for the iron. The merchants, men like Arturo Villareal, who owned the hardware store, and Frederico Oldenburg, a German who made soft drinks by mixing sugar, water, and aniline dyes, moved freely among the three countries, loyal only to the river.

The Indians lived as they had for a generation, fishing and farming, working as boat men, porters, rubber tappers, and servants. The colonists tried not to work and sought diversions from their isolation in gossip and ceremony. Each morning at eight an honor guard raised the Colombian flag over the plaza, and for a minute all movement in the town ceased. Soldiers and sailors came to attention, schoolchildren stood still, fishermen quieted their canoes at the landing. Putting their coffee aside, government officials struggled out of their chairs, placed their hands on their hearts, and whispered, *"Dios y patria,"* God and country. Even the priest played a role, pausing in the middle of the morning Mass, breaking off the confessional on those rare mornings when the events overlapped.

This exaggerated display of patriotism, initiated a decade earlier in the wake of Peru's failed attempt to retake the town by force, was by 1944 no longer really necessary. By then the townspeople, roughly equal numbers of Brazilians, Peruvians, and Colombians, had come to terms with their differences and discovered in their mixed loyalties an unexpected benefit. Rather than fight, the town decided to celebrate the national holidays of all three countries. These added up, and combined with the numerous church holidays, virtually guaranteed that work could be kept to a minimum.

When Schultes arrived in Leticia, he already had something of a local reputation. In Manaus, on his way upriver, the Brazilian authorities had invited him to join an official party for a tour of the famous Opera House, the pride of the city. Schultes refused, telling the delegation that the building had been built with the blood of Indians. When word of the slight reached Leticia, the Colombians were delighted. They had never been fond of Brazil and in recent months had grown especially irritated by the antics of the Brazilian consul, a ridiculous little man who at the slightest provocation or rumor of political strife raced about town offering anyone who would listen the protection of his government or, alternatively, the threat of aerial bombardment by a Brazilian air force everyone knew did not exist.

Schultes did his best to avoid the consul, at least for the first weeks. He preferred the company of Colonel Pedro Monroy, the commander of the Colombian military post, and Padre Luis de Garzón, the Capuchin priest who offered him a room at the mission. His work naturally took him into the forest, and before long he was living upriver, at the mouth of the Loretoyacu River, on a farm that belonged to Rafael Wandurraga, a man who would become one of his closest friends. A Colombian of Basque origins, Wandurraga had left Huila as a youth to seek his fortune in the Amazon. Living first at La Pedrera, the center of the balata rubber trade, he had later moved south to Leticia to establish a rubber operation along the north bank tributaries that drain the Trapezium.

Wandurraga was a merchant, but his abiding interest was the welfare of the Indians and the protection of their forests. An honest and decent man, he operated with a fixed markup of 10 percent, refused to barter in liquor, and provided for the wives and children of all the rubber tappers who worked for him. In return, the Witoto and Bora, Mirañas and Tikunas, did what they had vowed never again to do: gather latex from trees that a generation before had been the reason for the misery and torture of their parents. When Wandurraga had his accident, falling from his boat and shearing the pin of the propeller with his face, his life was saved by his Indian workers who carried his mangled body overland to Leticia. There they had stood silently all night long while the military doctor sewed his nose back onto his face, repaired the gash in his jaw, and stitched together the two halves of his tongue.

Wandurraga's farm and plantation at the mouth of the Loretoyacu was an ideal base for Schultes. Located at the edge of the forest where Hans Sorensen had made his observations, with rubber trails radiating in several directions, it offered a chance to work closely with *seringueros* who each day exploited the trees that had shown such promise. Schultes's primary task was to identify and select the individual rubber trees with the greatest potential in regard to both resistance and yield of latex. Seeds and budwood from these collections would then be forwarded to nurseries and plantations, both in Colombia and Costa Rica, where Sorensen's hypothesis could be fully tested by a range of field and laboratory experiments. For a botanist, his assignment represented a rare opportunity to live intimately with a complex group of plants that, despite their economic importance, had yet to be fully understood. Indeed, as Sorensen had told him in Bogotá, some of the most fundamental aspects of their biology and classification remained elusive. Though hevea was the basis of one of the world's most important industries, with global sales of crude rubber alone generating more than a billion dollars in 1940, no botanist knew even how many species were in the genus.

There were two main reasons for this uncertainty. In defining species, taxonomists attempt to identify a unique set of morphological, genetic, and chemical traits that together characterize a discrete entity that breeds true through the generations. Within any species there will be variation, and one key element of the art and practice of taxonomy is the ability to distinguish such differences from characteristics that are sufficiently distinct to warrant the delineation of a separate species. To know a form of life, to understand its particular spectrum of traits, a taxonomist prefers to examine as many specimens as possible. Since variation is often correlated with ecological adaptation, this generally implies studying representative collections from across the entire geographical range of the species. Problems arise when the botanical record is sparse. One sometimes does not know whether two collections with similar but divergent traits represent separate species or just two extremes of a biological continuum. Like many groups of tropical plants, the genus *Hevea* grows over a vast area, only a small part of which had been visited by botanical explorers.

In first thinking about this problem, Schultes deferred to Adolpho Ducke, whom he had met in Manaus on his way upriver. One of the great Amazonian plant collectors of the century, Ducke was a strange character with a mysterious past, an immigrant from Europe who throughout his long and illustrious career refused to reveal any information about his education and early life. Born in Trieste in 1876, he had come to the Amazon as an entomologist and only later, after many months in the field, was he drawn to botany. Schultes described him as "highly irascible, and sarcastically critical, he seemed often to suffer from a kind of persecution complex. His scientific publications were peppered with caustic criticism, leveled especially against botanists who worked . . . without benefit of field work."

A bitter man, Ducke was nevertheless a splendid botanist. For more than fifty years he collected plants throughout the Amazon. He knew individual trees as he might have known friends, had he been so inclined. He visited and revisited localities for the same reasons that ordinary people travel great distances to see family. Botanists believe that the entire Amazon supports roughly 25,000 species of trees. In his solitary wanderings Ducke found and described an astonishing 762 new arboreal species as well as 45 new genera of flowering plants. To put these figures in perspective, a botanist working in the temperate forests of North America would be thrilled to find a single new species. A new genus would require an act of God.

Yet even Ducke was stumped by *Hevea.* Of the ninety-six names applied to various species during the long history of taxonomic studies of the genus, nearly half were published by Ducke himself. Though the master of Amazonian botany and the scientist who had spent more time in the forests than any other, he could not make up his mind about this impossibly complex set of plants. His advice to Schultes was to live with the trees, ignore all distractions, and pay attention to the slightest evidence of variation. Schultes accepted this counsel and added an element certain to have escaped the misanthropic Ducke. Schultes elected to listen as well to the advice of the Indians whose lives and destiny had been so deeply affected by the plant they knew as the weeping tree.

Throughout the fall of 1944, Schultes lived on Wandurraga's farm and followed the tapping circuits. Each *seringuero* rose before dawn, entered the forest, cut into the bark of perhaps 100 widely scattered rubber trees, and then returned home for breakfast. For the next hour or two, as the sun warmed the forest, latex flowed into the *tejelinas,* the small tin cups tacked onto the trees at the base of the incisions. Schultes would begin walking around eight, examining the contents of each cup before the Indians returned at nine to collect the latex. Any trees showing unusual yields would be flagged. Then in the afternoon he and an assistant would return to collect voucher specimens and examine their branches for signs of blight—black lesions on the surface of the leaves. His goal was to find the most promising trees and then track their yields through three complete tapping seasons, October to December, 1944 through 1946. The entire Loretoyacu drainage supported perhaps 120,000 rubber trees. Schultes identified and monitored over 6,000 of the very best. Of these he selected 120 clones to be dispatched as budwood to Costa Rica, where they were grafted onto healthy rootstocks at the Turrialba research station.

During the long days and weeks in the forest, Schultes came to know these trees through the eyes of the *seringueros.* For them, of course, the term *Hevea brasiliensis* had no meaning. They distinguished rubber trees by habitat and bark color, dividing what Schultes viewed as a single species into three types: White rubber, or *seringueira branca,* with smooth, thin, brownish gray bark and milky white latex, grew in forests flooded only during the height of the rainy season. Black rubber, *seringueira preta,* was found in lower areas, wet and inundated for much of the year. Its bark is thick and soft, with a purplish hue. The third and least abundant was red rubber, *seringueira vermelha.* Often found growing between the white and the black, it has a creamy, almost yellowish latex and smooth bark the color of terra-cotta. In general, black rubber occurs in the upper Amazon; the other two are commonly found farther downriver. All material collected on the Tapajós, and thus all the rubber of Southeast Asia, is either white or red. The *seringueros* agreed that black rubber produced by far the best quality of latex. Only a fool, one of them told Schultes, would attempt to create a plantation with material from the Tapajós.

Schultes made other observations in the forest. He already knew that the flowering season for all rubber species lasted only a few weeks. It was one of the aspects of their biology that made his job difficult. Yet now he learned that the flowering of individual trees was even more ephemeral, often lasting just a day or two, an adaptation that no doubt hindered hybridization. Studying the flowers, he discovered that they are pollinated by tiny midges. He also was intrigued by the color of the leaves. In most populations of *Hevea brasiliensis,* the foliage is decidedly discolorous, with very different tones on either side of the leaves. The leaflets of the trees of the Leticia area were equally green on both sides, and the bottom surface was glossy. So pronounced were these traits that both Schultes and Ducke toyed with the idea of describing the tree as a new variety.

Throughout this time on the Loretoyacu, Schultes's main guide and companion was a young man of seventeen named Francisco "Pacho" López. The son of a Miraña mother and a white *balatero,* Pacho had come to Leticia to work for Wandurraga, who had known his family at La Pedrera. He and Schultes were a perfect match. A brilliant youth with a passion for plants, Pacho could climb any tree, follow any path, find food in the midst of a swamp. Hardworking, strong, and honest, he lived by his intuitions and never complained. More than once he would save Schultes's life. The first time was soon after they met when Schultes heeded Pacho's warning not to board a plane at Araracuara, a Colombian penal colony isolated on the Caquetá.

They had been working for some weeks at the base of the southern extent of the Chiribiquete uplands, on flat hills that had not been collected since von Martius explored the region in 1823. The whole time Pacho had been uneasy. Colombia has no death penalty. Araracuara, though named for the scarlet macaw, was home to a hundred of the worst criminals and murderers in the country. With rapids on all sides and a boundless forest, escape was impossible, and the convicts were free to wander as they pleased. Armed with pistols and shotguns, and guided by two Indians, Schultes and Pacho made their collections, avoiding whenever possible the streams and gardens where the prisoners gathered.

On the last day of the expedition, as Schultes organized the specimens and Pacho made preparations to return by river to La Pedrera to visit his mother, the warden, a former army major, informed them of a problem. The plane that made the run to Tres Esquinas, a trimotor

Fokker on floats, could carry four passengers. At La Pedrera the pilot had picked up a sick nun desperately in need of medical attention. There would be room on the plane for Schultes or the specimens, but not both. Either he or the plants would have to hang around Araracuara waiting for the next plane that was due within the month. Schultes agreed to send the specimens, much to Pacho's relief. As it turned out, the plane carrying the nun passed over the headwaters of the Caquetá and almost reached Tres Esquinas before it fell into the forest, killing all on board.

In the late fall of 1944, as the onset of the rains brought the tapping season to a close, Schultes returned from Wandurraga's farm on the Loretoyacu to Leticia to lay the groundwork for the next phase of his research. The material that Sorensen had brought out two years before had continued to show immense promise. The researchers at Villarteaga and Turrialba wanted more of it, lots more. Schultes's assignment was to collect three tons of rubber seeds and find a way to get them to the experimental plantations.

Logistically it was a nightmare. Rubber seeds, as the British had discovered, are remarkably short-lived. In the heat of the lowlands they survive barely three weeks before the embryo is destroyed by fermentation. Shipment by land and water was out of the question. Even if a vessel were available in Leticia, the voyage downriver and up the Putumayo to the head of navigation on the Orteguaza took twenty-five days. From there it would take at least six weeks on foot, along narrow trails that meandered through the forest and rose across the frigid heights of the Andes. The only airplane available was the Catalina, but the entire town depended on that monthly flight. What's more, its capacity was only 2,500 pounds. Even had he been able to commandeer three flights in a row, it would have done him no good. The fruiting season lasts only two months. Schultes had to be able to gather and store the seeds, prevent their germination, sterilize the entire collection to avoid fungal contamination, and somehow dispatch the shipment in a single consignment.

In January 1945 he flew to Bogotá, met with the Colombian agency in charge of purchasing the seeds, and obtained its permission to charter a larger airplane. He then returned briefly to Washington and secured the cooperation of Earle Blair and Carl Utz, senior officials of the Rubber Development Corporation. They agreed to make available for a single flight the old transport that the company had been using to supply the remote rubber camps of Miraflores and Mitú. Back in Colombia,

Schultes arranged for the Colombian Ministry of War to ferry the seeds from Bogotá to Medellín. From there they could be moved by road to the plantations at Urabá. By the time he returned to Leticia in the last days of January, the fruits were ripe and the capsules ready to explode and drop the seeds.

With no time to lose, Schultes set in motion a whirlwind of activity the likes of which Leticia had rarely seen. With the help of Wandurraga and Pacho, word spread overnight that a gringo was paying good money for seeds that anyone could gather for free. Within days women and children, idle soldiers, and dozens of rubber tappers with time on their hands headed for the forest. Schultes, meanwhile, needed a center of operations. Padre Luis de Garzón offered to lend him the old church, which had recently been converted to a theater. The building was ideal. Large and well ventilated, it had a clean cement floor, wooden walls, and a good thatch roof.

The next problem was storage. The first seeds collected would need to be kept sterile yet moist for nearly a month. The answer was sawdust, and fortunately the military garrison had a small sawmill upstream just out of town at Granja Caldas. Colonel Monroy happily offered 125 enormous sacks of the stuff, which Schultes transported to Leticia by canoe. But how to sterilize several hundred pounds of sawdust? Once again Colonel Monroy came to the rescue, lending him an enormous cauldron that the army used to feed its garrison when the troops went on maneuvers. For the next days Schultes perfected his technique, partially filling a burlap sack with sawdust, submerging it for ten minutes in boiling water, and then sun-drying the contents on sheets of plastic until the sawdust was just moist enough to preserve the seeds without stimulating germination. The entire town, meanwhile, wanted to know why the gringo was cooking sawdust.

In the meantime, the seeds were beginning to arrive. As it turned out, the winter of 1945 was an ideal year for the job. Normally in Leticia the floods begin by the middle of January, and by February much of the forest is inundated. When the fruits mature in early March, most of the seeds fall on water. Some sink, many are eaten by fish, most drift away to rot in the current. Only a few reach the isolated islands of dry land. This is one reason so few rubber seedlings survive despite the enormous production of seeds. By chance in 1945 the floods were a month late, most of the land remained dry, and the seeds could be readily gathered.

Schultes's main concern was not quantity but quality. It was essential that all the seeds be perfectly ripe. If they had fallen naturally from the trees, they were hard and had a brilliant shine. Those that had been prematurely harvested were dull and quickly blackened. Many arrived already destroyed by fungal infections. Others had been deliberately sabotaged. Early on he discovered that some of his Brazilian collectors, anxious to prevent another "rubber theft" by a foreigner, were boiling the seeds before delivery. These were soft and dull and easily rejected. Still, every single seed had to be personally inspected by Schultes. Those produced by white and red rubber trees weighed in at 150 to the kilo; black rubber seeds were slightly smaller, some 200 to the kilo. In gathering 3 tons, Schultes had to examine over 600,000 seeds.

By the time the plane arrived on March 23, he had more than secured his reputation in Leticia. As Wandurraga and the men worked throughout the night, sifting the sawdust and packing the seeds into hemp sacks for shipment, he and Pacho wandered up and down the narrow streets of the town, planting all the hevea seeds that had germinated while in storage. When he left the next day for Bogotá, the young boys were delighted. After two months they would again have their movies in the theater, provided they did not upset the Padre. As for their parents, they could not have known that Schultes had transformed their town. Forty years later the seeds that he and Pacho planted would grow into great trees, and instead of walking beneath fruit trees imported from Asia, their grandchildren would live beneath the canopy of the tree from which their history had been forged.

It was now April and time to travel. Schultes had five months before he was due back in Leticia. As soon as he had delivered the seeds to the plantation at Villarteaga, he flew south to Peru, where he crossed the Andes by road to visit the newly created rubber station at Tingo Maria on the upper Huallaga. Located in the foothills of the mountains, Tingo was the center for all USDA rubber operations in the central Andes and adjacent parts of the Amazon basin. After consulting with a leading Peruvian rubber agronomist, Manuel Sanchez del Aguila, who had just returned from Turrialba, he joined a party for a journey down the Río Tulumayo to visit the plantation of a Victor Langemek. They never arrived. Taken by the current, the canoe overturned in a rapid and was swept beneath a log jam. Most of the supplies were lost. Returning to Tingo, Schultes decided to push on for Lima. His real goal lay farther south, beyond Cuzco in the tropical lowlands of the southwestern Amazon. There in a vast expanse of rolling forest where the borders of Brazil, Peru, and Bolivia come together, another young American botanist had identified the most important and promising population of *Hevea brasiliensis* in all of South America.

Born in Chicago and trained at the Missouri Botanical Garden, Russell Seibert had arrived in Peru early in 1944 charged, as Schultes had been, with the task of finding and collecting noteworthy strains of rubber. He spent the first six months in Iquitos on the Ucayali, nosing around the bars and waterfront shanties, visiting the small farms where retired *seringueros* scratched a living from tired soil and old rubber trees swollen with wounds. Each of the rubber tappers had different memories of the boom, of the era when Iquitos came close to rivaling Manaus, and Julio Arana and his thugs ruled the city. On one issue, however, Seibert found the rubber tappers in complete agreement. To know the greatest of the trees—the *lecheros,* the real bleeders, as they put it—one had to travel upriver and beyond to the forests of Madre de Dios.

Seibert did so, flying in a Catalina from Lima across the mountains 600 miles to Puerto Maldonado, a small jungle settlement located at the confluence of the Tambopata and Madre de Dios rivers. From there he continued through the forest on foot, with his gear loaded on the backs of oxen. Traveling 140 miles in ten days, he crossed a dozen river drainages to reach a rubber depot known as Iberia. Following the tapping circuits just as Schultes had done on the Loretoyacú, Seibert confirmed that the rubber trees yielded extraordinary amounts of latex. What's more, though the blight was present, many trees showed clear evidence of resistance. The trees themselves were enormous, with broader leaflets and larger, more elongated seeds than usual in the species.

Botanists, including Schultes, had known about this rubber for some time. *Acre fino,* as it was called, was not just the finest rubber of the Amazon, it was far superior in quality to anything produced by the Far East plantations. Light and highly pliable, it was particularly in demand for specialty items such as surgical gloves and condoms that require great strength and elasticity. Even in the wake of the boom when the rubber industry all but disappeared in the Amazon, the market for *Acre fino* remained strong. In 1988, Chico Mendes would be murdered eighty miles away in Xapuri, Brazil, for defending the rights of the rubber tappers of Acre. What made their business viable and their way of life possible was this unique population of *Hevea brasiliensis.*

High yields, the best product, and the possibility of resistance made the trees of Acre a top priority of the rubber program. Anxious to see these forests for himself, Schultes met Seibert in Lima in late May 1945, and together they flew east, over the mountains and across the face of the eastern *montaña.* Reaching the lowlands, their airplane dropped beneath the clouds and passed low over the canopy of the forest. Even from the air Schultes could see that the emergent rubber trees were highly unusual. The leaves of most *Hevea brasiliensis* are deep green and normally very glossy on the surface. These trees had a bluish, almost aluminumlike cast, which Seibert had not noticed. Landing at Iñapari, a small village located on the Río Acre north of Iberia, the two botanists spent five weeks in the forest, much of the time puzzling over the proper classification of the trees. Seibert believed that they were basically a geographic race referable to *Hevea brasiliensis* and that the unusual traits had resulted in part from hybridization with another species, *Hevea guianensis* var. *lutea.* Schultes maintained that the trees were in fact unique, though the distinguishing characteristics were not constant enough to warrant the taxonomic delineation of a distinct variety or species. He suggested that the trees be considered an ecotype, identified in the literature by the common name, *Acre fino.* One can imagine these two plant explorers, slogging through the mud or huddled around damp campfires, debating these arcane points of botanical classification and nomenclature.

On the practical matter at hand, Schultes and Seibert were in total agreement: The rubber trees of Acre were the finest and largest specimens that either of them had ever seen. They pondered what might have become of the Southeast Asian plantations had Wickham's seeds come from Acre instead of the Tapajós. The plantations they were in the process of building would be based on the superior stock, and the potential seemed limitless. By the time they left the forest at the end of June, Schultes felt a renewed confidence in the rubber program. Seibert was certain that it would succeed, provided they secured the very best material. His plan (which he carried out) was to remain in the forests of Madre de Dios for three years. During that time, while Schultes was back in Leticia, Seibert made over 350 clonal selections of high-yielding and resistant strains, all of which eventually made their way to the research station at Turrialba.

With two months before the onset of the tapping season in Leticia, Schultes had time for one more expedition. From Lima he traveled overland to Cuzco and then by railroad to the shores of Lake Titicaca and across the Bolivian altiplano to La Paz. There he spent a weekend with an American schoolteacher, Wayne White, before flying five hundred miles east across the mountains to Guayeramerín, a small lowland outpost on the Mamoré River, just across from the Brazilian frontier. Finding the customs agents hopelessly drunk, he left Bolivia illegally, slipping across the caiman-infested river in a dugout paddled by Indian boys. Once in Brazil he took passage on the old Madeira-Mamoré railway, a steam locomotive that ran two hundred miles through the forest to Porto Velho, then a small Brazilian trading center on the upper Madeira River. Built between 1907 and 1912, this odd little railroad, isolated in the middle of the lowlands, bypassed nineteen major rapids on the Mamoré and Madeira, and ostensibly provided Bolivia with an outlet to the Atlantic. Financed by Brazilians as compensation for the annexation in 1903 of Bolivia's Acre territory, its construction had cost over six thousand lives, one death for every fifty yards of track.

Arriving in Porto Velho on July 26, Schultes met up with Edgar Cordeiro, a Brazilian agronomist from the Instituto Agronomico do Norte in Belém, and together they made immediate plans for the river. Embarking by motor launch on the 28th, they journeyed seven days downriver, making brief stops to collect wild rubber at Calama, Humaitá, Boca Tres Casas, and various points in between. Their immediate goal was the mouth of the Río Marmelos, one of the many affluents that make the Madeira, though but a tributary of the Amazon, a larger river than the Mississippi. Schultes's aim was to reach the savannahs that lie between the headwaters of the Marmelos and the adjacent drainage of the Río Manicoré.

Twenty years before, a lone rubber tapper had brought out material of a rare species of hevea, native to the patches of light forest found on the grasslands. Based on one leaf, one capsule valve from a fruit, and a single seed, Adolpho Ducke had described it as a new species, *Hevea camporum.* Schultes hoped to collect complete material of the plant, which he suspected might represent but a dwarf form of *Hevea pauciflora* var. *coriacea,* just as his discovery of the strange little shrub on the mountains of Chiribiquete had turned out to be a variety of the tree *Hevea nitida.* To find this out he was happy to travel four hundred miles by river, dragging a canoe through a dozen rapids and fighting off malarial fevers as he went.

For a month he and Cordeiro tried to get up the river. Unfortunately, the rains were late, and the water level was so low that they ran aground well short of their goal. To continue on foot would have meant abandoning the work in Leticia. Reluctantly, they turned back and began the long paddle downstream. They did collect *Hevea nitida,* a strange and highly variable species. In swampy habitats it had yellow latex and grew to ninety feet. On dry rocky hillsides the latex was the color of milk, and the tree never surpassed the height of a medium-sized palm. The plant had no value except for those unscrupulous tappers who added its latex to that of the commercial species. Once this was done, the entire mix would be ruined, for the latex of *Hevea nitida* prevented rubber from coagulating. Such adulteration was a serious problem, one that did not show up until the latex was mixed with acid or smoked over a fire.

Coming down the Marmelos, Schultes and Cordeiro also took note of the unusual abundance of a second wild species, *Hevea spruceana.* Though its watery and highly resinous latex was no good for rubber and its only use was as wood to make matches, the foliage had shown resistance to blight at Fordlandia. The expedition made a number of important collections, including budwood for propagation. Still, it was small consolation for having failed to reach the savannahs. Three years later Schultes would try once again to get up the river, only to have his effort thwarted by beriberi that numbed his legs and made walking impossible. Eventually in 1961 two Brazilian botanists, aided by the military, would helicopter to the savannahs and confirm Ducke's identification. *Hevea camporum* was indeed a plant new to science.

In September, when Schultes returned to Colombia, he found that his status in Leticia had changed. Before he had been seen by the townspeople as a distinguished though eccentric foreign scientist. Upon his return he was immediately absorbed as a bona fide member of the community, with all the incumbent social obligations. It fell to him, for example, to mediate the growing battle between the two arms of the Christian faith.

For almost a year now, since their arrival in November 1944, two Protestant missionaries, Orville and Helen Floden, had been challenging the monopoly long enjoyed by Padre Luis. The two parties, though obedient to the same God, could not have been more different. Like most Catholic clergy working in the lowlands, Padre Luis had a quiet patience for the fragility of the human spirit and the ease with which men in the tropics drifted into sin. The Flodens, though good and decent people, had a somewhat more dogmatic interpretation of the Scriptures. They especially disapproved of the amount of time Padre Luis spent in the local bar. They preferred to wait out the long tropical evenings at home, in a simple house decorated with furniture purchased from the Sears catalog and brought in from the United States. Their time was devoted to prayer or various home improvement projects, such as the brick fireplace Orville built so that on one night a year, Christmas Eve, he and Helen could sit by the fire.

Naturally, in those years in Colombia, a Capuchin priest and an evangelical Protestant could not speak directly to each other, and that's where Schultes came in. Floden, an excellent carpenter, owned the best set of tools in town. Padre Luis, who could break a hammer driving a nail, was always pestering Schultes to borrow one thing or another. Floden was happy to oblige but eventually had to insist that Schultes show the Padre something about building. Thus Schultes found himself teaching carpentry, among his many duties.

This was only the beginning. When political unrest in the highlands left Leticia exposed to the threat of another Peruvian attack, Colonel Monroy rolled out the cannon and placed his eighty troops on alert. He wanted to deploy them upstream in a line of defense that ran from Leticia into the forest. But his soldiers were from the Caribbean coast, terrified of the jungle and certain to get lost. So the Colonel turned to Schultes, gave him a uniform, and ordered his soldiers to follow the doctor unto death. Schultes, naturally, turned the outing to his advantage, collecting several novel plants during the week the troops sat in the forest. Once back in Leticia, Schultes was surprised to find himself addressed by everyone as *el señor ministro.* While he was away, a visiting botanist from Bogotá had gotten drunk, stood on a table in the bar, and nominated him for the position of minister of agriculture. The people of Leticia assumed that their eminent *botánico* had become a member of the cabinet of President Alfonso López. To them it made perfect sense.

Schultes escaped his newfound notoriety by moving back upriver to Wandurraga's farm. In his time away the Indians had finished a house for him. Located on the crown of a hill in an open field overlooking the river, it was a simple structure of thatch, built off the ground, with a floor of palm wood and three rooms divided by walls of bamboo. He still took his meals at Wandurraga's, but the house provided a quiet place to work and a semblance of privacy. At a total cost of thirty dollars, it was a bargain. But Schultes was seldom there. Within a week he was again on expedition, heading north overland from the Amazon to the Río Putumayo. The goal was to discover the northern limit of *Hevea brasiliensis* and find out whether it would be useful for Wandurraga to expand his operations beyond the headwaters of the Loretoyacu into the Putumayo drainage.

By Schultes's standards it was an uneventful passage. Thirty miles by canoe up the Río Amacayacu, fifty miles overland through the forest to the head of the Río Cotuhé, and then one hundred miles downriver on a raft built on the spot from palm trunks and balsa. There were no trails and, even worse for Schultes, no coca. The Tikunas never used it, and once beyond the headwaters of the Cotuhé, there were no Indians on the river. For two weeks the party lived on bread, three cans of tuna, two cans of sausages, two kilos of rice and sugar, a kilo of butter, a pound of onions, one pineapple, two papayas, and a kilo of dried *pirarucu* fish. Padre Luis, who came along for the walk, struggled through the mud, refusing to remove his long woolen robes that, with the rain, soon weighed more than forty pounds.

Schultes had one close call. The others had gone ahead, leaving him to survey the wild rubber. As it turned out, there was no *Hevea brasiliensis* beyond the Amacayacu, but several of the other species grew in some abundance. It had been raining, and his bifocals had steamed up. He took a step, stumbled, and his foot landed on what felt like a large rubber hose. Coiled on the ground was an enormous anaconda. Fortunately it had just eaten. His Indian guide, who believed in the sacredness of the snake, was still talking about Schultes's narrow escape a week later when the party reached the military post at Tarapacá. There Schultes discovered two pieces of good news: The plane that would carry them back to Leticia was due within days. More important, Major Gustavo Rojas Pinilla, the commander of the post when Schultes and Nazzareno went down the Putumayo in 1942, had been promoted and was serving in Bogotá. Schultes would not have to play any chess while they waited.

For Schultes the next year would be much like the one that had just passed. He spent the fall of 1945 with the rubber tappers on the Loretoyacu, and then returned to Leticia in January 1946 to organize the gathering and dispatch of another enormous collection of rubber seeds. On March 6 a ton went off to Villarteaga. Five hundred pounds were sent to a new experimental plantation located on the Río Calima, just north of Buenaventura on the Pacific coast. Four shipments, totaling almost half a ton, headed east for Brazil. The last consignment, more than three tons destined for the Peruvian plantations at Yurimaguas and Tingo Maria, was dispatched on April 1. Having inspected and processed more than 700,000 seeds, Schultes was again free to move.

The middle of May found him on the west coast of Colombia, inspecting the plantation on the Río Calima, examining some of the ten thousand seedlings that had sprouted from his Leticia collections. Then, eager for a respite from the lowlands, he returned for two weeks to Sibundoy. The beginning of June had him back in Pasto, collecting on the flank of the Galeras volcano. On June 5 his sojourn in the mountains was interrupted by an urgent message from the embassy in Bogotá. The president of Bolivia, Major Gualberto Villarroel, who had come to power in a military coup in December 1943, was interested in establishing rubber plantations in the eastern lowlands. The embassy wanted Schultes to fly to La Paz to talk things over with him.

Schultes left immediately, spent a week at eleven thousand feet fighting off malaria, and finally felt well enough to pay a call. He directed his driver to take him to the Presidential Palace. Erect and resplendent in a new uniform, Villarroel welcomed Schultes in a cavernous office. The meeting went well; the President seemed genuinely interested in the potential of the plantations, but the project went nowhere. A protest march within a month of Schultes's visit turned into a revolution, and the people of La Paz burst into the palace, dragged Villarroel to the street, and strung him up from a lamppost in the plaza. Battered, cut, and naked, his lifeless body hung throughout the night.

By then Schultes was one thousand miles away, readying himself for a third and final season in Leticia. With him this time was a Californian named George Black. Schultes had first met him in Belém where Black had been working at the Instituto Agronomico do Norte, mounting specimens in the herbarium. A wild character, fluent in Portuguese and so tall and thin you could practically see through him, Black turned out to be one of the toughest field men Schultes would ever know. What's more, he knew his plants. The first thing they did together was a second traverse of the Trapezium, completely on foot, on a compass bearing through the forest. Halfway across, Black lost one of his shoes in a muddy river crossing. Without a complaint he discarded his remaining shoe and walked barefoot to Tarapacá. Unwilling to wait a month at the military post for a plane, he and Schultes decided to return to Leticia by water, the long way. First they paddled a dugout canoe two hundred miles down the Putumayo—or Içá as it is known in Brazil. From its confluence with the Solimões, the main trunk of the Amazon, they had another thousand miles to go to reach the city of Manaus. Hitching rides with passing riverboats, they arrived without incident and by the end of September 1946 were back in Leticia toasting their good fortune with beer at Oldenburg's bar. Black would not always be so lucky on the rivers. Several years later, hunting for plants on the Río Madeira, he and two Indians would lose their lives attempting to shoot a rapid. Their bodies were never found.

By the end of the fall of 1946, Schultes had been in South America almost continuously for four years. His work in Leticia, together with what Seibert had accomplished in Peru, had done more to advance knowledge of wild rubber and the future of American plantations than that of any other explorer. In December, when he finally returned to Boston and throughout the spring of 1947, when he toured Europe, visiting herbaria in England, the rubber program stood poised on the edge of success. By then, however, the war had been over for almost two years and events had taken a turn that in the end would leave all of his efforts, as well as those of his colleagues, in vain.

• • •

In 1935 at the Seventh Nazi Party Congress in Nuremberg, Adolf Hitler announced rather calmly that "the problem of producing synthetic rubber can now be regarded as definitely solved. The erection of the first factory in Germany for this purpose will start at once." Fortunately for the free world, the statement was a lie. It was true that Germany for some time had been at the forefront of the scientific effort to find a chemical substitute for rubber. During World War I, with the British blockade strangling the country and German trucks running on tires of wood and rope, the Bayer Company of Germany turned out 2,500 tons of a substance known as methyl rubber. In 1929 scientists working for I. G. Farben invented Buna-S, the co-polymer of butadiene and styrene, which in the end would prove the most useful of all synthetics for making tires. Still, despite Hitler's assurances, immense technical problems remained, many of which the Germans never solved. Indeed, despite having the highest priority during the war, their synthetic rubber program never achieved full design capacity. Production peaked in 1943 at 109,000 tons. In terms of both mileage and tread wear, the best tires they could build lasted a tenth as long as what the Allies had. In a war that raged over thousands of miles, this weakness took on strategic significance.

In the United States a truly commercial synthetic rubber, known as neoprene, was first produced by DuPont in 1932. It found a limited market, but for the most part efforts to forge a full-scale synthetic industry went nowhere in the decade before the war. With the price of natural rubber dropping to three cents a pound, there was little economic incentive. The first plant capable of producing Buna-S rubber, which everyone knew to be the future of the industry, did not begin operations until April 1941. In that year synthetic rubber accounted for only 1.4 percent of national consumption. With a complacency difficult in hindsight to believe, America stumbled toward the abyss. As late as 1940 the country exported 125,000 tons of processed scrap rubber, 60,000 tons of which went to Japan. Standard Oil of New Jersey continued to work in partnership with I. G. Farben long after the onset of the European conflict.

The fall of France in May 1940 stunned the United States and finally prompted the government to lay plans for four plants capable of producing a total of 40,000 tons of Buna-S synthetic rubber. In September the National Defense Advisory Committee recommended that this capacity be increased to 100,000 tons. Incredibly, the orders for the facilities that would make the vital precursors, styrene and butadiene, did not go in until November 1941. It took the trauma of Pearl Harbor to galvanize the nation. In January 1942, Jesse Jones, head of the Reconstruction Finance Corporation, the agency that controlled the Rubber Reserve Company, ordered the capacity of the four plants to be increased immediately to 120,000 tons. In April, after a string of Japanese victories, he boosted that figure to 800,000 tons and in a symbolic gesture changed the name of the rubber from Buna-S to GR-S (Government Rubber—Styrene).

It was, of course, one thing for a government official to anticipate requirements and quite another for a fledgling industry to increase its capacity a hundredfold in twenty-four months. To make matters worse, rubber was not the only industrial challenge facing the nation. In 1940 the country's gross national product was $99.7 billion. During the first half of 1942 government agencies alone placed more than $100 billion in new orders, more than industry could fill in two years. Throughout the land there was intense competition for construction materials, equipment, and labor.

In the midst of this crisis the rubber industry raced against time to establish its infrastructure. Expediency and innovation ruled. When one of the sources of butadiene was needed for aviation fuel, the chemical engineers had to figure out how to make it from grain alcohol. There was no opportunity to worry about perfection of product or process. The use of existing oil refineries to make butadiene was costly and inefficient, but the plants were in place and had to be used. Since the shipping of butadiene required pressurized tank cars, which were in short supply, the rubber factories were built close to the oil refineries instead of near the markets or the tire manufacturing plants in Akron, Ohio. Over half the capacity ended up in Texas and Louisiana. Significantly, while the major tire companies controlled the production of GR-S, the factories making the precursors—575,482 tons of butadiene in 1945, 180,106 tons of styrene—remained under the umbrella of the emerging petrochemical industry. Altogether fifty-one major facilities were constructed. Full capacity was achieved by 1944. Incredibly, production of GR-S reached 830,780 tons by 1945. At a cost of nearly $673 million, the United States had pulled off one of the most outstanding scientific and engineering achievements of all time. Had it not been for the Manhattan Project, the $2 billion investment that created the atomic bomb, the synthetic rubber program would be remembered as the greatest technological breakthrough of the war.

With the defeat of the Japanese in August 1945, the government began the slow process of disengaging itself from the rubber industry. By the terms of their agreements with the Defense Plant Corporation, the companies that had run the plants had six months to decide whether or not to purchase them. Anticipating enormous profits with the end of rationing and the opening of new car markets, the major companies—Goodyear, Firestone, U.S. Rubber, and B. F. Goodrich, each of which had run three major facilities during the war—were happy to buy. After all, synthetic rubber had transformed the industry. In 1941 natural rubber accounted for 99 percent of American consumption. By 1945 the ratio was almost reversed, and petrochemical plants supplied over 85 percent of domestic use.

The rubber industry of the Far East, however, was not about to fade away. Fears that the Japanese would sabotage the plantations by deliberately introducing leaf blight proved unfounded, and once the war was over, production was quickly restored. Natural rubber still had two major advantages: It was cheaper and better. None of the synthetic compounds could match its flexibility, durability, and strength. As the men who made the tires used to say, "Synthetic rubber is a great material. It will mix in any proportion, and the more natural you put in, the better it is."

Tires made in America from synthetic rubber, though the best in the world, were still hopelessly inferior. In the first months after the war, the rationing board found that it had to allocate extra supplies to the Rocky Mountain States. Returning servicemen driving east from Pacific ports invariably blew their tires by the time they reached Denver. By 1947 natural rubber had recaptured 50 percent of the American market. Between 1946 and 1950 the production of the Far East plantations doubled each year. In 1948 natural rubber once again dominated world consumption, and the production in the United States of synthetic rubber fell to 540,000 tons and would have gone lower had it not been for government regulations that mandated a minimum synthetic content in rubber goods.

Caught between the resurgence of the Far East plantations and the technical breakthroughs of the synthetic industry were the efforts of men like Schultes and the agronomists of the USDA to establish an independent natural rubber industry in the Americas. Back in 1943, in an administrative reshuffling that in the end would be of monumental consequence, the financing of the Latin American rubber program passed to the control of the State Department, which allocated funds to the Bureau of Plant Industry, still a part of the Department of Agriculture. Initially this was a useful arrangement because it insulated the program from members of Congress, especially those from farming states, who might have questioned the wisdom of spending money on a crop that could never be grown in the United States. Eventually, however, the arrangement proved disastrous, both for the program and the country.

Almost from the moment the war ended, Robert Rands, Mayer's immediate superior and the head of rubber investigations at the Bureau of Plant Industry, became embroiled in a bureaucratic struggle with State Department officials who questioned the usefulness of the rubber program. Natural rubber supplies, they maintained, could be met with a national stockpile. Actual production from Latin America remained low, and it would be some time before the region could be a meaningful source. True independence from Asia would be found not in Latin America but in Louisiana and Texas. This was the heart of the critique. Synthetic rubber represented the future. The infrastructure was in place, the investment had been enormous, and the plants had to be used or they would be ruined by corrosion. America could control the entire process, from the extraction of petroleum to the final pressing of tires. What's more, the nation retained an absolute monopoly. As late as 1954, 92 percent of the world's synthetic capacity lay in the United States, with most of the rest in Canada.

The invention of Cold Rubber processing in 1946, an innovation that significantly increased the flexibility and tread wear of synthetic tires, appeared to foreshadow a series of technical improvements which promised to render natural rubber obsolete. In a confidential memo dated December 13, 1951, a mid-level State Department employee, Lester Edmond, dismissed the importance of the Latin American rubber program. "Even if we assume sufficient rubber will be available in 15 to 20 years to supply a significant portion of this country's defense needs for new rubber, it is extremely likely that by that time the technical development of synthetic rubber will have progressed to the extent where it will be completely interchangeable with natural." Faith in synthetic rubber was both absolute and politically convenient. There were far more votes in Texas than in Costa Rica.

Against this tide, Rands did his best to remind people of the real issues at stake: the susceptibility of the Far East plantations to leaf blight, the superior quality of natural rubber, the unproved potential of synthetic, and the tremendous cost of maintaining a national stockpile. This strategic reserve represented an investment of half a billion dollars. Most of it had to be rotated every three years. The annual maintenance cost alone was in excess of $25 million. If plantations could be established, Rands argued, they would serve as a living stockpile. Between 1946 and 1952 the government spent over $40 million on synthetic rubber research. In fourteen years the plantation effort had cost a mere $2.8 million. In an era when companies such as United Fruit were spending up to $20 million a year to expand production of bananas, it represented a piddling amount.

Yet the results had been impressive. Plant explorers had located and brought back incredible collections of germplasm, high-yielding and resistant strains that provided the building blocks of the program. At the research station at Turrialba in Costa Rica, a forest now grew of the very best material selected from the wild. At experimental plantations throughout the Americas, controls of the blight had been developed and their effectiveness extensively tested. The new director at Turrialba, a young pathologist from Illinois named Ernest Imle, had perfected nursery practices and budding procedures that in two years could produce a commercially viable sapling with a vigorous rootstock, a high-yielding trunk, and a blight-resistant crown. A simpler process—a single graft of stems with resistant foliage onto a high-yielding stock—could be accomplished in twelve to eighteen months at a cost of just three cents a tree. It would be many years before a breeding program would produce an ideal clone, the seeds of which would consistently grow into high-yielding and resistant trees. But all the material for the work had been gathered, the basic science was well advanced, and the team was ready to go.

On February 1, 1952, the financing of the rubber work was taken over by a branch of the State Department known as the Institute of Inter-American Affairs (IIAA). On March 31, Rands outlined the progress of the program in a memo sent to the manager in charge, Rey M. Hill, director of the Division of Agricultural and Natural Resources. "Our research program," he wrote, "has reached a crucial stage. It has provided methods and plant material that make rubber production in leaf blight areas of Latin America economically feasible." Two months later, on May 21, Rands followed up with a warning: "Our strategic and critical needs for rubber can be satisfied only partially by the development of synthetic substitutes, since these materials cannot meet our requirements without significant admixture of the natural product. Strategically, the development of sources of rubber in the Western Hemisphere is urgent to hedge against our being cut off entirely from sources of supply in the East by war, plant disease, or ideological developments."

Rands was not alone in recognizing the importance of the rubber work. A memo dated June 10, 1952, and circulated by Hill's second in command, Lyall Peterson, assistant director of the Division of Agricultural and Natural Resources, stated unequivocally that "the evidence we have points to the conclusion that Latin America can compete successfully with the Far East." On July 17, at a meeting of the Rubber Advisory

Panel held at the State Department to discuss the future of the rubber work, Paul Litchfield, CEO of Goodyear, a man who had lived through the crisis of 1941, called for an expansion of the program.

It was not what Rey Hill wanted to hear. The wrong man in the wrong place for the wrong reasons, Hill had already decided on largely political grounds that rubber was not for Latin America. He had said as much to Ernie Imle while on a perfunctory tour of the facilities at Turrialba. Imle, having spent twelve years of his life on the project—ten as director of the research station—disagreed. They argued, with Hill openly threatening to use his influence with officials in San José to have the entire operation shut down. Hill knew nothing of rubber and showed little interest in the experimental plantings that preserved four specimens of every rare clone from the Americas. He was more concerned about the political affiliations of the local workers hired by Imle. There was civil unrest in Costa Rica at the time. Hill wanted to know why all of Imle's employees were supporters of the Popular Front, a political party that the State Department had branded communist. Imle explained that everyone living in the countryside around Turrialba voted for the Popular Front. He had been hiring gardeners, not conducting a political witch hunt. Only years later did Imle discover that his own file at the State Department had been subsequently flagged, his character smeared, and his patriotism called into doubt.

Declassified documents at the U.S. National Archives reveal something of the process by which Hill and his allies dealt with the rubber program. In March 1953 Robert Rands, who had been working on the rubber program for more than twenty years, was pushed aside and replaced as head of the Division of Rubber Plant Investigations by Marion W. Parker, who nonetheless became an ardent supporter of the rubber effort. On March 19 Parker sent a memo to Hill reminding him that the unanimous endorsement of the Rubber Advisory Panel "ought to obviate discussion of the continuing need of research." It was time, in other words, to stop the bureaucratic infighting and get on with the work. Hill would have none of it. On April 3 he wrote his colleague Richard Cook, chairman of the Project Committee of the IIAA, seeking funds for an external review of the program. Cook raised the issue on May 20 at a meeting of the IIAA board of directors. There it was decided to table Hill's request pending the outcome of an exchange of letters between Harold Stassen, director of the Foreign Operations Administration, and Arthur Flemming, director of the Office of Defense Mobilization. The IIAA board was not willing to move against the rubber program until it had received an assessment of the strategic consequences.

On June 23, Stassen, whose agency at State was about to take over the rubber program, sent a letter downstairs to Flemming's office in the old Executive Office Building. In it he confessed, "We are at a loss to evaluate this program from the viewpoint of U.S. domestic interests. . . . Your advice and suggestions would be greatly appreciated." Ten days later, in a follow-up letter, he added, "I would very much appreciate your opinion . . . on whether an increase of natural rubber production in Latin America would significantly serve the security needs of the United States." When there was no response, Stassen had his deputy send yet a third note, on July 21.

Finally, on August 13, Flemming responded. His support of the rubber effort was unqualified. "We believe," he wrote, "that this research and development program should be continued, since a successful conclusion to the program would considerably enhance our security interests in natural rubber." He went on to remind Stassen of the vulnerability of the Asian plantations. "The accidental or deliberate introduction of this disease could within two years devastate nearly all the closely planted areas in the Far East. Since it takes seven years to bring a rubber tree into full production, the outbreak of leaf blight in the Far East could precipitate a natural rubber crisis at any time, as the Far East supplies over 70 percent of the world's rubber needs." Flemming ends his letter by reminding Stassen that the costs of the program were modest, the potential benefits immense. The endorsement of the rubber industry had been both consistent and absolute. How many times, he seems to ask, do we need to go over this same ground?

Industry was indeed behind the effort. In May 1953, Paul Litchfield of Goodyear had sent Rey Hill economic forecasts which showed that the nation would continue to be dependent on natural rubber for years to come. On July 24, W. E. Klippert, the man in charge of Goodyear's plantation program, wrote to Stassen and reminded him that "there has been remarkable progress in developing new high yielding and blight resistant strains of Hevea rubber for use in this hemisphere. . . . Over the next few years, the long period of fundamental research will begin to pay off." The letter then comments on the administrative hierarchy that had placed the rubber work under the jurisdiction of the State Department. "This rather unusual organizational arrangement has given rise to certain operational conflicts." Industry, in other words, was aware of the efforts to undermine the program and wanted the State Department to know. In his August 11 response, Stassen mentions only that "the questions you raise are deserving of most careful consideration prior to taking any definitive action with regard to the curtailment of this program."

By this time it was clear to all concerned that the reviews and meetings had been mere window dressing. The Foreign Operations Administration had no intention of keeping the rubber program going. They were just waiting for the appropriate moment to pull the plug. On September 2, Secretary of Agriculture Ezra Benson sent an angry letter to Stassen: "Step by step, that program has gotten over into your technical assistance program. . . . This work of unusual technical nature carried on for many years now finds itself so enmeshed in administrative entanglements with your Institute of Inter-American Affairs that the operation of the program is being seriously hampered. . . . The rubber industry of the United States has repeatedly and wholeheartedly endorsed the conduct of the natural hevea work. . . . I urge you to designate this rubber work as a distinct field project activity and give the Department of Agriculture the responsibility and latitude necessary to conduct it for you. . . . This rubber program is most urgent."

On September 25, a meeting of the Rubber Advisory Panel again brought together representatives from Agriculture, Commerce, State, Treasury, and all the major rubber companies. G. M. Tisdale, president of the U.S. Rubber Company, later Uniroyal, offered the most chilling assessment. "This company," he wrote in his submission, "operates sizable plantations in the Far East and development of strains of Hevea which can resist the South American leaf disease could some day prevent the destruction of the plantations. . . . We all know what a blight would be to the economies of the natural rubber growing countries and this company urges that the Research and Development be carried on." Reading the testimony of Paul Litchfield of Goodyear, one senses that he cannot believe he is being asked the same questions yet again. "This program is proving invaluable," he wrote, "in the development of higher yielding strains, methods of disease control, and locally effective production techniques in this hemisphere. . . . The purpose of this letter is to reaffirm our keen interest in the continuance of this highly important work."

By now, it was too late. On October 12, 1953, Rey Hill in a memo to D. W. Figgis, president of the IIAA, proposed eliminating the program altogether. Ten days later Harold Stassen wrote to each of the rubber executives who had made submissions to the Rubber Advisory Panel: "A thorough review of the regional research phase of the program has resulted in my recommending to the Departments concerned that this activity be terminated as far as the administration of FOA is concerned as of June 30, 1954." This was the program's death sentence. Stassen then went out to create a paper trail to rationalize his decision. On December 2, 1953, a five-page secret report calling for the end of the program was presented by a group known as the Research and Evaluation Division, Office of Research, Statistics and Reports. In the entire document there is only one reference to the leaf blight: "There is only a remote possibility that further developing a strain capable of resisting South American leaf blight might have significant long-term benefits in protecting major world sources of supply in S. E. Asia in the event that the blight struck there." In other words, since the development of resistant strains could not prevent the destruction of the Far East plantations should the blight break out, there was really no reason to develop disease-resistant plantations in the Americas. Such was the logic that doomed the rubber effort. On December 9, 1953, Harold Stassen circulated a secret memo to the secretaries of State, Treasury, Defense, and Agriculture announcing the cancellation of the program. It was, he said, of marginal importance to the military security of the United States.

In the summer of 1953, six months before the program was finally killed, the Rubber Division recommended that Schultes be reassigned to the United States for two years to prepare a botanical monograph on *Hevea* and its wild relatives. Even Rey Hill could not deny the merit of the proposal. In a letter to his superior he wrote, "As a result of [his] 12 years of exploration work, Schultes knows more about wild rubber, the various species and their relation to other plant life than any other man. The information he has accumulated over this 12 years is available only in Schultes' mind." To ensure that this knowledge was not lost, Hill supported the special assignment, provided that his approval not have any bearing on the "continuation, or discontinuation of the present rubber development program." It was an ironic endorsement. A monograph would be of great importance, but the real legacy of Schultes's experience grew in the fields of Turrialba and Los Diamantes.

Several times—first in 1946 and then in the spring of 1951—Schultes had gone to Costa Rica to see his trees. Ernie Imle, then director of the Research Station, remembers two incidents from the visits. The first was a minor accident—their jeep rolled in a ditch while en route to the farm at Los Diamantes. Schultes had stunned the workers with his ability to swear fluently in four languages, including Witoto. Then there was the luncheon on the day that he arrived in 1951. Knowing that this legendary figure was coming right from the jungle and would be starved for company, Imle's wife, Portia, put on an elaborate spread and invited several women friends. Schultes had indeed just returned from the Río Guaviare, and he arrived in the parlor of the cottage thin and gaunt, with what Imle remembers as a distant, even haunted look on his face.

When his eyes fell upon an old gramophone, sitting on a stand in a corner of the room, Schultes politely asked Portia if she had any part of Haydn's *Die Schöpfung* oratorio. She had never heard of it. He then reached into his knapsack, pulled out a recording, and asked if he might put it on. When he was away from the sounds of the forest, he explained, this piece of music was his only source of tranquillity and comfort. While the others chatted, he sat alone in a chair, eyes closed, listening to the music. The next morning he climbed to the top of the Irazú volcano and then spent the day collecting in the oak forests at ten thousand feet.

On July 17, 1953, Rey Hill wired the United States Embassy in Bogotá: "Schultes should prepare to permanently leave Colombia when coming to U.S. since not returning to Foreign duty. . . . Assignment to prepare monograph appears assured." It was not. Authorization for the two-year project never came through. In the fall of 1953, Schultes returned to his beloved Harvard and accepted a job as curator of the Orchid Herbarium of Oakes Ames at the Botanical Museum. The endowment that had established the position carried with it the stipulation that the curator work exclusively on orchids. With typical zeal Schultes plunged ahead, devoting himself to this one family of plants and publishing two books, including one on the species of Trinidad and Tobago. His rubber monograph was put on hold indefinitely and has yet to be completed.

During the five years Schultes worked on orchids at the Botanical Museum, the forest slowly reclaimed the plantations at Villarteaga and the other research gardens established by the rubber program. Soon after the program was terminated, agents dispatched by the State Department to Turrialba shut down the station and left with all the records. Within a year nearly all the rubber trees at Turrialba and Los Diamantes had been cut to the ground. The clonal garden that had once served as a repository for the germplasm of an entire continent was replaced by a field of sugarcane. The last of the Schultes and Seibert collections, a copse of rare rubber trees growing adjacent to the old laboratory, was destroyed by a forgettable botanist from Scotland who didn't want to have to walk too far to reach his experimental plantings of manioc.

The destruction of the rubber program, foolish as it was, might have taken its place in history as nothing more than another example of bureaucratic idiocy had it not been for a development that no one involved could have foreseen. In the decades after the war, demand for rubber products of all kinds rose dramatically. Between 1948 and 1978 world consumption increased at an annual average rate of 6.3 percent. In thirty years production went up by a factor of six. Though improvements on the plantations more than quadrupled yields, the supply of natural rubber could not keep pace with demand. For each of the ten years after the rubber program ended in 1954, synthetic rubber captured a higher and higher percentage of the market share. By 1964 it accounted for 75 percent of all production. Economists expected the trend to continue, and most predicted that in time natural rubber would be reduced to a historical footnote.

They were wrong. The first blow to synthetic rubber came with the 1973 oil embargo by the Organization of Petroleum Exporting Countries (OPEC), which more than doubled the price of raw materials for the industry. Plantations consume roughly half a ton of oil to produce a ton of rubber, whereas a petrochemical plant needs 3.5 tons to make the same volume. The increase in oil prices also affected consumers, making them far more conscious of gas mileage. This prompted the second and far more serious challenge to the synthetic industry: the rapid and widespread adoption of the radial tire. Until 1968 over 90 percent of cars in America ran on simple bias tires, the same fundamental technology that had been in place since 1900. The radial tire was a radical departure. By placing the cords within the fabric of the tire at 90 degrees to the direction traveled, and later adding a steel belt for extra strength, engineers at Michelin created a tire that gave better fuel consumption, better handling, and lasted twice as long. Once radials caught on in America, they quickly captured the market and by 1987 accounted for virtually 100 percent of sales. This, in turn, provided an immense boost to the plantation industry, for only natural rubber has the strength required for the sidewalls of radial tires. Here was a technological breakthrough that no one had predicted.

By 1993 natural rubber had recaptured 38 percent of the domestic market, and the United States was more dependent on it than at any time in the previous forty years. The promise of synthetic rubber had been only partly fulfilled. There is today no product that can match natural rubber's resilience and tensile strength, resistance to abrasion and impact, and capacity to absorb impact without generating heat. Today the tires of every commercial and military aircraft, from the 747 to the B-2 bomber and the space shuttle, are 100 percent natural rubber. Half the rubber in every pickup tire in America still comes from a tree. The enormous tires of industrial machinery are 90 percent natural. Nearly half of the rubber in every automobile tire originates on plantations located thirteen thousand miles away.

Each year the United States spends nearly a billion dollars a year importing natural rubber. In 1993 global consumption surpassed 5.5 million tons. Eighty-five percent of it was produced in Southeast Asia. Decades ago, scientists warned of the danger of leaf blight being deliberately introduced into the plantations. "None of the planting material used in establishing millions of acres of plantations in the east has any appreciable degree of resistance to the disease," wrote Loren Polhamus, a leading rubber authority at the USDA. Fundamentally, the situation has not changed. To this day a single act of biological terrorism, the systematic introduction of fungal spores so small as to be readily concealed in a shoe, could wipe out the plantations, shutting down production of natural rubber for at least a decade. It is difficult to think of any other raw material that is as vital and vulnerable.

The decision to end the rubber work was made in a miasma of ignorance, unaccountability, and arrogance for which modern bureaucracies have become renowned. As an act of folly it has few equals in botanical history. After a century the threat of South American leaf blight and the vulnerability of the Far East plantations continue to hang like a Damoclean sword over the neck of the industrial world. Had the rubber program continued, one can be almost certain that today there would be healthy, blight-resistant plantations growing close to home in the Americas. Had that happened, many of the trees would have been the descendants of wild seeds gathered years before by a solitary explorer in the forests of the Amazon. Instead, his dream of twelve years was wiped out by the stroke of a pen.

Chapter Twelve

The Blue Orchid,

1947–48

THE FIRST TIME Schultes felt the dull ache of malaria was on the afternoon of May 23, 1942, as he and his Italian companion Nazzareno Postarino paddled up the Río Karaparaná on their way to El Encanto. It was the height of the rainy season, and both riverbanks were flooded. Still, they had no choice but to make camp and rest until the fever passed. After stringing their hammocks above the boggy ground and kindling a fire from moss and bark, they lay in the rain for three days as the paroxysms of chills and night sweats convulsed Schultes's body. On the morning of May 27 the fever broke, and Schultes awakened to a blue sky, a cool breeze coming off the river, and sunlight falling through the forest. Still weak, he rose slowly from his hammock and cautiously made his way down to the river to bathe. He stumbled and fell against the muddy bank. Looking up he saw a solitary orchid growing on the surface of a half-drowned, mossy trunk. He went closer and reached for the delicate inflorescence. The petals and sepals were light blue, the lip somewhat darker with pale veins, and the back and wings of the column were streaked with red. He had never seen such a perfectly pure shade of blue. Teasing a blossom with his finger, Schultes knew that he held in his hand the legendary blue orchid. "Never," he wrote years later, "could a doctor have prescribed a more effective tonic! I had found my friend. . . . I was happy and could almost have believed that destiny had led me in these lowest of days to that one bright jewel of the jungle."

It was indeed an extraordinary prize. In three centuries the delicate plant *Aganisia cyanea,* originally named for the lover of Apollo, the Greek nymph Acacallis, had been found in the wild by only four explorers. In 1801, Alexander von Humboldt and Aimé Bonpland collected it at the base of Cerro Duida, the mountain of the Lost World that soars above the beginning of the Casiquiare, the natural canal that runs between the headwaters of the Río Negro in Brazil and the Orinoco in Venezuela. Fifty years later the British botanist Richard Spruce found it growing one thousand miles to the south, on a tree trunk by a forest stream near Manaus, just above the confluence of the Negro and the Solimões. In January 1853, Spruce collected it once more, on the upper Uaupés River, below the great cataract of Ipanuré. The better part of a century went by until the next collection occurred in 1939, a single plant found at the headwaters of the Uaupés by Schultes's friend José Cuatrecasas. This was one of the specimens that Schultes had asked to see when he arrived in Bogotá in 1941.

Schultes first became aware of the mysterious orchid as a student at Harvard when he came upon a color illustration in an issue of the *Botanical Magazine* published in 1916. It was less the novelty of the plant than one of the men who had found it that most intrigued him. To understand his affection for Richard Spruce, a humble Yorkshire botanist of the nineteenth century, one must accept the possibility that the seed of one generation can be born in the next and that the spirit of one long dead can reach across time not merely to inspire but to mold the dreams of another. They were not mentor and student, for Spruce had died twenty years before Schultes was born. Schultes's love of Spruce, a raw atavistic association bordering at times on obsession, became his strength, allowing him to endure, encouraging him always to achieve more, and providing his closest experience of spiritual certainty. Asked in recent years whether it might be possible that he had, subconsciously or unconsciously, modeled his life and career on that of Spruce, Schultes replied, "Neither. It was conscious."

They first met in 1922 in a hospital room in Boston when seven-year-old Richard Evans Schultes lay immobilized for several weeks with a childhood affliction. Each evening his father sat by his side, reading from a book entitled *Notes of a Botanist on the Amazon and Andes: Being Records of Travel on the Amazon and Its Tributaries, the Trombetas, Río Negro, Uaupés, Casiquari, Pacimoni, Huallaga and Pastasa; as Also to the Cataracts of the Orinoco, Along the Eastern Side of the Andes of Peru and Ecuador, and the Shores of the Pacific, During the Years 1849–1864*. The two well-worn volumes contained Spruce's letters and journal extracts, edited after his death by his close friend, the naturalist Alfred Russel Wallace. The book's subtitle provides a sense of the breadth of Spruce's journey, a fifteen-year odyssey that to this day represents one of the pinnacles of botanical exploration in South America.

Richard Spruce was born in the Yorkshire hamlet of Ganthorpe, near Malton, on September 10, 1817. The only son in a family of nine children, he grew up on the moor and before he was sixteen had compiled a list of more than four hundred local plants. His closest friend was an older tinsmith, Sam Gibson, an amateur botanist who kept by his forge not a Bible but a copy of Hooker's *British Flora* so caked in soot and grit as to be scarcely readable. When Charles Darwin's *Journal* first appeared in 1839, Gibson bought a copy for Spruce and fired his imagination with the thought that he might accomplish for botany what the great naturalist had achieved for zoology. It was a powerful dream, and one that sustained Spruce during the dreary years when he was forced to earn his living as a schoolteacher in the nearby city of York.

His release came in late 1844, when illness and the unexpected closing of his school sent him south to London where he was befriended by the botanists George Bentham and the great William Hooker himself, then director of the Royal Botanic Garden at Kew. Both had been impressed by several of Spruce's technical papers that had appeared in *The Phytologist,* a botanical journal first published in 1841. Bentham, who had just returned from Spain, suggested that Spruce travel to the Pyrenees, and fund his expedition by selling his collections as he went along. So in May 1845, Richard Spruce left England for the first time. He stayed in the mountains for a year, learned Spanish, and collected nearly five hundred species of mosses, including seventeen that were new. So significant were these collections that Bentham eagerly suggested a far more ambitious expedition to South America. In exchange for receiving one complete set of specimens for Kew, Bentham promised to act as Spruce's botanical agent, processing his collections and securing paid subscriptions from the major herbaria of Europe. With no money of his own and nothing more than this tenuous commitment from Bentham, Spruce sailed for the Amazon in the early summer of 1849. He had no idea where he would go or when he might return.

Spruce's journey would be unique, but the sentiments that propelled him to South America were shared by a remarkable cadre of natural historians, a new type of English scholar, inspired by Darwin yet free of the constraints of class and property that for so long had hampered the natural sciences in Britain. Travel would open their imaginations and provide the space to think new thoughts and the raw material from which to forge new theories of existence. What resulted from these journeys was a watershed of scientific discovery. Society was in flux, and so was biology, and each informed the other. A new plant from the Americas, the stratigraphy of a bed of fossils, the shape of the beaks of birds, the life cycle of a barnacle, a beetle found on the Scottish heath —these and a hundred other moments of illumination inspired the slow gestation of a set of ideas that would shake the intellectual and social foundations of the world.

Darwin's evolutionary theory—the idea that life was ever changing, that species were mutable, with natural selection favoring the survival of the most fit—came to him slowly and drew from many influences. But the inspiration occurred on board the *Beagle* during a long and sometimes lonely voyage that carried him thirteen thousand miles away from home. But what finally prompted him more than twenty years later to go public with his theory and write *On the Origin of Species* was a letter he received on June 18, 1858, from a distant correspondent on the Spice Islands. The letter, posted the previous February, read like an abstract of his own unpublished papers. Stranded en route to New Guinea, Alfred Russel Wallace had, in a fit of malarial fever, independently conceived an evolutionary theory so similar to Darwin's that their results were hastily published two weeks later in a joint paper read before an audience of thirty fellows at the Linnean Society of London on July 1, 1858. Wallace was able to make this famous contribution in part because six years previously, on another continent and suffering from another bout of fever, his life had been saved by his friend Richard Spruce.

Like Spruce, Wallace had started as a young schoolteacher of modest origins, enamored of Darwin and driven by a desire to explore the tropics. In 1848, at the age of twenty-five, he wrote his friend Henry Bates and audaciously proposed that they mount an expedition to solve "the problem of the origin of species." Bates was twenty-three, a kindred spirit and passionate bug collector, employed at the time as an apprentice to a hosiery manufacturer in Leeds. The industrial grime of the Midlands held little attraction, and Bates enthusiastically signed on. After securing the necessary commissions for their collections, these two self-taught naturalists left England in the last days of April 1848 bound for the Amazon. They arrived in Belém on the morning of May 28, just over a year ahead of Spruce.

For four months Wallace and Bates remained together, journeying up the Tocantins and exploring the forests surrounding Belém. Then, for reasons unknown, they went their separate ways—Wallace to the island of Mexiana in the north of the delta, Bates to the estate of a Scottish planter at Caripí, twenty miles upriver from Belém. Other outings followed, and for seven months the two wandered on their own, intoxicated by the scent of the tropics, overwhelmed by a river the width of a sea, and an estuary defined by forested islands the size of entire European countries. They came together once more in the summer of 1849 when Bates returned to Belém from Cametá and the Moju river on the evening of July 19. Wallace had been in town since June, awaiting the arrival of his brother Herbert, who had sailed from England on the brig *Britannia.* The ship docked in Belém on July 12. Traveling with Herbert Wallace was Richard Spruce.

It is not known whether Spruce, Alfred Wallace, and Henry Bates ever went into the forest together. For a naturalist the very thought of such an excursion is sublime. If they did, it could only have happened during the eleven days between Bates's return to Belém and Wallace's subsequent departure upriver at the beginning of August. Many writers have suggested incorrectly that the three came together four months later in November, at the town of Santarém at the mouth of the Tapajós, where Wallace had taken up residence. In fact, Bates reached Santarém on October 9 and left the next morning for Obidos, a village farther upstream. Spruce arrived in Santarém on November 19, the same day Bates embarked from Obidos for Manaus. This was to be the pattern of their travels, each following his own course, vaguely aware of the whereabouts of the other two, occasionally crossing paths as Spruce and Wallace did in Manaus in 1851 and later that year at São Joaquim on the Uaupés. What these men felt about each other is difficult to know, for their sentiments are buried in the formality of the Victorian age. One thing is certain: Taken together, their explorations transformed knowledge of the Amazon. Bates remained for eleven years, collecting over fourteen thousand species of insects, of which some eight thousand were new to science. Wallace stayed for just four years, before disease and the death of his brother sent him home to England. En route he met more misfortune: Seven hundred miles from Bermuda his ship caught fire. Drifting in a longboat, with the men all around him praying for rescue, he watched as the flames consumed his entire collection.

Long after Wallace had overcome this tragedy and moved on to Southeast Asia, long after even Bates was safely back in Europe, Richard Spruce remained in the forests of South America: a year at Santarém, a year at Manaus, then four years on the Río Negro. In a journey unmatched in the annals of natural history, he paddled over four thousand miles, collected twenty thousand specimens, and recorded the vocabularies of twenty-one previously unknown tribes. On this one expedition to and beyond the source of the Río Negro, Spruce described the making of coca powder, identified the botanical source of *yagé,* and collected for the first time the seeds of *yopo,* a powerful hallucinogenic snuff derived from the tree *Anadenanthera peregrina.* This was only a beginning.

In March 1855 he booked passage on a steamer that carried him up the Amazon to Iquitos, Peru. From there he headed overland toward the Andes, reaching the small town of Tarapoto, where he stayed two more years in the foothills of the mountains, amassing a vast collection including more than 250 species of ferns. Then toward the end of 1857, on his thirty-ninth birthday, an official letter arrived from home. With British soldiers having died by the droves from malaria in Crimea, Her Majesty's secretary of state for India wanted Spruce to proceed directly to Ecuador to secure seeds and supplies of cinchona bark, the only known treatment for the disease. A sensible itinerary might have carried him back down the Amazon and then south by sea around Cape Horn to Guayaquil on the Ecuadorian coast, and thence overland across the Andes to the cinchona region on the eastern side of the mountains. But because the commission instructed Spruce to go directly to the fields, he elected to proceed west on foot, a five-hundred-mile journey so perilous that it literally drove his dog mad. Living on wild berries and salted meat, suffering from attacks of catarrh so virulent that blood flowed from his mouth and nose, he somehow made his way through the dense jungle, and crawled out of the Andean cloud forest. Broken in body, though never in spirit, Spruce spent two years in the cinchona fields and gathered the seeds that became one of the foundations of the successful quinine plantations of the East.

• • •

Schultes spent the first months of 1947 in England, much of the time in the herbarium at Kew, studying collections that Spruce had made nearly a century before. Many of these specimens were *Hevea,* for rubber had been very much on Spruce's mind. A witness to the early days of the boom, Spruce saw the price rise in Manaus from 3 cents a pound to $1.50 in just the four years he spent on the Río Negro. As early as 1850, a quarter century before Wickham's famous shipment, Sir William Hooker had asked Spruce to obtain viable seeds. Given the transport available at the time, it proved an impossible task, but it did focus Spruce's attention on the plant, and throughout his travels he collected it wherever he could. His contribution was immense. Today ten species of hevea are recognized. When Spruce began, only two species were known. He found another six, of which four have held up to this day, with one other being reduced to varietal status. No one, with the possible exception of Schultes, has done more to advance botanical knowledge of the genus.

The distribution of Spruce's collections suggested to Schultes that the origin of *Hevea* lay in the remote reaches of the upper Río Negro. Naturally, he would have to go there. Upon his return to America he obtained authorization for his most ambitious expedition to date, a yearlong effort expected to begin in the fall of 1947. The purpose of the trip was to examine the wild relatives of rubber in the field, determine their range, and study their relationships—all with the goal of assessing the role they might play in the breeding program underway at Turrialba in Costa Rica. In particular, Schultes wanted to find *Hevea rigidifolia,* a highly unusual species endemic to the Río Uaupés and collected only twice since it was first discovered by Spruce in 1852. This was the scientific rationale. What motivated Schultes personally was the thought of following in Spruce's footsteps after a century in which no other botanist had strayed so far into the heart of the Amazon.

At Oxford in February, Schultes finally obtained his own copy of Spruce's journals. Flipping through the pages he found nothing he could remember from his youth, or that reminded him of the hours his father had spent by his bedside. Still, there were passages that had an odd resonance. Perhaps because he had shared similar emotions, they read like memories. In his first days on the Amazon, Spruce struggled to express his awe. "The largest river in the world," he wrote soon after arriving in Belém, "runs through the largest forest. Fancy if you can two millions of square miles of forest, uninterrupted save by the streams that traverse it. The natives think no more of destroying the noblest trees than we the vilest weeds; a single tree cut down makes no greater gap, and is no more missed, than when one pulls up a stalk of groundsel or a poppy in an English cornfield. . . . Here are grasses sixty feet high. . . . Instead of your periwinkles, we have handsome trees exuding a most deadly poison. . . . Violets the size of apple trees. Daisies born on trees like Alders!"

Later in the first volume, Spruce seems in retrospect to be speaking directly to Schultes, laying out the course of his imminent travels. "At the highest point I reached on the Uaupés, the Jaguaraté Caxoeira, I spent a fortnight, in the midst of heavy rains. . . . From that point upwards, one may safely assume that nearly everything was new, and I have no doubt that the tract of country lying eastward from Pasto and Popayán where are the headwaters of the Japuré, Uaupés, and Guaviare . . . offer as rich a field for a botanist as any in South America. But I have made enquiries as to the possibility of reaching it, and I find that it will be necessary to cross *páramos* of the most rugged and inhospitable character, and afterwards risk oneself among wild and fierce Indians, so that I fear its exploration must be left to some one younger and more vigorous than myself." Schultes had already crossed the Andes from Pasto and Popayán. His goal now, in the spring of 1947, was to enter the unknown lands from the south and east, traveling up the rivers that drained into the headwaters of Río Negro.

When South America was young and the continent was still attached to the landmass that became Africa, the predecessor of the Amazon flowed east to west, draining an arc of massive uplands, the remnants of which are now known as the Brazilian and Guiana Highlands. The river reached the Pacific somewhere along the western shore of contemporary Ecuador. Then 100 million years ago the two southern continents split apart. A mere 15 million years ago the birth of the Andean Cordillera dammed the river, creating a vast inland body of water that covered much of what is now the Amazon basin. It became, in effect, the largest swamp the world has ever known. In time these waters worked their way through the older sandstone formations to the east and formed the modern channel of the Amazon. Only then did the forest come into being.

The Río Negro and the Río Solimões, the two great branches of the Amazon that come together at Manaus, are legacies of these staggering geological events. The Solimões and its affluents are fed by water from ten thousand precipitous mountain valleys of the high Andes. Rich in sediments, these are the fabled milk rivers of Indian mythology, the source of earth and nutrients that each year replenish the floodplain of the lower river. The Río Negro, by contrast, drains the northern half of the Amazon basin, rising in the white sands and ancient soils eroded from the Guiana Highlands. The water is pure, almost sterile, and its dark color suggests its most compelling feature: a high concentration of humic matter, very little silt load, and a tannin content equal to that of a well-brewed cup of tea. The forests along its shores are magnificent, but the plants grow on soils leached of nutrients, and the animals are those that can survive in a land known to the Indians as the River of Hunger.

Not far from Manaus the Río Solimões and the Río Negro meet, and for several miles the black waters run beside the white, curling and twining in the current, slowly fusing into each other. The Río Negro alone is five miles wide at the mouth. By itself on any other continent it would be considered the second largest river on earth. The color of the Solimões is mocha, and there are floating islands of grass and hyacinth, emerald and blue. Dolphins thrive a thousand miles from the sea. One is the *tucuxi,* dark and in appearance very much like its marine relative. The other is the *boto,* one of a unique group of primitive freshwater cetaceans. Virtually blind, the creatures spend most of their lives upside down, navigating through echo location, emitting faint clicks as they hover in the murky water. Six to seven feet long, with pink skin, tiny eyes, and a prominently humped back, they swim along the river bottom using the sparse, tactile hairs covering their long snouts to locate food. When they come up for air, they breach the surface slowly as if afraid of the sun. To the Indians they appear otherworldly and thus sacred.

For Spruce, the river dolphins were just one of the wonders of this world of water, forest, and sky: the braided channel of the Río Negro, studded with thousands of islands, the Solimões falling away to a distant horizon. Water warmer than the air. A voyage across the confluence where for a week he never saw land. "It is impossible," he wrote soon after arriving in Manaus, "to behold such immense masses of water in the centre of a vast continent, rolling onwards toward the ocean, without feeling the highest admiration, and, when viewed under the setting sun and afterwards the descending and deepening gloom blends all into an indistinguishable mass, though the tumult of the contending waters is still distinctly audible, I felt it difficult to tear myself from the spot." Several months later Spruce's sentiments would be somewhat more subdued. After a year he would write, "The Río Negro might be called the Dead River—I never saw such a deserted region."

• • •

In Manaus on September 4, 1947, Schultes had no time to contemplate either the beauty of the river or the perils of the journey at hand. His launch, a riverboat belonging to Higson & Company and named the *João,* was set to sail for the Río Negro at six in the afternoon. Everything seemed to be going wrong. His field assistant Pacho López had made it safely from Leticia, but their aluminum canoe and botanical supplies shipped upriver from Belém had been delayed and would not arrive in Manaus until the evening of the 7th. Mechanics had failed to repair a second and smaller launch belonging to the Instituto Agronomico do Norte. That meant their companion, a young Brazilian botanist, Dr. João Murça Pires, would have to remain behind awaiting repairs. At noon came worse news. Padre McCormack, an American priest, telephoned to confirm that the money for the expedition had been wired from the United States Embassy in Rio de Janeiro to the U.S. consular agent in Manaus. Unfortunately, there wasn't one. So with the town sealed shut for lunch, Schultes had to find a lawyer and make out a power of attorney so that Murça Pires could cash the money order if it ever arrived at a local bank. He barely had time for a farewell drink with the Colombian consul, Herrán Medina, who escorted him and Pacho to the waterfront just before eight, two hours after the official departure time but a full hour before the *João* actually puttered out into the current.

Things continued to go badly. In the middle of the night, just as the launch struggled past the last of the large islands at Santo Antonio, a tremendous thunderstorm cracked open the night sky. In an instant the drop in air pressure became palpable. Violent winds buffeted the deck and threatened to sink the two smaller barges tied up alongside the hull of the *João.* The vessels were bound together by thick ropes spun from piassaba fiber, yet still one of the ropes snapped and a surge of water drove the barge into the hull of the boat. For an hour the captain fought the storm before finally giving up and steering the launch into the shelter of a shallow bay, protected from the wind by the forest. There was a sudden calm, but it did little good. The passengers, startled awake in their hammocks, could not stop speaking of their adventure. In time the conversation shifted to other topics but did not die down. All the commotion, the noise of the repairs, together with the mosquitoes that rose from the swamp, made for a sleepless night.

The next several days passed uneventfully. Like Spruce before him, Schultes had no specific itinerary. The Río Negro flows in a gentle arc roughly six hundred miles from Manaus away to the northwest and the settlement of São Gabriel, located just below the confluence where the river divides, with the Río Uaupés coming in from the west, and Río

Negro heading off to the north. Between Manaus and São Gabriel were a handful of villages, small and isolated, much as they had been in Spruce's day. Three hundred miles upstream from Manaus was the Salesian mission of Barcelos, and beyond, roughly halfway to São Gabriel, was Santa Isabel, another small Catholic mission. In between were a number of clearings and landings where riverboats stopped and fishermen gathered to trade. Schultes expected to collect on the lower river, but his ultimate goal lay beyond São Gabriel in the isolated forests of the headwaters.

Traveling slowly upriver, Schultes took notes about the vegetation and escaped the heat by working quietly through the middle of the day, translating several of Adolpho Ducke's technical papers on rubber. In the evenings he read Spruce or chatted idly with Padre Colombo, a garrulous Italian priest who ran the Salesian school at Barcelos. The Padre pointed out the rock carvings below the mouth of the Río Branco, the celebrated Ilhas de Pedras that had attracted Spruce, and he boasted of the coconut palms his mission had planted along the shores. At the Yurupari-roka, a massive granite boulder rising in the middle of the river, he spoke of the devil and questioned what force would draw a people to the edge of an endless swamp. He called the people of the river *cabocolos* and described them as a peasantry of the flood forest. What Indian blood remained on the lower reaches of the Río Negro had long been absorbed by this mixed race of men and women whose lives were tied to the cycle of the river. In the dry season they did well, living on fish and what produce they could grow on the riverbanks. The seasonal rains brought a period of trial. The river rose by as much as fifty feet, flooding homes and fields, scattering the fish, and forcing the cattle onto floating rafts or into distant fields beyond the flood line.

Four days out of Manaus the launch reached the mission at Barcelos. It was Independence Day, oppressively hot, and the ship's crew and passengers had been drinking *cachaça,* raw cane liquor, since noon. And apparently so had the entire town. Not one man could be induced to work, so Schultes spent three and a half hours unloading cargo, including a three-month supply of food for the two hundred children of Padre Colombo's school. The next day the crew recovered and slowly the *João* motored upriver, reaching the small village of Santa Isabel just after midnight on September 9. Schultes and Pacho slept on board and the following day worked through the morning moving their gear to a small two-room adobe house lent to them by the local representative of J. G. Araujo & Company, a trading concern based in Manaus. In the afternoon they paid a call on the priest of Santa Isabel, Padre Schneider, a German from the Sudetenland.

Schultes's initial plan was to borrow a canoe and make his way back downriver, visiting a number of important localities before meeting his friend Murça in Barcelos. But that afternoon word came that Padre Schneider's own launch would be leaving the next morning, 150 miles upriver for São Gabriel. It was a tough choice. They had almost no money and had already been obliged to break into their rations. Dinner that night was a can of sardines and a piece of bread. If São Gabriel proved to be as poor as Santa Isabel, there would be no place to eat. Still, Schultes was eager to move higher on the river. The forests around Santa Isabel were flooded, and it would be a month before the rubber tappers returned to their circuits. What's more, São Gabriel was the heart of Spruce country, his home for much of 1852 and the base from which he had explored the headwaters of both the Negro and the Orinoco. Thus Schultes elected to push on. Dispatching a note downstream alerting Murça to the change of plans, he boarded the priest's riverboat on the afternoon of September 11.

Once again it was an uncomfortable passage. Tied alongside the thatch-roofed launch was an old covered *batelão,* a dilapidated barge thirty feet long and twelve feet wide, piled high with cargo. Crammed on top of the gear were fourteen people and a dog. Among the passengers was a young girl, a leper whose hands and face were dreadfully affected by an advanced case of the disease. Pacho lay between her and an older man, the police chief of São Gabriel, and Schultes watched as the three shared bunches of grapes that Padre Schneider had managed to grow in his garden at Santa Isabel.

In the evening the sun was beautiful, and they came upon the three peaks of the Serra de Jacamín, the mountains of the Tapir, bathed in scarlet light. Lying still in his hammock, Schultes tried to sleep, but his rest was broken by the constant scraping of tin on wood. The barge leaked alarmingly, and throughout the journey a young Indian boy sitting alone toward the stern did nothing but bail. During the day the sound was not noticeable, but at night, even with the sounds of the forest at hand, the noise kept Schultes awake at least until the early morning hours. He must have eventually dozed off, because he was asleep at five when the launch ran too near the shore, and the trunk of a leaning tree smashed into the roof of the barge, crushing the tattered cabin and nearly sweeping the passengers to their deaths. Schultes jumped out of his hammock and grabbed a flashlight. Amazingly, no one was hurt. The light fell on a broken branch of the tree. There were young fruits, at least a number of recently fertilized ovaries. In a daze he broke off a few specimens and put them into his press. When dawn came, he discovered that the tree was *Micrandra minor,* an important relative of *Hevea* and a species that he had been eager to collect.

"Thus far have I advanced into the bowels of the land without impediment," wrote Richard Spruce on December 28, 1851, from the clearing of Uanauacá at the base of the cataracts below São Gabriel. In little over a month since leaving Manaus he had collected over three thousand specimens, many of them new species. Above Barcelos virtually every plant was unknown. "And so many things were in flower," wrote Spruce, "that I was obliged to confine myself to those which presented the greatest novelty of structure. Nothing like this has ever happened to me before. I was obliged, for instance, to shut my eyes to Myrtles, Laurels, Ingas. . . . I counted no fewer than fourteen species of Lecythis in flower, and all but one new to me!"

His luck changed the following morning. The rapids increased, and even with the wind filling the sails of his boat, it was impossible to make any headway. Yard by yard the Indians worked the *batalão* up the river. As ropes broke and men fell beneath the waves, anxiety replaced pleasure, and fear for the fate of his boat and crew distracted Spruce from the wonders of the forest at his side. After a terrible day on the river, his body exhausted, Spruce confided in a letter to a friend, "It may be true as Humboldt says that 'perils elevate the poetry of life' but I can bear witness that they have a woeful tendency to depress its prose."

A hundred years later Richard Schultes was having his own troubles as he approached the cataracts below São Gabriel. On the evening of September 13 his launch reached Jucabí, a Baptist mission at the mouth of the Río Curicuriarí. There the missionaries—a kindly American couple named Ross and their colleagues the Babcocks—received him well, filled his pack with food and old copies of the *Saturday Evening Post,* and invited him to return, both to await the arrival of his friend Murça and to climb the beautiful mountains of the Serra Curicuriarí, which rose from the forest several days' walk away from the mission to the south. Schultes was delighted to accept the offer but wanted first to get established at São Gabriel.

The following morning he departed Jucabí, one of twenty-four passengers crammed aboard the riverboat. The barge and most of the cargo had to be left behind in a quiet back eddy, for it was far too unwieldy to negotiate the ten miles of rapids that lay above the mission. Thus Schultes and Pacho arrived in São Gabriel with little food and nothing to cook it with. They approached the town's leading merchant, Señor Goncalves, whose son had available a one-room, tin-roofed house on the far side of town. The Salesian priest, Padre Miguel, was less helpful and had no interest in feeding the two strangers. Until the barge arrived five days later they lived on cans of sardines, bread, and cold cocoa. Both became weak and nauseated with headache, and as always when he was ill, Schultes grew homesick. "I really wish," he confided in his notebook, "that something would open up for me in New England."

None of this discomfort kept him from collecting. The season was not far advanced, and comparatively little was in flower. Nevertheless, he managed to find several novelties, including a new species of *Herrania,* a wild relative of the chocolate tree, growing on top of Serra de São Gabriel, a height of land that rose three hundred feet above the village. Wandering about the forest, he was thrilled to rediscover a number of other plants first found by Spruce: a rare species of *Sapium,* another *Micrandra,* the beautiful *Pouteria elegans.* From his notes it appears almost as if Schultes felt the presence of the Yorkshire botanist, as if Spruce were there to make the formal introductions. "I saw some wonderful examples of *Cunuria spruceana* in the forest," Schultes wrote, "and met, for the first time, the *Monopteryx uauçu,* which we shall endeavor to collect tomorrow."

Schultes liked São Gabriel, especially after he finally had a hot meal on September 17; and he described it as the "prettiest little Amazonian town I have ever seen." It was built on several hills that rose toward the Serra, and the church and mission school, whitewashed and covered in fresh thatch, stood out in rolling fields of tall grass. The town was immaculately clean, and there were a number of lovely cement houses, decorated in wood and painted in bright pastel colors. There were even two trucks, one an old Model T, the other a new Chevrolet brought in by barge from Venezuela during the war by the Rubber Development Corporation. The only road, if you could call it that, was a flat track that led half a mile to a beach where the riverboats landed during the dry season. From there you could look across the river and watch the sun set over the distant peaks of the Serra Curicuriarí.

Still, after a week, Schultes was growing increasingly worried about both his friend Murça and the bulk of the expedition gear. On the 20th he and Pacho decided to head back downstream to find out what had happened. Their launch was scheduled to depart São Gabriel at eight in the morning. It finally left at three and did not reach the Baptist mission at Jucabí until well after dark. They stayed there for five days, sleeping in the schoolhouse, feasting on the fine cooking of Mrs. Ross, and exploring the surrounding forest, which proved to be exceedingly rich in *Hevea* and related plants. When there was still no sign of Murça by the 26th, Schultes borrowed a twenty-foot canoe from Mr. Ross, rigged up a cloth tent over the stern, and headed off downriver with Pacho and one of the boys from the mission. Floating all night, they woke to a beautiful sky. On a pure white sandbar at the edge of the forest, they stopped to eat and bathe before continuing on for Santa Isabel.

A fever hit the following afternoon around three while they were still on the water. By evening Schultes was vomiting blood and had to ask to be taken in by a family at the village of Bom Jardim, at the base of the Serra Jacamín. For the next three days he lay in a hammock, nursed by a man with the improbable name of William de So Amazonas Belleza. Schultes believed he had influenza. His host, "William of the Beautiful Amazon," thought otherwise and administered injections of quinine. Whatever it was, the illness passed, and the patient slowly recovered. Though still weak, Schultes was ready by the first of the month to head for Santa Isabel.

Arriving late in the afternoon, Schultes found no sign of Murça Pires. Nor was his companion on the Araujo Company launch that arrived from Manaus later that night. No one aboard knew what had become of the Brazilian botanist, but rumor had it that he was still in Barcelos, hung up with motor trouble. By this time Schultes was completely out of money and desperate to connect with Murça. On credit he hired a boat and crew and left immediately for downriver. He did not get far before his own engine broke down, stranding him for seven hours on a sandbar while the *motorista* made repairs. By traveling all night he and Pacho López reached Barcelos on the afternoon of October 4.

There they found Murça, but the news was not good. The mechanic in Manaus who had worked on the Institute's launch had installed the engine off-center, and the resulting strain had broken the drive shaft. Fortunately, an old Portuguese blacksmith in Barcelos had managed to weld the two pieces together and was just finishing the painfully slow process of turning out a new shaft on a hand-driven lathe. While they waited, Schultes visited a Salesian priest, a former physician, who assured him his ailment had not been malaria, only a severe cold. Schultes was delighted by the diagnosis, Pacho highly skeptical.

The following evening they finally headed upstream, towing the Institute's boat with the launch that Schultes had hired at Santa Isabel. In addition, they now had the aluminum canoe and outboard, which allowed Schultes to dart in and out of the flood forest, making collections while the larger vessels made their way slowly against the current. On the morning of October 7 they reached the island of Xibarú, and Schultes spent several hours examining the local type of *Hevea micro-*

phylla, the most unusual of all the species of rubber. First discovered in 1902 by a German botanist Ernst Ule, it is a slender, sparsely crowned tree, often found growing in dense colonies along creek margins or around the shores of islands on land subject to extreme flooding. It is not uncommon on the Río Negro. Schultes would run into it at a dozen places. What he could never understand and what left him amazed as he examined the long, pointed capsule of a young fruit was how Spruce had somehow missed it. Though the Yorkshireman had paid particular attention to *Hevea* and found many new species, he never collected the most peculiar of them all. It was, Schultes believed, one of the great enigmas of Amazonian botany.

The wonder of it was lost on the crew, who had more immediate concerns. The river was falling, and if there was any chance of passing the Institute's launch through the rapids at São Gabriel, they would have to hurry. After four days at Santa Isabel, they finally had the motor properly installed, and the boat was ready to proceed under its own power. They left on the evening of October 9, hoping to reach the Baptist mission at Jucabí in a day or two. Instead, two hours after getting under way, the engine died. "This trip," Schultes wrote that night, "seems to be one disappointment after another, and I am wondering if I shall ever see *Hevea rigidifolia* in flower."

Much more was to come. Murça fought with the engine for a day before they finally accepted a tow from a passing launch, which dropped them at Jucabí on the evening of the 12th. By then Murça had the engine working fitfully, but neither he nor Schultes trusted it sufficiently to run the rapids below São Gabriel. Instead they arranged for the launch that had towed them upstream to return in two days to accompany them through the worst of the cataracts. It was just as well, for the next morning, on a short excursion from Jucabí, the engine failed yet again, stranding Schultes and Murça on a sandbar in the open sun for a day and a night. By the time they finally reached São Gabriel on the 14th, the river was low and the water fierce. It took thirty-two men working all day just to pass the launch through the first of three major rapids. "It was very difficult and dangerous," Schultes wrote, "and I feared for a while we wouldn't get it up."

In the following days there was one setback after another. First came the discovery that the formaldehyde Schultes had purchased in Iquitos to preserve his plants had been adulterated, likely cut with water. It was a disastrous revelation. His specimens, pressed in moistened newspaper and tightly sealed in rubber bags, should have lasted at least two months. Instead, when he inspected the plants that he had left in São Gabriel, all were destroyed. "I was never so downhearted," he wrote.

"Nearly all my collections for the last month and a half are rotten!" Then came the realization that all his food and, more important, many of his blotters and drying cartons had been stolen. When Murça had broken down at Barcelos, he had shipped the equipment on the Araujo launch upriver to Santa Isabel. When he and Schultes were unable to locate the gear there, they assumed it had been sent on to São Gabriel. They were wrong. Without the canned rations, reaching the remote peaks would be that much more difficult, and the loss of the drying materials meant that they would have to collect on a much reduced scale. As if this was not enough, the kerosene stoves belonging to the Institute were acting up, and Schultes had an impossible time drying the few specimens that had not been destroyed.

Murça, meanwhile, tended to the launch. A local machinist tore apart the engine, found a number of broken parts, welded various things together, shattered an irreplaceable spark plug, and then pronounced the motor cured. The following morning, October 17, the expedition continued up the Río Negro, powered by a 3.5 horsepower outboard and a 5 horsepower inboard motor, with a spark plug bound together by wire. The river current was unusually strong, and when the launch rolled in a rapid, flooding the outboard, they were forced to pull the launch upstream with the aid of ropes. Progress was very slow.

Not far above São Gabriel is a fork in the river that defines the entire basin of the upper Río Negro. Coming in from the west is the Río Uaupés, which runs through a series of extraordinary cataracts to and beyond the Brazilian border with Colombia, where the river is known as the Vaupés. The Río Negro proper rises in the north, at the confluence of the Casiquiare Canal and the Río Guainía, a river that flows in a great arch out of Colombia and forms for a good part of its length the border between that country and Venezuela. Draining into the upper Río Negro, above the confluence with the Uaupés and below the Casiquiare, are several fascinating affluents, few of which had ever been explored botanically. Among these were the Xié, Dimití, and Içana. Schultes hoped in time to explore all of these tributaries, in addition to the Guainía, the Río Negro itself, and, of course, the entire upper drainage of the Uaupés, at least as far as the Colombian frontier. His first goal was the Içana and an unknown mountain close to its headwaters known as the Cerro Tunuhí.

On the morning of October 18 the Institute launch reached the Ilha das Flores, near the mouth of the Uaupés. Schultes paused to collect but felt poorly, and immediately upon returning to the boat, took to his hammock. By afternoon he was once again awash in fever, and shaken by violent attacks of vomiting. By nightfall there was pain in every limb, as if a jackhammer within his bones was pounding to get out. Pacho stayed close by him throughout the night, as did Mr. Ross, who was traveling with them in the hope of establishing a mission on the upper Içana. In the morning it was obvious that the priest at Barcelos had been wrong. It had been malaria, not influenza, that had stricken Schultes three weeks before.

Now the disease was back with an intensity he had never known. All day on the 19th, as the launch slowly pushed up the Içana, Schultes lay convulsed by nausea and high fever. For five days he remained in an unnatural rest, with Pacho always by his side, encouraging him to turn back for São Gabriel. Schultes refused, saying that he would be well by the time they reached Tunuhí. Indeed, the mountain did pull him out of his hammock on the 23rd, and for a day he felt somewhat better. On the 24th he collected four plants, but the exertion proved too much. By the evening of the 25th he had drifted into delirium, and the bile in his vomit had been replaced by blood. The following day he was so weak that Murça, Pacho, and Mr. Ross had to carry him in his hammock from the Indian house where they were staying to the launch. Things were falling apart. Coming downriver that morning, Pacho and Murça attempted to run a stretch of rapids in the aluminum canoe. They capsized, losing prized specimens of *Hevea rigidifolia* as well as an oar, scissors, shoes, and, as Schultes later wrote, "almost their lives."

That night, October 26, they finally abandoned the expedition and started back down the Içana for São Gabriel. They left at 8:30, continued in the darkness throughout the night, and motored all the following day and night without stopping. Arriving at São Gabriel at dawn on the 28th, Schultes was taken immediately to the mission hospital where he lay for a week, his life saved by timely injections of quinine and atropine, two drugs derived from South American plants he knew so well.

On January 15, 1852, Richard Spruce approached São Gabriel with low spirits and an anxious heart. Weeks on the river had culminated in fourteen days of struggling through fearsome cataracts, his men yoked to ropes, his launch on occasion spinning like flotsam in the current. The stress was difficult to bear. "Rose this morning with a sensation of weariness and disgust scarcely conceivable," he confided in his journal. "The idea of having still another day to pass through like the two last was most depressing." Adding to his burden was the recent news about his friend Alfred Russel Wallace, who lay seriously ill with fever at São Joaquim, a small settlement on the Río Uaupés, just beyond its confluence with the Río Negro.

The two naturalists had last seen each other three months before in Manaus. Wallace had just returned from over a year on the Río Negro, a journey that had taken him beyond the headwaters, overland to the Orinoco. There his Indians had deserted him, and he had worked his way back 1,200 miles to Manaus, where he ran into Spruce on September 15, 1851. Wallace planned to remain in town just long enough to mount another expedition, this time to the Río Uaupés. Spruce, who had been in Manaus for almost a year, was himself eager to move. There was talk of their traveling together, but it was impossible to find a launch that could accommodate both parties, together with their gear and particular needs. Reluctantly, the two men parted ways in the first days of October, with Wallace returning to the river. Spruce would follow in six weeks, beginning his travels on November 14, 1851.

By then the fate of Wallace's brother Herbert had been confirmed. Upon his return to Manaus in September, Alfred had learned that his brother was gravely ill, struck down by yellow fever in Belém on the eve of his embarkation for England. But the letter that had brought the bad news was three months old, so Wallace had gone up the Río Negro uncertain whether his brother still lived. In fact, he had died the previous May, and Spruce now had the difficult task of delivering the bad tidings to one who was himself on the edge of death. "I had sad news two days ago from my friend Wallace," wrote Spruce on December 28, 1851. "He is at São Joaquim . . . and he writes me by another hand that he is almost at the point of death from a malignant fever, which has reduced him to such a state of weakness that he cannot rise from his hammock or even feed himself. The person who brought me the letter told me that he had taken no nourishment for some days except the juice of oranges and cashews."

Few in the settlement of São Joaquim expected Wallace to survive. For two months he lay in delirium, his body wracked by fever and perspiration, his spirit dampened by the deep depressions that invariably accompany severe cases of malaria. "I could not speak intelligibly," Wallace later reported in his journal, "and had not the strength to write, or even turn over in my hammock. . . . In this state I remained till the beginning of February, the ague continuing, but with diminished force; and though with an increasing appetite, and eating heartily, yet gaining so little strength, that I could with difficulty stand alone, or walk across the room with the assistance of two sticks. The ague, however, now left me, and in another week, as I could walk with a stick down to the river-side, I went to São Gabriel, to see Mr. Spruce, who had arrived there, and had kindly been to see me a short time before."

Neither Spruce nor Wallace record in their notes any further details of these two encounters. Incredibly, within a fortnight, Wallace embarked on yet another expedition, returning to the Uaupés where, among other discoveries, he first observed among the Tukano the feast of Yuruparí and the parading of the sacred flutes and trumpets that women, on pain of death, were forbidden to see. He admired the sound of the instruments but called it devil music, thus becoming the first of many Europeans to misunderstand this extraordinary ritual of initiation and remembrance.

Exhausted in spirit and anxious for home, Wallace nevertheless continued up the Uaupés, crossed the Colombian frontier, and paddled as far as the mouth of the Kuduyarí, just below the present town of Mitú. There, a week short of his goal, the great falls of Yuruparí, he turned back. Having passed "over fifty *caxoeira,* great and small, some mere rapids, others furious cataracts, and some nearly perpendicular falls," Wallace had had enough. Returning downriver with his collection of ethnographic artifacts, skeletons, and skins of hundreds of creatures, together with some fifty-two live animals, he reached São Gabriel on April 28, 1852. There he stayed but a day before continuing on for Manaus. In his journal he noted simply that he had "enjoyed a little conversation with my friend Mr. Spruce." Nothing more is said of this, his last meeting in South America with the man who had saved his life.

Six months after their parting, Spruce journeyed to the Uaupés, the river of Wallace's disappointment, and established residence at Ipanoré, a small village located at the foot of four miles of cataracts. Beyond the broken river, at the head of the rapids, was a Tukano longhouse, a maloca called Urubú-coará, the "buzzard's nest." It was there in November 1852 that Spruce himself witnessed a ceremony of the Yuruparí cult. "We reached the mallóca [*sic*] at nightfall," he remembered, "just as the *botútos* or sacred trumpets began to boom lugubriously within the margin of the forest skirting the wide space kept open and free of weeds around the mallóca. At that sound every female outside makes a rush into the house, before the *botútos* emerge on the open; for to merely see one of them would be to her a sentence of death." Unlike Wallace, who appreciated Indian life but viewed Indians as animals, Spruce passed no judgment. "The old Portuguese missionaries called these trumpets juruparís, or devils," he noted, "merely a bit of jealousy on their part."

Before the enormous façade of the longhouse, built of palm and painted with figures born of visions, three hundred men had gathered to dance. Red beads circled their ankles, their bodies were streaked in red and blue-black dye, and long strands of glass beads and jaguar teeth hung from their necks. Most wore tall headdresses fashioned from egret and parrot feathers, and many had eagle down glued to their chests. In his journal Spruce says very little about the actual ritual. As a botanist he appears to have been far more interested in the plants employed in the ceremony, particularly a strange bitter drink, brownish green in color and known to the Tukano as *caapi.* Throughout the night in the intervals between the dances, Spruce watched as a Tukano elder ran the length of the longhouse, singing and chanting, bearing in each hand a cup of the preparation. Approaching the dancers, the shaman dropped to the ground, his chin touching his knees as his arms lifted first one and then the other cup to whichever young man was waiting to receive it. Within two minutes the potion took effect. "The Indian turns deadly pale," Spruce reported, "trembles in every limb, and horror is in his aspect."

Spruce was fascinated by this strange preparation, and though he had yet to take the drug himself, he had certainly heard of the effects from traders and Indians on the Río Negro. "White men who have partaken of *caapi* in the proper way," he wrote, "concur in the account of their sensations under its influence. The sight is disturbed and visions pass rapidly before the eyes, wherein everything gorgeous and magnificent they had heard or read of seems combined. . . . A Brazilian friend said that when he once took a full dose of *caapi* he saw all the marvels he had read of in the *Arabian Nights* pass rapidly before his eyes as in a panorama. The Indians say they see beautiful lakes, woods laden with fruit, birds of brilliant plumage. . . . Soon the scene changes; they see savage beasts preparing to seize them, they can no longer hold themselves up, but fall to the ground."

The feast at Urubú-coará was to have been Spruce's chance finally to sample the drug. Unfortunately, it was not to be. "I had gone," he recalled years later, "with the full intention of experimenting with *caapi* on myself, but I had scarcely dispatched one cup of the nauseous beverage which is but half a dose, when the ruler of the feast—desirous, apparently, that I should taste all his delicacies at once—came up with a woman bearing a large calabash of *caxirí* (mandioca beer), of which I must needs take a copious draught, and as I know the mode of preparation, it was gulped down with secret loathing. Scarcely had I accomplished this feat, when a large cigar, 2 feet long and as thick as the wrist, was lighted into my hand, and etiquette demanded that I should take a few whiffs of it—*I,* who never in my life smoked a cigar or pipe of tobacco. Above all this, I must drink a large cup of palm-wine, and it will readily be understood that the effect of such a complex dose was a strong inclination to vomit, which was only overcome by lying down in a hammock and drinking a cup of coffee."

Needless to say, Spruce missed the rest of the party. In the morning, once back to his senses, he pursued the botanical identity of *caapi* and found to his astonishment that it was an unknown species, a liana in a family of plants that had never before been known to have narcotic or even medicinal properties. He called it *Banisteria caapi,* a name later changed to *Banisteriopsis caapi,* more commonly known in other parts of the Amazon as *yagé* or *ayahuasca.* His was to be the only complete botanical specimen of the plant for nearly a century. Eager to determine the chemical constituents of the drug, he dried a good quantity of stems and dispatched the collection downriver for England in March 1853. Unfortunately, the shipment was delayed, and by the time the boxes were finally opened by Bentham at Kew, mold and rot had ruined the specimens. "The bundle of Caapi would presumably have quite lost its virtue from the same cause," Spruce later wrote, "and I do not know that it was ever analysed chemically; but some portion of it should be in the Kew Museum at this day. . . . This is all I have seen or learnt of caapi or aya-huasca. I regret being unable to tell what is the peculiar narcotic principle that produces such extraordinary effects. . . . Some traveller who may follow my steps, with greater resources at his command, will, it is to be hoped, be able to bring away materials adequate for the complete analysis of this curious plant."

A hundred years later such a traveler did follow in Spruce's wake. But not only did Schultes find *yagé* again in the forest, taking it with Indians on more than twenty occasions, he searched the backrooms of Kew and eventually located the stems of the very plant that Spruce had collected at Urubú-coará, the morning after the feast of Yuruparí. In 1969 he had the material analyzed and found that after more than a century the stems had retained their potency. By then, of course, the chemical constituents of this most famous of Amazonian hallucinogens were well known. Schultes had the analysis done partially for science but mainly for Spruce.

The nuns who nursed Schultes while he lay sick and delirious at São Gabriel fully expected that, once he was able to travel, the botanist would return to Manaus for proper medical care. Predictably, he never considered such an option. After six days of rest his only thought was of the forest. It was the beginning of November, and soon the trees would be in flower. His companion Murça Pires had but a month before he was due back in Belém. Despite all the setbacks of the previous weeks, there was still time to salvage the expedition. Though weak and heavily medicated, Schultes left the hospital on November 2 and began preparing for a journey two hundred miles up the Río Uaupés. In his kit he packed two syringes and several ampoules each of quinine, atropine, atabrine, and various liver extracts, all gifts of the mission. His goal was a stretch of white sand beyond Ipanoré, the small settlement where Spruce was living the night he discovered *yagé.* Though Ipanoré appeared on no map, Schultes knew it as the place where *Hevea rigidifolia* had been discovered, and he hoped to collect from the very trees that Richard Spruce had found over a century before.

As Schultes headed up the Río Negro toward its confluence with the Uaupés, a tremendous weight lifted from his spirit. A week cooped up at the mission, enveloped with fever, away from his plants, was for him a kind of purgatory. He lived for his work. Every moment not in the forest was a lost dream, a species denied to science, a botanical mystery left unsolved. To understand his frustration—indeed, the source of his drive and ambition—one must appreciate just what kind of botanist he had become.

For most travelers the face of the tropical rain forest appears surprisingly monotonous, especially when experienced in the flat light of midday. An experienced naturalist readily discerns the basic growth forms of plants, distinguishing the foliage of a climbing liana from the leaves of an emergent tree, the flowers of an epiphyte from the blossoms of its host. The botanist perceives another level of complexity, recognizing families and genera, taxonomic categories which by their very alignment reveal evolutionary relationships that are meaningful and true. Yet even the most highly trained botanists are humbled by the immense diversity of the Amazonian forests. Confronted with the unknown, they collect specimens and do their best to identify a plant to family or genus. Only later, in the comfort of the herbarium and invariably with the assistance of a colleague specializing in that particular group of plants, will they figure out the species and obtain a complete determination.

In other words, most botanists working in the Amazon must come to peace with their ignorance. When they look at the forest, their eyes fall first on what is known and then seek what is unknown. Schultes was the opposite. He possessed what scientists call the taxonomic eye, an inherent capacity to detect variation at a glance. When he looked at the forest, his gaze reflexively fell on what was novel or unusual. And since he was so familiar with the flora, he could be confident that if a plant was new to him, it was likely to be new to science. For Schultes such moments of discovery were transcendent. He was once in a small plane that took off from a dirt runway, brushed against the canopy of the forest, and very nearly crashed. A colleague who was with him recalled years later that throughout the entire episode Schultes had sat calmly by a window, oblivious to the screams of the terrified passengers. It turned out that he had spotted a tree, a new species of *Cecropia,* and had scarcely noticed the crisis. What all this meant was that Schultes could resolve botanical problems in the moment, write descriptions in the field, realign species and genera just by holding a blossom to the light. In the entire history of Amazonian botany, only a handful of scientists have possessed this talent.

Thus for Schultes each journey was like a new morning, full of a day's potential for wonder or disappointment. Away from the mission, once again on the water, entering country first explored by Spruce, he felt drawn into a land that he believed had given rise not just to *Hevea* but to an entire complex of related plants derived from some distant and common ancestor. The lineage of *Hevea* was a large family, the Euphorbiaceae, but the immediate cousins were three genera known by the Latin names *Joannesia, Micrandra,* and *Cunuria.* Only a botanist could tell them apart, but for Schultes each had a history that he could understand, in the same sense perhaps that an anthropologist, holding in his hands a set of bones, can envision the descent of our own species. Years later, when Schultes was again on the banks of the Apaporis, he would stumble upon a tree at the edge of a rapid and know in an instant that it was a new genus, intermediate between *Joannesia* and *Micrandra* —a single tree, a new branch of the family, a plant that would forever bear the name *Vaupesia cataractarum.*

In the afternoon of the first day on the river, the expedition paused at the base of Serra Carangrejá so that Schultes could collect one hundred specimens of *Cunuria spruceana.* The next morning on the Ilha das Flores, the island of flowers, he found its relative *Cunuria crassipes* in full bloom. Things were definitely looking up. For once the engine was working well, and as the launch entered the mouth of the Uaupés, weaving past the studded islands and along the rocky shore, Schultes felt strong and confident. They stopped for the night at São Joaquim, the village where Wallace had been stricken with fever, and found it a strange place—a dozen or more thatch houses and a chapel, all in good condition yet completely uninhabited. A storm broke, and the hard rain made them glad for the shelter.

The next days brought more of the same. A steady movement upstream, a little collecting along the shore, then nightfall and a landing at some clearing—the old mission of Cunurí and later the half-abandoned village of Bela Vista. There was something eerie about these broken-down settlements, built on sterile white sands, empty and silent save for the odd cry of a hungry child. It made all of them, but especially Pacho, anxious for the river.

They left Bela Vista on the morning of November 7. Their attention was arrested by the sudden appearance on the right bank of a small granite mountain, a few miles in from the shore, and identified by one of the crew as the Serra Tukano. Schultes decided to explore the summit. Anchoring the launch into a quiet backwater, he, Pacho, and Murça, together with a young Tukano encountered on the riverbank, headed into the forest, slashing their way toward the base of the ridge. After several hours they came to a beautiful *caatinga,* an expanse of white sand in the midst of the forest. Strewn over the ground was a cover of thick and dry leaves, all of a single type. Looking about, Schultes realized he was in a pure stand of a slender, unbuttressed tree of the genus *Cunuria.* There were no flowers, but from the fruits and leaves alone he knew it was a new species, which he named on the spot *Cunuria tukanorum,* in honor of the mountain and the Indians of the region.

It was a most curious plant. He held a branch in his hand and slowly made his way across the *caatinga,* the fallen leaves crunching beneath his feet. He sensed that something was peculiar—just an intuition. Deep in thought, he walked directly into an enormous wasp nest and was stung more than twenty-five times in the neck, face, and hands. The mountain climb was off. Aching with pain and fever, he made his way slowly back to the launch, arriving after dark, tired but happy to have in his possession a new botanical treasure.

The next day brought even greater delights. Motoring all night, they reached by early morning the mouth of the Río Tiquié and the Salesian mission of Taracuá, a beautiful and vibrant community, quite unlike anything they had seen on the river. On a height of land stood large and imposing buildings of wood and brick, and skirting the mission were rich and extensive plantings of manioc and corn. The houses of the Indians lay across a small plaza, to one side of which rose the scaffolding that marked the walls of the enormous brick church then under construction. The priest was a German from Bavaria, Padre Lorenzo, a kind and friendly man who worked hard, treated the Indians well, and somehow, in the midst of the jungle, managed to keep himself supplied with enough barley, sugar, and hops to brew an excellent beer. It had a slightly bitter taste and was, as Schultes discovered that night at dinner, "indescribably good."

Schultes remained in Padre Lorenzo's care for a week. The next morning he encountered for the first time *Hevea rigidifolia,* and though the plant was not in flower, he took note of its smooth yellowish bark and secured with considerable effort a liter of latex for analysis. He also collected a century set, one hundred specimens, of *Cunuria crassipes,* a wonderful find that more than compensated for the four-inch gash he sliced into his leg with a machete as he gathered the plant. The following day brought a more serious problem. In the morning Schultes and Murça had set up their dryer in one of the small thatch storerooms belonging to the mission. Schultes had placed one press over a small primus stove. The bulk of their collections, including all three of Murça's presses, were mounted on a tripod over a different stove. That evening, after a long day exploring the forests around Taracuá, they returned in time to see smoke pouring through the roof of the storeroom. Fire had consumed four of the presses. The collections were ashes, as were the irreplaceable blotters and cardboards.

To lose what little equipment they still had was a terrible blow, but there was nothing to be done about it. The next morning Padre Lorenzo and a number of Indians from the mission piled aboard the launch, and the party headed up the Igarapé da Chuva, a small affluent that drained the flank of a distant mountain. As the creek narrowed, they continued by canoe, reaching around noon the head of a trail that climbed toward the height of land. At the foot of the mountain Schultes paused, his attention drawn to some leaves scattered in great abundance on the forest floor. They came from a tree of medium size with a perfectly columnar, unbuttressed trunk covered in smooth bark of a yellowish hue. He cut a branch and saw from the leaves and fruit that it was a *Micrandra.* A closer search revealed a single inflorescence, with just two open flowers. They were the blossoms of a *Cunuria.* Not only was this a new species, it was intermediate between the genera, a tree so novel that in an instant the genus *Cunuria* disappeared from the botanical record, absorbed by its cousin *Micrandra.* In his delight Schultes drifted to the summit of the mountain, utterly oblivious to the cold rain that soaked the rest of his party and made their outing miserable. Years later he would name the plant *Micrandra rossiana,* after Mr. Ross, the American missionary who was so helpful to him at Jucubí. As for *Cunuria tukanorum,* the plant that had brought him face-to-face with the wasps, it would become, once he had a chance to see it in flower, *Micrandra lopezii,* named for Pacho López, who first saw it on the *caatinga* at the base of Serro Tukano.

The discovery turned out to be the highlight of this phase of the expedition. Ipanoré, the site of so many of Spruce's collections, was for Schultes a disappointment. They reached the squalid settlement in the late evening of November 13, half a day after leaving the mission at Taracuá. Even in the fading light Schultes could tell that the only reason for the place to exist was its location at the foot of the cataracts that blocked the river. The village itself was a shambles of plaited palm shacks, falling down and abandoned. There was not an ounce of food to be had, and in the morning they could see why: Ipanoré was built on an expanse of completely sterile white sand in which not a bean could be grown. To make things worse, nothing was in flower. The few Indians hanging about were a sorry-looking lot and by reputation the only skilled thieves on the river. Though he collected a number of interesting plants and examined some fifty individuals of *Hevea rigidifolia,* Schultes left Ipanoré after a day, scarcely able to believe that Spruce had put up with the place for four months.

By then the expedition was ready to return downriver. Murça, in particular, was discouraged by the loss of equipment in the fire, which left him unable to process his specimens. To be on the river yet unable to collect was an unhappy fate for a botanist, and one he had tried most of his career to avoid. At any rate his leave from the Institute was almost up, and it was time to move. Schultes naturally would accompany him as far as São Gabriel but remain in the field, waiting for the forest to bloom. Then, free of the Institute's launch and all its mechanical woes, he and Pacho would continue their explorations by canoe. There were still things to do on the way downstream, a mountain to climb at Bela Vista, a few localities to revisit, accounts to settle, but for Schultes his time with Murça came to an end on the banks of the Uaupés, at the foot of the cataracts that had held Spruce hostage for so many lonely months. Within two weeks Murça Pires would be on a riverboat bound for Manaus. Schultes would be in São Gabriel poised to continue his explorations north to the mouth of the Casiquiare. Their last day together in the forest had them exploring a small creek that ran along the base of the Serra Wabeesee. Among the collections was a solitary epiphyte, a delicate plant growing on a fallen log. Murça had never seen it, but Schultes had. It was, he explained to his friend, the blue orchid, something that Spruce had found at Ipanoré.

It was June 27, 1853, the fourth day of the Feast of San Juan, and Richard Spruce had several things on his mind as he scratched out a letter to Sir William Hooker, director of the Royal Botanic Garden at Kew. "The gratification I naturally feel," he wrote, "at finding myself in *terra Humboldtiana* is considerably lessened by various untoward circumstances." He had come down the Río Uaupés in late March and headed directly north on the Río Negro, reaching San Carlos, the Venezuelan town at the mouth of the Casiquiare Canal, on April 11. Since then his every effort had been devoted to searching for food, and he had been lucky to eat every other day. When Humboldt went through the country in 1800, San Carlos had been a thriving mission, one of many along the waterway that linked the Amazon and the Orinoco. By 1853 there had not been a priest in the region for twenty years. The only outsiders were a pair of Portuguese traders, and it was one of these men who had first warned Spruce that the Feast of San Juan would be the signal for the local natives, mostly disgruntled Kuripakos and *mestizos,* to rise up and massacre all the whites. Even as he wrote his letter to Hooker, Spruce found himself under siege, barricaded in a thatch house with provisions and weapons at hand.

For three days the defenders remained on guard, each man standing watch in turn, with all conversation suspended in a state of intense anxiety. The Portuguese, who had lived in San Carlos for some time, were certain that an attack was imminent. Spruce had his doubts and remained sanguine throughout the ordeal, even taking time to work on his plants between watches. The threat was serious. "Of their ultimate success against us there can be little doubt," he confided at the time, "for they are 150 against three. My firm resolve, in case of being attacked, was not to allow myself to be taken alive, and so suffer a hundred deaths in one."

Fortunately, the locals, deterred by Spruce's guns and exhausted by a week of dancing and drinking, never attacked. The days passed and the sound of the gathering shifted to the far side of the river. With the supply of rum coming to an end and the Indians having spent their powder firing salvos at the moon, the crisis ended, and Spruce was able to return to his work. Untouched by fear, he remained around San Carlos for nearly five months, making a number of excursions and acquiring what provisions he could for his next major expedition: a traverse of the Casiquiare to the Orinoco and the flank of Cerro Duida, the mountain of the Lost World, an isolated massif that rises more than eight thousand feet above the forest floor. It was not until the last days of November 1853 that he directed his launch up the Río Negro and into the yellow mouth of the Casiquiare. As a guide he carried with him a worn copy of Humboldt's *Personal Narrative.*

The weeks and months that followed passed as in a dream. The Casiquiare meanders for some 150 miles through dense hanging vegetation, along winding shores that give way to long stretches that appear as if carved into the earth by an engineer. In origin the canal is an arm of the Orinoco that branched off from the main channel, flowed away from the river, and became lost on the low flat plain that stretches west and south toward the basin of the Amazon. Only by chance and an accident of topography did the waters of the two drainages come together. Along the course of the Casiquiare are isolated mountains of stone, like the Roca de Guanári, where Humboldt determined the longitude and latitude of the canal and thus proved its existence, but all of these are dwarfed by the Cerro Duida. For the final seven days of the approach to the Orinoco by launch, this mountain dominates the sky. It was for Spruce the most compelling sight he had experienced in South America. "At sunset," he wrote, "the mountain was very grand, the ridges assuming a purple hue, while the interstices were veiled in an impenetrable gloom, and a stratum of white fleecy cloud was floating beneath the summit."

Humboldt had written that Cerro Duida could not be climbed. Spruce, who had long resolved to prove him wrong and "rifle its botanical treasures," was himself humbled when finally confronted by its cliffs and impenetrable ravines, where the "salient edges glitter like silver." He stayed at the foot of the mountain on the savannah that spreads around its base, at the settlement of Esmeraldas. There, on Christmas Day, 1853, he became consumed by a strange melancholia that drained his spirit. "You will credit me," he wrote a friend, "when I say that to the sight Esmeralda is a Paradise—in reality it is an Inferno, scarcely habitable by man." A warm wind blew the sand across an empty plaza but brought no sound of life. Not a bird or butterfly or even a dog was to be seen. The few huts of the old mission had roofs of rotten leaves, and straw doors that sealed away the inhabitants who, "bat-like, drowsed away the day and only steal forth in the grey of the morn to seek a scanty subsistence." Lost in a whirlwind of isolation, Spruce came to hate the place. "If I passed my hand across my face," he wrote, "I brought it away crushed with blood and with the crushed bodies of gorged mosquitoes. . . . If I climbed the cerros, or buried myself in the forest, or sought the center of the savannas, it was the same. . . . With me the wounds bled considerably."

Spruce made no effort to climb Duida. It would have been a rich prize. When the mountain was finally explored by two Americans in 1928, it yielded no fewer than seven hundred different plants, of which over two hundred were new to science. Instead, Spruce pushed on toward the headwaters of the Orinoco, lands never before seen by a European. For three months he lived among the Maquiritare Indians, exploring many of the affluents that drain the Serra Parima, the source of the Orinoco, before finally retracing his journey through the Casiquiare to San Carlos.

He remained on the Río Negro only until the end of May, when he decided to return to the Orinoco, this time by way of the Río Guainía and the overland route taken before him by Wallace and Humboldt. It was not a difficult journey. From the head of a small affluent of the Guainía, a two-day walk brought him to the Río Atabapo, which flowed into the Orinoco. Continuing north, he passed the mouth of the Guaviare, the largest tributary of the Orinoco, and reached the great cataracts at Maypures. There, encamped on the savannah, he met, "a wandering horde of Guahibo Indians, from the river Meta."

Among them was an old man, sitting by a fire. Around his neck hung the dried stem of a plant as well as a snuffbox of sorts: the leg bone of a jaguar, closed at one end with pitch and at the other plugged with a cork of marima bark. Along its length dangled aromatic rhizomes and roots. In one hand and balanced on his knee, the Indian had a wooden platter; in the other was a small pestle, also of wood, that he used to grind a small pile of roasted seeds. Every so often he took the stem from around his neck and tore off a thin strip, which he chewed with obvious satisfaction. When Spruce made inquiries, the old Guahibo responded in broken Spanish. "With a chew of *caapi* and a pinch of *niopo,* one feels good! No hunger—no thirst—no tired!" Spruce knew what *caapi* was. *Niopo* turned out to be a snuff derived from *Anadenthera peregrina,* a tree of the open grasslands first described as a hallucinogen by Humboldt. It was also known as *paricá* or *yopo.* Spruce collected seeds for analysis and purchased a full set of implements, including the Y-shaped snuffing tube, made of bird bones and pitch, with which the shaman snorted the ritual powder.

He did not have a chance, however, to partake of the drug himself. Already he could feel at the base of his spine the first ache of malaria. In haste he retreated from the open savannah of Maypures, hoping to reach the Río Negro before the full onset of disease. Five days in an open canoe was his undoing. By the time he reached San Fernando on the Atabapo, he lay prostrate and helpless, and knew that if he continued his journey, he would die. As it was, he nearly did. Taken into the care of a covetous woman named Carmen Reja, he lay in a sleepless trance for thirty-eight days. By night violent attacks of fever racked his body; by day he suffered vomiting and chills. For food he managed only a biscuit of arrowroot dissolved in water. His thirst was unquenchable. He could barely breathe. Anticipating death, he arranged with a local official for the disposal of his plants, and then he waited for the end in a state of complete apathy.

His Indians did not desert him. Instead, they traded all his equipment for rum, which kept them so drunk that they were unable to attend to his needs. He was fully at the mercy of his nurse, whose smile he later described as a "demoniacal scowl." As he lay dying, she gathered her friends in the adjoining room, counted her spoils, and abused him with foul language. Her favorite line, Spruce later reported, was "Die, you English dog, that we may have a merry watch-night with your dollars!"

When, after nineteen days, he stubbornly remained alive, one of Carmen Reja's neighbors suggested that it might be necessary to try a little poison.

Often in the upper Río Negro in the last days of June, at the time of the Feast of San Juan, a cool wind blows over the forest. The temperature drops as low as 70 at night, with rain and mist making the air seem much colder. In 1948 the change of weather came early and caught Schultes and Pacho, at the end of May, alone on the river in an open canoe, heading upstream for San Carlos. They had not eaten for three days. Wet and chilled to the bone, Schultes with some reluctance broke open a wooden case lying at his feet. Inside were twenty-four bottles of the cheapest Brazilian brandy, a consignment that Schultes was delivering from the mission at São Gabriel to Padre Antonio at San Carlos. On the assumption that something in their stomachs was better than nothing, Schultes twisted off a cap, took a drink, and passed the bottle to Pacho. The brandy tasted like gasoline. Still, it warmed them up and for a few fleeting minutes provided an immense sense of well-being. With the sun in his face, Schultes drifted into thought.

The river had held them for ten months—not as long as Spruce but long enough. Christmas, 1947, had found him on the granite summit of the Piedra de Cocuí, the Hawk's Rock, 1,200 feet above the forest floor. At his back was Venezuela, at his feet the frontier of Colombia and Brazil, boundaries that had come to mean nothing to him. He moved up and down the river at will. In January he explored the Río Xié and found at its mouth the first flowering specimens of *ucuquí,* a delicious fruit that had defied scientific description since first reported by Wallace over a century before. Its secret had been known to the Indians. An enormous tree with immense buttresses on the trunk, it flowers in a single day: Literally hundreds of thousands of small ephemeral blossoms burst forth at dawn and fade by twilight. Schultes and Pacho came upon it by chance at six in the morning, their attention drawn by the patter of flowers falling like rain in the forest. He collected a century set and named the species *Pouteria ucuquí.*

The end of January brought them to the headwaters of the Río Curicuriarí, at the base of Serra Cujubí, where Schultes found a novel variety of *Micrandra lopezii.* Then it had been back to the Uaupés, Serro Tukano, Ipanoré, and the valley of the Tiquié to secure flowering material of the new species they had found the previous fall with Murça Pires. In March 1948, Schultes headed briefly back downriver to Manaus, where he dispatched his specimens and reprovisioned the expedition before returning to São Gabriel. April meant the Río Içana, Jucabí, and the summit of Serra Curicuriarí, where he found *Abuta rufescens,* an arrow poison first collected by von Martius 130 years before. The three-week exploration of the botanically unknown Río Dimití consumed all of May and left both Schultes and Pacho physically depleted far more seriously than either of them imagined at the time.

Now, in June, worn and exhausted, they were making their way back to San Carlos and the mouth of the Casiquiare for the final phase of their sojourn in the basin of the Río Negro. Unlike Spruce and Wallace, Schultes's goal was not the Casiquiare but the headwaters of the Río Guainía, the river that flows from the northwest and joins the Casiquiare to form the Río Negro above San Carlos. There was a mountain to know there, the Cerro Monachi, and along the Guainía lived the Kuripáko, a people whose knowledge of medicinal plants was believed to be both extensive and shared by all the members of the tribe, not held by the shaman alone. In São Gabriel, Schultes had seen samples of their trade goods: great wooden manioc graters with quartz chips set in resin and exquisite pottery with designs painted in red pigment extracted from the leaves of a vine and glazed with the sap of an unknown tree.

Sometime around five in the afternoon the brandy kicked in. In one moment Schultes had been deep in thought, relishing the prospects of the journey. In the next, or so it seemed, he woke to an ugly morning, his head pounding and his mouth feeling as if it had housed for the night every one of the mangy dogs spread out around his legs on the floor of the hut. He opened one eye. He was inside an Indian shelter, and sprawled across their gear, deep in sleep, was young Pacho. Schultes stood up, stumbled beneath the thatch, and emerged into a blinding light. Shading his face with his hand, he made his way down a narrow path to the river and found to his immense relief that he had managed to tie up the canoe before passing out. Turning and walking up the bank, he came upon three young boys standing arm in arm, gently laughing. Schultes smiled weakly and asked them where he was. They told him, and he made a mental note to warn the Padre about the liquor.

San Carlos appeared as dismal to Schultes as it had to Spruce. The long rows of mud houses were built against one another, thatch roofs intertwined so that if one kitchen caught fire, the entire town would burn—as indeed it had, twice in recent history. The land was unplanted, there was no market, and the streets of the town were like open sores, littered with filth and human waste. Schultes could not understand how the people survived. There was so little food. At seven in the morning the men hung around the plaza while the women begged for soap and tobacco. Only with the help of Padre Antonio, and for an exorbitant price, was Schultes able to obtain a five-gallon can of *farinha* for the journey. When Schultes saw the miserable wooden shack that served as school, dormitory, and church of the mission, he understood why the priest needed the brandy. In fact, the Padre never touched the stuff; he used it only to influence the mayor. After only one night at the mission, Schultes and Pacho continued upstream, passing into the mouth of the Guainía at midday on May 28.

The Río Guainía was a land of hunger, a river of few fish and less game. At Tomo, a small settlement where Spruce had dried his plants for four days, living exclusively on the meat of a pair of toucans, Schultes and Pacho found no food at all. By the time they reached the Raudal del Sapo, the Frog's Cataract, they had exhausted their emergency rations and were considering breaking into their last supplies: four cans of Boston baked beans that Schultes had carried wherever he went, less as food than fetish. After four days on the river they came upon what appeared to be an apparition: a small dugout canoe paddled by a solitary white woman wearing a straw bonnet. She was an American, Sophia Müller, an evangelist for the New Tribes Mission, who had been living for years among the Kuripako at a small outpost known as Sejal. The Catholic priests, unhappy with the competition, had been trying to run her out of Colombia for a decade. On the rare occasions when a military patrol came up the river, she slipped into the forest and found refuge with the Kuripako. Schultes knew of her reputation, and she of his.

That night at dinner—three bowls of his *farinha*—she did her best to recruit him, arguing that anyone who knew the wild as he did ought to be saving souls, not plants. Besides, she suggested, when the final trumpet sounds, everything will be revealed, including all knowledge of plants. Why not just wait until then? He replied that by then he might be deaf and unable to hear the trumpet. He wanted to know now. For three days she kept at him, with the conversation becoming more and more surreal. Finally, on the morning of June 2, they parted ways, with Schultes and Pacho continuing upriver, destined for the flank of the Serranía de Naquén. As she saw them off at the landing, the poor woman appeared so desperately thin and sickly that Schultes made the ultimate sacrifice. Of his four cans of beans, he left two behind.

It proved a hasty decision, for in the following days he would need the food. That evening he first noticed the tingling in his fingertips. Initially he thought it a consequence of using formaldehyde to preserve his specimens, but within days the sensation had spread to his toes. As he approached the headwaters of the Río Guainía, he realized that a thousand miles from the nearest city, in the midst of a sterile forest where not a tree was in fruit, he was coming down with beriberi. Caused by a deficiency of vitamin B_1, or thiamine, in the diet, the disease is progressive. Early symptoms of lassitude and numbness in the limbs and extremities gradually give way to the complete degeneration of the nerves, resulting in the loss of reflexes, muscular atrophy, a complete inability to move, and ultimately death. The only treatment is thiamine—massive doses injected repeatedly over time. Schultes was a long way from a pharmacy.

Padre Miguel, the Salesian priest at São Gabriel, had already saved Schultes's life once. Now he had an idea that might do it again. By the time Schultes reached São Gabriel on June 21, his symptoms were pronounced, and it was imperative that he get proper medical attention as soon as possible. One option was to continue downriver six hundred miles to Manaus. Padre Miguel suggested another route. By traveling up the Uaupés, Schultes could reach the mission at Taracuá and the confluence of the Río Tiquié in twenty-four hours. From there it was only four hours up the Tiquié to the Ira-Igarapé, a narrow tributary at the head of which was a small encampment occasionally occupied by nomadic Indians. The whites knew the spot as Puerto Macú, named after the tribe. From Puerto Macú a trail ran across a height of land to the headwaters of another small river, the Abiú, which flowed into the Taraíra, a larger river that for much of its length formed the boundary between Colombia and Brazil. The Taraíra, Schultes knew, entered the Apaporis just twenty miles above the confluence of that river with the Río Caquetá. If he could get to the Caquetá and still have enough gasoline left to motor upstream for three hours, he would reach the Colombian military post at La Pedrera. From there he could fly directly to Bogotá. The total distance by land and water would be half that of going downriver to Manaus.

For Schultes a choice between Brazil and Colombia was no choice at all. The next morning he and Pacho departed São Gabriel at first light in a hired launch that ran all day and night, reaching Taracuá at midday on June 23. From there, the following afternoon, they left for the Ira-Igarapé. Padre Miguel had implied that they would reach Puerto Macú in a day. In fact, it took five, consuming much of their gas and leaving both Schultes and Pacho exhausted by the time they finally approached the landing at the edge of the forest.

By chance the nomadic Macú were there, a dozen men and women, furtive and uncertain what to make of the outsiders. For generations they had moved at will through the forests of the headwaters, living on wild fruits, herbs, and meat that they killed with darts coated with the most powerful curare known in the Amazon. The other forest dwellers, the Tukanoan peoples who grew food and dwelt in longhouses surrounded by planted palms, did not consider Macú to be people. They were not born of the anaconda but were breathed into being from dust and dirt by the first human, Yeba, the child of Jaguar Woman and Father Sun. They were subhuman, mediators between the realm of the living and the spirits of the forest, representatives of the female principle, objects of sexual desire and loathing, creatures useful only as slaves or servants. Yet the Macú lived among the sedentary tribes, and it was they who tended the ritual fires and prepared the sacred tobacco, smoked in enormous cigars two feet long and balanced on a carved fork driven into the ground. At the root of all the hatred and contempt lay a singular distinction between the tribes: Though technologically primitive, the Macú were the masters of the forest. Their knowledge of plants, their ability to secure skins, teeth, and feathers, and their dexterity with the poisons were unsurpassed. Few in number, fearful and timid, virtually unknown, the Macú had eluded Schultes for some years. Now, if only for a day, he would have a chance to be with them.

As the Indians moved slowly toward the shore, Schultes stepped tentatively from his canoe, careful not to stumble. His feet were numb, and he could walk only with the aid of two sticks, but he tried desperately to appear safe, ordinary, without affliction. He welcomed an old woman who came closest to him, and he smiled as she ran her hand over the side of the aluminum canoe. She mumbled a few words that Pacho could understand.

"She wants to know," Pacho said with a smile, "how many pots you can make from this."

"Pacho," Schultes said, "I think you had better dig out the trade goods."

Trade was something the Macú were very good at, and before long Schultes and Pacho had unloaded a bolt of cloth, two machetes, a pair of scissors, as well as one large aluminum pot that they had been saving for some time. It was the perfect moment. The Macú had something in mind for that evening—a party of sorts, nothing ceremonial, just a few men hanging out and taking *caapi.* Schultes and Pacho had arrived just in time.

The preparation was a simple cold-water infusion of the bark of a liana. No admixtures were employed, and the color of the final potion had a yellowish hue, quite unlike the coffee-brown color typical of *yagé.*

The effects, however, were unmistakable. "I learned experimentally," Schultes later reported, "that it had strong hallucinogenic properties." In fact, the intoxication was as intense as any he had ever experienced. It lasted all night and into the dawn. His eyes still painted with color, his mind swirling in the cool mist and scent of the forest, his body awkward and ashamed, he worked his way slowly along a trail to a small clearing where the old man who had danced all night revealed the source of the drug. It was a liana, clearly in the same family as *yagé,* but as Schultes lifted a leaf in his tingling hand, he recognized immediately that it was a new species in a genus of plants never before known to be psychoactive. He named it *Tetrapterys methystica* and with the help of the shaman gathered specimens as well as material for chemical analysis. These bulk collections were unfortunately lost in the following days when the canoe was spun over by a rapid, but the herbarium specimens survived and later confirmed that in the midst of his darkest hour, with beriberi progressing through his body, Schultes had indeed discovered a new hallucinogen. It was a relative of *Banisteriopsis caapi,* the sacred plant first found over a century before by Richard Spruce in his own period of trial on the banks of another of the wild streams of the Río Negro.

The short portage from the head of the Ira-Igarapé overland to the drainage of the Taraíra turned out for Schultes to be a three-day slog on feet that felt like stumps. Guided by the Macú, they finally arrived at the headwaters of the Abiú on the morning of July 6. The Indians hastily built a shelter for Schultes and then, after a simple meal, walked back over the trail to the Ira-Igarapé to get the canoe and the remainder of the equipment and supplies. For three full days, in the midst of the most terrific rainstorms he would ever experience in the Amazon, Schultes and Pacho remained alone, waiting. At no point did he consider what might have happened had the Macú not lived up to their promise to return. Though he had been with them for less than a week, his trust was absolute.

The Indians did come back, at midday on July 10, and by dusk Schultes and Pacho had paddled two hours down the Abiú. By this time, they had no gas and very little food. In their condition, getting up the Caquetá to La Pedrera without a motor would be almost impossible. How they would manage and what they would do if they couldn't were questions both agreed to worry about later. For the time being the river flowed in their favor, and they were glad of it. Another day on the Abiú brought them to its mouth on the Taraíra. The next morning, their food completely exhausted, they set out at dawn, and reached the first of two major sets of rapids by midday. A difficult portage, another stretch of river, a second portage, and then two days floating beneath a burning sun until finally, on the afternoon of July 14, they drifted from the Apaporis onto the main channel of the Río Caquetá. There, in a small back eddy on the left bank, they came upon a squad of Brazilian soldiers swimming lazily in the river. They had arrived by chance that morning and, fortunately, had gasoline and food to spare.

Schultes and Pacho wasted no time in going on to La Pedrera. The planes arrived only once a week, and Schultes was not about to risk missing a flight and having to wait around the military post and mission. The last time that had happened, at the end of the Apaporis expedition in 1943, he had painted the entire interior of the church. After five years it probably needed another coat, and he was not interested. All he could think about as they motored upriver in the dark was the cool mountain air of Bogotá, the fruit stands by the Plaza Santander, and Mrs. Gaul's English cooking at the Pensión Inglesa. They stopped for the night around ten, slept fitfully on the riverbank, and were back on the water by seven the following day. An hour later the sandstone mountain of La Pedrera loomed above them, and they could see the white statue of the Virgin Mary, luminescent in the morning light.

At the landing, before even leaving the canoe, Schultes asked a guard when the next plane was due. The soldier began to laugh. Schultes repeated the question. The soldier still did not believe the question was serious.

"Have you not heard¿" he asked in Spanish.

"What¿" replied Schultes.

"La Violencia," the soldier said, as if the word could explain all. Only later did Schultes learn that since the assassination of Jorge Gaitán on April 9, a civil war had paralyzed the nation. Much of Bogotá had been burned to the ground. There had not been a military flight to La Pedrera since the beginning of May, nor, given the state of emergency in the highlands, was one expected. Schultes and Pacho had traveled three hundred miles to reach a remote outpost that, now cut off from the air, was even more isolated than the place they had begun their journey.

With the disease each day making more of his body numb and useless, Schultes faced a river trip of more than seven hundred miles. There was no option. Abandoning his aluminum canoe in favor of a larger river launch, he and Pacho left La Pedrera after a day, bound for Manaus by way of the Río Japurá and the Solimões. The passage took nine days of steady running. When finally they arrived in Manaus, on the night of July 24, Schultes could barely walk and had to be supported by Pacho on his way to the hospital. There Schultes got hold of a syringe, learned how to use it, and bought enough thiamine to get him through a few weeks.

Schultes naturally had no intention of hanging around Manaus nursing himself with injections. On his way into town from the docks he had noticed a fine riverboat belonging to the American Chicle Company. That night he dispatched Pacho to make inquiries, and by morning everything was arranged. Within three days of arriving in Manaus, Schultes was back on the river headed for the Río Madeira. His goal was the savannah at the headwaters of the Marmelos. Three years before, when he had come down the Madeira from Bolivia, low water had prevented him from reaching the home of the rare endemic *Hevea camporum*. He was not about to let beriberi get in the way of a chance to secure flowering specimens. On July 31, their first day on the river, he and Pacho made collections from twenty-four different trees of *Hevea spruceana*.

Inevitably, after a month on the Madeira and Marmelos, Schultes was forced to give up. Once they reached the headwaters and had to drag a small launch through the shallows, his legs could not take it. He brought Pacho back to Manaus and then, in reward for his service, invited him to attend a botanical congress at Tucuman in Argentina. From there the two of them flew to Lima and on to Bogotá in a whirlwind tour beyond the wildest dreams of a young Miraña lad from the forests of La Pedrera. Before Schultes returned to Boston in December 1948, he made sure that Pacho would have a job waiting for him in Leticia. It was from there, seven months later, that Schultes received the terrible news that on July 6, 1949, Pacho López had died of tuberculosis. Schultes was certain that he had caught the disease during their time together on the Río Negro. Pacho was only twenty-two. For five years they had been inseparable, and for Schultes, losing Pacho was like losing a son.

Schultes remembered Pacho as he stood in a small English churchyard at Terrington in the early spring of 1950. For some months he had been in Europe visiting herbaria and working on his collections at Kew. Just that morning he had come north by train to Yorkshire. On the estate of Castle Howard, in a hamlet called Coneysthorpe, he had found the small stone cottage in which Richard Spruce had spent the last twenty years of his life. Living in the cottage was an old woman whose mother had been Spruce's housekeeper. As a little girl she had brought him his fiddle in the evening and stuffed woolen blankets beneath the sills to protect him from the cold. Schultes had tea with her and sat quietly listening to her tales as his thoughts drifted into every corner of the cottage.

Now, in the soft light of an English spring, he had come to pay homage at Spruce's grave. He still felt immense sorrow over the loss of young Pacho, yet at the same time he knew that his friend's memory was secure. He had seen to it. There were now more than a dozen plants that bore the name "lopezii." He knelt by the grave and read the simple inscription on the stone: RICHARD SPRUCE—TRAVELLER AND AUTHOR OF MANY BOTANICAL WORKS. BORN AT GANTHORPE, SEPTEMBER 10, 1817. DIED AT CONEYSTHORPE, DECEMBER 28, 1893. He looked around the grave and saw dandelions and foxgloves, lichens growing on stones, and the thick tufts of grass along the base of the wall of the church. From such simple beginnings, he thought, did this modest man arise. He remembered a line from Juvenal: *Scire volunt omnes, mercedem solvere nemo.* Everyone wants to know, no one to pay the price. A few months later, once again in a canoe on the remote Apaporis, Schultes would recall this moment and write his own epitaph for Spruce.

"I have felt it in the deep shadows of the Amazonian forests and in the blinding brightness of the Amazonian waters; I have felt it in herbaria; I have felt it while standing before Spruce's humble cottage in the hamlet of Coneysthorpe; I felt it again as I stood reverently before Spruce's grave in the churchyard at Terrington; but there, under the lowering sky of a Yorkshire April, I *knew* it to be true: Richard Spruce still lives, and will live on to fire the heart and shape the thoughts of many a plant explorer as yet unborn, who will tread Spruce's trail to carry forward his great, unfinished work."

CHAPTER THIRTEEN

The Divine Leaf of Immortality

IN THE MORNING Tim and I awoke to the fragrant air of the rain forest and the sound of banana leaves moving in the breeze. By evening we were camped on desert sand, watching fishermen land *corvina* from balsa rafts with billowing cotton sails. In one hundred miles the moist slopes of the Ecuadorian lowlands had yielded to the barren coastal plain of Peru, a land where by day nothing breathes or moves save the dissolute officials who police the roadblocks and the trucks that roll down the narrow ribbon of the Pan-American Highway. The Humboldt current like a great ocean river flows along the Peruvian coast, cooling the humid winds that blow from the sea, causing rain to fall over the water and fog to envelop the shore. On land there is almost never rain, and the earth is parched. From Puenta Pariñas in the north to beyond the border with Chile in the south, between the blue sea and the windswept western flank of the Andes fifty miles inland, stretches one of the driest deserts in the world.

Even in the best of times the northern land route into Peru is dismal: mostly wretched towns surrounded by acres of shanties, roofless shelters woven from straw, split bamboo, and cane; truck stops black with grease and petrol; restaurants where soup costs more if the chickens have been plucked; cemeteries where for want of wood the names of the dead are laid out in dark stones. At the garrison town of Tumbes, close to the Ecuadorian frontier, a corpulent officer held us up for an hour until Tim gave *una cosita,* a little something, for the Guardia's "retirement fund." Farther south we were stopped once more. Beside the *retén* was a large billboard threatening fines and imprisonment for those found urinating in public. In front of the notice, illuminated by the incandescent light of the police station, stood a dozen bus passengers calmly pissing.

When we traveled south to Lima, in early December 1974, the usual air of melancholy and corruption was tempered by actual tragedy. The fishing industry, at one time the largest in the world, had collapsed. In towns like Chimbote, which a decade before had processed millions of tons of *anchoveta,* the fleet lay idle in the bay. Though the scent of fishmeal still hung in the air, the processing plants were shut down. Along the beaches the wind was cold, from the south, and the salt chuck churned with the remains of dead fish and birds. In most years this carnage, a natural phenomenon, is localized far offshore, in a graveyard in the sea where the cold Humboldt current rising from the depths of the Pacific meets the warm waters of the tropics. As the waters fuse, nearly every form of life is extinguished, from microscopic plankton to the soaring albatross.

In some years, such as this one, the warm equatorial current reaches south, wedging itself along the coast and deflecting the cold waters out to sea. Since this usually occurs around Christmas, it is known as El Niño, the Christ Child. It is an ironic name for a condition that invariably brings disaster. For a season the world turns over. Torrential rains fall on the desert and floods scour the coastal towns. For one thousand miles the beaches are lined with rotting fish, seals, sea lions, and the corpses of creatures unknown and unimagined. Each gust of wind is putrid, and with fog rolling over the shore and nothing growing on the barren earth save the silvery gray shadows of tillandsia, a plant that lives on dew, the land appears doomed, as if haunted by the past. After six months in Colombia and two in Ecuador, where we had made but a single collection of coca, Tim and I had looked forward to Peru. After a week on the coast we longed for the Sierra.

In Peru they say that Lima was once a beautiful city, and, the expression goes, they've being saying that for five hundred years. The austere Spaniards, with their habit of building cities that defy geography, planted their capital in the desert—on the banks of a river through which flowed too little water and just far enough inland to ensure that their descendants would not benefit from the relief and pleasure of a sea breeze. Most of the year the city is enveloped by coastal fog that blots out the sky. In December and January, when the summer sun burns away the mist, the air is choked with diesel fumes, smoke, and exhaust. Mounds of garbage litter the beaches, and the ocean is awash in effluent.

Perhaps Lima was beautiful a generation or two ago—in 1919 when the population was only 170,000 or in the 1940s when Schultes passed through and the Limeños numbered but 600,000. Today the population is over 7 million, and one Peruvian in three lives in the capital. Most are urban refugees who have fled the poverty and political chaos of the hinterland and settled in the *pueblos jovenes,* the "young towns," a convenient but cruel euphemism for the squalid slums that dominate the city. For the poor, scattered in roofless huts built of grass and reeds, the Rimac River is now both lifeline and sewer. They drink its water, use it to wash their clothes and children, and depend on it to carry away the waste of a city where there is seldom rain. During the course of a day the color of the river changes from tan to black and then back again to brown.

Tim and I avoided the center of Lima and stayed in Miraflores, a beachfront suburb that until the 1940s was separated from the capital by haciendas and miles of open ground. The city now reaches to the sea, but the back streets of Miraflores, lined with mulberry trees, remain quiet, even peaceful. We found a small pension run by a kindly old woman who owned the place and treated her guests as family. A few blocks from the ocean, it was an ideal refuge, especially as we ended up in Lima for nearly a month, repairing the Red Hotel, establishing contacts in the various ministries, studying the coca collections at the Universidad San Marcos, as well as the living plants growing at the Botanical Garden at La Molina, the agricultural university located south of the city. It was a curious and instructive time. Though the cocaine wars had yet to shake the country and the revolutionary terrorism of the Shining Path was some years away, coca and cocaine were on the minds of everybody. Yet with few noted exceptions such as the eminent neurosurgeon Fernando Cabieses or the botanist Edgardo Machado at La Molina, it was astonishing how little anyone knew of the plant. Nearly every person we encountered—biologists at the universities, narcotic agents at the American embassy, bleary-eyed coke freaks on the beaches at Punta Hermosa—spoke as if coca leaves and the pure chemical extract were one and the same. When I mentioned coca to a Californian staying in the pension, the same fellow who had kept us awake half the night snorting cocaine like a Hoover in the adjacent room, he thought I was talking about chocolate.

Cocaine was unknown until 1860, when it was isolated at Göttingen by the German chemist Albert Niemann. In 1846 the archaeologist Johann Jakob von Tsudi, having observed the traditional use of coca leaves in the highlands, wrote, "I am clearly of the opinion that moderate use of coca is not merely innocuous, but that it may even be very conducive to health." The praise of the influential Italian neurologist Paolo Mantegazza, whose work especially inspired Sigmund Freud, was more effusive. "I prefer a life of ten years with coca," he wrote a year before Niemann's discovery, "to one of a hundred thousand without it."

The Corsican chemist Angelo Mariani agreed. In 1863 he patented Vin Tonique Mariani, a combination of coca extract and red Bordeaux wine that became an overnight sensation. Indeed, Mariani holds the curious distinction of being the only person responsible for two U.S. presidents, one pope, and at least two European monarchs turning on to coca. Pope Leo XII habitually carried a flask of the wine on his hip and was so enamored of the drink that he presented Mariani with a gold medal of merit. In America the ailing Ulysses S. Grant received a teaspoon of it each day in milk for the last five months of his life. Among the other noted enthusiasts who provided Mariani with written testimonials were President William McKinley, the czar of Russia, the Prince of Wales, Thomas Edison, H. G. Wells, Jules Verne, Auguste Rodin, Henrik Ibsen, Emile Zola, and Sarah Bernhardt.

A serious student of the plant as well as a promotional genius, Mariani created a full line of products. In addition to Vin Mariani there was Elixer Mariani, a stronger version of the wine; Thé Mariani, a coca extract without the wine; a throat lozenge known as Pâte Mariani, and Pastilles Mariani, the same lozenge spiked with pure cocaine. To sell these various preparations this enterprising chemist secured the endorsement of the French Academy of Medicine and a list of over three thousand physicians who swore by his products. One prominent doctor, J. Leonard Corning, described Vin Mariani as "the remedy par excellence against worry." Before long this "wine for athletes" was being taken by everyone from the Bavarian army and the French championship lacrosse team to professional singers and Gibson girls seeking longevity and eternal youth. Advertisements in America described it as a modern panacea and the perfect cure for "young persons afflicted with timidity in society." In time Vin Mariani became the most popular prescribed medicine in the world.

The wave of popularity peaked in 1884, the year Sigmund Freud published his misguided paper, "On Coca," and Carl Koller discovered the anesthetic properties of cocaine, which led to the first use of local anesthesia in surgery. This medical breakthrough in particular transformed the practice of ophthalmology, allowing for the first time the painless removal of cataracts. A pamphlet put out by Parke-Davis suggested that cocaine might be the "most important therapeutic discovery of the age, the benefit of which to humanity will be incalculable." The drug company, which then controlled the American cocaine market, had more than eye surgery in mind. By the 1880s Parke-Davis was marketing cocaine in candies, cigarettes, sprays, throat gargles, ointments, tablets, over-the-counter injections, and a cocktail known as Coca Cordial. Articles in learned medical journals recommended coca and cocaine for everything from seasickness to stomach pain, hay fever, mental depression, and, more ominously, the treatment of addiction to alcohol and opium. For that scourge of the nineteenth century, female masturbation, one physician recommended "a topical dose to the clitoris for prevention."

The *British Medical Journal* enthused in an editorial that coca represented "a new stimulant and a new narcotic: two forms of novelty in excitement which our modern civilization is highly likely to esteem." The American public certainly did. In 1885 a patent medicine manufacturer in Atlanta named John Pemberton registered the trademark for a preparation called French Wine of Coca: Ideal Nerve and Tonic Stimulant. A year later he removed the wine and added the kola nut of Africa, rich in caffeine, as well as citrus oils for flavoring. Two years after that he replaced the water with soda water because of its association with mineral springs and good health, and began to market the product as an "intellectual beverage and temperance drink." In 1891 Pemberton sold his patent to Asa Griggs Candler, another pharmacist from Atlanta, and a year after that the Coca-Cola Company was launched. Sold as a treatment for headache and advertised by Candler as a "sovereign remedy," Coca-Cola soon found its way into every drugstore in the land. The soda fountain, a kind of poor man's health spa, became an institution, and all over the country men and women were strolling into their pharmacies and asking for the drink that only years later became known as "the pause that refreshes." In the early days you ordered a bottle by asking for "a shot in the arm."

By the turn of the century there were some sixty-nine imitations of Coca-Cola on the market, all of them containing cocaine. In 1906, aware of growing public concern and anticipating passage of the Pure Food and Drug Act, which would prohibit the interstate shipment of food or drinks containing the drug, Coca-Cola removed cocaine from its formula. It continued, however, to rely on the source plant as a flavoring agent. To this day coca leaves are brought into the United States by the Stepan Chemical Company of Maywood, New Jersey, the only legal importer in the country. Once the cocaine has been removed and sold to the pharmaceutical industry, the residue containing the essential oils and flavonoids is shipped to Coca-Cola. The company is not especially proud of this fact, but it ought to be, for it is the essence of the leaves that makes Coca-Cola the "real thing."

Even as cocaine was being enjoyed by the public at large, medical opinion was slowly turning against the drug. Inevitably, exaggerated claims of its therapeutic value brought on a backlash of disappointment. Sigmund Freud clearly loved cocaine, which he saw as a miracle drug. In a letter to his wife, Martha, he teased, "You shall see who is stronger, a gentle little girl who doesn't eat enough or a great wild man who has cocaine in his body." Perhaps blinded by euphoria, Freud recommended cocaine as a treatment for a host of ailments including morphine addiction and alcoholism. Between 1880 and 1884 the *Therapeutic Gazette* of Detroit published sixteen reports of cures of opium addiction by cocaine. Parke-Davis advertised the drug as the only successful treatment.

It soon became apparent, however, that the cure could be as bad as the disease. By 1886 cases of cocaine psychosis with tactile hallucinations—the notorious illusion of bugs crawling beneath the skin—began to haunt the profession. By 1890 the medical literature contained four hundred cases of acute toxicity brought on by the drug. Albrecht Erlenmeyer, recognizing the dangers inherent in chronic use of cocaine, pronounced it "the third scourge of mankind," after alcohol and morphine. In but a few years cocaine went from being described as the most beneficial stimulant known to man, the drug of choice of presidents and popes, to being perceived as a modern curse, the embodiment and cause of every social ill. In the United States a series of laws increasingly circumscribed its use and availability. In 1922 cocaine was condemned as a narcotic, which it is not, and within a decade the public became convinced that it was a dangerous addictive drug, used only by musicians, artists, and assorted degenerates.

In Peru, meanwhile, the medical establishment watched this reversal of fortune with some interest. During the short history of European and American fascination with the drug, virtually no one drew a distinction between cocaine and coca. In the medical literature, the popular press, and even the advertising copy of Mariani, the terms were used interchangeably. As sensational tales of cocaine addiction began to circulate late in the century and the medical profession came to consider cocaine and morphine as equally dangerous, coca became associated with opium, and the public was led to believe that the ruinous effects of habitual opium use would inevitably befall those who regularly chewed coca leaves. Thus a mild stimulant that had been used with no evidence of toxicity for at least two thousand years before Europeans discovered cocaine came to be viewed as an addictive drug.

This was just the opening that a number of Peruvian physicians had been looking for. For the most part these men were liberal, with a concern for the plight of highland Indians that in its intensity was matched only by their ignorance of Indian life. When they looked up into the mountains from Lima, they saw only abject poverty, poor health and nutrition, illiteracy, and high rates of infant mortality. With the blindness of good intentions they searched for a cause. Since political issues of land, power, oppression, and raw exploitation struck too close to home, forcing them to examine the structure of their own world, they settled on coca. Every possible ill, every source of embarrassment to their bourgeois sensibilities, everything keeping the nation from progressing, was blamed on the plant. Dr. Carlos A. Ricketts, who first presented a plan for coca eradication in 1929, described coca users as feeble, mentally deficient, lazy, submissive, and depressed. Another leading commentator, Mario A. Puga, condemned coca as "an elaborate and monstrous form of genocide being committed against the people." Referring in 1936 to Peru's "legions of drug addicts," Carlos Enrique Paz Soldán raised the battle cry: "If we await with folded arms a divine miracle to free our indigenous population from the deteriorating action of coca, we shall be renouncing our position as men who love civilization."

In the 1940s the push for eradication was led by Dr. Carlos Gutiérrez-Noriega, chief of pharmacology at the Institute of Hygiene in Lima. Considering coca "the greatest obstacle to the improvement of the Indians' health and social condition," Gutiérrez-Noriega established his reputation with a series of dubious scientific studies, conducted exclusively in prisons and asylums, which concluded that coca users tended to be alienated, antisocial, inferior in intelligence and initiative, prone to "acute and chronic mental alterations," as well as other reputed behavioral disorders such as "absence of ambition." The ideological thrust of his science was blatant. In a report published in 1947 by the Peruvian Ministry of Public Education, he wrote, "The use of coca, illiteracy and a negative attitude towards the superior culture are all closely related."

It was largely as a result of Gutiérrez-Noriega's lobbying that the United Nations dispatched a team of experts in the fall of 1949 to look into the coca problem. Not surprisingly their findings, published as the 1950 Report of the Commission of Enquiry on the Coca Leaf, condemned the plant and recommended a fifteen-year phasing out of its cultivation. Such a conclusion was never in doubt. At the press conference held at the Lima airport when the commission arrived to *begin* its investigation, the chairman, Howard B. Fonda, then vice president of the pharmaceutical company Burroughs Wellcome, announced that coca was without question "absolutely noxious," "the cause of racial degeneration . . . and the decadence that visibly shows in numerous Indians," and he promised the assembled journalists that his findings would confirm his convictions. Eleven years later both Peru and Bolivia signed the Single Convention on Narcotic Drugs, an international treaty that called for the complete abolition of coca chewing and the end of coca cultivation within twenty-five years.

Incredibly, in the midst of this hysterical effort to purge the nation of coca, none of the Peruvian public health officials did the obvious: analyze the leaves to find out just what they contained. It was, after all, a plant consumed each day by millions of their countrymen and women. Had they done so, their rhetoric might have softened. In June 1974, while he was back at Harvard, Tim and his immediate boss at the U.S. Department of Agriculture, Jim Duke, had obtained a kilogram of sun-dried coca leaves from the Chapare region of Bolivia and arranged for the first comprehensive nutritional assay. Just before Christmas a letter from Duke arrived at the embassy in Lima confirming the results, which were astonishing. All along Tim had maintained that coca was benign, that the amount of cocaine in the leaves was small and absorbed in association with a host of other constituents which no doubt mediated the effect of the alkaloid. It was, he suggested, analogous to coffee or tea. Pure caffeine, extracted from the plants and injected, could not be compared to a cup of tea taken in the morning. He often quoted the physician William Golden Mortimer, who as long ago as 1901 reminded his profession that the effect of cocaine no more represents the effect of the leaves than prussic acid in peach pits represents the effect of peaches.

Still, even Tim was amazed by Duke's letter. Coca had been found to contain such impressive amounts of vitamins and minerals that Duke compared it to the average nutritional contents of fifty foods regularly consumed in Latin America. Coca ranked higher than the average in calories, protein, carbohydrate, and fiber. It was also higher in calcium, phosphorus, iron, vitamin A, and riboflavin, so much so that one hundred grams of the leaves, the typical daily consumption of a *coquero* in the Andes, more than satisfied the Recommended Dietary Allowance for these nutrients as well as vitamin E. The amount of calcium in the leaves was extraordinary, more than had ever been reported for any edible plant. This was especially significant. Until the arrival of the Spaniards there were no dairy products in the Andes, and even today milk is rarely consumed. The high level of calcium suggested that coca might have been an essential element of the traditional diet, particularly for nursing women.

There was more important news. Andean peoples often take coca after meals, explaining that the leaves, which they consider a "hot" substance, balance the "cold" essence of potatoes, which form the basis of their diet. Intrigued by this association, an anthropologist at Indiana University, Rod Burchard, had decided to investigate. His studies, recently completed, suggested that coca helps regulate glucose metabolism and possibly enhances the ability of the body to digest carbohydrates at high elevation. This confirmed the essential point of Duke's letter: that coca leaves were not a drug but a food and mild stimulant essential to the adaptation of Andean peoples.

This evidence put in perspective the ravings of men like Gutiérrez-Noriega and the draconian recommendations of international bodies such as the 1950 U.N. Commission of Enquiry. The deadline for the elimination of coca established in 1961 by the United Nations expired in 1986. The Peruvian effort went nowhere. Today the eradication campaign is being spearheaded by the American government, which has a new set of good intentions and even greater ignorance of Indian life. The heart of the debate, then as now, has not been the pharmacology of coca or the deleterious effects of cocaine. Efforts to eradicate the traditional fields began fifty years before an illicit trade in the drug even existed. The real issue is the cultural identity and survival of those who traditionally have revered the plant. In the Andes to use coca is to be *Runakuna,* of the people, and the chewing of the sacred leaves is the purest expression of indigenous life. Take away access to coca, and you destroy the spirit of the people.

Shortly after Christmas, Tim and I decided to separate for a few days. I wanted to do a little climbing in Ancash, a mountainous province eight hours by road northeast of Lima. He would proceed in the Red Hotel to Ayacucho and drop into the valley of the Apurimac and the coca fields of San Francisco. The plan was to meet up in Cuzco in a fortnight and head down the Sacred Valley to Quillabamba and beyond to the Río Santa Maria, an affluent of the Urubamba and one of the world's richest coca-producing areas. Mine was to be a hasty trip to one of the most beautiful regions of Peru, the Callejón de Huaylas, a narrow valley of brown earth and lush green fields of maize and alfalfa running between two mountain ranges, the stark and forbidding Cordillera Negra to the west and the luminous Cordillera Blanca, snow-capped and soaring to the east. A spectacular massif one hundred miles long and just over twelve miles wide, Cordillera Blanca has more than fifty peaks above eighteen thousand feet, including Huascarán, at twenty-two thousand feet the tallest mountain in Peru and the highest summit in the tropical world.

Rain and mist kept me for the most part below the ice, camped at the base of Huascarán. After a week, with the clouds still clinging to the summit ridge, I abandoned the mountain and made my way back to the coast, eager to move south and rejoin Tim. From Lima the train to Huancayo and the central Andes heads due east, speeding along the banks of the Rimac River to Chosica, a faded resort town perched just high enough above the coastal plain to remain untouched by winter fog. Its elevation is 2,400 feet, the distance from Lima twenty-five miles. Beyond Chosica stands a wall of mountains, broken only by a narrow canyon of pink rock. Clinging to the edge of cliffs, with the bed carved into the granite, the tracks enter this opening, and the train slowly climbs above the river. Gradually, over the course of a long day, a dramatic passage unfolds, an engineering feat of such stunning magnitude that one almost forgets the construction of the Central Railway cost the lives of over seven thousand workers, mostly Chinese coolies imported specifically for the project. In seventy-five miles the railway passes through sixty-six tunnels, crosses fifty-nine bridges, and zigzags twenty-two times as it rises fifteen thousand feet to the Ticlio Pass and the dizzying heights of the continental divide. Surveyed by an American railway engineer, Henry Meiggs, and built between 1870 and 1893 by the Peruvian Ernesto Malinowski, the route is the highest passenger line in the world, with a rate of ascent so precipitous that oxygen is dispensed in the first-class cars. Those traveling at the back of the train fight off the headache and nausea of *soroche* by chewing coca, the standard remedy for altitude sickness in the Andes.

The old man sitting beside me had not said a word since the train pulled out of Lima's Desamparados station. In the desert heat, in a train car stuffy with smoke and the earthy scent of unwashed bodies, he wore wool, homespun knee breeches, knitted cap and poncho, and rubber sandals. His clothing was threadbare, and his face weathered and deeply lined. Most of the day he slept, and I watched him as the train climbed steeply into the mountains, passing into the jagged landscape beyond Matucana and above the deep cut at Casapalca where the Rimac is stained with runoff from the mining operations that poison the upper reaches of the drainage and ruin the drinking water of Lima.

On the empty seat between us I spread out several small bags of coca, and throughout a long afternoon, as the train inched its way over grades too steep to be believed, I sampled leaves from each one, using a variety of lime sources that I had found in the market in Lima. In Colombia the Indians of the coast had used seashells, whereas those of the interior preferred raw limestone, dug from the earth and fired in a variety of ways. Here in Peru the people used a hardened paste, known as *llipta* or *tocra,* derived from the ashes of various plants, mixed at dawn with dew, potato water, and a sprinkling of human urine. In the mountain the sources of ash include the roots and stems of haba beans, and the fruits of a number of columnar cacti. Farther east in the *montaña,* where the Andes fall away to the lowlands, the people use corn cobs, cacao pods, and the roots of bananas.

The rock-hard *llipta,* white and caustic, burned my tongue. I tried it first with excellent leaves sold in the Lima market as *Cuzco verde.* They were light green, fresh, and of the highest quality. A second type came from the Huallaga valley beyond the Andes to the east and was known as *gringuita* because of the delicate flavor and light color of the leaves. Harvested by hand, sun-dried, and then allowed to sweat just long enough to become pliable, the leaves had been shipped to the capital in sixty-kilo bales, wrapped in a special fabric of woven wool known as *jerga.* Great care had been taken every step of the way. Coca must always be able to breathe, and if at any point in the drying process the leaves become wet, they ferment, turn brown, and lose all value. A third commercial variety was *Cuzco negro,* also known as *coca pisada,* or trampled coca. In this case the fresh leaves are spread out and deliberately walked on. Once dried in the sun they develop a dark brownish color and a special flavor, the result of a mild fermentation, quite different from the foul taste of leaves spoiled by moisture.

Yet a fourth bag contained the finest of all, the small and delicate, light green, and somewhat brittle leaves of the coca of Trujillo. Found only in the river valleys of the north coast and on the arid slopes of the upper Río Marañón farther to the east, this drought-resistant variety is grown in irrigated fields beneath the shade of spreading *Inga* trees.

Unlike the other commercial types, which botanically were all *Erythroxylum coca,* Trujillo coca was unique, closer to Colombian both in habit and morphology. Having examined living plants at the botanical garden at La Molina, Tim was certain it was *Erythroxylum novogranatense,* the coca of Colombia, though quite possibly an endemic variety. Harvested three times a year and sun-dried on large cement patios, the leaves have the scent and taste of wintergreen, an essential oil lacking in all other coca. Though only 6 percent of Peru's total production, the harvest of Trujillo leaves is the most closely controlled by ENACO, the National Coca Company of Peru, the government corporation that holds the monopoly on the domestic production and international distribution of the leaves. All coca destined for legal export is taken to the ENACO warehouses in Trujillo, where leaves from different areas are mixed together and packed in eighty-kilo bales for shipment to the United States. In the early 1970s over five hundred metric tons of Trujillo coca went out each year, most of it destined for Coca-Cola.

The train shuddered to a halt. Just ahead was the station at Ticlio and the mouth of the Galera tunnel that traverses the divide, running four thousand feet through the summit of the mountain. The old man woke with a start, glanced out the window toward a dark lake nestled at the base of the crater that swept around the station, and then looked the other way, his eyes falling on the bags of leaves between us. His smile revealed teeth worn from a lifetime of eating dry corn and lime. From a bundle at his feet he withdrew an aluminum pot, which contained a thick gruel of barley, and a cloth wrapped around a dozen or more boiled potatoes. These we shared, carefully peeling away the skin, and after a suitable pause I reached into one of the small plastic bags and withdrew a handful of leaves.

"Hallpakusunchis, Taytáy," I said in broken Quechua. "Let us chew coca together, Father." He nodded, somewhat surprised by the greeting, and with great deliberation reached for a finely woven coca bag that lay on the top of his belongings. With a faint sigh he slumped back on the seat and slowly withdrew from the *chuspa* one leaf, then another, and finally a third. Satisfied that each was unblemished and whole, he placed the leaves one on top of the other to form a *k'intu,* a small offering, which he held out to me. I noticed his worn hands for the first time; forefinger and thumb etched with dirt, the skin with the sheen of copper. I accepted the *k'intu* and made one for him. After we had completed the customary exchange, he held his *k'intu* a few inches from his mouth and slowly blew over the leaves, whispering words I could not understand. This was the *phukuy,* the ritual invocation that sends the essence of the leaves back to the earth, the community, the sacred

places, the souls of the old grandfathers. The exchange of leaves is a social gesture, a way of acknowledging a human connection, no matter how tenuous or transitory. The blowing of the *phukuy* is an act of spiritual reciprocity, for in giving selflessly to the earth, the individual ensures that in time the energy of the coca leaves will return full circle, as surely as rain falling on a field must inevitably be reborn as a cloud.

One by one the old man removed leaves from his bag, stripped away the midrib with his teeth, and added what remained to the quid slowly growing inside his cheek. Every so often he reached into his coca bag for a broken clump of *llipta* and bit off a small piece to add to the center of the quid. When a trickle of green saliva formed at the edge of his mouth, he discreetly wiped it away with his hand. Two ounces a day, I thought; more than a ton of leaves chewed over a lifetime—not chewed but slowly formed into a neat quid that, once exhausted, is reverently removed from the mouth, never spat, and placed on the earth, a rock, a field, again as an offering for Pachamama and the good ancestors.

For forty minutes or more we sat together without speaking, quietly enjoying the leaves as the train zigzagged down the steep slope to Yauli, another small mining town marred by slag heaps yet isolated on a boundless and breathtaking tableland where llamas grazed and small rock cairns stood like sentinels on the skyline. Another slow descent and the train reached La Oroya, a cold and desolate mining center, all black hills and slate skies overlooking land long ago poisoned by arsenic and lead. My old companion rose and stepped off the train. Through the dim light, as the train pulled away from the station, I watched him walk away toward a row of wooden barracks of Dickensian squalor.

After the bleak mining towns carved and blasted into the rock of the Andes, the basin of the Mantaro River appeared like an oasis, fertile, temperate, verdant, a perfectly flat valley surrounded by rounded hills on a soft and undulating horizon. I left the train at Huancayo—at eleven thousand feet the major commercial center of the valley—and spent the evening and following morning exploring the famous markets of the city. Coca was readily available, sold mostly in small stores known as *estancas,* many of which carried nothing else. A pound of leaves went for 60 soles, roughly $1.50. I made several collections, including an aromatic *gringuita* from Huanaco, which I sampled throughout the morning as I wandered among the market stalls and the dazzling array of fruits and vegetables, delicately carved gourds, hides, and great piles of llama and alpaca wool, some of it dyed with the soft browns and reds of barberry, walnut, and madder root. A large section of the market was dedicated to the healing arts and, like a medieval apothecary, hung with dried roots and leaves, lizards, toads, snakes, animal entrails, insects, and great sacks of magic powders. One woman was selling *Cantua buxifolia,* the national flower of Peru, a beautiful plant often found in hedgerows in the high Andes. Surprised, I stopped to ask what it was used for and only then remembered that long before the Spaniards took the plant as their symbol, the red bell-shaped blossoms, known in Quechua as *qantus,* were consecrated by the Inca as the flowers of the dead.

In the afternoon I caught a local bus headed for the crossroads of Huanacachi, ten miles out of town. There the large transport trucks stopped, and with any luck I would find a ride south for Cuzco. Sitting beside me on the bus was a handsome young man, a university student returning to his studies at the Universidad Huamanga in Ayacucho. In the idle way one chats with a stranger, I asked about his family and background, the nature of his studies, his hopes and expectations. He answered with the gentle lilt of highland Spanish, saying very little but with exquisite style. His words hung in the air as the dilapidated bus bounced and rumbled along the narrow dirt road that ran out of Huancayo.

At the first roadblock south of the city, two policemen came onto the bus to check papers. Both were of Indian descent, elevated into the ranks of *mestizos* by uniforms and rations that had left them plump and sweaty. They insulted the Indian women, abused an old man who could not read, and ignored the foreign travelers. I handed the first of them my passport, acknowledged their perfunctory nod, and turned just in time to see my companion torn from his seat and pistol-whipped across the face. The next moment he was off the bus spread-legged against a mud wall. The police beat him savagely and laughed as he fell limp to the ground. No one else in the bus noticed. No one saw him get up and sprint down the alley of mud bricks. No one heard the gunfire and the shouts of rage from the police. In a moment it was all over. The youth was gone. The police appeared once again in command. The placid tone of the Andes had returned, as innocent as a guidebook.

Stunned by the violence and disturbed by the reflexive passivity of the other passengers, I stepped off the bus at the crossroads at Huanacachi, where I waited for five hours before flagging down a large transport destined for Cuzco. Climbing out of the valley, the road passed for hours through a remote and barren land, arid and cold. By the time we reached Huancavelica and pulled up in front of the cathedral for the night, it was well past midnight and the plaza was deserted. I slept in the back of the truck, as did the driver, who was again behind the wheel at dawn, slowly negotiating the narrow streets, inching his way toward the *salida* for Ayacucho and the road south.

Morning in Huancavelica revealed a small colonial town surrounded by barren rocky hills. In addition to the cathedral, beautiful in the soft light, there were seven ornate churches, each betraying a painful history. It was here that the Spaniards found silver and, more important, the mercury they required to smelt the mountain of silver they had found at Potosí in Bolivia. By 1564 Indian slaves from all parts of Peru were being marched into the Huancavelica mines. Beaten with iron bars, torn by whips, they remained underground for a week at a time, strung together in iron collars, working rock faces to extract the daily quota, twenty-five loads per man, one hundred pounds to the load. To their overseers they were not people but "little horses," and when after a few months they stumbled from the pit, chests burning from the toxic dust and vapors, muscles contorted with shakes and tremors, unable to walk, let alone maintain their production, they were simply killed to make room for the next worker. In time mothers began to bury their infant sons alive to save them from the horror of the mines.

In a perverse way the mines played a significant role in the history of coca. The early Spaniards had written glowingly about the plant. In his *Royal Chronicles,* Garcilaso de la Vega wrote that the magical leaf "satisfies the hungry, gives new strength to the weary and exhausted, and makes the unhappy forget their sorrows." Pedro Cieza de León, who traveled throughout the Americas between 1532 and 1550, noted that "when I asked some of these Indians why they carried these leaves in their mouths . . . they replied that it prevents them from feeling hungry, and gives them great vigor and strength. I believe that it has some such effect." The first botanical description of coca, written by the Spanish physician Nicolas Monardes, appeared in a book entitled *Joyfulle News Out of the Newe Founde Worlde, Wherein Is Declared the Virtues of Herbs, Treez, Oyales, Plante and Stones,"* translated in 1582 by John Frampton.

Predictably, the Catholic Church attempted to outlaw the plant. At ecclesiastical councils held in Lima in 1551 and 1567 the bishops condemned its use as a form of idolatry and secured a royal proclamation declaring the effects of the leaves an illusion of the devil. By then it was too late even for the Church. Too many Spaniards were making fortunes growing and trading coca, and those running the mines had found that without leaves Indians would not work. A face-saving compromise allowed the Church to reverse its position. The cultivation and selling of coca was deemed acceptable, but the use of the leaves in religious ceremonies remained punishable by death. In 1573, Viceroy Francisco de Toledo removed all controls on secular commerce, and for the next two hundred years, as thousands died in the mines and on the plantations, coca became a mainstay of the colonial economy. Production soared by a factor of fifty, and by the end of the sixteenth century, taxes on coca were providing the Church with much of its revenue.

The dramatic rise in coca production in the years following the Conquest has kept alive one of the most enduring misconceptions about coca; the notion that during the century of Inca domination, the leaves were available only to the ruling elite. Garcilaso de la Vega was perhaps the first to make the claim, and it has since been parroted in nearly every popular account of pre-Columbian Peru. Though the idea has obvious appeal, especially to those critical of the contemporary use of the leaves, evidence for such a monopoly is, in fact, rather sketchy. The Inca did exert control over the production and distribution of various agricultural items including coca, according to many of the early chroniclers who visited the imperial capital. But it is less clear whether their observations of court life in Cuzco, based largely on testimony of Inca nobles, reflect what was going on in the outlying areas of the empire. The history of coca would suggest otherwise.

It is true, of course, that the Inca revered coca above all other plants. For them it was a living manifestation of the divine; its place of cultivation a natural sanctuary approached by all mortals on bended knee. Unable to grow the leaves at the high elevation of Cuzco, successive rulers ordered plantations to be replicated in gold and silver, in delicate gardens enclosed by temple walls. Coca figured prominently in every aspect of ritual and ceremonial life. Before a journey, priests tossed leaves into the air to propitiate the gods. At the Coricancha, the Court of Gold, the Temple of the Sun, sacrifices were made to the plant, and supplicants only could approach the altar if they had coca in their mouths. The future was read in the venation of leaves and in the flow of green saliva on fingers, by soothsayers and diviners who had acquired their knowledge by surviving a bolt of lightning. At initiation young Inca nobles competed in ardous foot races, while maidens offered coca and *chicha.* At the end of the ordeal each runner was presented with a *chuspa* filled with the finest of leaves as a symbol of his new manhood.

Long caravans carrying as many as three thousand large baskets of leaves regularly moved between lowland plantations and the valleys leading to Cuzco. Without coca, armies could not be maintained or marched across the vast expanse of the empire. Coca allowed the imperial runners, or *chasquis,* to relay messages four thousand miles in a week. When the court orators, or *yaravecs,* were called upon to recite the history of the Inca at ceremonial functions, they were aided only by a system of knotted strings, or *quipus,* and coca to stimulate the memory. In fields priests and farmers offered leaves to bless the harvest. A suitor presented coca to the family of the bride. Official travelers lay spent quids of leaves on rock cairns dedicated to Pachamama and placed at intervals along the paths of the empire. The sick and dying kept leaves at hand, for if coca was the last taste in a person's mouth before death, the path to paradise was assured.

But just as the Inca venerated the plant, so, too, did the other peoples of the Andes. The earliest archaeological evidence suggests that coca was used on the coast of Peru by 2000 B.C. Actual leaves that can be botanically identified as Trujillo coca have been dated to A.D. 600. Lime pots, lime dippers, and ceramic figurines depicting humans chewing leaves have been found at virtually every major site from every era of pre-Columbian civilization on the coast, Nazca, Paracas, Moche, Chimu. The very word coca is derived not from Quechua but from Aymara, the language spoken by the descendants of Tiwanaku, the empire that predated the Inca on the altiplano and in the basin of Titicaca by five hundred years. The root word is *khoka,* a simple term meaning bush or tree, thus implying that the source of the sacred leaves is the plant of all plants. Evidence suggests that an active trade in coca was established in the Bolivian highlands as early as A.D. 400, a thousand years before the dramatic expansion of the Inca.

The unique genius of the Inca, and the key to the power of their rule, was their ability to incorporate a remarkably diverse population by assimilating local rulers, absorbing religious ideas, manipulating regional animosities, and establishing new relationships of authority based on ancient notions of exchange—all the time working with local institutions to promote the expansion of their empire. When necessary they could respond ruthlessly, annihilating rebellious tribes, shifting populations to distant lands, and pacifying new conquests by transplanting entire colonies of loyal subjects. Rarely, however, did they violate ancient traditions or institutions that posed no threat to the integrity of the empire. Given this pattern, it seems extremely unlikely that they would have prohibited or been able to suppress a cultural practice as vital and fundamental as the use of coca. The production and distribution of the plant no doubt came under control, and its sacred role within areas close to the capital may well have been circumscribed. On the fields and slopes of the Andes, however, the plant almost certainly continued to be employed as it always had been: as a mild and essential stimulant in a harsh and unforgiving landscape.

• • •

We rode up mountains and down valleys, one after another, for five days. In the dry bottomlands morning glories grew as trees, and castor beans spread weedlike and wild on dusty trails running through land flattened by erosion. On the wetter slopes, fields were laid out like checkerboards, lined with eucalyptus and agave, and those that had been left fallow were thick with cosmos and zinnias, helianthus, bidens, and calceolaria, all yellows and golds and reds.

Under the Inca, fields such as these colored the land. Famine was unknown. The year Pizarro arrived in Peru, there was more soil under cultivation in the Andes than there is today, and the quantity of food produced was far higher. Though one crop in three failed due to frost, hail, or drought, the extremes of the Andean climate—hot sunny days and cold nights—made possible preservation techniques that allowed basic foods to be stored almost indefinitely. Jerky, the English term for dried meat, is derived from the Quechua word *ch'arki. Ch'uñu,* dried potatoes that have been repeatedly frozen by night, stomped, and sun-bleached by day, was a fundamental staple, and the world's first example of freeze-dried food. A few pounds could sustain an Incan soldier for a month or keep a family alive through the lean weeks of the rainy season.

To preserve and distribute the surplus of empire, the Inca established an extraordinary network of depots, or *qollqas.* Built usually outside of populated areas, often along imperial roads, these repositories held vast stores of food, weapons, fine cloth, and coca, precious goods and everyday items in such quantities that armies could be maintained on the move, and no part of the realm need ever fear starvation. The scale of the system was astounding. At the ancient center of Huánuco, archaeologists have uncovered the remains of five hundred storehouses with a total capacity of over a million cubic feet. At Vilcashuamán, the Sacred Falcon, the crossroads where the route from Cuzco to the coast intersected with the royal road following the spine of the Andes, there were over seven hundred *qollqas.* Sixteen years after the Conquest a Spanish army encamped for more than seven months beside an Incan storehouse in the Jauja Valley of central Peru, feeding itself with just some of the 800,000 bushels of food discovered within a single complex of *qollqas.*

These storage facilities, containing such enormous reserves of food and goods, were just one feature of a dynasty known less for its art and culture than the sheer brilliance and efficiency of its administrative organization. The Inca ruled with absolute power but understood that a people well fed and treated fairly produced more and worked harder than those suffering from injustice and privation. The entire system was based on the enlightened self-interest of a nobility that derived its authority from the gods and its power from the control of the most formidable armies of the known world. Demands on the individual were severe but predictable and consistent. The state guaranteed freedom from every sort of want and in exchange demanded heavy tribute in labor. This fundamental reciprocity allowed the ruling elite to harness tremendous levies of workers to build the great public works of the empire. Evidence of their tireless efforts is found everywhere in Peru today: rivers channeled and straightened to liberate precious agricultural land, massive boulders carved into ceremonial pools or decorated with wild façades of iconography, remnants of royal roads, bridges, palaces, and way stations. Traveling down the Andes, along mountainsides covered with banks of agricultural terraces and the thin traces of stone aqueducts that carried water for hundreds of miles, one cannot imagine how so much could have been accomplished in but a century of Inca rule.

When the Spaniards asked about the origin of the empire, they were told of a time of darkness and a great flood sent to the world by the creator Wiracocha. Taking pity on the earth, Wiracocha stood on an island in Titicaca and flung into the sky the sun, moon, and all the stars. Then he ordered these heavenly bodies to populate the land. The Sun, lord of the universe, dispatched his son Manco Capac, and the Moon gave her daughter Mama Occlo to be his bride. Together they emerged from the waters of Titicaca, from the islands of the Sun and the Moon, and bearing a golden staff, began a great odyssey. Their instructions were to search the world, probing the earth with their staff, until they found a place that would accept it. There they were to establish their kingdom.

For years they wandered, escorted each day by wild geese, each night by condor. Finally, at the base of a mountain known as Wanakauri, the ground swallowed the staff and a rainbow rose in the sky. Manco Capac called upon the local people to abandon their nakedness, their diet of wild seeds, their wretched lives to follow him into the valley where the imperial capital of Cuzco would be built. Manco Capac, son of the Sun, became the first Inca, taking as his *coya,* or queen, his sister, the daughter of the Moon. Thus from its mythological inception the empire was inspired by the gods, and the Inca, his family, and all his offspring were known to be divine. As the royal family grew, it divided into lineages, each composed of the descendants of a former ruler. Each ruler had his own palace in Cuzco, and after his death, his deified remains were preserved there, cared for by an enormous retinue of servants and loved ones, all members of his own *ayllu,* or clan. Incan history, as told to the Spaniards, revolved around the heroic exploits of the twelve men who in turn inherited from Manco Capac the mantle of the Inca. Their lineage began at the beginning of time and ended with the betrayal and murder of Atahualpa in 1533 at Cajamarca.

In fact, as late as the 1430s the people who became the Incas were still a relatively small mountain tribe, controlling just the land between the canyon of the Apurimac and the deep valley of the Vilcanota-Urubamba. In 1438 there was an attack by the Chanca, blood enemies from beyond the Apurimac, and Cuzco was nearly overrun. The ruling Inca fled, but the people rallied around his son, who, inspired by a vision of the sky god and aided by stones that sprang to life to fight by his side, regained control of the battlefield and vanquished the Chanca. In victory the prince took over the kingdom, changing his name to Pachacuti, Reformer of the World. This was the seminal moment in the history of the empire, the actual beginning. The eight rulers who had come before, and who were forever memorialized in Incan mythology, most likely were not successive leaders at all but rather the heads of the various clans who fought beside Pachacuti on the slopes above Cuzco.

Following his triumph, Pachacuti transformed his capital, draining the marshes, redirecting the flow of the Saphi-Huatanay and Tullumayo streams that passed through the town, and laying down the outline of a new city that would conform to the feline shape of a puma, the nocturnal embodiment of the sun. He built a palace for his lineage and restored the glory of the Coricancha, the Temple of the Sun. He then created the Acllahuasi, the House of the Chosen Women, erected temples to the sky gods, and instituted laws and the worship of the sun, the moon, stars, thunder, and the rainbow. He declared himself the Son of the Sun and set out to conquer the world.

With Cuzco as the axis, the navel of the universe, Pachacuti conceived Tawantinsuyu, the Four Quarters of the World. To the northeast lay Antisuyo and beyond the formidable forests of the Amazon. To the southeast was Collasuyo, land of origins, the altiplano and the fertile basin of Titicaca, richest domain of the empire. To the northwest was Chinchaysuyo, the central Andes, the desert kingdoms of the north coast, the mountains of Ecuador. The final quadrant, Condesuyo, lay to the southwest, across the Andes to the desolate wastes of the Atacama Desert.

Beginning in 1438, Pachacuti moved steadily north and south, his power radiating in all directions before him, his armies victorious, his conquests consolidated through the imposition of religion and language, the creation of infrastructure, the incorporation of local elites, the savage destruction of all who resisted. By 1463 the Inca controlled the central Andes, from the present location of Ayacucho in the north to Titicaca in the south. Subsequent campaigns by Pachacuti and his son Topa Inca Yupanqui extended the empire north into Ecuador. In 1465 Topa Inca marched south from Quito, invaded northern Peru, and defeated the Chimu, a dynasty that had ruled six hundred miles of the desert coast for two centuries. Six years later Topa Inca succeeded his father as Inca and continued his conquests, taking all of highland Bolivia, invading modern Chile through the Atacama, and reaching as far south as the Maulé River, two thousand miles south of Cuzco.

In 1493, the same year that the Catholic Church divided the New World between Spain and Portugal, Topa Inca died, and his place was taken by his son Huayna Capac. By 1520, less than a century after Pachacuti's victory over the Chanca, this grandson was fighting in southern Colombia, and stones taken from the quarries of Cuzco were being used to build his palaces in Quito. The empire stretched over three thousand miles, the largest ever forged on the American continent. Within its boundaries lived nearly all the people of the known world. There were fourteen thousand miles of paved road, rich and bountiful fields, immense temples dedicated to the sun. There was no hunger. All matter was perceived as divine, the earth itself the womb of creation. Then came smallpox, the death of Huayna Capac, and the beginning of the cataclysm.

There was still no sign of Tim when I woke, stiff and cold, just before dawn at the Pensión San Jorge in Cuzco. The shutters of my room opened onto a small park beyond which the terra-cotta roofs of the city piled one atop the other, their color the same as the red earth I had seen the evening before as the truck crested the height of land above the Anta Valley and the lush fields gave way to adobe walls and mud tracks leading into the heart of the ancient capital. At night the city had appeared like a carnival, particularly the Plaza de Armas where foreign travelers outnumbered locals. The only Indians to be seen were sitting in a long row beneath a colonnade, hawking trinkets and weavings. Across the way the baroque façade of La Compañía, the church of the Jesuits, was lit up like a birthday cake. There had been music in the air, mostly rock and roll, all of it coming from garbled speakers set up in the streets by cafés and bars eager to attract customers. Outside the restaurants ragged children huddled, waiting for a chance to dart in and steal small bits of food. At a corner of the plaza I came upon a German couple haggling with an old woman over the price of a handwoven belt, intricately detailed with figures in flight. The Germans were unwilling to pay 60 soles, the price then of a cup of coffee in Munich.

The morning sky was overcast, but the clouds were clearing with the dawn. The air smelled of eucalyptus smoke and rain on whitewash and old stones. Below my window, a young girl hurried by, dragging behind her a black llama. She glanced up and I saw her face, round and full. Just a decade earlier it was common to see herds of llamas being watered at the fountains of the city. Now only the odd one appeared, usually as an attraction for tourists. But in the countryside the animals are still revered. Once a year, on a special day in August, their owners join them in their corrals to drink and chew coca together. The llamas are decorated with bright tassels and given a special concoction of barley mash, *chicha,* medicinal herbs, and cane alcohol. The men make the offering, holding the animals down as bottles of the brew are poured into their throats. By the end of the day both man and beast are completely drunk, and together they stagger out of the corral, following their other companions, laughing, singing, dancing.

At seven the bells of the cathedral rang out, calling the people to Mass. The sound was beautiful, and I followed it into the streets, joining a dozen or more women, all dressed in black, as they slowly made their way toward the Plaza de Armas. I hesitated at the base of the stone steps and then slipped into the cathedral, staying just long enough to see the opulence, the dark vaults, a row of priests in red robes, old Indian women lighting candles, dust drifting in shafts of sunlight. The nave lies on the foundation of the Quishuarcancha, the palace of Wiracocha, father of Pachacuti. An Inca prince is said to be imprisoned within the walls of one of the towers, waiting for the masonry to decay that he may emerge to reclaim his land. In 1950, when a powerful earthquake destroyed much of Cuzco, thousands gathered in the plaza, watching the belfry, expecting it to crumble. It was badly damaged, but it did not fall.

Outside again in the cool air, I felt the sun on my face and looked over the plaza, trying to imagine a time when Inca priests gathered to greet the day, spreading their hands and washing their skin with sunlight. At the Coricancha, the Court of Gold, more than four thousand priests lived, each with ritual duties. The sun temple itself was draped entirely in gold, with the chambers dedicated to the moon, thunder, and stars sheathed in silver plate. There was a garden of golden maize, tended by the *Mamaconas,* the wives of the sun, and a life-sized effigy of a young boy, again made out of solid gold. A cornice of gold a foot wide surrounded the building, and within the main chamber was a golden altar and a ritual font that together weighed over three hundred pounds. The centerpiece of the Coricancha was an immense golden disk, positioned so that its surface would catch the morning light and reflect the sun rays into the temple. This sacred object, the most famous lost treasure of the Andes, has never been seen by Spaniards since the day three of Pizarro's men arrived to collect Atahualpa's kingly ransom. With their own hands they desecrated the Coricancha, removing seven hundred sheets of gold, each weighing several pounds, and dozens of figurines of animals and plants, sacred vessels and fountains. For the Inca priests, for whom gold had no monetary value, the violation by these vandals was a sacrilege beyond imaginings.

I left the plaza and made my way slowly up the steep streets that led away from the center, toward the hillside that rises above the city, and the ruins of Sacsahuaman, the Royal Hawk. When Pachacuti laid out his capital in the shape of a puma, he envisioned the Tullumayo stream as the animal's spine, a river of bone, its confluence with the Saphi-Huatanay as the tail. The head of the cat was to be the height of land that divided the two streams to the northwest. With cliffs falling away on three sides toward the city below, the outcrop formed a natural fortress. All that was required was a row of battlements on the northern side, a zigzag of ramparts that the Inca perceived as teeth, sharp, powerful, invincible. To build the fortress Pachacuti, and later his son and grandson, maintained a constant workforce of twenty thousand for sixty years, rotating the workers through the seasons so that in the end countless thousands of royal subjects gave of their labor for the benefit of the Inca.

Although the fortress, having served as a quarry for four centuries, is today a shadow of what it once was, it remains the single most impressive public work of the Inca. The giant ramparts, sixty feet high and running in three rows for a third of mile, look down on a broad, artificially leveled parade ground, beyond which lies another fortified redoubt, the Rodadero Hill. The lower terrace wall is made up of the finest Inca masonry, as perfectly formed as the most delicate walls of the Coricancha. Yet here the individual megaliths are of a breathtaking scale. The largest is twenty-five feet high, ten feet across, and is believed to weigh 361 tons. Others weigh between 50 and 200 tons. Many were cut from the earth on site, others were dragged overland, twelve miles from a quarry southeast of Cuzco at Rumicolca.

The Spaniards who saw Sacsahuaman in its glory, when it housed five thousand imperial troops, when its water towers stood and the lodgings and religious sanctuaries of the Inca had yet to be pillaged, could not believe that the fortress was the work of men. Nothing created in the history of the Old World could compare. The Church declared the stonework, in particular, to be the product of demons, an assertion no more fantastic than many more recent attempts to explain the enigma of Inca masonry. The ill-fated explorer Colonel P. H. Fawcett was the first to express the idea, since repeated by every tour guide, of a secret plant capable of dissolving rock. A slew of writers have argued for an extraterrestrial origin, a suggestion not only silly but demeaning, implying as it does that the ancestors of the highland Indians were incapable of executing what was in fact their greatest technical achievement.

The actual explanation is far simpler and more elegant than fantasy would allow. To quarry the stone the masons sought natural weaknesses in the rock, small fissures that could be widened by planting a wooden wedge and soaking it with water. Once a block of stone broke free, it could be worked with harder rocks, by a series of abrasive blows that in time would transform its surface. Experiments have shown that, even without iron tools, a shapeless lump of andesite can be turned into a smooth cube a foot in dimension in just two hours. To work a larger stone with complex angles would presumably involve dusting the edges with chalk or colored powder, setting the rock in place, removing it to see what needed to be ground off, and then setting it in place once again. Excavations have shown that this, in fact, is what the Inca masons did. Load-bearing surfaces fit with exquisite precision across the entire surface of the stones. The vertical joins, by contrast, are tight on the exposed face but ragged behind and filled with mortar. Clearly, there was no magic technique. Only time, immense levies of workers, and an attitude toward stone that most westerners find impossible to comprehend.

For the Runakuna, the people of the Andes, matter is fluid. Bones are not death but life crystalized, and thus potent sources of energy, like a stone charged by lightning or a plant brought into being by the sun. Water is vapor, a miasma of disease and mystery, but in its purest state it is ice: the shape of snowfields on the flanks of mountains, the glaciers that are the highest and most sacred destination of the pilgrims. When an Inca mason placed his hands on rock, he did not feel cold granite, he sensed life, the power and resonance of the earth within the stone. Its transformation into a perfect ashlar or a block of polygonal masonry was service to the Inca and thus a gesture to the gods, and for such a task, time had no meaning. This attitude, once harnessed by an imperial system capable of recruiting workers by the thousand, made almost anything possible.

If stones were dynamic, it was only because they were part of the land, of Pachamama. For the people of the Andes, the earth is alive, and every wrinkle on the landscape, every hill and outcrop, each mountain and stream has a name and is imbued with ritual significance. The high peaks are addressed as Apu, meaning Lord. Together the mountains are known as the Tayakuna, the Fathers, and some are so powerful that it can be dangerous even to look at them. Other sacred places, a cave or mountain pass, a waterfall where the rushing water speaks as an oracle, are honored as the Tirakuna. These are not spirits dwelling within landmarks; rather, the reverence is for the actual place itself.

A mountain is an ancestor, a protective being, and all those living within the shadow of a high peak share in its benevolence or wrath. The rivers are the open veins of the earth, the Milky Way their heavenly counterpart. Rainbows are double-headed serpents that emerge from hallowed springs, arch across the sky, and bury themselves again in the earth. Shooting stars are bolts of silver. Behind them lie all the heavens, including the dark patches of cosmic dust, the negative constellations that to the highland Indians are as meaningful as the clusters of stars that form animals in the sky. Lightning is concentrated light in its purest form. Places struck by lightning receive offerings of coca and are never forgotten; objects that have been hit become imbued with regenerative power. A person killed by lightning is buried on the spot and instantly becomes a Tirakuna. One who survives a lightning strike receives the gift of divination, the ability to read the future in the pattern of coca leaves tossed to the ground.

These notions of the sanctity of land were ancient in the Andes, but under the Inca they became formalized in a remarkable way. From the ruins of Sacsahuaman one can see the routes of the royal roads, running away to the four quarters of the empire. On the horizon one hundred miles to the southeast, Apu Ausangate soars to twenty thousand feet, guarding the approaches to Collasuyo. From the city below it is difficult to pick out the outline of a puma, but looking east, beyond the outstretched arms of the massive cement statue of Christ placed atop the ruins, the land rises to the height of Pachatusan, the Fulcrum of the Universe. Viewed from the temple of Coricancha, the sun on the winter solstice rises directly over Pachatusan. In the time of the Inca, great monoliths stood on the mountain, all part of a scheme that turned the land itself into a template for the royal astronomers.

If Cuzco was the navel of the world, Coricancha was its axis. The Court of Gold, the Temple of the Sun, was the hub of a conceptual wheel with forty-one imaginary spokes radiating to and beyond the horizons, their alignments determined by the rise and fall of the stars and constellations, the sun and the moon. Along these sightlines, or *ceques,* were 327 sacred sites, each with its own day of celebration, each revered and protected by a specific *ayullu,* or community. In this way each person and every clan, though rooted to a specific locality, was bound to the cosmological framework of the empire.

These shrines, or *huacas,* were stations on holy paths that existed in both a literal and metaphysical sense. One of the most sacrosanct of Inca memories was the birth of the nation, the moment at the foot of Wanakauri when the earth had at last accepted the golden staff of Manco Capac. On the summer solstice, during the Festival of Inti Raymi, priests traveled to Wanakauri to pay homage to the setting sun. A second group of pilgrims celebrated the birth of the sun by following the path of the *ceque* that led southeast to the sanctuary of Willkanuta, the House of the Sun, one hundred miles away from Cuzco on the continental divide. This imaginary line continued south through Titicaca to the ancient ruins of Tiwanaku. Tracking the same line to the northwest, the *ceque* reached Ecuador and the northern limits of the empire. It was this trajectory that Manco Capac followed on his mythical journey north from the islands of the sun and the moon. Thus in making the pilgrimage the priests reaffirmed the primordial linkage between Cuzco and Titicaca, birthplace of the world and origin of the first Inca.

Throughout the year, ritual activities such as these bore witness to mythology and confirmed the cosmic order of the empire, often in astonishing ways. During solstice celebrations, for example, or during moments of great portent such as the passing of a king or a serious defeat in battle, the priests would call for the sacrifice of Qhapaq Hucha. The victims, children and animals blessed by the Sun, were called to Cuzco from all parts of the empire. Some were killed in the capital; others were chosen to carry portions of the sacrificial blood back to their communities, where in due course they, too, would be killed. The entourages arrived in Cuzco by royal road, but on leaving they followed the sacred routes of the *ceques,* walking in a straight direction over mountains and across rivers, sometimes for hundreds of miles, visiting local shrines, paying homage to the perfection of their fate. These journeys, as much as the sacrifice of the children, realigned the people with the Inca and represented symbolically the triumph of empire over the imposing landscape of the Andes.

The Spaniards did all in their power to crush the religious spirit of the Inca. In a manual published early in the seventeenth century as a guide for missionaries, Pablo José de Arriaga offered simple advice for priests confronting idolatry. "Everything that is inflammable is burned at once," the priest wrote, "and the rest broken into pieces." Arriaga himself spent a year and a half in Peru during which time he alone was responsible for the destruction of 603 *huacas,* 3,418 household shrines, 617 ancestral mummies, and 189 field shrines. Yet, despite his zeal, even Arriaga realized the impossibility of his task. For it was not a shrine that the Indians worshiped, it was the land itself, the rivers and waterfalls, the rocky outcrops and mountain peaks, the rainbows and stars. Every time he or a fellow priest planted a cross on top of an ancient site, they merely confirmed in the eyes of the people the inherent sacredness of the place.

In the wake of the Conquest, when the last of the temples lay in ruins, the earth endured, the one religious icon that even the Spaniards could not destroy. Through the centuries the character of the relationship between the people and Pachamama has changed, but its fundamental importance has not. Some years after I first visited Cuzco, I spent a month engaged in ethnobotanical work with several colleagues in the small village of Chinchero, fifteen miles by road from the ancient capital and the site of the summer palace of Topa Inca Yupanqui. Though the flora was spectacular and the agricultural skills of these highland farmers nothing short of genius, what had impressed me most was the daily round, the accumulation of gestures that together spoke of an intimate and profound reverence for the very soil on which the village lay. The village, of course, was not merely the warren of adobe and thatch houses clustered around the small church. It was the totality of the existence of the people—the ancient ruins that ran away from the village and hung like memories to the edge of the cliffs overlooking the river, the fields cut into the precipitous slopes of their sacred mountain Antakillqa, the lakes on the pampa where the sedges grow, and the waterfall where no one went for fear of meeting Sirena, the malevolent spirit of the forest.

For the people of the village every activity was an affirmation of continuity. At dawn the first of the family to go outside formally greeted the sun. At night when a father stepped back across the threshold into the darkness of his small hut, he invariably removed his hat, whispered a prayer of thanksgiving, and lit a candle before greeting his family. In the morning before the labor in the fields began, there were always prayers and offerings of coca leaves for Pachamama. The men worked together in teams forged not only by blood but by reciprocal bonds of obligation and loyalty, social and ritual debts accumulated over lifetimes and generations, never spoken about and never forgotten. Sometime around midday the women and children would arrive with great steaming cauldrons of soup, baskets of potatoes, and flasks of *chicha.* The families feasted together every day, and in the wake of the meal the work became play, the boys and girls taking their place beside their fathers, planting, hoeing, weeding, harvesting. At the end of day the women scattered blossoms on the field, and the oldest man led the group in prayer, blessing the tools, the seed, the earth, and the children.

Every February at the height of the rainy season the fastest young man in the village dresses up as a woman and, pursued by virtually the entire population, races around the boundaries of the community's land. It is an astonishing physical feat. The distance traveled is only twenty miles, but the route crosses two soaring Andean ridges. The runners first drop one thousand feet below the village to the base of Antakillqa, then ascend four thousand feet to the summit of the mountain before descending to the valley on the far side, only to climb once more to reach the high pampa and the trail home. It is a race but also a pilgrimage, and the route is defined by sacred places, crossroads and rock cairns, waterfalls and trees where the participants must stop to make ritual offerings. Warmed by alcohol and fueled by coca leaves, the runners fall away into trance, emerging at the end of the day less as humans than spirit beings who have fought off their adversaries and have affirmed for yet another year the boundaries of their land. It is their way of defining their place, of proclaiming their sense of belonging.

The sun was high over the valley, and the first of the tour buses had arrived. Young guides hovered on the great stones or regaled tourists with standard accounts of tunnels leading from Cuzco to the ends of the world. The silence of dawn had been displaced by the whirl and clicking of cameras. On the path back to Cuzco I passed two young French travelers. Both were chewing coca, stuffing leaves into their mouths, like horses eating hay. I walked back to town and had breakfast at the Restaurante Azul, in a room hung with faded posters of the Beatles. Later, on my way to see the remains of the Coricancha, I dropped by the *pensión* where I found a note waiting for me from Tim, who had just arrived in Cuzco. Evidently his trip had gone well. The note read:

> Willi, I'm at the Hotel Montecarlo. Everything's great. Pogo sends a lick. P.S. I saw valleys full of coca and found it wild in the forest. I think I can trace it to its source. Timote

The full significance of what Tim had found did not come clear until the following day when, after a night in the bars of Cuzco, we drove out of town, once again destined for the lowlands. There are two main routes from Cuzco to the Sacred Valley of the Vilcanota. One drops two thousand feet in an hour to cross the river at the village of Pisac. The other road climbs to the northwest, crossing a high massif before falling to the town of Urubamba. We had decided on the latter and had driven in the rain over the pampa of Yanacona, past Chinchero and the lakes of Piuray and Huaypo.

"There were two lovers," Tim was saying. "One was a virgin of the Sun, the other a lowly peasant boy. When the Inca discovered that the young girl had broken her vows, he ordered the two of them buried alive, face up." He paused to light a cigarette. He had lost weight in the forest, and his skin was brown from the sun and wind.

"But when night fell all sorts of strange things happened. The stars moved, rivers dried up, and the fields became poisoned. Only the earth around those lovers remained fertile. So the priests demanded that their bodies be dug up and burned. But all they found was a pair of tubers."

"So those were the first potatoes."

"That's the myth. But it's interesting. They never say that they came from seeds. Nearly all the cultivated plants are like that—an image of the flesh being torn away and planted, the food coming from the body of the plant."

The Red Hotel passed over a height of land, and for an instant we seemed to float, suspended between the moment and a vista of unexpected scale and beauty. Across the surface of a vast upland plateau stretched the plains of Maras, fallow fields of gold and russet, broken by hedgerows of agave and patches of brown earth where oxen and ploughmen had been. In the distance, where the edge of the escarpment fell away to the Vilcanota River, mist hung over the valley and skirted the flanks of the mountains that rose on the other side, beyond the snows and into the clouds.

"What about coca?" I asked.

"That's the strange thing. The Indians say that the plant was discovered when the first mother lost her child. They mean, of course, the primordial mother, the essence of Pachamama. She is Maria, but she is also Mamacoca. In a sense she is both woman and the plant itself. The legend is pretty simple. Sad, lonely, stricken by grief, she wanders through the forest. Casually she tastes a leaf, only to find that it relieves her hunger and eases the pain. And that's probably what really happened. Some hunters on the move, hungry, perhaps lost, came upon a coca plant. The leaves are relatively tender, especially the new growth. Certainly edible, an obvious famine food."

"How do you know what it looked like?"

"What do you mean?"

"The actual plants. How do you know what they looked like? It was five thousand years ago."

"At least," Tim said. "That's the incredible thing. What I saw in the forest could have been wild. I still don't know if it was or if the seeds had just been cast there by birds. But it doesn't matter. The plant was thriving in the middle of the forest. You understand what I'm talking about."

"A cultivated plant able to survive on its own."

"More than that. A cultivated plant virtually unchanged for thousands of years. Compare it to potatoes or corn or any other domesticated plant. Most have been worked and reworked by humans for so long that they bear no resemblance whatsoever to their ancestors. Look at these fields. This was the birthplace of potatoes. Some of these farmers have one hundred different varieties growing in a single field, each a little different, each with a name. Each completely dependent on man."

"Unlike coca."

"Yes, unlike coca."

The implications were extraordinary. In the mountains of the southern Andes we were passing through one of the world's great centers of plant domestication, equal in significance to the Middle East or China. The development of potatoes began here some six thousand years ago when hunters and gatherers found that by eating certain clay soils they could detoxify the small poisonous tubers of a weedy annual. Over the centuries eight species of wild solanums were brought into cultivation, a feat of agricultural ingenuity that resulted in as many as three thousand distinct clones of potatoes, recognized today in the Andes by more than one thousand different native names. Through observation the people became masters of potato cultivation. They learned, for example, to rotate crops every seven years, a practice that confounded the early Spaniards who could not understand why farmers insisted on having but one field under production, with six lying fallow. Such a cycle was in fact essential to prevent potatoes from being destroyed by a common pest, a nematode whose cysts could survive in the ground for only six years.

Experimentation was ongoing both before and during the time of the Inca. Just to the west of the road that was carrying us toward the Sacred Valley were the ruins of Moray, three large circular depressions carefully excavated from limestone sinkholes and lined with perfectly formed agricultural terraces. They appear as three giant arenas sculpted from the earth. The largest is 100 feet deep, with a basal diameter of nearly 150 feet. The entire complex turns out to have been an agricultural research station, with each terrace reproducing the growing conditions of different ecological regions of the empire. Vertical stones marked the limits of afternoon sun at the equinoxes and solstices. Silver or gold plate may have lined the terraces to concentrate sunlight. Protected from the elements, the base of the massive excavation is consistently 15 degrees centigrade warmer than ground level, a temperature range equivalent to the average annual difference between London and Bombay. Each terrace thus corresponded to 3,000 feet of altitude, allowing the replication in miniature of twenty distinct ecological zones. Though Moray was built above 11,000 feet, it encapsulated the empire, enabling Inca officials to both anticipate yields from various regions and experiment with new crops.

Certainly potatoes were grown at Moray, and quite possibly coca. But while experimentation at this site and throughout the Andes had over the centuries utterly transformed the potato, what had happened to the most revered plant of the Inca? If Tim was right, if coca today existed in the wild or even if feral populations were able to survive and reproduce in the forest, it implied that artificial selection had barely affected the plant despite its long association with humans. That, in turn, would provide a vital clue to its geographical origins.

We knew there were at least three different cultivated varieties of *Erythroxylum.* One was the coca of Colombia, the Sierra Nevada, and the mountains of the Paez Indians. A second was found in the desert valleys of northern Peru, inland from the city of Trujillo. This variety we had seen growing in Lima at La Molina, the agricultural university. The third was known as Bolivian, but it referred to coca found along the eastern slope of the Andes in both that country and Peru. It was this variety that Tim had examined on his recent trip to the Apurimac and which we expected to find below Machu Picchu, beyond Quillabamba in the valleys of Urubamba-Vilcanota affluents.

Just by having seen the plants Tim knew that Colombian and Trujillo were more closely related to each other than they were to Bolivian, an observation later confirmed experimentally in the United States and Canada. This part of the puzzle was solved through two avenues of research. Chemical analysis revealed that Colombian and Trujillo shared certain compounds, including a rare flavonoid unknown in the coca of Bolivia. Second, breeding experiments showed that Colombian and Trujillo produced fertile hybrids, whereas Bolivian did not hybridize with Colombian, and when crossed with Trujillo, it yielded abnormal offspring that could not survive in nature. This eliminated the possibility that Trujillo coca, located geographically between the ranges of the other two varieties, was itself a simple hybrid.

Yet based on chemical evidence as well as the breeding experiments, Trujillo was clearly intermediate, leaving only two possibilities. One evolutionary scenario would have Trujillo giving rise to the other two, but this was unlikely. Trujillo is a true cultigen, grown in the hot desert under irrigation, utterly dependent on humans for its survival. Under the best of conditions its seeds, like those of all coca plants, are viable for only two weeks. In the hot sun they last barely a day. This trait alone suggested that coca had arisen in the moist forests of the *montaña.* When combined with Tim's discovery of wild or feral coca in the Apurimac, the evidence proved overwhelming. It was highly unlikely that a dependent cultigen could be the ancestor of a species that was still capable of surviving under natural conditions. According to Tim, Trujillo coca was the intermediate stage of an evolutionary sequence from Bolivian to Colombian.

This suggestion was confirmed by one further botanical observation. Plants, like people, have to worry about inbreeding, and they have various mechanisms to ensure they do not pollinate themselves. One means of encouraging cross-fertilization is to have the length of the styles—the elongated part of the female organ—vary relative to the length of the stamens, the male or pollen-bearing structure, within the flowers of a population. Technically, this adaptation is known as heterostyly. Both Trujillo and Bolivian coca are strongly heterostylous and do not self-pollinate. Colombian coca, by contrast, is self-compatible. In other words, a single shrub is fully capable of producing viable seeds on its own, a trait useful for the individual plant but less helpful in an evolutionary sense for the species. The breakdown or loss of the mechanism that prevents plants from fertilizing themselves is universally recognized in botany as a derived or more recent trait. Thus Colombian coca most certainly evolved after both Trujillo and Bolivian.

For Tim this was a revelation. Bolivian coca was clearly *Erythroxylum coca.* The coca of Colombia was certainly distinct and warranted recognition as a different species, *Erythroxylum novogranatense.* In order to acknowledge the close affinities of Trujillo and Colombian, he decided to describe them as varieties of the same species, *E. novogranatense* var. *novogranatense* and *E. novogranatense* var. *truxillense.* This may seem unimportant, but to Tim it was a defining moment, the endpoint of months of speculation and inquiry. It was not just an opportunity to position these remarkable species on an evolutionary tree. It was a chance to codify the past, to provide anthropologists and historians with a framework for understanding the domestication and diffusion of a sacred plant that had inspired every civilization of the Andes.

• • •

Rain had begun in earnest by the time we dropped into the valley of the Vilcanota, and in the town of Urubamba the market women had retreated beneath the spreading branches of an immense *pisonay* tree. We stopped for dinner—spicy trout, tamales, and frothy *chicha* served from great stoneware jugs—and then drove on through peach orchards and lush fields of maize to Ollantaytambo, the fortified temple that once guarded the northern end of the Sacred Valley from incursions from the lowlands beyond. Just outside of town we pulled over for the night and fell asleep with the roar of the river dominating all sounds.

By morning the weather had cleared, and we woke to bright sunlight on the cliffs and high mountain walls that enclose the narrow valley. The air was cold, fresh with the scent of eucalyptus and willow. Just beside our camp an old woman had kindled a twig fire and was waiting to sell us hot glasses of *ponche,* a delicious drink of toasted and ground *haba* beans mixed with water, sugar, cinnamon, and cloves. With her were three schoolgirls neatly dressed in gray uniforms and a young shepherd boy in a vermilion poncho and trousers of coarse wool. The girls scattered at the sight of Pogo and then returned, still giggling, to hover over Tim as he began to prepare our collections from the day before.

The children appeared to know everything about plants and were somewhat taken aback by our ignorance. Shyly at first and then in great bursts of enthusiasm they explained that plants were like people, each with its own mood and story. Cacti sleep by night. Mushrooms grow when they hear thunder, lichens only in the presence of human voices. The solitary blossoms of the open field have no feelings for others. Delicate gentians fold up their petals in shame. The late-blooming plants of the hedgerows are simply lazy, unwilling to work for the good of the community. All plants have names and are useful, the young shepherd remarked, but his personal favorites were found along the banks of the river: lemon grass, mint, and *mora,* the wild blackberry of the Andes.

For most of the day Tim and I explored the ruins, climbing the monumental terraces that rise toward the temple of the sun and later crossing the Patacancha River to scramble up the slope of Pinkuylluna, the steep height of land that hovers over the village. Despite the recent rains, the earth was dry and loose, dusty by noon. The mountain slope had the aspect of a desert garden, with great cascades of Spanish moss, flowering stalks of puya and aloe, and cacti of many species growing alongside bunch grass, lycopodiums, and hardy orchids. We climbed as far as the most prominent ruin, an ancient barracks or prison by local accounts, perched above a precipice. There we rested in the shade of its immense walls. From such a vantage the valley floor appeared as an oasis, winding like a serpent between the barren flanks of the mountains. The strategic significance of the town was obvious. To the east the valley widens toward Pisac and the approaches to Cuzco. To the west the Vilcanota passes into a series of increasingly precipitous gorges, eventually running beneath Machu Picchu and reaching the lowland forests of Quillabamba. Any army moving up or down the valley would have to take Ollantaytambo, a lesson that was not lost on Manco Inca, the prince who raised the standard of revolt in 1536 and nearly succeeded in crushing the Spaniards at Cuzco and driving them from Peru.

The uprising initially had gone well. The Inca army occupied Sacsayhuaman, laid siege to Cuzco, and set much of the city afire. In a desperate assault Spanish cavalry broke through and mounted the heights to attack the fortress. A ferocious battle ensued and thousands died, including 1,500 native troops slaughtered in captivity by the Spaniards. Once the fortress fell to his enemies, Manco Inca released his levies for the annual harvest and retired with the nucleus of his army to Ollantaytambo. Eager to defeat the rebellion before the native armies regrouped, Hernando Pizarro marched on the temple fortress but was repulsed with heavy losses. The following year, however, an influx of reinforcements allowed the Spanish to reestablish control of the Andes, and Manco Inca was forced to abandon Ollantaytambo and retreat down the valley into the remote mountain wilderness of Vilcabamba. There, at the fortress city of Vitcos and on the slopes of Salcantay and hundreds of other jagged peaks that reach west as far as the Apurimac, he established a new Inca state, a base from which he planned to wage guerrilla war until his forces were strong enough to reconquer the Americas. Thus began the astonishing saga of resistance and betrayal that marked the final end of the Inca lineage.

The first blow came in July 1537 when a Spanish force surprised the Inca at Vitcos. Manco escaped, carried in the arms of five of his swiftest runners. The Spaniards captured twenty thousand of his followers, including his wives and children, as well as the sacred mummies of his ancestors, the holy relics of the temple, and vast herds of llamas and alpacas. When the Inca refused to surrender, Francisco Pizarro had the Inca's principal wife and sister-queen stripped and flogged at Ollantaytambo. Riddled with arrows, her naked body was bound to a raft and floated down the Vilcanota past the eyes of the rebels of Vilcabamba.

Manco fought on, taking advantage of the civil war that broke out among the Spaniards, building the strength of his own forces until his influence reached well beyond the Vilcabamba. Then in 1542, following the victory of the Pizarro brothers over the Almagro faction, he made the fatal mistake of providing refuge for six fugitives from the defeated side. For three years these Spaniards found a sanctuary at Vitcos. In 1545 they repaid Manco Inca by murdering him during a game of bowls. The Inca was succeeded by three sons. One ruled until 1560, essentially as a Spanish puppet. The second, Titu Cusi, is best known for the memoir he dictated, the only record of the Conquest as seen through the eyes of the Inca. The third son, Tupac Amarú, inherited his father's spirit, and waged war in the Vilcabamba. In 1572 he, too, was betrayed and captured, alone and unaided, on a riverbank in the lowland forests far beyond the homeland of his ancestors. Taken to Cuzco, he was beheaded in the public square, and with his death came the end of the Vilcabamba resistance.

In time the exploits of Manco Inca and Tupac Amarú drifted into legend, and the remote lands where they found refuge became infused with mystery. Tales were told of a lost capital, a city never visited by Spaniards, with storehouses still full of gold and Inca treasures. The location of this fabled city captivated the imaginations of many explorers and adventurers, not the least of whom was Hiram Bingham, the Yale historian who, in a remarkable two-week period in July 1911, discovered not only Machu Picchu but also the ruins of Vitcos and the white rock of Yurac Rumi, site of Chuquipalta, the principal shrine of the Vilcabamba. Continuing deeper into the mountains and dropping to the lowlands beyond, Bingham made a fourth major discovery that summer: the overgrown ruins of Espíritu Pampa. Unfortunately, by then the expedition was running out of food. Bingham made only a cursory examination of the site and never appreciated the significance of what he had found. Until his death he believed that Machu Picchu, perched on a rocky saddle at nine thousand feet, high in the cloud forest above the Vilcanota, was the lost capital of Manco Inca. He was wrong. The actual site was indeed Espíritu Pampa, a fact not finally confirmed until a series of expeditions in the mid-1960s.

It was a plant that drew me to the Vilcabamba, albeit in an indirect way, and long after Tim and I had stared into the mountains from the slope above Ollantaytambo. In 1976 a native collector told Tim of a curious liana he had seen used by Campa Indians in the Chanchamayo valley of eastern Peru. Known as *chamairo,* the bark was reportedly added to coca leaves to sweeten the quid. Unfortunately, no specimens were collected, and Tim was unable to find any reference to the plant in the botanical literature. Later that year while examining coca paraphernalia at the Ethnografiska Museum in Göteborg, Sweden, he stumbled upon an unidentified piece of bark collected from the Campa by the great ethnographer Erland Nordenskjöld in 1922. Described as a coca admixture and labeled *yarnayru,* it was almost certainly *chamairo.*

Returning to Peru in 1978, Tim rummaged the markets of Lima and found *chamairo* being sold by an herb dealer. The bark was fibrous, tough in texture, and reddish brown in color. Astringent and extremely bitter, it seemed the opposite of a sweetening agent. A subsequent herbarium search at the Field Museum in Chicago, then the world's largest collection of Peruvian plants, uncovered but a single specimen, collected by an anthropologist more than a decade earlier and identified as *Mussatia hyacinthina* of the Bignoniaceae. Eager to secure further specimens as well as bulk collections for chemical analysis, Tim had urged me to keep an eye out for the liana during my travels in the lowlands.

The spring of 1981 found me living among the Chimane, a riverine people who inhabit the gallery forests and sandbars of the Río Maniqui in the remote eastern foothills of the Bolivian Andes. The Chimane use coca, which they call *sa'si,* mixing the entire leaves with ash derived exclusively from the burned spathe of a common palm and small pieces of stringy bark that turned out to be *chamairo.* Alone, the bark was bitter, as Tim had reported, but taken in combination with the ash, the effect was quite the opposite. It was as if a teaspoon of sugar had been added to the quid.

Chamairo did not grow in the Chimane homeland. They acquired it from wandering merchants, and to secure voucher specimens I had to retrace the trade routes to the source. After a month in the forest I traveled by dugout canoe five days down the Maniqui to the outpost of San Borja, where I hired a small plane to fly to Rurrenabaque, a settlement on the Río Beni. There I contacted a merchant who traded with the Chimane and obtained both bulk samples of the bark and a great deal of information concerning the range and habits of the plant. For another week I followed various leads until finally, on a Sunday, I found it growing outside a Tacana Indian village some fifty miles into the forest from Rurrenabaque. It was indeed *Mussatia hyacinthina.*

A month later, having journeyed up the Río Beni by river launch and crossed back over the Andes to La Paz, I returned north to Cuzco where I ran into an extraordinary American by the name of John Tichenor. A keen botanist and superb white-water guide, Tichenor had pioneered routes down most of the major rivers of the Andes, including the formidable Apurimac, headwaters of the Amazon. His favorite trick was to strap collapsible rafts onto the backs of mules, traverse mountain ranges, and put in on rivers deemed by everyone else to be unnavigable. The potential for plant exploration was obvious. I had always wanted to see the dry intermontane basin of the upper Apurimac and follow the river as it fell through the cloud forest to the lowlands. John appreciated the challenge and offered to guide me and a small party across the Vilcabamba.

After crossing the Vilcanota at Chaullay, we traced Bingham's route up the Río Vilcabamba, passing over short stretches of Inca road until reaching Vitcos and the sacred stone of Yurac-rumi. From there Bingham had turned north and west for Espíritu Pampa; we continued south, crossing two high passes to reach the headwaters of the Río Choquetira, which flowed to the Mapillo, an affluent of the Apurimac. After following these drainages for five days we reached a broken-down hacienda isolated above the dusty canyon of the headwaters of the Amazon. The next morning we put into the river, and for another week John led us through rapids so spirited that at times the rafts folded in half, with paddlers in the bow knocking heads with those in the stern, above those in the middle.

One morning by the river, just as I was preparing a chew of coca, three Indians walked into our camp. Noticing my small bag of leaves, one of them suggested that for a really sweet quid I ought to try a piece of *chamairo.* I was astonished, for I had not expected to find the plant in this part of Peru. When I inquired of its whereabouts, the oldest of the three pointed to a high ridge that disappeared beyond the horizon. There was no chance of following him to the source, so I asked whether he might gather the plant and meet us downstream. A simple task, he replied, and we agreed to meet in two days at a hacienda just above the jungle clearing of Osambre, four days by raft above our destination of San Francisco. Much to my astonishment the old man was waiting there when we arrived, with an enormous bundle of leaves and stems in his arms. It was indeed *chamairo,* but a quick examination left little doubt that it was a different species, and one that turned out to be new to science.

Encouraged by this discovery, John and I returned to Peru two years later and followed the Vilcanota River to the forested lowlands of the Machiguenga Indians. There we also found the curious plant, and in doing so passed through the Pongo de Mainique, the rocky gorge that marks the last range of the Andes.

• • •

I had wanted to linger a few days at Ollantaytambo, but Tim was anxious for the lowlands, and understandably so. In 1963 a botanist had estimated that the lower valleys of the Vilcanota supported some 80 million coca bushes. Production that year reached nearly 7 million pounds, and this at a time when there was no illicit market for cocaine. By 1975 demand had soared, and the lands around Quillabamba had become one of the most notorious coca-growing regions in the world. Once we left Ollantaytambo, it was not long before we saw the first plantings.

The road rose out of the valley and climbed past glaciers to a high pass before falling away to cloud forest and the upper reaches of the Río Santa Maria, an affluent of the Vilcanota, or the Urubamba as it is known below Ollantaytambo. Tea plantations covered the higher slopes, but below six thousand feet coca grew in small plots or as solitary bushes planted in gardens. The first extensive fields began three thousand feet lower and stretched along the valley floor as far as Chaullay where the Santa Maria flows into the Urubamba. From there the drainage narrows, and no other plantations are seen until Quillabamba. At that point the entire basin is alive with plant stimulants: tea up above, coffee and chocolate interplanted with coca in the hot bottomlands, coca alone in neat terraced rows and vast plantings that brighten the mid-elevations with a delicate hue of light green.

For several days Tim and I explored the lush valleys that converge at Quillabamba. In remnant patches of forest and along streambeds and hedgerows, we searched for feral populations of coca, finding as expected that the cultivated plants escaped the plantations with ease, dispersed by birds attracted to the bright red fruits. We also came upon several species of wild *Erythroxylum,* growing alongside or even within the plantations. Known as *monte coca* or *coca coca,* these were occasionally used to adulterate shipments of coca destined for the markets of Cuzco. One species, *Erythroxylum raimundii,* grew into an impressive tree more than twenty feet high, with thick dark green leaves and beautiful red fruits. We first found it on a steep hillside at twilight when we paused to chew coca. At first we took no notice. Then suddenly I laughed, realizing that we were sitting at the edge of an entire grove of the plant, a rare species and one that we had been seeking for some time.

Naturally we visited plantations, examining nurseries and seedbeds, collecting living material for propagation, studying the weeds and pests of the open fields, all the while sampling the finest coca grown in Peru. The farmers seemed delighted to sit with us on the stony ground, sharing leaves as they described the difficulties of their work. They propagated coca always by seed, harvesting the fruits just before maturity and planting them beneath a light covering of soil in fertile beds protected and shaded from the sun. The young seedlings are gradually exposed to increasing amounts of sunlight, and after four months are transplanted to the fields. After three years a plantation yields a small harvest of choice, delicate leaves that are ritualistically distributed to family and friends. The second harvest is somewhat larger, but it is not until the sixth year that maximum production is achieved. Although yields tend to fall off after a decade, a plantation once established may remain productive for forty years.

After only a week in the fields listening to the farmers, we realized that the crop substitution program, a key component of the American effort to eradicate coca, was a fantasy. Harvested by hand every four months or as often as six times a year if fertilized and sprayed, coca outperformed every other agricultural product in the valley, earning eight times as much per acre as coffee, twenty-five times more than cacao. Ideally adapted to the poor, well-drained, and highly eroded slopes of the *montaña,* and suffering from few natural pests or predators, it flourished where most other cultivated plants could not even grow. At a hacienda in the upper reaches of the Santa Ana valley, we asked a group of farmers whether it might be possible to substitute other crops for coca. The men laughed aloud and then inquired who would possibly want to do so. One of them bent to the ground and grabbed a handful of dry soil. Letting the dirt fall between his fingers, he said, *"Es imposible. ¿Qué es lo que podemos sembrar en este suelo cansado?"* What else could we possibly plant in this tired earth?

Tim had learned what he needed to know, and the pull of the lowlands, with the promise of new plants, took us away from Quillabamba, farther down the Urubamba to Sahuayacu and the valley of the Río Chalpimayo. There we stayed for another week in an orchard of mango trees, exploring the forest for medicinal plants and ornamentals before finally taking our leave and returning to the mountains, climbing in a few hours from dense jungle to a fifteen-thousand-foot pass dappled with ice, heading south for Bolivia.

CHAPTER FOURTEEN

One River

THERE COMES A moment in a journey when the allure of the unknown fades and memories of home suddenly appear exotic. Tim and I slipped over this threshold in highland Bolivia. After a final excursion to the coca fields of the Yungas of La Paz and beyond the Coroico Gorge to the lowland forests of the Alto Beni, I had been on the road for fifteen months. With the exception of one trip to Boston to confer with Schultes, Tim had been away for over a year. The Red Hotel had traveled thousands of miles back and forth across the Andes, and we had collected over three thousand plants—altogether some ten thousand individual specimens—in addition to hundreds of living collections, rhizomes, cuttings, tubers, and seeds. Both of us had plans that would keep us in South America for another month or two. Tim was to return up the coast to examine the coca fields at Trujillo. My goal was the Amazon and a slow passage north to Colombia. But as we drove out of La Paz heading for Peru, we could sense the pull of home.

From Titicaca we crossed the mountains to the sea, passing salt flats colored with wild flamingos and dropping into the shadow of El Misti, an immense snowcapped volcano. At dawn we slipped through Arequipa, an old Inca town transformed by the Spaniards and built entirely with pearly white stone, and continued west to Camaná and the coast. From there we sped north. It was the end of summer, and light fog hung over the desert, obscuring the sky and darkening the cliffs that ran along the shore. By evening we were drinking beer in Chala, a small fishing village that once sent runners into the mountains to Cuzco to deliver fresh fish for the table of the Inca. We slept by the roadside on the hard sand and woke before light. By early morning we were two hundred miles north, having passed through the ancient Nazca lines and paused for just a few minutes to rest in the orange groves of Palpas. Just south of Lima we pulled off the road once more. A flight of seabirds brought us to the shore, and we came upon a rocky bluff that sheltered an isolated cove, a crescent beach of white sand curled around a small island speckled with guano.

It was noon, and the sun had burned off the sea mist. We waited for an hour or two until the light softened somewhat, and then we ate several handfuls of the dried flesh of a cactus we had found in the mountains of Bolivia. It was a wild relative of *huachuma,* or San Pedro, the Cactus of the Four Winds, a magical plant rich in mescaline used by *curanderos* on the northern coast of Peru. The species we had collected, *Trichocereus bridgesii,* had never been reported as a hallucinogen, but an old Aymara woman we had met on the altiplano had referred to it as *achuma* and said that it made one drunk with visions.

We sat quietly on the sand, watching the waves crash on the shore. Both of us soon felt queasy, a faint trace of nausea that could have been the first sign of intoxication or the onset of poisoning. This uncertainty slowly gave way to an unmistakable sensation, a warmth in the belly, a faint intimation of what lies ahead. The wind breathing in the air, a bird in flight, silent and composed. The waves falling on the sand, spitting up white froth. Suddenly the wave was within, the ebb and flow, pulsing, moving physically into the body.

We stood up and walked along the beach toward a headland where the ocean swell broke over a series of tidal pools, alive with starfish, urchins, and crabs. A surge of energy carried us up the face of a cliff, bare feet touching rock, dark stones bursting into blossoms. The wind blowing off the sea lifted us onto a broad promontory beyond which lay the entire desert. Every gesture brought forth a reaction in the dry air, a wave of color that ran away to the horizon. The air took form, became tactile. It was like swimming in a pool of soft pastel light. I turned and saw Tim silhouetted against the sea. Overhead was the confusion of a dazzling sky. Around the sun were figures flying in circles, creatures with red breasts, blue serpents for hair, and eyes like saucers of light spinning in tighter and tighter circles. A vortex of memory. A song, a luminous spiral. Tim's voice, a vision of blinding light, a tapestry of pearls embroidered with gold and silver thread, a blanket to rest on.

The sky opened. A dome of the deepest blue gave way to black with small crystals of light flaring on all sides. I looked down and saw the brown earth receding. We were caught on the wings of birds, passing through space, through emptiness, over lands of purple sand and rivers of glass running to the sea. From the desert shapes emerged, castles and temples, enormous lizards draped over dunes, totemic figures etched onto the sand, a mere semblance of known things. Flying along the wild face of mountains, in the wind the touch of clouds on feathers. They were our feathers, sprung from the skin. The eye of a hawk. The beauty of water carving veins in the earth. The wind carrying us away into the night sky and beyond the scattered stars. Nothing to fear.

Suddenly, a distant voice came from far below. A well of darkness. The pale face of a smiling child. I turned and saw a raptor arched across a morning sky, flying on, its beak aimed at the center of the sun. There was no sound, just the image of a soaring bird heading for oblivion. And then a slow spiraling descent that seemed to draw the earth to my feet. And once again the ground. Slowly I stood up and walked to the point overlooking the sea. The sun was down. Hours had passed. I looked back and saw Tim sitting on a stone at the center of a pool of velvet light. Pogo darted in and out of shadow. We had no idea where we were, and for a moment Tim seemed uncertain.

"Are you okay?" I asked. The words came awkwardly. We both laughed.

"Did you see the sun?" Tim said.

"Yeah."

In that moment all things seemed possible. A collective vision, movement through time and space, metamorphosis—these were no more illusory or wondrous than the beauty of a dry blade of grass sprouting from beneath a rock in that barren desert.

Tim followed me to the edge of the bluff. We walked blindly but managed to find a way down the rocks. The moon had yet to rise, and beyond the far side of the cove, headlights passed in the darkness. There were ships landing, small parties of men delivering contraband. We rested on the beach and then continued up the shore, retracing our steps until we came upon a strange sight: an enormous whale bone stranded in the sand. I looked back and saw waves of color flowing from Tim's brow. I walked on, climbing slowly up a steep slope that rose from the sea. Pogo darted ahead, chasing a fox. I paused at the crest of the hill and waited. To the east the mountains lay bare, a dark ridge on the night horizon. Overhead stretched the Milky Way. And here below, parked on the desert floor, was the Red Hotel.

The moon had turned slowly through the sky, and the desert was coming alive. Colors softened. The light changed, and silver traces flashed by in the rough wind. The surge of waves on the shore, the deep breath of exhaustion. We collapsed on the sand, with Pogo hovering by our side. Clouds rolled by, time was suspended. The smugglers worked through the night. Foxes yelped, the odd bird cried. Gradually the eastern sky lightened, and the first intimations of dawn took us both by surprise.

"Listen to that," Tim said.

There was a low drone in the earth, deep, unmistakable. An impulse, resonant and complete.

"That is the sound of life," he said. "I'm not speaking in metaphor. I mean the actual sound of life. The tone of energy within your cells."

The dawn was fully upon us, and the clouds to the east took on a luminous tone in an empty sky. Every color of the sunset returned in an infinitely more subtle hue. A great rolling wall of mist swept over the shore, and by the time it dissipated, there were fishermen on the beach, combing the surf for bait.

Two days later Tim and I parted ways in Lima. I flew to Pucallpa, a jungle town on the upper Ucayali, a place so dismal that within hours of arriving I had booked passage to Iquitos, a larger city five hundred miles downriver. Iquitos, as it turned out, held me for nearly three weeks. First I ran into Adriana de Vaughn, an old friend of Tim's who still worked for what was left of the Amazon Natural Drug Company, the outfit that had sponsored Tim and Dick Martin in the late sixties. Adriana led me to Fernando Tina, a Yagua Indian described by Tim as the finest botanist he had ever known. I found Fernando selling *empanadas* in front of the Cine Atlantico. For ten days he astonished me with his knowledge of the forest. From the scent of bark alone Fernando had constructed a system of plant classification every bit as complex as anything I had been taught at Harvard. On the banks of the Nanay and Itaya rivers he identified ferns used to treat whooping cough, orchids effective in the treatment of boils, and a rare variegated calathea known as *tigrepanga* employed as an admixture to *ayahuasca.* In the forest we lived on melastom fruits, medicinal teas brewed from *sacha ajo,* the garlic vine, and *chicha* made from the fruits of *miriti* and other palms. At night we drank *chuchuhuasi* and rum while Fernando regaled me with stories of anacondas the width of rivers and electric eels that had lit up the primordial night sky.

From the floating market of Belém we motored up the main channel of the Amazon and for a week explored ox bow lakes and stagnant back eddies searching for palm fruits and ripe seed of the giant water lily *Victoria amazonica.* This beautiful plant, with its enormous leaves capable of supporting the weight of a small child, grows in side channels and standing bodies of water throughout much of the Amazonian floodplain. One evening we slept by the side of a swamp, just to watch the flowering of the lotuslike blossoms. Precisely at sunset, from every corner of the swamp, massive flower buds rose slowly above the surface of the water. Triggered by the falling light, the flowers, easily the size of a human face, opened with a speed that could be readily seen. The brilliant white petals stood erect, and the flower's fragrance, which had been growing in strength since early afternoon, reached a peak of intensity.

As Fernando gazed over the water, his eyes falling on a pair of *jacanas* skipping over the surface of a leaf, I explained that the same process that generates the scent also increases the temperature of the central part of the blossom by exactly 11 degrees Celsius above whatever the air temperature happens to be. The combination of color, odor, and heat attracts a swarm of beetles, which converges on the center of the flower. As night falls and temperatures cool, the flowers begin to close, trapping the beetles within the blossom. By two in the morning the flower temperature has dropped, and the petals begin to turn pink. By dawn the flowers are completely closed, and they remain so for the rest of the day. In the early afternoon the outer sepals and petals alone open. By then a deep shade of reddish purple, they warn other beetles to stay away. Last night's beetles, meanwhile, remain trapped in the inner cavity of the flower. Then, just before dusk, the male anthers release pollen, and the beetles, sticky with the juice of the flower and once again hungry, are finally allowed to go. In their haste to find yet another opening bloom with its generous offering of food, the beetles dash by the anthers and become covered with pollen, which they then carry to the stigma of another flower, thus pollinating the ovaries.

Fernando listened to this explanation without comment, accepting it easily, as if it was just another story. In the morning all words were lost in the mundane task of wading through the spiny petioles of the leaves and retrieving the ripe fruits found only on the bottom of the swamp. It was miserable work made pleasant by Fernando's laughter. With each cautious step he joked of yet another hazard of the murky waters: stingrays, piranhas, snakes, and catfish that chew holes directly through the flanks of their victims. For the rest of the day, as we rode a river launch back to Iquitos, he teased me about another fish, the needle-sized candiru, a parasitic catfish attracted to the scent of urine and known to lodge itself with painful spines inside a man's urethra.

After twenty days on the rivers around Iquitos, the thought of a slow passage to Leticia lost its appeal, and when time came to leave, I decided to fly. Fernando and his family saw me off at the airport. From the small window I waved to his young girls. Dressed in bright calico, with pink and blue ribbons in their black hair, they fluttered at his side. As the battered DC-3 rose above the forest and passed over the serpentine flow of the Amazon, I thought of Fernando returning to his *empanada* business. Within a few years he died of leprosy, and his knowledge of plants was lost.

The river spread beneath the plane, and from the air one could sense its pulse. The water did not flow but eased its way through the forest, as if responding to a distant tremor. At the confluence of the Río Napo, just minutes by air downstream from Iquitos, the Amazon absorbed the rivers of Ecuador. The scale was daunting. When Francisco Orellana sailed down the Napo in 1541 and reached the Amazon, he went temporarily insane. Coming as he did from the parched landscape of Spain, he could not conceive that a river on God's earth could be so enormous. Little did he know what awaited him two thousand miles downstream where the river becomes a sea and the riverbanks, such as they are, lie one hundred miles apart.

There was a sudden shudder off the right-hand side of the plane, and I looked up to see the gray edge of a tropical storm, hung like a sheet over the basin. Smoke was pouring from the engine, and the propeller had stopped. Buffeted by winds and lashed by rain, the plane lumbered along, stranded over the forest. With each gust the fuselage shook. Passengers screamed. Luggage broke loose. A chicken darted up the aisle of the plane. When at last the port city of Leticia came into view, the pilot banked sharply at high speed. Twice he overshot the runway.

On his third approach, with the left wing not two hundred feet above the canopy of the forest and the plane tilted at 45 degrees, he landed halfway up the slippery, rain-soaked field. Sirens wailed, and a fire crew raced for the plane. Fortunately, the pilot had shut down the engine in time. Only later did we learn that on his first approach he had come within one hundred feet of colliding with the observation tower.

Leticia offered no temptations. Having paid Fernando for his help, I was nearly broke. When I learned that a cargo flight was leaving the next day and that a few dollars would land me in the copilot's seat, I made immediate plans for Bogotá. That night I slept on the floor of a shack and the next dawn rode another DC-3 over the forests and plains that reach to the edge of the Andes. Sharing the flight were a priest and two nuns, who sat in the back on top of a load of dried fish. I was certain that the Colombian capital was within reach and that in a few days I would be back in Boston. What I didn't know was that Boston had come to Bogotá and that waiting at the Canadian embassy was a letter from Professor Schultes letting me know that he would be in Bogotá on the day of my arrival.

I found him that evening at The Halifax, the most recent incarnation of Mrs. Gaul's Pensión Inglesa, the boardinghouse that had been Schultes's home in Bogotá for twelve years. Mrs. Gaul was long gone, and the current owner, Joan Hodson, another Yorkshire woman, had moved uptown to Calle 93. But the essence of the place remained unchanged. It was simple, comfortable, stodgy, an island of English reserve in a city perpetually on the edge of disaster. I felt a little self-conscious ringing the bell. I was staying at the Hotel Kayser, a dive in La Candelaria, and I had not called in advance. I knew that Schultes hated telephones. When he first came back from South America, a journalist asked what he had found most difficult about adapting to life in Boston. "Parking meters, appointments, meetings, and telephones," he answered. "In the Amazon, time has no meaning."

A maid welcomed me at the door, and I followed her past a circular staircase into the dining room, where Schultes sat at the head of a long table shared with three other foreigners. I smiled to see his red tie, blazer, and the same rimless bifocals that had first greeted me at Harvard so many months before.

"Hello, Professor." He looked up and spoke as if I had been expected.

"Wade. How good to see you. Come and let me introduce you to these people."

I reached across the table and shook hands with a Swedish couple and a young Scottish botanist named Brinsley Burbidge.

"You must be starving," Schultes said. "Sit down and eat. Have you ever seen anything like it? Roast beef and Yorkshire pudding. Shepherd's pie. In Bogotá, of all places. It's wonderful."

I filled my plate. After months of yuca and plantains, the food was delicious.

"What brings you to Bogotá?" the young woman asked.

"I'm just passing through on my way back to Boston."

"He's a lumberjack from British Columbia who has walked the Darien Gap," Schultes explained. The woman looked puzzled.

"And you?" I asked.

"We're adopting a baby."

"The paperwork is horrendous," her husband said. "We've been here over a month."

I felt Schultes's hand on my arm. The others noticed and slipped off into another conversation.

"Have you heard from Tim?" he asked.

"No, I was out of touch."

"He's going to be all right, but he's been very sick."

"What do you mean?"

"Hepatitis. Serum hepatitis. Usually one catches it from dirty needles. Tim hasn't . . . he doesn't—"

"No, of course not," I interrupted.

"I hardly thought so. The good news is that he'll quit smoking. With hepatitis you lose your taste for tobacco."

"But how is he? How did he get it?"

"You tell me. The incubation period is about ninety days. Think back."

"I was with him the whole time except for the last month in Lima. We were around Cuzco for six weeks, Quillabamba and the lower Vilcanota. Then there was Bolivia, Titicaca, and the Yungas of La Paz. At least a month."

"Did you try the coca?"

"Yes, we did."

"Excellent, isn't it?"

"Yes, it is."

"Perhaps my favorite. Now, go on."

"Coroico and Coripata for a week or two. And later we— Oh, my God. The roadblock."

Leaving Cuzco for Bolivia, I remembered, we had dropped over the

Andes to search for wild coca in the lowland forests of Madre de Dios. There had been a serious outbreak of yellow fever, and officials from the Peruvian Ministry of Health had set up a roadblock near Shintuyu. Only those who could prove that they had been inoculated were allowed to proceed. I had my papers, but Tim had left his medical certificate in Cuzco. His only option was to accept an inoculation on the spot with a needle that looked and felt as though it had been used as a plow.

"A foolish mistake," Schultes said casually. "I always carried my own."

"What do you mean?" I asked.

"My own syringe. I always carried my own equipment. That way you could be sure. Did I ever tell you about that time in Miami?"

"With the customs agent?"

"Yes. You know that story?"

"I think I remember. But go ahead."

"That son of a bitch." He began to laugh, and his laughter drew in the others at the table.

It was one of his favorite tales, and I enjoyed watching him tell it. On his way through Miami in 1948, sick with malaria and beriberi, barely able to walk, he had been in no mood to indulge an officious customs agent. Miami at that time was a small airport, and Schultes knew all the authorities, including Mr. O'Brien, an Irishman from East Boston who ran the place. But in the year he had been away exploring the Río Negro, O'Brien had retired and a number of new agents had been hired, including the one whose misfortune it was to start rummaging through Schultes's luggage. In a shoe the agent found a suspicious-looking tin box, which he opened.

"What's this?" he demanded. Schultes leaned over on his two canes.

"It looks to me like a hypodermic syringe."

"Of course it is."

"Then why did you ask me?"

"What do you use it for?"

"I take injections with it."

"What kind of—" Before the agent could finish the sentence, Schultes cut him off, ordered him to close the bag, and demanded to speak with his superior. Within minutes Mr. Lomes appeared, O'Brien's replacement as head of the airport.

"Hello, Doc. I've been looking forward to meeting you. O'Brien told me all about your work. How can I help?"

"Just by letting me know," Schultes replied, "why my tax money is supporting an employee who doesn't recognize a hypodermic syringe. If you want to know if I am a drug addict, for God's sake, just ask."

"I wouldn't try that today," I said as he wound up the story.

"No." He laughed. "It's a different time."

After dinner we moved into the study and sat by the fireplace to have a drink. Professor Schultes poured from his own bottle, a fine Colombian rum that we took on ice. For a while he continued to reminisce in an idle way—amusing stories and anecdotes that for the most part I had heard before. He was not by nature a modest man, but he never spoke about his achievements, and as for adventures, he maintained that he had never known one. If Schultes had contempt for anything, it was for latter-day explorers who court disaster as an end in itself. He never took deliberate risks, and what others saw as moments of peril and daring, he viewed as simple impediments to his work, inconvenient yet unavoidable. He lived in a world of plants. For those around him, especially his students, this nonchalance was mesmerizing.

"I do understand your wanting to get back to Boston," he said. "I always did after a year away, especially in the winter. I always missed the winter. But there is one thing I would like you to do before you go."

"What is it?" I asked.

"Well, Tim seems to have done quite well with coca."

"I think he has worked it out."

"Yes, I'm sure he has. But there is one final issue. Amazonian coca. Tim is going to need some living material, and he won't be able to get it himself. So I'd like you to go to Mitú. It's a lovely town. You can give my regards to the Monsignor and then find one of the Cubeo boys to take you up the Vaupés. You can be back here in a month."

"Okay," I said, forgetting for the moment that I barely had enough money to get back to my hotel. Then I remembered.

"Professor, there is one slight complication. I'm a little short of—"

"Don't worry," he interrupted. "I think I can tide you over."

"That would be great."

"Good. Now. A couple of things. The Tanimukas flavor their coca by blowing balsamic incense through the warm ash. It's very good. Keep your eye out for it."

"What's the source plant?"

"The Colombians call it *pergamín,* or *brea.* It's the resin of *Protium heptaphyllum.* In Tanimuka the name is *hee-ta-ma-ká."*

"And what else?" I asked.

"Pardon me?"

"You said there were a couple of things."

"Yes, of course. What was it?" Schultes paused and took a sip from his drink. His eyes brightened. "I remember now. Have you tried *yagé?"*

"Yes." I laughed.

"How was it?"

"Pretty awful."

"I just saw colors."

"I just threw up."

"That often happens. Perhaps you should try it again. In fact, perhaps you should spend a little more time in the Vaupés. There is one river I would like you to see—the Piraparaná, the river of the Barasanas and Makunas. Wonderful people. You'll find a way to get there."

Three days after meeting Schultes, and having passed over the Andes by road, I found myself flying southeast from Villavicencio to Mitú. It was an empty cargo plane, yet another DC-3, only this time without a door. Fortunately, the weather was fair, with little turbulence, and I was able to spend most of the flight perched beside the opening in the fuselage, a few thousand feet above the undulating earth. It was like riding the back of a pickup truck through the sky. There were great clouds on the horizon, and on the ground the roads and ranches of the *llanos* soon gave way to open grasslands with no sign of human habitation, a vast expanse of savannah etched with streams and rivers, all flowing toward the east. By the time the plane passed over the Río Guaviare the gallery forests of the river margins had spread out and coalesced, and the land had disappeared beneath an unbroken cover of jungle. Only the rivers remained to remind anyone living on the ground of a world beyond the canopy.

Of the people who lived in that forest, I knew only what I had learned from Schultes and what I had managed to pick up over the last days in Villavicencio. Before we parted in Bogotá, Schultes had given me a letter of introduction to an old friend of his, Dr. Frederico Medem, a Latvian nobleman who had fled the Russian revolution, settled in Colombia, and become one of the world authorities on Amazonian snakes and crocodilians. In Villavicencio, Medem welcomed me warmly and for two days shared both his memories of Schultes and his profound knowledge of the Northwest Amazon.

He was an enormous man, of impeccable grace and dignity, with a delicate goatee and the thick hands of one who had spent his life in the forest. His home resembled the quarters of an old rubber trader. It had wooden floors and a tin roof, an open verandah hung with hammocks, and walls decorated with jaguar and bushmaster skins. In every room were rare collections, casually displayed: exquisitely carved medicinal curing sticks from Bahia Solano, breastplates from the Kuna, snuff tubes from the Orinoco, the skull of an eighteen-foot caiman killed on expedition in 1956. His office had walls of books and rare manuscripts, wooden barrels filled with rolled-up charts. Overhead a ceiling fan cast faint shadows across the desk as he caressed an artifact or ran his fingers over a fading map drawn by hand a century before.

His most treasured possession was an old shaman's necklace, a single strand of palm fiber threaded through a six-inch quartz crystal, very similar to what I had seen used among the Ingano on the night that Tim and I took *yagé* with Pedro Juajibioy. I had mentioned this to Medem and recalled our experience.

"So you know something about this?" he said with a smile.

"It's the lens," I said. "The shaman uses it to look into the patient's body."

"Yes," he replied gently, "but it is so much more. Here, before you go into the forest, you must read this."

He reached behind him and pulled from the shelf a white paperback, which he handed to me. Written in Spanish, the book's title was *Desana: Symbolism of the Tukano Indians of the Vaupés.* The author was anthropologist Gerardo Reichel-Dolmatoff.

"My old friend will teach you that this quartz is compressed solar energy. The Indians see it as the penis of Father Sun, as crystallized semen. In the colors are thirty hues, all distinct energies that must be balanced. But it is still more. It becomes the shaman's house. When he takes *yagé,* this is where he goes. Inside. And from within he looks out at the world, to the territory of his people—the forests, the rocky hillsides and streams—watching and watching the ways of the animals. That's his vision. But his enemy is also there, doing the same. So they meet in the spirit world, each encased in an armor of crystal, each standing on a hexagonal shield, each struggling to unbalance the foe. It is battle at close quarters."

Long after Medem had retired for the night, I remained reading in his office. It was from Reichel-Dolmatoff that I first learned of the symbolic importance of the rivers. For the Indians of the Vaupés they are not just routes of communication, they are the veins of the earth, the link between the living and the dead, the paths along which the ancestors traveled at the beginning of time. Their origin myths vary but always speak of a great journey from the east, of sacred canoes brought up the Milk River by enormous anacondas. Within the canoes were the first people, together with the three most important plants—coca, manioc, and *yagé,* gifts of Father Sun. On the heads of the anacondas were blinding lights, and in the canoes sat the mythical heroes in hierarchical order: chiefs, wisdom keepers who were dancers and chanters, warriors, shamans, and finally, in the tail, servants. All were brothers, children of the sun. When the serpents reached the center of the world, they lay over the land, outstretched as rivers, their powerful heads forming river mouths, their tails winding away to remote headwaters, the ripples in their skin giving rise to rapids and waterfalls.

Each river welcomed a different canoe, and in each drainage the five archetypal heroes disembarked and settled, with the lowly servants heading upstream and the chiefs occupying the mouth. Thus the rivers of the Vaupés were created and populated, with the Desana people coming into being on the Río Papuri, the Barasana on the upper Piraparaná, the Tukano on the Vaupés, the Makuna on the Popeyacá and lower Piraparaná. In time the hierarchy of mythical times broke down, and on each of the rivers the descendants of those who had journeyed in the same sacred canoe came to live together. Still, they recognized each other as family, speakers of the same language, and to ensure that no brother married a sister, they invented strict rules. To avoid incest, a man had to choose a bride who spoke a different language.

When traders and missionaries first penetrated the Vaupés early in this century, they attempted to differentiate the various peoples by assigning tribal names according to language. Hence in an area the size of New England, with a total population of perhaps ten thousand, twenty-five tribes were identified. To a great extent these designations had no meaning. In most parts of the world, especially among small indigenous societies, the boundaries of language and culture are the same. To speak Waorani is to be Waorani. In the Vaupés there are societies organized in this way. The nomadic Macú live apart, marry among themselves, and remain monolingual. The Cubeo of the Kuduyarí and Kubiyú have three exogamous patrilineages, all speaking the same tongue. But in the Vaupés these are the exceptions. Among most peoples of the region, sharing a culture and way of life does not imply sharing a common language.

When a young woman marries, she moves to the longhouse of her husband, who by customary law must speak a different language. Their children will be raised in the language of the father but naturally will learn their mother's tongue. Their mother, meanwhile, will be working with their aunts, the wives of their father's brothers. But each of these women may come from a different linguistic group. In a single settlement as many as a dozen languages may be spoken, and it is quite common for an individual to be fluent in as many as five. Through time there has been virtually no corrosion of the integrity of each language. Words are never interspersed or pidginized. Nor is a language violated by those attempting to pick it up. To learn, one listens without speaking until the language is mastered.

One inevitable consequence of this unusual marriage rule—what anthropologists call linguistic exogamy—is a certain tension in the lives of the people. With the quest for potential marriage partners ongoing, and the distances between neighboring language groups considerable, cultural mechanisms must ensure that eligible young men and women come together on a regular basis. Thus the importance of the gatherings and great festivals that mark the seasons of the year. Through sacred dance, the recitation of myth, and the sharing of coca and *yagé,* these celebrations promote the spirit of reciprocity and exchange on which the entire social system depends, even as they link through ritual the living with the mythical ancestors and the beginning of time.

For over three hours the plane moved slowly over the forest. There were no towns or roads, no natural features to break the rhythm of the land. At times the plane seemed to hang in the air like a motionless ship on an endless sea. From the length of the flight alone one might have assumed that a town of some size waited at the other end, but from Medem and Schultes I knew that Mitú was still little more than a name on the map, a cluster of thatched houses, the odd tin roof, an airstrip of white sand.

In the late forties and early fifties when Schultes was there, the Vaupés was virtually unknown to the outside world. The first permanent mission was only established in 1914 at La Pedrera on the Río Caquetá. Mitú, farther north on the Río Vaupés, was a small trading post and missionary station until 1936, when it became the administrative center of the newly formed *comisaría* of the Vaupés. In 1950 the entire frontier region of Colombia and Brazil, lands bounded in the north by the Río Guaviare and in the south by the Caquetá, was as remote and wild as any place on earth. Yet Schultes came to know it quite literally like the back of his hand. On one occasion in 1958 he was flying from Bogotá direct to Mitú. As the Catalina dropped out of the Andes, it came upon a blanket of cloud stretching south as far as the pilots could see. They flew on, hoping for a break in the cover. After two hours, well beyond radio contact, they were lost, with neither the fuel to return to Bogotá nor any means of locating their position. In desperation they called Schultes to the copilot's seat. With fuel running low, he told them to fly south. Suddenly the plane veered, dove through a small opening in the clouds, and leveled off over the forest. They could have been anywhere. Schultes looked around, recognized a bend in a river, and directed the astonished pilots to Mitú.

He had first seen the country in 1943 when the Rubber Development

Corporation had ordered the survey of the Apaporis River. That expedition, though successful, had been for him exceedingly frustrating. In nearly a year he had collected only 350 plants, yet 20 had turned out to be new to science. "I deeply regret," he wrote in 1945, "that it was impossible to make a larger and more representative collection. Carried out at a cost of a life and innumerable difficulties imposed by distance, inaccessibility, absence of population, treacherous rapids and falls, this trip had as its chief purpose tasks so time-consuming that extensive botanical collecting was out of the realm of feasibility. A very preliminary examination of this small collection, however, has brought to light so many new or rare plants that we can say without fear of exaggeration that it is imperative that a thorough botanical survey of the Apaporis drainage be made. Only with an understanding of the flora of the Apaporis can we hope to arrive at an accurate understanding of the composition, distribution and history of the western Amazonian flora."

His opportunity came in the fall of 1950 when, after more than a year away in Europe and the United States, he returned to South America. It was a unique time. He remained on the payroll of the USDA, and his work with rubber was ongoing. Funding was still plentiful, and the political maneuverings in Washington that would eventually kill the program had yet to be felt by the men in the field. But with the end of the war, the atmosphere of crisis had passed, and oversight of his activities slackened. For the next three years he and a handful of explorers were essentially free to go anywhere their scientific intuitions took them.

Mostly they returned to their favorite places, seeking clues that might allow them to understand the floristic diversity that had overwhelmed them in their earlier travels. Of particular interest were the savannahs and the isolated massifs and mountains scattered throughout the basin. Already botanists had discerned patterns in their vegetation. Schultes's friend and colleague Paul Allen had surveyed the savannahs of the Kuduyarí and discovered palms that were absent farther south at Chiribiquete and Cerro Campana on the Apaporis. Schultes went to Cerro Circasia on the Vaupés and found plants closely related to those of the savannah. On a high sandstone plateau at Araracuara on the Caquetá he noted a flora more similar to that of Circasia than to what he had seen on the mountains of Chiribiquete, yet unique in its own right with a high level of local endemism. The ultimate goal of these expeditions was to understand the relationship between the flora of the Andes and that of the ancient highlands of the Guiana Shield, and thus reconstruct the geological and evolutionary history of the entire Northwest Amazon. The key was plant exploration, securing enough collections in enough places to begin to fill in the botanical map.

In December 1950 and continuing through the early spring of 1951, Schultes mounted a series of Colombian expeditions, crossing the Río Guaviare to explore the Mesa La Lindosa before turning west and entering the Serranía de la Macarena, the ancient ridge of uplands adjacent to the flank of the Andes. Unfortunately, it was the height of *La Violencia,* and even in the most remote reaches of the country there was no escaping the horrific carnage of civil war. On the eve of a major battle, Schultes was forced out of the Macarena and back to Villavicencio.

In the following weeks two encounters occurred that would define the remainder of his years in the Amazon. Word had reached him of a young naturalist, Isidoro Cabrera, a *llanero* whose father had been killed and family home destroyed when the fighting swept through San Martin. Schultes had been looking for a field assistant ever since the death of Pacho López, but he had concerns about hiring a lad from the open savannah to work in the rain forest. Cabrera offered to serve for free until Schultes was satisfied with his performance. He was paid from the start. Over the next three years Schultes and Cabrera would collect over ten thousand specimens, and their collaboration would go down as one of the most significant in the history of South American botany.

The second meeting of note took place over *tintos* in a small café in Bogotá. Don Miguel Dumit was a businessman from Antioquia who had purchased a Catalina and inaugurated the first commercial flights over the Vaupés. With him was his German pilot, Captain Liebermann. Dumit was interested in rubber and wanted Schultes's help in establishing a remote trading post that could be supplied by air. The location would be up to Schultes. All that mattered was the supply of wild rubber. There had to be enough in the immediate region to warrant commercial exploitation. Indian workers, if not available locally, could be brought in from elsewhere. They would, Dumit assured Schultes, be treated fairly and compensated properly for their labor.

Schultes knew just the place: a stretch of the Río Apaporis three hours by paddle above Jirijirimo, the waterfalls sacred to the Makuna. For ten miles the river disappears into the earth, and shamans descend to the depths to negotiate for days with the fish masters. It was there, he told Dumit, that he had found the highest concentration of rubber in all the Apaporis. Dumit gave his approval, and within two weeks Schultes and Cabrera were on their way back into the forest to establish Soratama. Named for the village where Dumit was born, it would be Schultes's wilderness base for three years. It was an extraordinary opportunity. The lure of employment meant that Indians from all over the Vaupés would be coming to him. With logistical support guaranteed and transport assured, he returned to his beloved Apaporis and entered the most productive phase of his career as an ethnobotanist. In three years he would collect over two thousand medicinal plants. Within a month he made one of his most important contributions to the study of hallucinogenic plants: the discovery of the identity of *yá-kee,* the intoxicating snuff known to the Indians as the semen of the sun.

The dust had hardly settled on the runway when I realized that Mitú was unlike any settlement I had known in the lowlands. Although it was the administrative capital of a region the size of Ireland, the sleepy little town consisted of one main street running a few blocks from the airstrip to the banks of the Río Vaupés. Along the way were neatly graveled lanes, painted cottages, and dozens of children in bright uniforms, laughing and playing in the churchyard. It was as close to idyllic as any town I had seen in the Amazon.

The only hotel was a small *residencia* on the waterfront, owned by a kind merchant and his wife, Doña Leja. It was just around the corner from the *comisario*'s home, a two-story wooden house with open verandahs overlooking the river on one side and the plaza on the other. Just by looking at the place one could tell that the most important business in Mitú was conducted over tall, cold glasses of sweet lemon juice and rum as the sun dropped over the forest. There was one bank in town, but it never had any money, and paper currency was so scarce that almost all transactions ran on credit, just as they had in Schultes's day. In fact, about all that had changed since his time was the view of the river. It was better, according to Doña Leja, before the saplings he planted grew into mature trees.

The river itself was a thousand feet wide, milky brown in color, with a slow and steady current. In the morning it was irresistible, warm and soothing, laden with silt. One could walk along the grassy riverbank to the landing above town and then swim slowly downstream, listening all the while to the quiet voices of the Indians who gathered along the waterfront—Cubeos and Tukanos for the most part, but also Tatuyos and Desana. The seasonal rains had begun, and though the skies were often blue, the river was rising. The Indians watched it carefully, sensing its modulations, examining bits of leaves and flowers as they floated by.

Several days after arriving in Mitú I joined a small party heading upriver for the Río Kubiyu. The destination was a Cubeo settlement two hours above the mouth of the affluent and close to a trail that led to the savannah of Guranjudá. The *motorista,* a young Cubeo named Ernesto, assured me that at the house of his sister there would be plenty of coca and that it would be possible to collect specimens and observe the preparation. Ernesto was a dashing youth. With his hair cut short and parted on one side, his khaki trousers and white short-sleeved shirt open at the neck, he reminded me of photographs Schultes had taken among the Cubeo in the early 1950s. Even then the Indians living close to Mitú had adopted the style of the *mestizo.* Men had given up the *guayuco,* the traditional loincloth, and women had become partial to calico cloth.

The Vaupés is by reputation a treacherous river, and indeed between Mitú and the Brazilian frontier at Yavareté there are more than sixty rapids. Heading upriver there are ten others including the great falls of Yuruparí. But the short stretch between Mitú and the mouth of the Kubiyu was slow and placid, as was the tributary itself. Seven hours of steady running, interrupted only by an early afternoon storm, and we arrived at a landing below a small clearing. A trail rose up the slippery bank, passed through a manioc field, and led to a simple thatch house standing alone at the edge of the forest. Like the other dwellings we had seen coming upriver, the house had two levels: a sleeping platform built on stilts, and on the ground a hearth and cooking area where the women and children gathered to escape the sun.

Ernesto's brother-in-law Noel was away gathering coca but was expected soon. The woman of the house, Ernesto's sister Magdalena, welcomed us warmly, and before we could unpack our gear, she brought each of us a large calabash of *chicha* and several slabs of warm cassava, the unleavened bread made from the poisonous roots of bitter manioc. No sooner had she slipped away to the fire than her daughter, a pretty child of ten, arrived with a sauce of hot chilies and fish, and a plate of roasted *mojojoye,* the fat and succulent grubs found only in the rotting hearts of certain species of palms. At the sight of this local delicacy, the other children gathered around—three boys and another small girl who was beside herself with giggles. I offered one of the grubs to the oldest boy and watched as he eagerly plucked it from the plate, bit off the head, and sucked out the oily contents.

As we waited for Noel, I climbed down into the kitchen and sat with Magdalena as she and her daughters prepared cassava. The basic food of the Amazon, equivalent to potatoes or rice, its elaboration is complex and so time-consuming that it is the organizing principle in the lives of all the women. The plant is a member of the spurge family, and its roots are full of toxic cyanogenic glycosides. In sweet manioc the poisons are concentrated in the root bark, which can be readily removed. But in bitter manioc, plants which are far better adapted to growing in the tropical lowlands, the poisons are spread throughout the flesh. In all of the Northwest Amazon, women use the same basic method to process the starch and remove the toxins.

The key is the *tipitipí,* an elastic length of basketry ingenious in its design. The women first peel the roots and then grate them on boards made rough by crystals of stone driven into the wood. The pulp is placed on a flat basket suspended horizontally beneath a triangle of poles. Water is poured over and through the mash, which is kneaded repeatedly. What collects below in a ceramic vessel is a liquid poison so toxic that a cup of it can kill a human. But overnight the potion separates, with pure tapioca starch forming a layer below and the venomous liquid hovering above. Heated by daylight over an open fire, this liquid is transformed into a soothing tea. The residue is pure starch. The pulp remaining on the surface of the basket is placed within the *tipitipí.*

On one level the *tipitipí* is simply an elaborate six-foot-long woven sieve. With a pull against the weave, the mouth of the *tipitipí* enlarges to perhaps four times its normal diameter. Stuffed with manioc pulp, it is hung from the rafters of the house, and a wooden pole is placed through a hoop woven into the other end. This becomes the fulcrum, and incredible pressure can be exerted simply by pushing down on one end of the pole. The force drives out all the moisture, and with it the venom. What remains in the *tipitipí* is a fibrous material that can be molded into cassava or simply dropped onto a hot clay skillet to create *farinha.* Thus the women create nourishment out of poison.

The *tipitipí* is also viewed as a serpent. If used improperly, it becomes an anaconda, which can devour its victims. This instrument, which creates the staple food, is a power not unlike that of the strangling coils of a snake. Sometimes a shaman will take on the image of the *tipitipí* and float on a river. As he uncoils in metaphysical space, he pays homage to the food mothers, knowing that manioc is a child, the offspring of women who give birth in gardens, returning by dusk with babies held tightly to their breasts.

What manioc is to women, coca is to men. Ernesto's brother-in-law Noel returned in the late afternoon with two large baskets of fresh leaves. He was a small man with an effusive personality, far more *campesino* than Indian by nature. But like most *mestizo* traders and rubber tappers of the Vaupés, he had adopted native ways, including the use of coca, or *patu* as he called it in Cubeo. While his wife prepared manioc by her hearth, Noel kindled a second fire and began the slow process of toasting his harvest on a large flat ceramic pan four feet in diameter, supported over the fire by five sizable rocks. With a fan woven from palm leaves, he stirred the coca with a steady rhythm, singing as he worked, tossing the leaves periodically into the air, and listening as they fell back onto the hot clay surface.

Ernesto, meanwhile, carefully swept a portion of the dirt floor and fired a large pile of dry cecropia leaves, two varieties known in Cubeo as *juakubu* and *opodokabú.* The flames flared well over his head, singed the thatch of the roof, and quickly died back, leaving a pile of white ash. I collected a sample of ash and then knelt beside Noel to examine his leaves. They had none of the delicacy of Colombian coca and were most certainly, as Schultes had suggested, aligned with *Erythroxylum coca,* the plants of eastern Peru and Bolivia. But the leaves were larger than those of the *montaña,* rounded at the tip and tougher in texture. The characteristic lines normally found running parallel on either side of the midrib on the underside of the leaf were faint or entirely lacking. Though clearly related, this was a different coca from what we had seen in the southern Andes.

I looked up as Ernesto positioned a heavy wooden mortar, about nine inches in diameter and three feet tall, carved from the trunk of a pupunha palm. When the coca was ready, the leaves brown and brittle, Noel placed several handfuls into the mouth of the mortar, and Ernesto began to pound them with a long wooden pestle. It was hard, steady work, and sweat soon dripped from his brow. Every so often he stopped and tipped over the mortar to examine the contents. When the leaves had been reduced to a bright green powder, he poured the contents of the mortar into a large calabash and mixed in some ash, roughly a handful for every two handfuls of coca. The color turned a rich gray-green.

The next step involved wrapping the powder in a piece of cloth, securing the bundle to a stick, and shaking the contents vigorously inside a covered vessel. As small clouds of green dust filled the air, Noel asked me if I had ever tried *patu.* I mentioned the coca I had used in Peru and Colombia. Ernesto cringed at the thought of chewing whole leaves.

"Qué bárbaro," he said.

He then handed me an open tin can and a spoon. Inside was the final product, a finely sifted powder the consistency of talc.

"Cuidado," Noel said. Be careful. I followed his instructions and placed a large spoonful gently on my tongue. Within seconds I coughed, and great puffs of green smoke blew out of my mouth and nostrils. Ernesto laughed aloud.

"Never, never try to talk," he exclaimed. "Just wait and let the *patu* come together on its own."

I tried again, carefully controlling my breath until saliva had moistened the powder, and I was able to form it into a pasty lump with my tongue. I had not yet mastered the technique but already could feel the coca trickling down my throat. The flavor was smoky and delicious. Within a few minutes the inside of my cheek was numb, and a sensation of well-being had spread throughout my body. It was a feeling that lasted the rest of the evening, through long conversations, and well into the night.

The next morning we left for the forest early, following a trail that ran away from the river to the *cocales,* the coca fields that Noel had established on dry ground, well beyond the reach of the seasonal floods. Fortified by a huge wad of *patu,* I moved effortlessly over the rough terrain and for the first time in my experience in the lowlands felt truly oblivious to the heat. It was no wonder, I thought, that Schultes had chewed coca every day during his years in the Amazon. According to Reichel-Dolmatoff, he once pulled out a can of the powder at a cocktail party in Bogotá. In measured tones he explained to anyone who would listen that the preferred ash came from the large palmate leaves of *Cecropia sciadophila,* not the decidedly inferior foliage of *Cecropia peltata.* He naturally had the good stuff.

Noel's coca field was unexpectedly small, little more than a patch of ground cleared in the midst of the forest. I stood for a moment and looked over the bright foliage. On the trail I had watched for evidence of feral populations, plants capable of escaping cultivation and surviving on their own. I had seen nothing, and now I knew why. Noel's plants were tall and spindly, with weak branches draped in lichen. Insects were rampant. I asked Noel about the seeds, and he assured me that they were worthless. Coca could be propagated only by cuttings, bits of stem driven into the ground. The enigma resolved itself. The very fragility of these plants revealed their origins.

Tim had been right. I realized this as I walked through the field, making a series of collections and listening as Noel described the categories recognized by the Cubeo. The most common of the plants was *kárika patu,* the sweet coca, the desirable one. The other two were known as *hoki patu,* stick coca, and *wehki patu,* coca of the tapir. Both were said to be too powerful for use. To me they all appeared to be the same species, clearly referable to *Erythroxylum coca.* But the flowers told the real story. With a hand lens I could see that all of the styles, the female part of the blossom, were short and of the same length. In other words, the mechanism built into the species to ensure cross-fertilization was absent. Though several of the plants were in fruit, I was certain that few of the seeds would be viable. If this population was typical of the Vaupés, it implied that Amazonian coca was a true cultigen, totally dependent on humans for its survival. Centuries ago someone had brought the plant out of the southern mountains, learned to propagate it vegetatively in the lowlands, and set in motion a sacred metaphor, the image of the divine leaf carried up the Milk River in the belly of the Anaconda. Over time local conditions had produced a novel form of the plant, a unique variety eventually described by Tim as *Erythroxylum coca* var. *ipadu.*

From what Noel told me about growing coca, I was quite certain that cuttings would root easily and, if kept moist, retain their viability for some time, perhaps even weeks. Still, the sooner I dispatched my collections to Bogotá the better. Ernesto was heading downstream for Mitú in two days, which was ideal because it permitted time to visit Guranjudá, the savannah where, in April 1953, Schultes discovered *Philodendron dyscarpium,* an oral contraceptive known to the Cubeo as *he-pe-koo-ta-ta.*

Like the mountains of Chiribiquete, the savannahs of the Northwest Amazon are remnants of the ancient uplands that once stretched continuously across the northern face of the continent. The eroded remains of sandstone mountains, they are like desert islands in the forest. There is almost no soil, and the ground is so porous that even torrential rains leave little mark. Only plants adapted to conditions of severe drought can survive, and thus the savannahs have become through time repositories of species often found nowhere else in nature. But on the day that I wandered above the twisted caves and tunnels of Guranjudá, collecting a wild coca and several other novelties, it was the flora at the edge of the savannah that caught my attention—in particular a number of medium-sized trees in the nutmeg family. I recognized them as virolas by their bloodred resin. To the Tukanoans they are the source of a powerful hallucinogen acquired at the beginning of time when the daughter of the Sun scratched her father's penis and ground the dried semen into snuff. Today shamans take the drug to journey beyond the Milky Way to contact the spirit master, *Viho-mahse.* The first to experience the intoxication and write about it in a botanical sense was Schultes.

Schultes had several things on his agenda in June 1951 when he returned to the Apaporis to establish Soratama. His rubber work was ongoing, and he had ambitiously taken on the task of translating into English the

Andean journals of Hipólito Ruiz and José Antonio Pavón, eighteenth-century Spanish botanists whose papers, lost for two centuries, had recently been found in a bombed-out wing of the British Museum of Natural History. Then there were the practical challenges of carving a trading post out of the forest, maintaining the flow of supplies, and overseeing the work and morale of the men. In this he had considerable help from Isidoro Cabrera as well as José Restrepo, an enterprising merchant from Antioquia whom Dumit had hired to manage the station.

Tall and lanky, with a high forehead and long aquiline nose, Restrepo worked hard by day but was given to flights of fantasy. He was an odd character with unusual tastes. His home in Medellín was painted jet black, as was every room within it. At Soratama he spent most of his time discussing Colombian politics or concocting impossible schemes to make his fortune. Schultes, having no interest in money and viewing discussions of current affairs preposterous in an outpost cut off for months from the latest news, found companionship in his work.

For years he had been aware of conflicting reports describing the use of hallucinogenic snuff in lowland South America. Richard Spruce had found *yopo* at Maypures on the Orinoco in June 1854 and identified the source as the tree *Anadenanthera peregrina.* The famous German ethnologist Theodor Koch-Grünberg, traveling in the Northwest Amazon between 1903 and 1905, told of a "magical snuff . . . prepared from the bark of a certain tree. . . . The sorcerer blows a little through a reed into the air. Next he snuffs, whilst he absorbs the powder into each nostril. . . . Immediately the witch doctor begins singing and yelling wildly, all the while pitching the upper part of his body backwards and forwards."

In 1945, when the Smithsonian Institution published the seminal five-volume *Handbook of South American Indians,* an anthropologist produced a map showing *yopo* distributed among tribes throughout the Northwest Amazon. This map left Schultes utterly perplexed. In his years in the field he had consumed copious amounts of tobacco snuff and had taken *yagé* more times than he could remember. He had never come upon *yopo,* and he knew that the source tree grew only on the open savannahs of the Guaviare and Orinoco, with scattered populations thriving at the mouth of the Río Branco on the lower Río Negro in Brazil and on the open savannahs of the Río Madeira. It was never found in the rain forest. What's more, Koch-Grünberg, who wrote exclusively of the wet tropics, had mentioned a bark being employed to make the drug, whereas Richard Spruce had quite specifically identified seeds as the source of *yopo.*

The mystery was very much with him on the morning of June 26, 1951. For a week he and a group of rubber workers had been surveying the forest on the north bank of the Apaporis between the mouth of the Río Pacoa and the Kananarí. Among them was a young Puinave, the son of a shaman who had been flown in from the Río Inírida, the highest Colombian affluent of the Orinoco. As they moved through the forest making general collections, they happened upon a small flowering tree, which Schultes recognized as *Virola calophylla.* The lad examined the bark and innocently turned to Schultes.

"This is the tree that gives *yá-kee.* My father uses it when he wants to talk with the little people."

Schultes froze, momentarily stunned. He knew that the homeland of the Puinave was one of the regions where *yopo* did grow, and no doubt the boy's father employed the hallucinogen in shamanic rites. But what on earth was *yá-kee?*

"Does he use the seeds?" Schultes asked, thinking that *yá-kee* might just be another name for *yopo.*

"No," the boy answered, "just the bark."

Schultes quietly led him away from the others to a clearing on the riverbank where they could be alone.

"Do you know how to make it?" he asked.

"Of course."

The next morning they left their encampment early and returned to the tree they had found the day before. The Puinave youth insisted that the bark had to be removed at first light, before the sun penetrated the canopy and warmed the trunk. With a quick slash of the machete, Schultes found that it came off easily in long strips. Within minutes a thick red liquid oozed from the inner surface.

The Puinave youth soaked the bark for thirty minutes and then rasped it, placing the shavings in an earthen pot. He added a small amount of water and then kneaded and squeezed the material, straining it several times through a piece of cloth into a second and smaller vessel. Once satisfied, he discarded the shavings and placed the liquid over a low fire, allowing it to simmer for several hours. When the preparation had been reduced to a thick dark syrup, he removed the pot from the heat and set it in the sun so that the residue would solidify slowly into a dry crust. With a pestle of polished stone, he ground this deposit into a fine powder, mixing in equal amounts of ash derived from the bark of wild cacao. The final product was sifted and placed in a large snail shell. Attached to the shell were two bird bone tubes, each plugged with a stopper of bright feathers. The Puinave then handed the works to Schultes. By then it was afternoon, and Schultes had been waiting all day to try the snuff. Following his young companion's instructions, he put the end of one tube in his nostril, placed his lips on the other, and blew hard through his mouth.

"It may be of interest," he wrote the following day, "to append a few observations which I was able to make personally after taking *yá-kee.* I took about one-third of a level teaspoonful of the drug in two inhalations using the characteristic V-shaped bird-bone apparatus by means of which the natives blow the powder into the nostrils. This represents about one-quarter the dose usually absorbed. . . . Within fifteen minutes a drawing sensation over the eyes was felt, followed very shortly by a strong tingling in the fingers and toes. The drawing sensation in the forehead rapidly gave way to a strong and constant headache. Within one half hour, there was a numbness of the feet and hands and an almost complete disappearance of sensitivity of the finger-tip; walking was possible with difficulty, as in the case of beri-beri. Nausea was felt until about eight o'clock, accompanied by a general feeling of lassitude and uneasiness. Shortly after eight, I lay down in my hammock, overcome with a heavy drowsiness which, however, seemed to be accompanied by a muscular excitation, except in the extremities of the hand and feet. About 9:30, probably, I fell into a fitful sleep which continued, with frequent awakenings, until morning. The strong headache over the eyes lasted until noon. A profuse and uncomfortable sweating, especially of the armpits, and what might have been a slight fever lasted from about six o'clock all through the night. There was a strong dilation of the pupils during the first few hours of the experiment. No food was taken and no tobacco was smoked from the time the experiment began until one o'clock in the afternoon—that is, for twenty hours during the course of the experiment.

"Since this experiment was performed under primitive conditions in the jungle, all observations had to be made by myself. In spite of its many and serious shortcomings, the experiment indicates the narcotic strength of the snuff. The witch doctors see visions in color, but I was able to experience neither visual hallucinations nor color sensations. The large dose used by the witch doctor is enough to put him into a deep but disturbed sleep, during which he sees visions and has dreams which, through the wild shouts emitted in his delirium, are interpreted by an assistant."

For Schultes this innocent discovery was both a culmination and a beginning. He had, in fact, stumbled upon *yá-kee* in the early spring of 1942 in the Putumayo while walking from El Encanto to La Chorrera. His field notes for May 31, 1942, record the collection of a small unidentified tree known to the Witoto as *oo-koo'-na,* with "red resin in bark.

Intoxicating." But the significance of the collection was lost on him at the time, and the specimen soon forgotten. Now after nine years, with a positive identification in hand and certain knowledge of the efficacy of the snuff, any number of curious leads presented themselves.

Within a week he had found a second psychoactive species, *Virola calophylloidea.* A month later, while working among Taiwanos on the Río Kananarí, he came up with a third, *Virola elongata.* Over the next months he was able to document the preparation of the snuff among the Cubeo and Tukano on the Vaupés, the Barasana and Makuna on the Piraparaná, and the Kuripakos far to the north on the Río Guainía. Then in 1953 a report reached him from an American missionary working in Venezuela among the Waiká, or Yanomami, at the headwaters of the Orinoco.

According to this account the Waiká used a snuff called *ebéna* to contact the *hekula,* spirits of rocks and waterfalls. For the next decade tantalizing but indefinite reports suggested that *ebéna* was derived from *Anadenanthera,* the source of *yopo.* Schultes had his doubts. Finally, in the summer of 1967, he journeyed to two Waiká villages widely separated in the remote upper reaches of the Río Negro in Brazil. At the Salesian mission of Maturacá, he found something he had never seen in thirty years of studying hallucinogenic plants: the casual, essentially recreational use of drugs. In every house hung a large bamboo tube of snuff. The Indians took it formally during rituals and festivals, but any man was free to dip into the stash if he felt the urge. It was not uncommon to see an individual high on *ebéna,* dancing and singing alone, while the rest of the village went about the daily round. Schultes confirmed that the source of *ebéna* was, as he had always suspected, a virola, in this case *Virola theiodora.*

At the second village he arrived by chance during the annual festival of the dead. Sitting on the ground, he watched as a shaman lifted a four-foot grass tube to his mouth and placed the other end into Schultes's nostril. He endured the blast of snuff that ripped into his sinus and transformed the world before his eyes. But he was not prepared for what came next. At the height of the feast, the Indians ran out of snuff. Schultes waited expectantly, anxiously, as they opened several bamboo quivers and removed dozens of poisoned arrows. Having scraped off the venom and ground it into dust, they placed the powder into the blowing tube. Before Schultes could say anything, the shaman drove the venom into his nose. What amazed him was that nothing changed. He felt the same surge, the same magnetic release from the world. Poison and hallucinogen were one. It was something he would never understand and never forget.

This was not the end of the virola story. Upon his return from the field Schultes had several samples of *ebéna* analyzed. An astonishing 11 percent of the dry weight of the snuff was composed of a series of potent tryptamines, including 5-methoxy-N, N-dimethyltryptamine, arguably the most powerful hallucinogen known from nature. More curious than the strength of the snuff was the fact that the chemical constituents were both in kind and concentration almost identical to what had been found in *yopo.* The source trees were botanically unrelated. One drug was derived from seeds, the other from inner bark. Yet somehow the Indians had recognized both plants and discovered ways to exploit their remarkable chemical properties. The botanical identity of *yopo, Anadenanthera peregrina,* had been determined by Richard Spruce in June 1854. It was only appropriate that the sources of *yá-kee* and *ebéna,* various species of *Virola,* were discovered by Schultes a century later.

Two years after the Venezuelan expedition, Schultes's work with virola took yet another serendipitous turn, again sparked by an unexpected encounter in the forest. The unusual chemical properties of the resin had shown considerable promise as an anti-inflammatory drug. Commissioned by one of the major pharmaceutical companies to gather one hundred kilograms of bark, Schultes found himself in February 1969 on the banks of the Río Loretoyacu, near Leticia. His assistant was Rafael Witoto, a native of El Encanto on the Karaparaná.

As they stripped the bark from the trees, the young man turned to Schultes and said, "This tree is the one my father made little pellets from. He ate them when he wanted to speak with the little people."

The remark startled Schultes, who had been down this path before.

"You say he *ate* it."

"Yes."

It was impossible. Schultes understood the pharmacology of the drug. He knew that tryptamines had to be absorbed through the nose or injected. Taken orally they are rendered inert by monoamine oxidase, a naturally occurring enzyme in the stomach. Yet experience had taught him that no information could be dismissed out of hand.

"What do you call it?" he asked.

"Oo-koo'-na," Rafael said. Witoto was one of two indigenous languages Schultes spoke fluently, yet the word had no meaning to him. He did not know that he had recorded it in his field notes twenty-five years before.

"Can you make it?"

The boy could, and the next morning the work began. The procedure was familiar: the inner bark was scraped away and the liquid boiled off until a dark brown syrup remained. But rather than allowing it to dry, Rafael formed the paste into small pellets and then rolled them in a vegetable salt derived from the bark of a tall forest tree. There were no other admixtures. Schultes ate four pellets. The effect was unmistakable. The mystery was how it had happened. Schultes assumed there had to be another compound in the paste that inhibited monoamine oxidase. No chemist believed him until one working in Sweden found trace amounts of beta-carbolines, compounds known to block the action of the enzyme. Once again Schultes was vindicated by his faith in the word of his informants.

For more than a week following my return from the Kubiyu I stayed around Mitú, hoping to find a way to reach the Río Piraparaná, the river that Schultes had encouraged me to explore. Much of the time I spent in the office of the *comisario,* poring over maps of the Vaupés, or visiting with the padre who had known Schultes well. Through his recollections I sensed more than ever the isolation of the rubber work, the image of Schultes and Cabrera living in the forest, drifting down these immense and unknown rivers, only to return after weeks or months to Soratama, which was itself but a handful of huts in a clearing on the bank of the Río Apaporis.

"Dreams spring from the horizon," the padre told me, "but they are stolen by the suffocating walls of the forest. It's a wonder he didn't go mad."

Throughout their time together Isidoro Cabrera kept a journal, mostly notes about the collections, enlivened with bursts of poetry, impassioned pleas for the protection of nature, expressions of bitter contempt for those who had destroyed his family and home. "How I feel for the good people," he wrote at one point, "and the old women and children, those who are defenseless and who suffer the most from the consequences of this group of imbeciles who poison our land. My poor countrymen, how long will we allow ourselves to be exploited and murdered¿" He was twenty-nine, seven years younger than Schultes. Of their first meeting, Cabrera noted simply, "He greeted me like I was somebody."

Their travels, compressed into the short passages of the journals, have a stoic quality rendered powerfully by the simplicity of Cabrera's writing. "Trouble at the Yuriparí rapids," reads one of the daily entries, "one canoe lost. Young Miguel pulled under and disappeared. Reached camp by noon." The end of July 1951 found them at Soratama challenging Alejandro Betancourt, a rubber trader of ill repute who had arrived with an agent of the law to seize several of the Indians for debt. Schultes confronted them to no avail. The women were taken away, but the event was never forgotten. Months later a young Makuna killed Betancourt. "The son of a bitch deserved it," noted Schultes, who immediately offered the Makuna a job.

In 1952 the success of Soratama inspired Miguel Dumit to establish a second rubber station on the Apaporis, farther downstream in the very heart of the Indian lands. On February 16, Schultes and Cabrera commenced a slow descent of the river, passing through the cataracts of Yayacopi and reaching an isolated group of Tanimukas who had fled the rubber atrocities to settle at the headwaters of the Igarapé Peritomé, far from their traditional homeland on the Río Popeyacá. Cabrera marveled at the shape of their longhouse: round, immense, with a great vaulted ceiling. Schultes was impressed by the flavor of the coca. The village chief, Tuemejí, spoke no Spanish, but he and Schultes communicated through an interpreter, Alirio Mejira. In the evening, as they gathered around the men's circle to make coca, the formula of the preparation was revealed. Rolled tubes of ischnosiphon leaves were tamped half full with small lumps of white resin, extracted from trees and aged for five months. Fired at one end, the burning resin tube was inserted into piles of cecropia ash and blown vigorously so that the myrrhlike aroma permeated the ash, thus flavoring the coca. Schultes found the powder pleasant but slightly irritating to the throat. Within three days, however, as he and Isidoro reached the mouth of the Popeyacá and came upon another Tanimuka village, he had learned to enjoy the taste.

On the morning of February 23 he left for the forest to gather *yagé.* That night he and Cabrera, along with all adult men and women, took the potion and danced, arms colored with feathers, the men chanting and forming an inner circle, the women enveloping them in a wider sphere. It was a simple dance: three steps forward, a blow to the earth with the right foot, four gentle steps, followed by eight of tremendous vigor. In the morning a shaman presented Schultes with a stone ax, a magical implement capable of killing the children of enemies.

The following day the journey continued downstream to the *maloca* of Capitán Pajarito, another chief, who lived just above the mouth of the Piraparaná. On February 26 a site nearby was chosen for the new station—high ground adjacent to a flat stretch of river suitable for the Catalina. Schultes named the camp Jinogojé, in Makuna, "Cave of the Anaconda." Here he would again make botanical history. For the next eighteen months, with the exception of brief excursions to the upper Río Negro and the Putumayo, all of his explorations would be within a one-hundred-mile radius of Jinogojé. He would not spend a great deal of time in any one village, but his movements would resonate throughout the entire region.

Within days of establishing Jinogojé, Schultes began planning an expedition to the homeland of the Yukuna, an Arawakan people inhabiting the Miritiparaná, a remote tributary of the Caquetá born in the lightly forested uplands south of the Apaporis. Three hundred miles long, the Miritiparaná begins as a shallow transparent stream, flowing over a bed of pure white sand, and then tumbles through seven major rapids. Doubling in size at the confluence with the Guacayá, its principal affluent, the river flows south, entering the Caquetá twenty-five miles above La Pedrera.

At the end of February 1952, Schultes and Cabrera walked from the Apaporis overland, nine hours of arduous portage to reach the headwaters of the Guacayá. There they spent a week among Tanimuka Indians, close allies of the Yukuna, and they encountered the last surviving Matapies, a people whose language had been lost a generation before. Of the Yukuna, Schultes learned of two clans: the People of the Eagle, who numbered perhaps 250, and the People of the Boar, of whom only 40 remained alive. It was the fruiting season, the time of the great annual celebrations, and Schultes and Cabrera were dazzled by the power of the Yukuna wrestlers, men who consumed pounds of coca a day as they squared off and jousted with clubs of knotted wood.

Unprepared for an extended stay, Schultes and Cabrera were forced to retreat to the Apaporis and Soratama. Schultes flew to Bogotá but returned on April 4, eager to reach the Miritiparaná. With Cabrera sick with malaria, their departure for the Guacayá was delayed until April 21. They finally were back among the Tanimuka on the eve of the *Kai-ya-ree,* the annual festival celebrating the maturation of the pupunha palm. For a month the Indians had accumulated vast stores of dried meat, coca, sacred snuffs, and fermented brews for the dancers. Emissaries of the *capitán* of the longhouse had scattered through the forest inviting all families of the Yukunas, Matapies, and Tanimukas to begin their movement toward the *maloca.* The guests had gathered in a dozen places, all within a day's walk of the ritual site, to await the sound of drums announcing that all was in readiness. The sound of the signal drums could be heard at a distance of a day's travel through the forest. When Schultes and Cabrera entered the village, they were greeted by the chief, who offered them coca and tobacco as well as the ritual salutation, a song recalling the events of the previous year.

The dancing began at midnight, and by then Schultes had been dressed as an animal with a shirt of coarse brown cloth hammered from the inner bark of the *llanchama* tree, anklets of hollow seeds, and a long skirt made of strips of bark that rustled like grass as he moved. His mask was also bark cloth, painted with black pitch and decorated with yellow and white designs. The opening dance was the *cha-vee-nai-yo,* an offering to the spirit of evil. Sixteen men in four groups wound in and out of each other, movements that climaxed in an intense flurry before ending on a note of calm, a line dance with a simple repetitive rhythm.

"The mask for the *cha-vee-nai-yo,*" Schultes later recalled, "is weird: a human face fashioned of blackened pitch, with eye holes through which the dancer peers, a wedge-shaped wooden nose, and a leering, toothless mouth. The eerie sight of so many hideously unreal devil masks and the weirdly monotonous minor chant with its far-off, hollow sound as it is sung through the mask and hood had an almost hypnotic effect upon me. And, as I took part in this dance and joined in the chant myself, it was not hard for me to imagine that such an unearthly ritual must be placating some unearthly force."

In turn, through dance, the Tanimuka saluted each animal of creation. First the young boys adorned as monkeys and carrying leafy branches mimicked the lithe movement of primates scattering through the canopy. The tapir dance was slow and lumbering, the one dedicated to the anteater startling in the realistic depiction of the costume. The movements of the deer dance were graceful and rapid, darting steps that perfectly captured the frightened and nervous character of the animal. The dance of the wild bee was accompanied by a low buzzing drone. That of the fruit bat was sung with unexpected beauty, a high chant, squeaky and shrill like the voices that emerge from caves. The most powerful dance of all, the one heard by Schultes just as he took a dose of *yá-kee* snuff, was that of the jaguar. As he watched the nimble dancers pouncing and whining, snarling like cats, he saw the masks of wooden teeth, glass eyes, and whiskers of black pitch come alive, move toward him, and disappear in a pool of colored light.

The festival lasted four days, with Schultes dancing throughout. Only years later would he understand what he had lived and experienced through movement. For the Yukuna, the dances of the *Kai-ya-ree* tell the story of life, expressing all that they believe about the origins and evolution of the natural world. In taking on the images of supernatural beings and wild creatures, in balancing the forces of good and evil, the dancers ensure the health and fertility not only of their people but of the earth itself.

• • •

After a month on the Miritiparaná, Schultes elected to return to the rubber station at Jinogojé. The simplest route was to retrace his steps across the Guacayá portage to the lower Apaporis. Schultes went in the opposite direction, up the Miritiparaná and overland through miles of forest to the headwaters of yet another major river, the Popeyacá. He reached the Makuna settlements there on June 4. Traders had cautioned him to be wary of the Popeyacá Makuna, a people said to be notoriously deceitful and dangerous. Schultes scoffed at the warning. "The Makuna are not treacherous," he wrote, "but when called upon to match the treachery and guile of an unwanted intruder, they are equal to the task of defending their homes and families with treachery." His way of establishing rapport was to take *yagé,* which he did on his first evening in the *maloca,* sitting patiently as the shaman stained his hands red and painted red dots on his face to facilitate the jaguar transformation. They attached rattles to his elbows and ankles, presented him with a shaman's wand, and taught him to play panpipes as the shaman sang the mythological history of the world. The clear notes carried over the forest, Schultes recalled, and he came to realize for the first time that the shaman's breath had a creative power of its own. The paintings on the face of the *maloca* were the same images as those carved into petroglyphs by the ancients, and these were the same as what he saw an hour after the shaman cleaned the ritual vessel with a scarlet feather and poured him a dose of the drug.

A week later he and Cabrera continued downriver, reaching the Apaporis just above the mouth of the Piraparaná and arriving at Jinogojé on June 13. There they remained for a month. On July 10 the Catalina appeared unexpectedly in the western sky. On board was a five-man delegation from the Colombian Ministry of Education. It included Enrique Gomez, son of then president Laureano Gomez, as well as a young journalist by the name of Belisario Betancur who would himself become president of Colombia thirty years later. Indeed, it would be Betancur who in 1983 would present Schultes with the Cruz de Boyacá, Colombia's highest civilian honor and one that had never before been awarded to a foreign naturalist. But the most intriguing member of the group was a young anthropologist making his first trip to the Amazon. In a letter written just before his death in 1994, Gerardo Reichel-Dolmatoff remembered their first encounter:

"A steady rain was falling and the dank smell of forest was in the air. The people surrounded us. . . . Most of them had malaria-worn faces; wiry little mestizos clad in tatters shook hands with us; in the background were some Indians, most of them naked, some with trousers. A tall stranger, gaunt and bearded, dressed in crumpled khaki, but unmistakably American, walked up to me and said: 'I am Richard Schultes.' "

That evening after dinner the anthropologist and the botanist spoke long into the night, a wide-ranging conversation that drifted from plants and natural history to botanical terminology, Greek etymologies, the Latin classics. At one point Reichel-Dolmatoff mentioned Richard Spruce. With a jerk of his head Schultes stared at his guest and said in a toneless voice, "He is my hero." The comment surprised Reichel-Dolmatoff. This was not idle sentiment, it was the raw invocation of lineage.

"There was something in his expression, in his eyes, that alerted me," Reichel-Dolmatoff would recall. "The next day I noticed the same expression on his face. We had gone a few miles upriver and now were standing on the riverbank, and in front of us, on the other side, the forest was rising like a wall. We looked in silence and then Schultes said, as if speaking to himself, 'I know every tree, every single tree one can see from here.' He was standing very quiet, eyes wide open, with updrawn eyebrows. 'That man's a fanatic,' I thought, but immediately I corrected myself; no, that was not the word. He had not been talking about trees; it wasn't about trees at all. The forest meant something else to him; the forest was a mediator. And sometimes I wonder what the forest came to mean to those Victorian naturalists, to Wallace, Spruce, Bates. And to the Victorian deep inside Richard Evans Schultes."

For Reichel-Dolmatoff, who would go on to become the world's foremost authority on the ethnology of the Vaupés, the chance encounter at Jinogojé was a pivotal moment. The Indians spoke of the Piraparaná, a river "where their people lived in painted longhouses and where hardly ever a foreigner had been seen." Schultes, too, had mentioned the river, and "although his quiet voice and sober-minded descriptions were unromantic, there was something that captivated me. . . . The Indians and Richard Schultes, each in their own way, had shown me a new dimension. I left Jinogojé with the conviction that this had been a liminal spot on my way."

Schultes, of course, had his own plans for the Piraparaná, but they would have to wait. He was feeling poorly, and when the delegation left on July 13 for Mitú and Bogotá, he went along, sitting beside Reichel-Dolmatoff behind the pilot. Overloaded with rubber, the Catalina struggled to leave the water. The forest loomed ahead. There was a moment of panic. The hatch flung open, and Indians dumped bale after bale of rubber as the plane ricocheted over the surface before finally rising sharply over the rocks and barely clearing the edge of the canopy. Reichel-Dolmatoff settled back to catch his breath. "Schultes," he later wrote, "had remained in full control, holding on to his field notes and presses." In fact, Schultes hardly noticed the crisis. His head pounded, and he was beginning to sense the convulsions of yet another attack of malaria, this time so severe that it would leave him prostrate in Bogotá for nearly a month.

From Doña Leja's *residencia* in Mitú I had moved to the waterfront and strung my hammock in an open shelter built for itinerant traders and Indians. It was there that word reached me of the arrival of the missionary pilot. From the padres I had learned of San Miguel, an abandoned Catholic mission located midway up the Río Piraparaná, roughly one hundred miles by air south of Mitú. Established only seven years before, it no longer had a resident priest, but the longhouse stood and the forest had yet to reclaim the grass airstrip. There were Barasana and Tatuyos still living there and other Indians in the area: Taiwanos farther upstream, Makuna twenty miles below on the Caño Komeyaká, and Barasana and Bará on the Caño Colorado, a small affluent that enters the Piraparaná just above the mission. With the rains beginning in earnest and my time and resources limited, the best chance I had of reaching the drainage was to drop in by air.

The pilot was a southern Baptist, tall and taciturn, who had no love for the Catholics and little interest in helping a young botanist collect coca. But he knew San Miguel and had landed there several times, most recently to evacuate an Indian who had been bitten by a poisonous snake. He was willing to take me to the mission with the understanding that the return flight would be at his discretion. We would leave in the morning, and he would try to be back in ten days. I agreed and spent the rest of the afternoon organizing my gear and purchasing trade goods —shotgun shells mostly, but also machetes, fish hooks, aluminum pots, flashlight batteries, and several bolts of brightly colored cloth. We left Mitú just after dawn. The small plane rose into the clouds and then burst over the canopy like a wasp, minuscule and insignificant.

The forest stretched to the horizon, with nothing to betray that people had ever set foot on the land. Schultes once wrote of his time on the Miritiparaná, of drifting downstream and reading in the patterns of the vegetation the history of the river, shadow places where settlements had been, their ancient presence revealed by the persistent growth of domesticated palms and fruit trees found nowhere in nature. From the air I saw only forest and rivers, and occasionally faint patches of bright foliage that might once have been fields. Then, an hour out of Mitú, I looked ahead to a small opening in the canopy. As we approached the mission of San Miguel, its isolation appeared complete. There was the *maloca,* a thatch structure the size of a warehouse, and around it a burnished clearing, bare and clean. Beyond the clearing were fields, simple and monotonous, and beyond the plantings the forest.

The plane flew low over the *maloca* and then banked sharply over the river before turning back to begin a steep glide to the short and dusty runway. By then there were a dozen Indians waiting in the grass, mostly women and children but also two men. As soon as the plane taxied to a stop, the pilot stepped out, waved a perfunctory greeting, hastily unloaded, and was off. Within minutes he was gone, and I was following my gear and the young boys along the narrow trail that led to the *maloca.* It was only when we were inside that I introduced myself to the men and explained my reasons for visiting. The younger of the two, the one wearing trousers and a tattered shirt, spoke broken Spanish. He was Rufino Vendaño, the *capitán.* The older man was his father, who became known to me as Pedro. His legs were painted with dark spirals, he wore a *guayuco,* a loincloth, and his face was stained with red *achiote.*

They piled my supplies and equipment in the front of the *maloca,* in the open area reserved for visitors, and then invited me to join them toward the rear where the women and children had gathered in a boisterous circle around a large ceramic vessel. Inside was a mud nest, and darting for the rim were dozens of termites. With great delight the boys and girls plucked them before they could escape, bit off the wings, and popped the rest of the insect into their mouths. Everyone was laughing, and the laughter increased as I accepted an invitation to join in. Rufino's father squatted beside me and made sure I was given the choicest morsels. After the last of the insects had been eaten, Rufino looked somewhat puzzled.

"It seems," he said, "that you and the pilot do not come from the same country."

"No," I replied truthfully, "we don't."

It was a simple gesture, sharing a meal with a host, but one that seemed to put everyone at ease. Just a few years earlier two British anthropologists, Christine and Stephen Hugh-Jones, had lived for twenty-two months in a neighboring longhouse on the Caño Colorado. I did not know them at the time, but I admired their work and was without doubt the beneficiary of the considerable goodwill they had established with the Barasana. By evening, when we had returned from the fields and bathed in the stream that ran alongside the clearing, I felt completely at home.

As darkness fell, Rufino lit a torch dripping in resin, and a red glow illuminated a small circle where the men had gathered. The women and children had retired to their family compartments, which ran along either side of the *maloca,* at the far end, each divided one from the other by a wall of thatch. Soft words were spoken, and every so often you could hear a mother gently chiding a child. But the night clearly belonged to the men. Close to where I had hung my hammock, Rufino kindled a fire to toast coca while his son reduced cecropia leaves to ash and prepared the mortar and pestle. Schultes always spoke of the hypnotic sound of night in a *maloca:* the deep thumping and pounding of coca, the voices of men around a fire, music played on deer skulls and shells. One could sense the comfort of shelter, the thought of sleeping beneath a protective canopy while the forest outside reverberated with nocturnal life.

That first night I took a great deal of coca, perhaps too much, for even as the last embers of the fire faded away, I lay awake in my hammock, unable to sleep. In the absence of people the *maloca* seemed strangely alive. The space was vast, perhaps one hundred feet long and sixty feet across, with a vaulted ceiling that rose thirty feet above the dirt floor. The symmetry of the structure was exquisite: eight vertical posts spaced evenly in two rows, with two smaller pairs near the doors, crossbeams, and pleated rows of thatch woven together over a grid of rafters. Reichel-Dolmatoff wrote that the longhouse is a model of the cosmos, with each architectural element charged with symbolic meaning. The roof is the sky, the house beams the stone pillars and mountains that support it. The mountains in turn are the petrified remains of ancestral beings, the Mythical Heroes who created the world. The smaller posts represent the descendants of the original Anaconda. The floor is the earth. The long ridge pole overhead represents the path of the sun that separates the living from the limits of the universe.

Beneath the ground runs the River of the Underworld, the destiny of the dead. The Barasana bury their people in the floor of the *maloca,* in coffins made from broken canoes. As they go about their daily lives, living within a space explicitly perceived as the womb of their lineage, the Indians walk above the physical remains of their ancestors. Yet the spirits of the dead drift away, and to facilitate their departure, the *maloca* is always built close to water. And since all rivers, including the River of the Underworld, are believed to run east, each *maloca* must be oriented along an east-west axis, with a door at each end, one for the men and one for the women. Thus the placement of the *malocas* adjacent to running streams is not just a matter of convenience. It is a way of symbolically acknowledging the cycle of life and death. The water both recalls the primordial act of creation, the riverine journey of the Anaconda and Mythical Heroes, and foreshadows the inevitable moment of decay and rebirth.

Outside the longhouse is a world apart, the place of nature and disarray. The owner of the forest is the jaguar, and the demon spirits have been transformed into animals that eat without thought and copulate without restraint. White people are like the animals, dwelling at the margins of the world, reproducing with such abandon that their numbers swell, spilling over into lands reserved from the beginning of time for the Barasana, Makuna, and the other people of the Anaconda. The world of the wild is a place of danger, the origin of disease and sorcery, the realm where shamans go in dreams and hunters walk each time they leave the protective confines of the *maloca* and surrounding gardens.

A wild coca grew near the rapids, and the Barasana maintained that it was the coca of their fathers. Rufino called it *coca de pescado* and said that it was so much stronger than the domesticated plant that the people avoided it even though it had been used by the ancestors. In order to collect it, we left San Miguel early, with the dawn mist still lying over the forest.

The river ran east through a region of stagnant backwaters and swamps and then fell suddenly through the first of a series of rapids. Rufino was in the rear of the canoe, and a man known to me as Pacho guided us from the bow. He was Tatuyo, originally from the upper Piraparaná, though he now lived in a longhouse south of San Miguel. He led us several miles downstream to the mouth of the Caño El Lobo, an affluent incongruously named in Spanish after the wolf. Its actual name was Caño Timiña. I knew it as the highest point that Schultes reached on the Piraparaná. It was one of the few times in his life as an explorer that he was unable to reach his goal.

Recovering from malaria in the summer of 1952, he and Isidoro returned to Jinogojé on August 14. His plan was to head up the Piraparaná and find a way overland to the headwaters of the Río Tiquié and then drift down into Brazil along the Vaupés to the Río Negro. It was a modest journey by Schultes's standards, three hundred miles as the crow flies, but it offered a chance to revisit areas he had not seen since 1948 when his work had been curtailed by beriberi. With any luck he might once again make contact with the nomadic Macú.

Unfortunately, the short dry season was upon them, and by the time he and Isidoro reached the Caño Timiña on September 5, water in the affluent had dropped so low that movement overland to the Tiquié proved impossible. Schultes elected to remain on the Piraparaná for a month and then proceed to the Río Negro by air. A terrible storm carried their plane to Mitú on October 14. A week later Captain Liebermann dropped them on the Río Negro, just below the mouth of the Casiquiare. After only two weeks in the area, including a brief trip up the Río Guainia, they returned to Mitú, where Schultes succumbed once again to malaria. It was his seventh attack, and he had no choice but to return to Bogotá. Within ten days he would head home to spend Christmas 1952 in Boston.

During his time on the Piraparaná, Schultes collected five hundred plants, including the wild coca that Rufino, Pacho, and I found growing quite commonly along the lower river. It turned out to be *Erythroxylum cataractarum,* a small tree first described by Richard Spruce. Rufino again insisted that the leaves could be eaten but with some risk of harming the stomach. This led to talk of other plants, including *yagé* and its various admixtures. One of these was *Sabicea amazonensis,* known to the Barasana as *kana.* Schultes had collected it twice. Usually found growing around the *malocas,* it is a scandent vine with opposite leaves and small pink berries that are said to sweeten the brew. Anthropologist Stephen Hugh-Jones observed its use during the Yuruparí, the sacred rites of male initiation. He likened the fruits to hearts on a string, each a symbol of a generation, all linked together by the vine, an umbilicus running away from the longhouse into the rivers and from there to the source of all humanity, the place in the east where the Sun is born. In drinking *yagé* for the first time, in eating this fruit, the young boys remember the ancestral origins of life. The shaman's rattle, itself a heart, contains seeds of the fruit, and thus by blowing over the rattle the shaman transforms the heart and soul of the initiate.

When we returned to San Miguel, I casually mentioned to Pacho that I would like one day to drink *yagé.* Nothing more was said, and for several days we continued our work in the forest, collecting for the most part along the Caño Colorado, a beautiful stream overhung with dense vegetation. Then one evening not long after sunset, Rufino asked me if I was interested in taking *yagé.* Evidently his father had prepared it earlier in the day during our absence. I accepted the offer happily and without concern. In the Vaupés the potion is always a simple infusion made by pounding the liana and soaking the shredded stems in cold water. Admixtures are used, but the combination is not boiled and concentrated. The result is a milder brew, at least by reputation.

• • •

We were sitting on four small wooden stools placed around the men's circle in the *maloca.* Rufino, like his father Pedro, was wearing a guayuco, as was Pacho. All three had carefully decorated their bodies, tracing lines of red and black dye on their faces and painting their legs with small wooden rollers that left geometric patterns on the skin. Each wore a seed anklet and a simple headdress—a corona of green and yellow parrot feathers, tufts of eagle down, and a long tail feather taken from a scarlet macaw. In the center of the circle was a large red ceramic vessel with swirling designs around the rim. Inside was a frothy liquid. Near the base of the pot were rattles, panpipes, and other musical instruments made from turtle shells and deer skulls. The men had already danced shoulder to shoulder in a line, singing as they circled the pillars of the longhouse. Now they waited quietly. The women and children had long since retired, and the only light came from a resin torch burning at the base of one of the house posts.

Rufino's father stood up and began a solemn chant. When it was over, he dipped a black calabash into the *yagé* and passed it to his son. Rufino grimaced as he drank the potion, as did we all. The taste was bitter and nauseating. There followed more singing and dancing, high tremulous voices and the sound of rattles and anklets. Then there was a hush of expectation as Pedro prepared the next allotment of the brew. It was after the fourth round of *yagé* that I realized the Barasana made up in quantity what their preparation lacked in potency. I had not seen it made, but I knew from Rufino that it included, in addition to the basic liana, the leaves of a plant known as *oco-yajé,* or water *yagé.* This was almost certainly the vine *Diplopterys cabrerana,* a tryptamine-containing admixture used throughout the Northwest Amazon to enhance the brilliance of the visions. Schultes had collected it twice in 1952—once on the Popeyacá and again among the Barasana on the Caño Timiña. Tryptamines are soluble in cold water, and from the number of leaves in his father's recipe, I gathered from Rufino that we were in for quite a ride.

I sat quietly among them, unable to participate yet conscious of the power and authority of their ritual. The plant took them first. In soft murmurs Rufino spoke of a red sun, a red sky, a red rain falling over the forest. Nausea came quickly and he vomited. Immediately his father offered another draft of *yagé,* which he took, spitting and gasping. Until then I had felt nothing, but the sound of his retching caused me to turn aside and throw up in the dirt. Pacho laughed and then did the same. We all took more *yagé,* several more cycles. An hour or more passed. I looked up and saw the edges of the world soften and felt a resonance coming from beyond the sky, like the intimation of a hovering wind pulsating with energy.

At first it was pleasant, a wondrous sense of life and warmth enveloping all things. But then the sensations intensified, became charged with a strange current, and the air itself took on a metallic density. Soon the world as I knew it no longer existed. Reality was not distorted, it was dissolved as the terror of another dimension swept over the senses. The beauty of colors, the endless patterns of orblike brilliance were as rain falling away from my skin. I caught myself and looked up, saw Rufino and Pacho gently swaying and moaning. There were rainbows trapped inside their feathers. In their hair were weeping flowers and trees attempting to soar into the clouds. Leaves fell from the branches with great howling sounds. The sky opened. There was a livid scar across the heavens, stars throbbing, a great wind scattering everything in its path. Then the ground opened. Snakes encircled the posts of the *maloca* and slipped away into the earth. One could not escape. The rivers unfolded like the mouths of blossoms. Movement became penetration. Then the terror grew stronger. Death hovered all around. Ravenous children and animals of every shape and form lay sick and dying of thirst, their nostrils plunged into the dry earth. Their flanks lay bare and exposed, and all around rose a canopy of immense sorrows.

I tried to shake away the forms from the luminous sensations. Instead, my thoughts themselves turned into visions, not of things or places but of an entire dimension that in the moment not only seemed real, but absolute. This was the actual world, and what I had known until then was a crude and opaque facsimile. I looked up and saw my companions. Rufino and Pacho sat quietly, heads down, hunched around a fire that had not been there before. Rufino's father stood apart, arms outspread as he sang. His face was upturned, and his feathered corona shone like the sun. His eyes were brilliant, radiant, feverish, as if focused into the very nature of things.

Slowly, as the night moved forward, the colors softened and the terror receded. I felt my hands running over the dirt floor of the *maloca,* saw dust tinged with green light, heard the voices of women laughing. Dawn was coming. I could hear it in the forest. My companions still remained by the hearth, but the fire had died and the air was cold. I stood and stretched my muscles. Tired but no longer afraid, I slipped into my hammock. For the longest time I lay awake, wrapped in a cotton blanket, like a drained child sweating out the end of a fever. The last thing I saw before drifting off to sleep was a placid cloud of violet light softly descend on the *maloca.*

Some hours later I was awakened by the roar of an airplane passing just over the roof of the longhouse. I looked up and saw narrow shafts of light cutting through the thatch. My head ached and I wanted to drink, but other than that I was fine. I felt clean, as if my body had been washed inside and out. Sitting up, I found myself surrounded by young boys; they followed me outside into the sunlight and down the path that led to the river. The water was cool and refreshing, delicious to drink. There was a shout, and one of the boys pointed to the riverbank. It was the missionary pilot. Beside him stood Rufino and his father. They had packed away their regalia, but their legs still bore decorative motifs, and black dye was smeared across their faces. The pilot had his hands on his hips.

"Gone native, have we?" he called out. "I wouldn't touch that water if I was you."

"You're early," I said.

"Actually, I'm two days late."

"Oh."

"Well, come on, then. I don't have all day. I have to be in Miraflores by noon."

It made for an awkward departure. I gathered my gear and specimens, left what remained of the trade goods with Rufino, and within twenty minutes was airborne, soaring above the *maloca* and over the forest toward Mitú. The sudden shift in perspective was startling. The streams fell behind, grew into rivers, and the rivers spread like serpents through a silent and unchanging forest. Rufino had likened *yagé* to a river, a journey that takes one above the land and below the water, to the most remote reaches of the earth, where the animal masters live and lightning is waiting to be born. To drink *yagé,* Reichel-Dolmatoff wrote, is to return to the cosmic uterus and be reborn. It is to tear through the placenta of ordinary perception and enter realms where death can be known and life traced through sensation to the primordial source of all existence. When shamans speak of facing down the jaguar, it is because they really do.

In the spring of 1953, the last year that he spent in the Amazon basin, Schultes returned to Colombia, met Isidoro in Medellín, and turned south, returning to the valley where he had first encountered the enchantment of a continent. From Sibundoy he traveled east, across the Andes to Mocoa, and from there down the Putumayo to the Sucumbíos, the river of the Kofán, the tribe that had embraced him more than a decade before. It was as if he were saying farewell to lands and peoples that had made him who he was.

In April he returned to the Vaupés for one final expedition. He turned east and followed the river through each of the sixty cataracts that barred the way to Brazil. Then, instead of proceeding downstream, he turned up the Río Papurí, working his way through impossible rapids to the land of the Desana, yet another people of the Anaconda. In June he was back in Mitú waiting for a plane. There followed a final visit to Sibundoy, to see Pedro Juajibioy and touch again the realm of mystery unleashed by *yagé.* On Sunday, July 19, 1953, he was back in Bogotá, en route for America. His last botanical collection occurred that day on the *páramo* of Chisacá, close to sunset as mist enveloped the mountain. The collection number was 20210. It was a simple composite with no apparent use or significance. It would be more than four years before he would collect another plant, and never again would he be free to spend months at a time in the forest.

Tim once said that science and myth were one, that the natural world was but the manifestation of thoughts and impulses occurring on endless metaphysical planes, all enveloped by the mind of the healer. Schultes said he didn't understand, but I know he did. Often I think of how it must have been for him all those years ago. And what comes to mind is my recollection of Tim, standing in the forest, hands reaching for some curious plant. When I look at photographs of Schultes in the field—in Oklahoma with Weston La Barre, straddling a mule in Oaxaca, with Pacho López on the Río Negro—what I see is Tim, his posture, his eyes, his sense of wonder and delight. Schultes is an old man now, and like all old men he has forgotten much of what he knew and did. But he never forgets Tim, and in his memories they have merged, student and teacher, father and son, like two branches of a river flowing into one.

Notes on Sources

I BEGAN THIS project by conducting some thirty hours of interviews with Professor Schultes in the spring of 1990. Consisting of reminiscences and somewhat more elaborate accounts of many of the same stories and anecdotes I had heard over the years at Harvard, the transcripts nevertheless provided a framework from which to proceed. Perhaps most significant, they revealed Schultes's strengths and weaknesses as a source of information about his life and times. His recollections and sense of chronology were vague, his capacity for introspection limited. Although proud of his achievements, it had never dawned on him to place his work in historical context. On the subject of plants, however, his memory proved almost eerily infallible. I soon learned that if I could anchor a question to a specific botanical collection, then, as if by magic, people, places, dates, and events would emerge.

With the exception of three months in the fall of 1947, Schultes never kept a journal during his many years in the field. But like all botanists he recorded the date and location, and assigned a personal number to each of his collections. Most of these data he kept in a series of notebooks. Between 1951 and 1953, however, while back on the Apaporis, he used yellow notepads, one inch by two inches in size, with each paper sheet numbered top and bottom so that he could write at one end and simply tear off the number and place it with the specimen. This saved him time in the field, though it was destined to frustrate a biographer who had to work through eight thousand small bits of paper. The effort was worthwhile, for these botanical notes, despite some inconsistencies, provided a wealth of information and were an essential aid in reconstructing chronologies and itineraries. The types of plants found, the number of collections made, the ecological settings, and the data on localities, collectors, and vernacular names together painted a portrait of his days in the forest. In five years of research I became familiar with most of his localities and many of the twenty-seven thousand or more plants that he collected.

Second only to his passion for botany was Schultes's interest in photography. He took thousands of photographs, mostly of plants but also of scenes and moments that captured his imagination. These he filed away, placing each negative in a small envelope and gluing a two-inch-square print to the upper left-hand corner on the outside; on the envelope itself he wrote a short caption and noted the date the picture was taken. Most of these images, including some of his finest photographs, have never been published. Without them I could never have written this book as I did. It was from the snapshots, in particular, that I culled all the small details of dress and deportment, of appearances and gestures, of landscape, forests, planes, riverboats, mission posts, and cities and tribes long since transformed and laid waste by half a century of violent change.

Although Schultes disliked working for the government, the fact that he did for most of his years in the Amazon provided another invaluable source of information. While many documents have been lost or discarded, several of his most important field reports are preserved at the U.S. National Archives. These provided maps, daily logs, itineraries, as well as lists of personnel and equipment. They proved especially useful in reconstructing the early years of the rubber effort, his exploration of the Apaporis, and the subsequent work at Leticia. It was in searching for these documents that I found the declassified papers outlining the scandalous course of events that led in 1954 to the termination of the rubber program.

A final element, and the one that reached beyond the notebooks, photographs, and displaced documents, was the cadre of men and women with whom Schultes worked between 1936 and 1953. In Colombia I interviewed Hernando García Barriga, Alvaro Fernández-Pérez, Isidoro Cabrera, Roberto Jaramillo, and Pedro Juajibioy. My earlier travels brought me to Jorge Fuerbringer, Frederico Medem, and Gerardo Reichel-Dolmatoff. At Harvard I spoke with Gordon Wasson, and I found Eunice Pike alive and well in Texas. Helen Floden, the missionary of Leticia, wrote from Florida, Irmgard Weitlaner from outside of Mexico City, William Burroughs from Kansas. In North Carolina I spent three days in a retirement home with Weston La Barre. Wayne White spoke from Arkansas. Russell Seibert and Ernie Imle, both of the rubber program, are retired in Florida and Maryland respectively. Albert Hofmann is eighty-nine and living in Switzerland.

Fuerbringer, Medem, Reichel-Dolmatoff, Fernández, Wasson, and, of course, Tim Plowman have since died. All the others, including Schultes, are well on in years. I have been fortunate indeed to have had the chance as a writer and student to know these remarkable individuals.

The following references are not exhaustive. They simply provide a road map and represent a few of the sources consulted in the writing of this book.

Chapter One. Juan's Farewell

The coca project at the Botanical Museum began with Dick Martin, who published a superb review paper, considered a classic in the field. See Martin, R. T., "The Role of Coca in the History, Religion and Medicine of South American Indians," *Economic Botany* 24(4): 422–38, 1970. After Martin dropped out of graduate school, the coca assignment passed to Tim Plowman. Of Tim's eighty published papers, forty-six are related to *Erythroxylum.* Of these the most important are "Botanical Perspectives on Coca," *Journal of Psychedelic Drugs* 11(1–2): 103–17, 1979; "The Identity of Amazonian and Trujillo Coca," *Botanical Museum Leaflets* 27: 45–68, 1979; "Amazonian Coca," *Journal of Ethnopharmacology* 3: 195–225, 1981; "The Identification of Coca (*Erythroxylum* Species): 1860–1910," *Botanical Journal of the Linnean Society* 84(4): 329–53, 1982; "The Ethnobotany of Coca," *Advances in Economic Botany* 1: 62–111, 1984; "The Origin, Evolution and Diffusion of Coca, *Erythroxylum spp.,* in South and Central America," in Stone, D. (ed.), *Pre-Columbian Plant Migration,* papers of the Peabody Museum of Archaeology and Ethnology, vol. 76: 125–63, 1984.

Richard Evans Schultes is the author of ten books and 496 scientific papers. His most important works are *The Healing Forest: Medicinal and Toxic Plants of the Northwest Amazon* (with R. Raffauf), Dioscoroides Press, Portland, Oregon, 1990; *The Botany and Chemistry of Hallucinogens* (with A. Hofmann), Charles C. Thomas, Springfield, Illinois, 1980, and *Plants of the Gods* (with A. Hoffmann), McGraw-Hill, New York, 1979. He has also published two books of photography: *Where the Gods Reign,* Synergetic Press, Oracle, Arizona, 1988, and *Vine of the Soul* (with R. Raffauf), Synergetic Press, Oracle, Arizona, 1992.

The species problem in cannabis is discussed in Anderson, L. C., "A Study of Systematic Wood Anatomy in Cannabis," *Botanical Museum Leaflets* 24(2): 29–36, 1974; Emboden, W. A., "Cannabis—a Polytypic Genus," *Economic Botany* 28(3): 304–10, 1974; Fullerton, D. S., and M. C. Kurzman, "The Identification and Misidentification of Marijuana," *Contemporary Drug Problems,* 1974, pp. 291–344; Schultes, R. E., "Random Thoughts and Queries on the Botany of Cannabis," in Joyce, C. R. B., and S. H. Curry (eds.), 1970, 11–38; *The Botany and Chemistry of Cannabis,* J. & A. Churchill, London, 1970; Schultes, R. E., et al., "Cannabis: An Example of Taxonomic Neglect," *Botanical Museum Leaflets* 23(9): 337–67, 1974. Schultes's main antagonist in the debate was Ernest Small. For his views see Small, E., *The Species Problem in Cannabis: Science and Semantics,* 2 vols., Canada Corpus, Toronto, 1979.

Tim's near fatal poisoning by *Brunfelsia chiricaspi* occurred at Santa Rosa on December 3, 1968, and is recounted in Plowman, T., *"Brunfelsia* in Ethnomedicine," *Botanical Museum Leaflets* 25: 289–320, 1977. For further details see his dissertation thesis, *The South American Species of Brunfelsia (Solanaceae),* Harvard University, Cambridge, Massachusetts, 1974. In reconstructing events from my travels with Tim, I used his field notebooks and journal as well as the more extensive daily entries in my own journals of the time.

Chapter Two. Mountains of the Elder Brother

For insight into the world of the Sierra Nevada, the publications of Gerardo Reichel-Dolmatoff are the fundamental source. His monograph, published in two volumes in 1950 and 1951, was reprinted in 1985. See Reichel-Dolmatoff, G., *Los Kogi: Una Tribu de la Sierra Nevada de Santa Marta, Colombia,* 2 vols. Procultura, Nueva Biblioteca Colombiana, Editorial Presencia, Bogotá, 1985. For other Reichel-Dolmatoff publications see:

———. "Contactos y cambios culturales en la Sierra Nevada de Santa Marta," *Revista Colombiano de Antropología,* 1: 15–122, 1953.

———. "Training for the Priesthood Among the Kogi of Colombia," *in* J. Wilbert (ed.), *Enculturation in Latin America: An Anthology,* Latin American Center, University of California, Los Angeles, 1976.

———. "Templos Kogi: Introducción al Simbolismo y la Astronomía del Espacio Sacrado," *Revista Colombiana de Antropología* 19: 199–246, 1977.

———. "The Loom of Life: A Kogi Principle of Integration," *Journal of Latin American Lore* 4(1): 5–27, 1978.

———. "Some Kogi Models of the Beyond," *Journal of Latin American Lore* 10(1): 63–85, 1982.

———. "The Great Mother and the Kogi Universe: A Concise Overview," *Journal of Latin American Lore* 13(1): 73–113, 1987.

———. "Análisis de un Templo de los Indios Ika, Sierra Nevada de Santa Marta, Colombia," *Antropologica* 68: 3–22, 1987.

———. *Notas Etnográficas Sobre los Indios Ika de la Sierra Nevada de Santa Marta, 1946–1966,* Universidad Nacional de Colombia, Bogotá, 1991.

———. *Indians of Colombia: Experience and Cognition,* Villegas Editores, Bogotá, 1991.

For a fine travel book with wonderful passages on the Kogi, see Moser, B., and D. Tayler, *The Cocaine Eaters,* Longmans, Green and Co., London, 1965. For a disturbing and evocative sense of the horror of the Conquest, see Eduardo Galeano's astonishing trilogy, *Memory of Fire* (translated by Cedric Belfrage), Pantheon Books, New York, 1985, 1987, 1988. See also Galeano, E., *Open Veins of Latin America,* Monthly Review Press, New York, 1973. For a discussion of the demographic impact of disease, see Bethell, L. (ed.), *Cambridge History of Latin America,* vol. 3, Cambridge University Press, Cambridge, England, 1984; Cook, N. D., and Lovell, W. G. (eds.), *Secret Judgements of God,* University of Oklahoma Press, Norman, 1992; Hemming, J., *The Conquest of the Incas,* Harcourt Brace Jovanovich, New York, 1970; Crosby, A., *Ecological Imperialism: The Biological Expansion of Europe, 900–1900,* Cambridge University Press, Cambridge, England, 1986; Crosby, A., *The Columbian Exchange: Biological and Cultural Consequences of 1492,* Greenwood Press, Westport, Connecticut, 1972.

Chapter Three. The Peyote Road, 1936

The photo album belonged to Weston La Barre and is deposited with his papers at the National Anthropological Archives, National Museum of Natural History, Smithsonian Institution, Washington, D.C. Also among his papers is an unpublished manuscript, *The Autobiography of a Kiowa Indian,* 184 pages. This oral history of Charlie Charcoal, recorded by La Barre in 1936, contains vivid descriptions of peyote ritual and visions. This document, together with photographs, La Barre's journal notes and early publications, and Schultes's detailed account of his own peyote experience presented in his undergraduate thesis, allowed me to reconstruct their ritual intoxication with some precision.

For Schultes's early family life I depended on interviews with him, his wife, Dorothy, and various family members. His sister, Clara Loring, kindly shared letters and the unpublished manuscript of her autobiography. On several occasions Schultes himself walked me through the streets of his old neighborhood. For a brief history of East Boston, see *East Boston: Boston 200 Neighborhood History Series,* Boston 200 Corporation, Boston, Massachusetts, 1976.

For biographical information on Oakes Ames, including correspondence between him and Schultes, see Plimpton, P. A. (ed.), *Oakes Ames: Jottings of a Harvard Botanist, 1874–1950,* Harvard University Press, Cambridge, Massachusetts, 1979. For Ames on cultivated plants, see *Economic Annuals and Human Cultures,* Botanical Museum of Harvard University, Cambridge, Massachusetts, 1939.

Schultes first read about peyote in Heinrich Klüver's *Mescal: The Divine Plant and Its Psychological Effects,* Kegan Paul, Trench, Trubner & Co., London, 1928. The book later appeared as *Mescal and Mechanisms of Hallucinations,* University of Chicago Press, Chicago, 1966. Schultes as a young student learned of Tarahumara and Huichol use of peyote in Lumholtz, C., *Unknown Mexico,* 2 vols., Charles Scribner's Sons, New York, 1902. Also available to him was Mooney, J., *Calendar History of the Kiowa,* Seventeenth Annual Report of the Bureau of American Ethnology, Smithsonian Institution, Washington, D.C., 1898. Weston La Barre's *The Peyote Cult* first appeared in 1938. An expanded edition that included several of La Barre's later papers on peyote was published in 1975 by Archon Books, Hamden, Connecticut.

For other sources on peyote, see Anderson, E., *Peyote: The Divine Cactus,* University of Arizona Press, Tucson, 1980; Furst, P., *Hallucinogens and Culture,* Chandler & Sharp, San Francisco, 1976; Myerhoff, B., *Peyote Hunt: The Sacred Journey of the Huichol Indians,* Cornell University Press, Ithaca, New York, 1974; Stewart, O. C., *Peyote Religion,* University of Oklahoma Press, Norman, 1987. For a sense of the wonder of Kiowa life, see Momaday, N. S., *The Way to Rainy Mountain,* University of New Mexico Press, Albuquerque, 1969.

Schultes's 1936 unpublished undergraduate thesis, *Peyotl Intoxication: A Review of the Literature on the Chemistry, Physiological and Psychological Effects of Peyotl,* is deposited at the Economic Botany Library at Harvard University, Cambridge, Massachusetts. His publications on peyote and the Kiowa include the following:

———. "Peyote and Plants Used in the Peyote Ceremony," *Botanical Museum Leaflets* 4(8): 129–52, 1937.
———. "Peyote and Plants Confused with It," *Botanical Museum Leaflets* 5(5): 61–88, 1937.
———. "Peyote and the American Indian," *Nature Magazine* 30: 155–57, 1937.
———. "The Appeal of Peyote as a Medicine," *American Anthropologist* 40(4): 698–715, 1938.
———. "Peyote—an American Indian Heritage from Mexico," *El Mexico Antiguo* 4: 199–208, 1938.
———. *The Economic Botany of the Kiowa Indians* (with Paul Vestal), Botanical Museum of Harvard University, Cambridge, Massachusetts, 1939.

Both Schultes and La Barre testified against Senate Bill 1399, as did Franz Boas, A. L. Kroeber, and other prominent anthropologists. For their submissions, see *Documents on Peyote,* Document no. 137817, pt.1, February 8, Mimeographed Bureau Report on S. 1399, 75th Congress, 1st Session, U.S. Bureau of Indian Affairs, Washington, D.C., 1937.

Chapter Four. Flesh of the Gods, 1938–39

The early Spanish reports on *teonanacatl* are reviewed by Schultes in *"Plantae Mexicanae II:* The Identification of *Teonanacatl,* a Narcotic Basidiomycete of the Aztecs," *Botanical Museum Leaflets* 7(3): 37–55, 1939, and *"Teonanacatl,* the Narcotic Mushroom of the Aztecs," *American Anthropologist* 42(3): 429–43, 1940. For Schultes on *ololiuqui,* see *A Contribution to Our Knowledge of Rivea Corymbosa: The Narcotic Ololiuqui of the Aztecs,* Botanical Museum of Harvard University, Cambridge, Massachusetts, 1941; and "The Devil's Morning Glory," *Nature Magazine* 49(9): 463–64, 1956.

For William E. Safford's theories on *ololiuqui* and *teonanacatl,* see Safford, W. E., "An Aztec Narcotic," *Journal of Heredity* 6(7): 291–311, 1915; and "Narcotic Plants and Stimulants of the Ancient Americas," *Annual Report of the Smithsonian Institution for 1916,* pp. 387–441, 1917.

A copy of Blas Pablo Reko's letter of July 18, 1923, to J. N. Rose is preserved on herbarium sheet number 1745713, United States National Herbarium, Smithsonian Institution, Washington, D.C. Reko first identified *nanacate* as a mushroom and *ololiuqui* as a morning glory in Reko, B. P., "De los Nombres Botanicos Aztecas," *El Mexico Antiguo* 1: 113–57, 1919.

Schultes's field itineraries for 1938 and 1939 are outlined in his unpublished doctoral dissertation *Economic Aspects of the Flora of Northeastern Oaxaca, Mexico,* 2 vols., Harvard University, Cambridge, Massachusetts, 1941. Deposited at the Economic Botany Library at Harvard, the thesis is illustrated with forty-six photographs taken by Schultes and ten by Eunice Pike. An interview with Ms. Pike, together with subsequent correspondence, provided many additional insights. She and her fellow missionary, Florence Hansen Cowan, discuss the difficulty of explaining the Christian faith to the Mazatec in Pike, E., and F. Cowan, "Mushroom Rituals Versus Christianity," *Practical Anthropology* 6(4): 145–50, 1959. For an account of her experience in Huautla and a sense of the town in the years when Schultes was there, see Pike, E., *Not Alone,* Summer Institute of Linguistics, Academic Publications, Dallas, Texas, 1964.

For Mazatec and Chinantec ethnography and landscape, see Bevan, B., "Report on the Central and Southern Chinantec Region," vol. 1, *The Chinantec and Their Habitat,* Publication 24, Instituto Panamericano Geografico y Historia, 1938; Cowan, G. M., "Mazateco Whistle Speech," *Language* 24(3): 280–86, 1948; "Mazateco House Building," *Southwestern Journal of Anthropology* (2): 375–90, 1946; Cowan, F. H., "A Mazateco President Speaks," *América Indígena* 12(4): 323–41, 1952; "Notas Etnográficas Sobre los Mazatecos de Oaxaca, México," *América Indígena* 6(1): 27–39, 1946; "Linguistic and Ethnological Aspects of Mazateco Kinship," *Southwestern Journal of Anthropology,* vol. 3: 247–56, 1947; Weitlaner, R. J., and Weitlaner, I., "The Mazatec Calendar," *American Antiquity* 11(3): 194–97, 1945.

Before his untimely death, Jean Bassett Johnson published several important papers. For an account of his discoveries, see: "The Elements of Mazatec Witchcraft," *Ethnological Studies* No. 9, pp.128–50, Gothenburg Ethnographical Museum, 1939; "Some Notes on the Mazatec," *Revista Mexicana de Estudios Antropológicos* 1(2): 142–56, 1939; "Note on the Discovery of *Teonanacatl,*" *American Anthropologist* 42: 549–50, 1940. In this brief communication, Johnson properly identifies R. J. Weitlaner as the scholar who first secured the mushrooms from José Dorantes. Weitlaner obtained the specimens during Easter week, 1936, and later passed them along to B. P. Reko. When Reko sent them to Schultes, he made no mention of Weitlaner. Thus, in his first paper on the subject ("Plantae Mexicanae II: The Identification of *Teonanacatl,* a Narcotic Basidiomycete of the Aztecs," *Botanical Museum Leaflets* 7[3]: 37–55, 1939) Schultes failed to acknowledge Weitlaner's contribution. This led to a misunderstanding, which was cleared up in a series of letters between Johnson and Schultes in late 1939. For details, see Schultes to La Barre, January 30 and February 3, 1940, letters among La Barre's papers deposited at the National Anthropological Archives, Washington, D.C.

Irmgard Weitlaner, widow of Jean Johnson and daughter of Robert Weitlaner, provided useful insights into the dramatic events that led to the identification of *teonanacatl.* Ms. Weitlaner explored Oaxaca with her father and later was a member of the Bevan expedition. She spent a month among the Chinantec in 1935, traversed the Chinantl in 1936, and was with Bevan, Johnson, and Louise Lacaud when they first witnessed the mushroom ceremony in Huautla in 1938.

Wasson's correspondence with Schultes, La Barre, Robert Weitlaner, and Robert Graves is among the papers of the Tina and Gordon Wasson Ethnomycological Collections deposited at Harvard University. For biographical notes on Wasson, see Riedlinger, T. J. (ed.), *The Sacred Mushroom Seeker: Essays for R. Gordon Wasson,* Dioscoroides Press, Portland, Oregon, 1990. Wasson's account of taking *teonanacatl,* "Seeking the Magic Mushrooms," appeared in *Life* magazine on May 13, 1957. See also Wasson, R. G., et al., *María Sabina and Her Mazatec Mushroom Velada* (New York: Harcourt Brace Jovanovich, 1974); Wasson, R. G., et al., *Persephone's Quest:*

Entheogens and the Origins of Religion, Yale University Press, New Haven, Connecticut, 1986. Also of interest is Wasson, R. G., "Notes on the Present Status of *Ololiuhqui* and the Other Hallucinogens of Mexico," *Botanical Museum Leaflets* 20(6): 161–93, 1963. For the impact of Wasson's work on the life of María Sabina, see Estrada, A., *María Sabina: Her Life and Chants,* Ross-Erikson, Santa Barbara, California, 1981.

In the early 1960s the work of Timothy Leary and Richard Alpert, together with the presence of Schultes and the affiliation of Gordon Wasson with the Botanical Museum, made Harvard a major center for the study of hallucinogens. In the summer of 1963, Andrew Weil, then an undergraduate student of Schultes, edited "Drugs and the Mind," *The Harvard Review,* vol. 1, no. 4, 1963. The articles include personal accounts of psychedelic experiences, visionary ramblings of Alpert and Leary ("The Politics of Consciousness Expansion," pp. 33–38), and more sober contributions from Schultes ("Hallucinogenic Plants of the New World," pp. 18–33) and Wasson ("The Mushroom Rites of Mexico," pp. 7–18). See also Schultes, R. E., "Botanical Sources of the New World Narcotics," *The Psychedelic Review* 1(2): 145–67, 1963.

Albert Hofmann describes his revelations in *LSD: My Problem Child,* McGraw-Hill, New York, 1980. See also Hofmann, A., "How LSD Originated," *Journal of Psychedelic Drugs* 11(1–2): 53–68, 1979; "The Discovery of LSD and Subsequent Investigations on Naturally Occurring Hallucinogens" *in* Ayd, F. J., and B. Blackwell (eds.), *Discoveries in Biological Psychiatry,* Lippincott, Philadelphia, 1970. For Hofmann's work with morning glories, see "The Active Principles of the Seeds of *Rivea Corymbosa* and *Ipomoea Violacea," Botanical Museum Leaflets* 20(6): 194–212, 1963. See also Hofmann, A., "History of the Basic Chemical Investigations on the Sacred Mushrooms of Mexico," *in* Ott, J., and J. Bigwood (eds.), *Teonanácatl: Hallucinogenic Mushrooms of North America,* Madrona Publishers, Seattle, Washington, 1978.

Chapter Five. The Red Hotel

An account of the Darien expedition appears in Sebastian Snow's *The Rucksack Man,* Hodder & Stoughton, London, 1976. For a history of Colombia, see Bushnell, D., *The Making of Modern Colombia,* University of California Press, Los Angeles, 1993. For a sense of Bogotá before the fall, see Mendoza de Riaño, C., *Así Es Bogotá,* Ediciones Gamma, Bogotá, 1988. For an oral history of the *Bogotazo,* see Alape, A., *El Bogotazo: Memorias del Olvido,* Planeta Colombiano Editorial, Bogotá, 1983. For studies of *La Violencia,* see Pécaut, D., *Orden y Violencia: Colombia 1930–1954,* 2 vols, CEREC, Bogotá, 1987. *La Vorágine* was translated into English by E. K. James. See Rivera, J. E., *The Vortex,* G. P. Putnam's Sons, New York, 1935. For Schultes's work as a translator in Mexico in the summer of 1941, see Stakman, E. C., et al., *Campaigns Against Hunger,* Belknap Press, Cambridge, Massachusetts, 1967.

The description of *Rhytidanthera regalis* was published in Schultes, R. E., "Plantae Colombianae XIV: Rhytidantherae Montis Macarenae Nova Species," *Botanical Museum Leaflets* 16(5): 106–11, 1953.

For Paez ethnography and use of coca, see Antonil, *Mama Coca,* Hassle Free Press,

London, 1978. Yohimbine, a stimulant said to prolong erections in men, is derived from the bark of an African tree, *Pausinystalia johimbe.* For therapeutic use, see Weil, A., *Natural Health, Natural Medicine,* Houghton Mifflin, Boston, 1990. For two wonderful books on the cocaine trade and the spirit of the open road in Colombia, see Nicholl, C., *The Fruit Palace,* William Heinemann, London, 1985, and Sabbag, R., *Snowblind,* Bobbs-Merrill, Indianapolis, Indiana, 1976.

Schultes published more than a hundred papers on hallucinogens and stimulants. For the best review papers, see "Antiquity of the Use of New World Hallucinogens," *Archeomaterials* 2(1): 59–72, 1987; "The Place of Ethnobotany in the Ethnopharmacological Search for Psychotomimetic Drugs," *in* Efron, D., et al. (eds.), *Ethnopharmacological Search for Psychoactive Drugs,* pp. 33–57, Government Printing Office, Washington, D.C., 1967; "The Plant Kingdom and Hallucinogens," *Bulletin on Narcotics* (United Nations), 21(2): 3–16, 1969; 21(3): 15–27, 1969; 22(1): 25–53, 1970.

For a discussion of *Iochroma fuchsiodes,* see Schultes, R. E., "A New Hallucinogen from Andean Colombia: *Iochroma fuchsiodes," Journal of Psychedelic Drugs* 9(1): 45–49, 1977. For reference to Guambiano use of *Brugmansia,* see Schultes, R. E., and A. Bright, "A Native Drawing of an Hallucinogenic Plant from Colombia," *Botanical Museum Leaflets* 25(6): 151–59, 1977. For more on the tree of the evil eagle, see Lockwood, T. E., "The Ethnobotany of *Brugmansia," Journal of Ethnopharmacology* 1: 147–64, 1979; Bristol, M. L., "Notes on the Species of Tree Daturas," *Botanical Museum Leaflets* 21(8): 229–48, 1966; Bristol, M. L., "Tree Datura Drugs of the Colombian Sibundoy," *Botanical Museum Leaflets* 22(5): 165–227, 1969. For the hexing herbs in European sorcery, see Hansen, H. A., *The Witch's Garden,* Unity Press, Santa Cruz, California, 1978; Heiser, C. B., *Nightshades: The Paradoxical Plants,* W. H. Freeman & Co., San Francisco, 1969. For use among the Jivaro, see Harner, M. J. (ed.), *Hallucinogens and Shamanism,* Oxford University Press, London, 1973.

For the archaeology of San Augustín, see Reichel-Dolmatoff, G., *San Augustín: A Culture of Colombia,* Praeger Publishers, New York, 1972. For discussion of pre-Columbian use of psychoactive mushrooms in Colombia, see Schultes, R. E., and A. Bright, "Ancient Gold Pectorals from Colombia: Mushroom Effigies?" *Botanical Museum Leaflets* 27(5–6): 113–65, 1979.

For Schultes and Burroughs, see Burroughs, W. S., and A. Ginsberg, *The Yage Letters,* City Lights Books, San Francisco, 1953; Harris, O. (ed.), *The Letters of William Burroughs 1945–1959,* Viking, New York, 1993; Morgan, T., *Literary Outlaw: The Life and Times of William S. Burroughs,* Henry Holt, New York, 1988.

Chapter Six. The Jaguar's Nectar

Melvin Bristol, a graduate student of Schultes, spent thirteen months in the valley of Sibundoy in 1962–63. See Bristol, M. E., "Sibundoy Ethnobotany," doctoral dissertation, Harvard University, Cambridge, Massachusetts, 1965. For healing practices, see Seijas, H., "The Medical System of the Sibundoy Indians of Colombia," doctoral dissertation, Tulane University, New Orleans, Louisiana, 1969. For *yagé* use by the Kamsa and Inga, see Bristol, M. E., "The Psychotropic *Banisteriopsis* Among the Sibundoy of Colombia," *Botanical Museum Leaflets* 21(5): 113–40,1966. For Schultes on hallucinogenic plants of Sibundoy, see *"Desfontainia:* A New Andean Hallucinogen," *Botanical Museum Leaflets* 25(3): 99–104, 1977; "A New Narcotic Genus from the Amazon Slope of the Colombian Andes," *Botanical Museum Leaflets* 17(1): 1–11, 1955; "A New Plant Source of Narcotic Drugs: *Methysticodendron Amesianum," Bulletin on Narcotics,* pp.1–4, October–December 1956; "A Powerful New Narcotic," *The Chemist and Druggist,* p. 200, August 25, 1956.

For the expedition of Pérez de Quesada, see Hemming, J., *The Search for El Dorado,* Michael Joseph, London, 1978. For an extraordinary analysis of the history of Sibundoy and the upper Putumayo, see Taussig, M., *Shamanism, Colonialism, and the Wild Man: A Study in Terror and Healing,* University of Chicago Press, Chicago, 1987. For a critical examination of the power and influence of the Capuchin priests, see Bonilla, V. D., *Servants of God or Masters of Men? The Story of a Capuchin Mission in Amazonia,* Penguin, Harmondsworth, England, 1972.

For Schultes on guayusa, see *"Ilex guayusa* from 500 A.D. to the Present," *Etnologiska Studier* 32: 115–38, Gothenburg, Sweden, 1972; "Discovery of an Ancient Guayusa Plantation in Colombia," *Botanical Museum Leaflets* 27(5–6): 143–60, 1979. See also Patiño, V. M., "Guayusa, a Neglected Stimulant from the Eastern Andean Foothills," *Economic Botany* 22: 310–16, 1968; Shemluck, M. J., "The Flowers of *Ilex guayusa," Botanical Museum Leaflets* 27(5): 155–60, 1990.

There is a vast literature on *yagé* and *ayahuasca (Banisteriopsis caapi).* Basic botanical and ethnobotanical sources include Gates, B., *Banisteriopsis, Diplopterys (Malpighiaceae), Flora Neotropica* (Monograph No. 30), The New York Botanical Garden, Bronx, New York, 1982; Rivier, L., and J.-E. Lindgren, "Ayahuasca, the South American Hallucinogenic Drink: An Ethnobotanical and Chemical Investigation," *Economic Botany* 26: 101–29, 1972; Schultes, R. E., "The Identity of the Malpighiaceous Narcotics of South America," *Botanical Museum Leaflets* 18(1): 1–56, 1957; Schultes, R. E., "The Beta-carboline Hallucinogens of South America, *Journal of Psychoactive Drugs* 14(3): 205–20, 1982.

For ethnological studies, see Reichel-Dolmatoff, G., *The Shaman and the Jaguar: A Study of Narcotic Drugs Among the Indians of Colombia,* Temple University Press, Philadelphia, 1975. For a collection of fine papers, including contributions from Schultes, Jean Langdon, Dennis McKenna, Bronwen Gates, Luis Luna, Anthony Henman, and others, see *América Indígena* 46(1): 5–256, 1986. For more on *yagé* use among the Siona, see the papers of Jean Langdon:

———. "Yagé Among the Siona: Cultural Patterns in Visions" *in* Browman, D. L., and R. A. Schwarz (eds.), *Spirits, Shamans, and Stars,* Mouton Publishers, The Hague, 1979.

———. "The Siona Hallucinogenic Ritual: Its Meaning and Power" *in* Morgan, J. (ed.), *Understanding Religion and Culture: Anthropological and Theoretical Perspectives,* University Press of America, Washington, D.C., 1979.

———. "Cultural Bases for Trading of Visions and Spiritual Knowledge in the

Colombian and Ecuadorian Montaña" *in Networks of the Past: Regional Interaction in Archeology,* Proceedings of the 12th Annual Conference of the Archaeological Association of the University of Calgary, Calgary, Alberta, Canada, 1981.

Chapter Seven. The Sky Is Green and the Forest Blue, 1941–42

Schultes described *yoco* as *Paullinia yoco* in Schultes, R. E., *"Plantae Colombianae II—*Yoco: A Stimulant of Southern Colombia," *Botanical Museum Leaflets* 10(10): 301–24, 1942. Schultes's discovery of *Herrania breviligulata* stimulated his interest in wild species of this close relative of cacao. Seventeen years later, in one of his most significant taxonomic contributions, he published a revision of the genus. See Schultes. R. E., "A Synopsis of the Genus *Herrania," Journal of the Arnold Arboretum* 39: 216–78, 1958.

Between 1967 and 1986, Schultes published a series of thirty-eight numbered articles on biodynamic and toxic plants, principally of the Northwest Amazon. Appearing under the weighty title *De Plantis Toxicariis E Mundo Novo Tropicale Commentationes* and published mainly in the *Botanical Museum Leaflets,* these papers consist for the most part of botanical descriptions and ethnopharmacological notes culled from his collections. The plants are listed alphabetically by genus, and in each case Schultes includes information on date, locality, altitude, and names of collectors. These papers proved of great value to my work. By cross-referencing with his field notebooks, I could trace the pattern of his discoveries as he moved throughout the Northwest Amazon. Schultes drew heavily on these obscure articles when he wrote *The Healing Forest: Medicinal and Toxic Plants of the Northwest Amazon,* Dioscoroides Press, Portland, Oregon, 1990.

In the 1960s and early 1970s, at a time when the major pharmaceutical companies had virtually abandoned natural products research, Schultes resisted the trend and wrote a series of review papers calling for increased attention to the medicinal potential of plants. See Schultes, R. E., "The Widening Panorama in Medicinal Botany," *Rhodora* 65(762): 97–120, 1963. In this paper, edited from a speech delivered at the University of West Virginia, October 29, 1962, he warns of a looming crisis that few at the time could anticipate.

"Civilization is on the march in many, if not most, primitive areas. It has long been on the advance, but its pace is now accelerated as the result of world wars, extended commercial interests, increased missionary activity, widened tourism. The rapid divorcement of primitive peoples from dependence upon their immediate environment for the necessities and amenities of life has been set in motion, and nothing will check it now. One of the first aspects of primitive culture to fall before the onslaught of civilization is knowledge and use of plants for medicines. The rapidity of this disintegration is frightening. Our challenge is to salvage some of the native medico-botanical lore before it becomes forever entombed with the cultures that gave it birth."

For Schultes on the potential of medicinal plants, see also:

———. "The Plant Kingdom and Modern Medicine," *The Herbalist,* pp.18–26, 1968.
———. "From Witch Doctor to Modern Medicine: Searching the American Tropics for Potentially New Medicinal Plants," *Arnoldia* 32(5): 198–219, 1972.
———. "The Future of Plants as Sources of New Biodynamic Compounds," *in* Swain, T. (ed.), *Plants in the Development of Modern Medicine,* Harvard University Press, Cambridge, Massachusetts, 1972.
———. "The Plant Kingdom: A Virgin Field for New Biodynamic Constituents," *in* Fina, N. J. (ed.), *The Recent Chemistry of Natural Products, Including Tobacco: Proceedings of the Second Philip Morris Science Symposium,* Richmond, Virginia, 1976, pp. 134–71.

For a history of curare, see McIntyre, A. R., *Curare: Its History, Nature and Clinical Use,* University of Chicago Press, Chicago, 1947; Thomas, K. Bryn, *Curare: Its History and Usage,* J. B. Lippincott, Philadelphia, 1963. For the travels of Raleigh, La Condamine, Humboldt, Waterton, Martius, Schomburg, Wallace, Bates, and Spruce, see Goodman, E. J., *The Explorers of South America,* University of Oklahoma Press, Norman, 1972; Smith, A., *Explorers of the Amazon,* Viking, New York, 1990; Von Hagen, V. W., *South America Called Them,* Knopf, New York, 1945. For Waterton's adventures, see Waterton, C., *Wanderings in South America,* Dutton, New York, 1895. For the saga of Richard Gill, see Gill, R. C., *White Water and Black Magic,* Henry Holt, New York, 1940; Humble, R. M., "The Gill-Merrill Expedition: Penultimate Chapter in the Curare Story," *Anesthesiology* 57: 519–26, 1982.

For the major medical breakthroughs, see Griffith, H. R., and G. E. Johnson, "The Use of Curare in General Anesthesia," *Anesthesiology* 3: 418–20, 1942; King, H., "Curare Alkaloids I. Tubocurarine," *Journal of the Chemical Society* (London) 2: 13–81, 1935; Wintersteiner, C., and J. D. Dutcher, "Curare Alkaloids from *Chondodendron tomentosum,*" *Science* 97: 467, 1943. For taxonomic studies of the botanical sources of curare, see Krukoff, B. A., and H. N. Moldenke, "Studies of American *Menispermaceae,* with Special Reference to Species Used in Preparation of Arrow Poisons," *Brittonia* 3(1): 1–74, 1938; Krukoff, B. A., and A. C. Smith, "Notes on the Botanical Components of Curare," *Bulletin of the Torrey Botanical Club* 64: 401–9, 1937; 66: 305–14, 1939; Krukoff, B. A., and J. Monachino, "The American Species of *Strychnos,*" *Brittonia* 4(2): 248–322, 1942.

For a discussion of *ayahuasca* admixtures, see Ott, J., *Ayahuasca Analogues,* Natural Products Co., Keenewick, Washington, 1994; Pinkley, H., "Plant Admixtures to Ayahuasca, the South American Hallucinogenic Drink," *Lloydia* 32(3): 305–14, 1969; Schultes, R. E., "Ethnotoxicological Significance of Additives to New World Hallucinogens," *Plant Science Bulletin* 18: 34–41, 1972. For the most insightful recent work, see McKenna, D., and G. H. N. Towers, "Biochemistry and Pharmacology of Tryptamines and Beta-carbolines: A Minireview," *Journal of Psychoactive Drugs* 16(4): 347–58, 1984; McKenna, D. J., et al., "Monoamine Oxidase Inhibitors in South American Hallucinogenic Plants: Tryptamine and Beta-carboline Constituents of Ayahuasca," *Journal of Ethnopharmacology* 10: 195–223, 1984; McKenna, D. J., and G. H. N. Towers, "On the Comparative Ethnopharmacology of Malpighiaceous and Myristicaceous Hallucinogens," *Journal of Psychoactive Drugs* 17(1): 35–39, 1985. For Schultes on folk classification of *yoco,* see Schultes, R. E., "Recognition of Variability in Wild Plants by Indians of the Northwest Amazon: An Enigma," *Journal of Ethnobiology* 6(2): 229–38, 1986.

Homer Pinkley, a graduate student of Schultes, lived for a year among the Kofán in 1965. See Pinkley, H. V. "The Ethnoecology of the Kofán," doctoral dissertation, Harvard University, Cambridge, Massachusetts, 1973. For ethnohistory and shamanism, see also Robinson, S., *Toward an Understanding of Kofán Shamanism,* Latin American Studies Program Dissertation Series, Cornell University, Ithaca, New York, 1979.

Schultes met Gaspar de Pinell in Sibundoy, and while staying at La Chorrera, he studied the padre's famous report. See Pinell, P. Gaspar de, *Excursión Apostólica por los Ríos Putumayo, San Miguel de Sucumbíos, Cuyabeno, Caquetá y Caguán,* Imprenta Nacional, Bogotá, 1929. See also Robuchon, E., *En el Putumayo y Sus Afluentes,* Lima, 1907; Whiffen, T., *The North-West Amazons: Notes on Some Months Spent Among Cannibal Tribes,* Constable, London, 1915.

For basic history of rubber, see Baum, V., *The Weeping Wood,* Doubleday, New York, 1943; Polhamus, L. G., *Rubber: Botany, Production and Utilization,* Leonard Hill Books, London, 1962; Wilson, C. M., *Trees and Test Tubes: The Story of Rubber,* Henry Holt, New York, 1943. For the horrors of the rubber era in the Putumayo, see Anon., *The Putumayo Red Book,* N. Thomson & Co., London, 1913; Collier, R., *The River That God Forgot,* Collins, London, 1968; Hardenburg, W. E., *The Putumayo: The Devil's Paradise,* T. Fisher Unwin, London, 1912; Taussig, M., *Shamanism, Colonialism and the Wild Man,* University of Chicago Press, Chicago, 1987. See also Weinstein, B., *The Amazon Rubber Boom, 1850–1920,* Stanford University Press, Stanford, California, 1983. For Schultes's survey of the rubber plants of the Putumayo, see Schultes, R. E., "Lacticiferous Plants of the Karaparaná-Igaraparaná Region of Colombia," *Acta Botanica Neerlandica* 15: 178–89, 1966.

Chapter Eight. The Sad Lowlands

For remarks on the psychoactive potential of *Coriaria thymifolia,* see Schultes, R. E., and A. Hofmann, *Plants of the Gods,* McGraw-Hill, New York, 1979. For the best modern source on the Conquest, see Hemming, J., *The Conquest of the Inca,* Harcourt Brace Jovanovich, New York, 1970. See also Hemming, J., and E. Ranney, *Monuments of the Incas,* University of New Mexico Press, Albuquerque, 1982. To my mind this beautiful book of photographs and text captures the spirit of the Inca better than anything else in print.

For the disappearance of coca from Ecuador, see León, L. A., "The Disappearance of *Cocaïsmo* in Ecuador," *Bulletin on Narcotics* 4(2): 21–25, 1952; León, L. A., *"Historia y Extincion del Cocaísmo en el Ecuador: Sus Resultados," América Indígena* 12(1): 7–32, 1952; Myers, T. P., "Formative Period Occupation in the Highlands of Northern Ecuador," *American Antiquity* 41(3): 353–60, 1976; Naranjo, P., "El Cocaísmo entre los

Aborígenes de Sud América: Su Difusión y Extinción en el Ecuador," *América Indígena* 34(3): 605–28,1974; Patiño, V. M., *Plantas Cultivadas y Animales Domesticos en América Equinoccial,* Imprenta Departamental, Cali, Colombia, 1967.

For ethnography of lowland Quichua, see Whitten, N. E., *Sacha Runa: Ethnicity and Adaptation of Ecuadorian Jungle Quichua,* University of Illinois Press, Urbana, 1976. For Siona-Secoya, see Vickers, W. T., "Cultural Adaptation to Amazonian Habitats: The Siona-Secoya of Eastern Ecuador," doctoral dissertation, University of Florida, Gainesville, 1976. For Jivaroan peoples, see Harner, M., *The Jívaro: People of the Sacred Waterfalls,* Doubleday/Natural History Press, New York, 1972. For Blomberg's misadventures, see Blomberg, R., *The Naked Aucas: An Account of the Indians of Ecuador* (translated by F. H. Lyon), George Allen Unwin, London, 1957.

For a critical look at the Summer Institute of Linguistics, see Stoll, D., *Fishers of Men or Founders of Empire?* Zed Press, London, 1982. For a very uncritical look at Rachel Saint and Dayuma, see Kingsland, R., *A Saint Among Savages,* Collins, London, 1980. For the Wycliffe/S.I.L. interpretation of events, see Hitt, R. T., *Jungle Pilot: The Life and Witness of Nate Saint,* Hodder and Stoughton, London, 1959; Wallis, E. E., *The Dayuma Story,* Harper and Bros., New York, 1960; Wallis, E. E., *Aucas Downriver,* Harper and Row, New York, 1973. The most revealing of the books written by those directly affected by the massacre is Elliot, E., *Through the Gates of Splendour,* Hodder & Stoughton, London, 1957. Betty Elliot also wrote *Shadow of the Almighty,* a biography of her deceased husband, published in London by Hodder & Stoughton, 1958. See also Elliot, E., *The Savage, My Kinsman,* Harper and Row, New York, 1961.

Chapter Nine. Among the Waorani

For ethnography of the Waorani, the papers of Jim Yost are the basic source. See Yost, J. A., "People of the Forest: The Waorani," *in* Ligabue, G. (ed.), *Ecuador in the Shadow of the Volcanoes,* pp. 95–115, Ediciones Libri Mundi, Venice, Italy, 1981; "Twenty Years of Contact: The Mechanisms of Change in Wao ("Auca") Culture," *in* Whitten, N. E. (ed.), *Cultural Transformations and Ethnicity in Modern Ecuador,* pp. 677–704, University of Illinois Press, Urbana, 1981; Yost, J. A., and P. M. Kelley, "Shotguns, Blowguns and Spears: The Analysis of Technological Efficiency," *in* Hames, R. B., and W. T. Vickers (eds.), *Adaptive Responses of Native Amazonians,* pp. 189–224, Academic Press, New York, 1983.

For medical studies of the Waorani, see Kaplan, J. E., et al., "Infectious Disease Patterns in the Waorani, an Isolated Amerindian Population," *American Journal of Tropical Medicine and Hygiene* 29(2): 298–312, 1980; Larrick, J. W., et al., "Snake Bite Among the Waorani Indians of Eastern Ecuador," *Transactions of the Royal Society of Tropical Medicine and Hygiene* 72: 542–43, 1978; Larrick, J. W., et al., "Patterns of Health and Disease Among the Waorani Indians of Eastern Ecuador," *Medical Anthropology* 3: 147–91, 1979; Theakston, R. D. G., et al., "Snake Venom Antibodies in Ecuadorian Indians," *Journal of Tropical Medicine and Hygiene* 84: 199–202, 1981.

Jim Yost and I published three papers based on our work together. See Davis,

E. W., and J. A. Yost, "The Ethnobotany of the Waorani of Eastern Ecuador," *Botanical Museum Leaflets* 29(3): 159–217, 1983; "The Ethnomedicine of the Waorani of Amazonian Ecuador," *Journal of Ethnopharmacology* 9(2–3): 273–98, 1983; "Novel Hallucinogens from Eastern Ecuador," *Botanical Museum Leaflets* 29(3): 291–98, 1983.

Chapter Ten. White Blood of the Forest, 1943

For an excellent review of the rubber program, 1940–54, see Rands, R. D., and L. G. Polhamus, *Progress Report on the Cooperative Hevea Rubber Development in Latin America,* Circular No. 976, U.S. Department of Agriculture, Washington, D.C., June 1955. For figures on world production of natural rubber, 1910–40, see data compiled by the Rubber Staff of the Bureau of Foreign and Domestic Commerce, U.S. Department of Commerce, July 24, 1942, U.S. National Archives, Washington, D.C.

The impetus provided by the wartime emergency led to the publication of a series of papers on rubber, many written by Schultes's superiors and associates at the USDA. These articles are fascinating both for their contents and for the style and tone in which they are written. Earnest and grave, cautious yet fundamentally hopeful, they confront the severity of the crisis and yet, like all good propaganda, leave the reader certain that in the end American scientific ingenuity will prevail.

For the strategic value of rubber in the war effort, see Blandin, J. J., "Why Rubber Is Coming Home," *Agriculture in the Americas* 1:1–7, 1941; Burkland, E. R., "Speaking of Rubber," *Agriculture in the Americas* 1(1): 7–12, 1941. For alternative sources of latex, see Brandes, E. W., "Go Ahead, Guayule!," *Agriculture in the Americas* 2(5): 83–86, 1942; Brandes, E. W., "Rubber from the Russian Dandelion," *Agriculture in the Americas* 2(7): 127–31, 1942; Loomis, H. F., "Castilla Rubber's Comeback," *Agriculture in the Americas* 2(9): 171–76, 1942. See also Schultes's *Report on Common Names, Lactiferous Plants, Colombia,* submitted to J. de W. Mayer, March 20, 1943, U.S. National Archives, Washington, D.C.

For general articles on the domestication and history of rubber, see Imle, E. P., "Hevea Rubber: Past and Future," *Economic Botany* 32: 264–77, 1978; Schultes, R. E., "The Odyssey of the Cultivated Rubber Tree," *Endeavor* (n.s.) 1(3–4): 133–38, 1977; Schultes, R. E., "The Tree That Changed the World in One Century," *Arnoldia* 44(2): 2–16, 1984.

For Schultes's *Cryptostegia* survey, see Schultes Report No. 2, December 10, 1942, submitted to C. B. Manifold, Rubber Reserve Company, December 11, 1942, U.S. National Archives, Washington, D.C. For Schultes's work on *Sapium* and the explorations of the Villalobos concessions, see Schultes Report No. 3, submitted to J. de W. Mayer, Rubber Reserve Company, Bogotá, February 6, 1943, U.S. National Archives, Washington, D.C.

For the Apaporis expedition, see *Survey of the Apaporis River Basin,* submitted to the Rubber Development Corporation by E. L. Vinton and R. E. Schultes, December 6, 1943, Bogotá. This document, deposited at the U.S. National Archives, Washington, D.C., includes a brief review by Mayer and four separate reports. Schultes Report No. 4, *Air Survey of the Apaporis River Basin,* submitted March 12, 1943;

Vinton and Schultes Report No. 1, *Survey of the Upper Apaporis,* submitted August 10, 1943; Vinton's *Report on the Airport at Puerto Hevea,* submitted September 28, 1943; Schultes Report No. 5, *Survey of Middle and Lower Apaporis,* submitted November 3, 1943. See also Schultes's itinerary reports for April to July, 1943, submitted July 13, 1943, to E. M. Blair, Rubber Development Corporation, Washington, D.C.

For Mayer on Vinton's illness and the importance of entering the Apaporis, see Mayer to E. M. Blair, January 16, 1943. For Blair's response, see Blair to Mayer, January 26, 1943. For Mayer's decision to dispatch both Schultes and Vinton to the Apaporis, see Mayer to Blair, April 7, 1943.

For Schultes on the Apaporis, with a review of his results on rubber yields and discovery of the new species, see "Terrain and Rubber Plants in the Upper Apaporis of Colombia," *Suelos Ecuator* 1: 121–37, 1958, and "Glimpses of the Little-Known Apaporis River in Colombia," *Chronica Botanica* 9: 123–27, 1945. For descriptions by Schultes of the new species found on top of Chiribiquete, see Schultes, R. E., "Plantae Colombianae IX," *Caldasia* 3: 121–30, 1944; Schultes, R. E., "Plantae Austro-americanae III: De Plantis Principaliter Colombiae Orientalis Observationes," *Botanical Museum Leaflets* 12: 117–48, 1946; Schultes, R. E., "Plantae Colombianae VIII," *Caldasia* 11: 23–32, 1944. For contemporary botanical studies of Chiribiquete, with a map indicating location of the mountain named for Schultes, see Estrada, J., and Fuertes J., "Estudios Botanicos en la Guayana Colombiana, IV: Notas Sobre la Vegetación y la Flora de la Sierra de Chiribiquete," *Revista del Academia Colombiana de Ciencias Exactas, Físicas y Naturales* 18(71): 483–87, 1993.

Chapter Eleven. The Betrayal of the Dream, 1944–54

For Mayer's defense of Schultes and Vinton, see Mayer to W. A. Stanton, November 30, 1943, Rubber Development Corporation, U.S. National Archives, Washington, D.C. For Paul Allen's report, see Allen, itinerary reports for May and June 1943, submitted July 13, 1943, to E. M. Blair, Rubber Development Corporation, U.S. National Archives, Washington, D.C. For Schultes's near miss with the Selective Service, see W. L. Clayton to Henry Linam, July 17, 1943, U.S. National Archives, Washington, D.C.

For wartime efforts to create plantations in the Americas, see Brandes, E. W., "Rubber on the Rebound—East to West," *Agriculture in the Americas* 1(3): 1–11, 1941 (an abstract of this paper appeared in *Indian Rubber World* as "A Plantation Industry for America"); Brandes, E. W., "The Outlook for Plantation Rubber in Tropical America," *Chronica Botanica* 7(7): 320–23, 1943; Brandes, E. W., "Progress in Hemisphere Rubber Plantation Development," *India Rubber World* 108: 143–45, 1943; Crane, B. V., "The Production of Rubber in South and Central America," *India Rubber World* 107(3): 259–64, 1942; Klippert, W. E., "The Cultivation of Hevea Rubber in Tropical America," *Chronica Botanica* 6(9): 199–200, 1941; Klippert, W. E., "Small-Farm Rubber Production," *Agriculture in the Americas* 2(3): 48–53, 1942; Polhamus, L. G., "War Speeds the Rubber Project," *Agriculture in the Americas* 2(2): 29–31, 1942;

Rands, R. D., "Hevea Culture in Latin America: Problems and Procedures," *India Rubber World* 106(3): 239–43, 350–56, 461–65, 1942.

For history of the Ford plantations at Fordlandia and Belterra, see Eidt, R. C., "Plantaciones de Caucho en el Brasil Fordlandia y Belterra," *Agricultura Tropical* 9(1): 17–28, 1953; Johnston, A., "The Ford Rubber Plantations," *India Rubber World,* May 1, 1941, pages 35–38; June 1, 1941, pages 199–202. For discussions of the South American leaf blight, see Rands, R. D., "South American Leaf Disease of Para Rubber," *USDA Department Bulletin* No. 1286, p.18, 1924; Rands, R. D., "Progress on Tropical America Rubber Planting Through Disease Control," *Phytopathology* 36: 688, 1946; Langford, M. H., *South America Leaf Blight of Hevea Rubber Trees,* U.S. Department of Agriculture Technical Bulletin 882, 31 pages, 1945.

Hans Sorensen's letter to T. J. Grant of March 20, 1944, is preserved at the U.S. National Archives, Washington, D.C. For Sorensen's scientific contributions, see Sorensen, H. G., "Crown Budding for Healthy Hevea," *Agriculture in the Americas* 2: 191–93, 1942; "Colombia's Plantation Rubber Program," *Agriculture in the Americas* 5: 106–8, 114–15, 1945.

For a sense of Leticia, circa 1943–46, I am deeply indebted to Helen Floden, who shared both her insights and her private papers and photographs. Of particular interest was a letter dated November 6, 1943, received by the Flodens from W. G. Scherer. In this long letter Scherer, a missionary then stationed in Iquitos, Peru, briefed the Flodens on what to expect on their arrival in Leticia. The detailed list of services and goods available, size and character of the population, means of transportation, and rhythms of the seasons provided a vivid portrait of the town.

Schultes described his work in Leticia in a number of articles written in the field and published in Colombia. See Schultes, R. E., "Aprovechamiento Científico de Una Riqueza Natural Colombiana," *Agricultura Tropical* No. 12, pp. 31–42, 1946; Schultes, R. E., "Esperanza Agronómica para la Amazonia Colombiana," *Agricultura Tropical* No. 2, Supplement No. 2, pp. 3–22, 1946; Schultes, R. E., "El Cauchero Abanderado del Vaupés," *Revista Nacional de Agricultura* No. 564, pp. 1–8, 1952.

For the 1923–24 USDA rubber survey, see La Rue, C. D., "The Hevea Rubber Tree in the Amazon Valley," *USDA Department Bulletin* No. 1422, 1926. For efforts to understand the complex taxonomy of rubber, see Schultes, R. E., "The History of Taxonomic Studies in *Hevea,*" *Botanical Review* 36: 197–276, 1970; Schultes, R. E., *A Brief Taxonomic View of the Genus Hevea,* Malaysian Rubber Research and Development Board, Monograph No.14, Kuala Lumpur, 1990. Schultes wrote a number of early papers on *Hevea,* based on his initial work in the Colombian Amazon. See "The Genus *Hevea* in Colombia," *Botanical Museum Leaflets* 12(1): 1–17, 1945; "Estudio Preliminar del Género Hevea en Colombia," *Revista del Academia Colombiana de Ciencias Exactas, Físicas y Naturales* 6(22–23): 331–38, 1945; "Studies in the Genus *Hevea I,*" *Botanical Museum Leaflets* 13(1): 1–11, 1947. Schultes's shipments of rubber seeds from Leticia are referred to in numerous documents on deposit at the U.S. National Archives, Washington, D.C. See Schultes to R. D. Rands, April 1, 1946.

For Seibert's work, see Seibert, R. J., "A Study of *Hevea* in the Republic of Peru," *Annals of the Missouri Botanical Garden* 34: 261–353, 1947, and "Searching the Jungles to Improve Rubber Trees," *Foreign Agriculture* 14: 153–55, 1950. For Schultes's inspection of the Calima plantations in western Colombia, see Schultes's report to C. Molina, May 31, 1946, U.S. National Archives, Washington, D.C.

For history of synthetic rubber and reviews of the wartime emergency, see Brenner, M. A., *The Outlook for Synthetic Rubber,* Planning Pamphlet No. 32, National Planning Association, Washington, D.C., 1944; Herbert, V., and A. Bisio, "Synthetic Rubber: A Project That Had to Succeed," *Contributions in Economics and Economic History* No. 63, Greenwood Press, Westport, Connecticut, 1985. For the role of natural rubber in the modern economy, see Barlow, C., *The Natural Rubber Industry,* Oxford University Press, Oxford, England, 1978; French, M. J., *The U.S. Tire Industry: A History,* Twayne Publishers, Boston, 1991; Grilli, E. R., B. B. Agostini, and M. J. Hooft-Welvaars, *The World Rubber Economy,* Johns Hopkins University Press, Baltimore, Maryland, 1980.

For postwar plantation efforts, see Bangham, W. N., "Plantation Rubber in the New World," *Economic Botany* 1: 210–29, 1947; Brandes, E. W., "Progress Toward an Assured Natural Rubber Supply," *India Rubber World* 116: 491–97, 507, 1947; *Rubber's Return to the Western Hemisphere,* Goodyear Tire and Rubber Company, Akron, Ohio, 1952, 24 pages; Litchfield, P. W., "A Living Stockpile for National Security," *Notes on America's Rubber Industry,* n.s. 16, Goodyear Tire and Rubber Co., Akron, Ohio, 1951; Schultes, R. E., and A. Uribe, "The Future of Rubber Growing in Colombia," *Agriculture in the Americas* 7(10–11): 127–30, October-November 1947.

The bureaucratic struggle that culminated in the termination of the rubber program is well documented in recently declassified papers at the U.S. National Archives. For Lester Edmond's memo, see Edmond to Armstrong, Comments on Attached Report by TCA on Rubber Projects in Latin America, December 13, 1951. For R. D. Rands's arguments in favor of the rubber effort, see Rands to Rey Hill, March 31, 1952; Rands to Hill, May 21, 1952; Rands to L.E. Peterson, June 4, 1952. For other supporting arguments, see L. E. Peterson, Reference Document on the Point IV Rubber Development Program in Latin America, November 1, 1950. See also M. W. Parker to Hill, March 19, 1953.

Industry support of the program was consistent and unanimous. See testimony of Paul Litchfield (Goodyear) and John Collyer (B. F. Goodrich), Report of the Department of State Rubber Advisory Panel Meeting of July 1952. See also Litchfield to Harold Stassen, July 24, 1953, and September 29, 1953; G. M. Tisdale (U.S. Rubber) to Harold Stassen, September 28, 1953; A. L. Viles (Rubber Trade Association) to Harold Stassen and Arthur Flemming, October 1, 1953; W. E. Klippert (Goodyear) to Harold Stassen, July 24, 1953.

For exchange between Harold Stassen and Arthur Flemming, see Stassen to Flemming, June 23 and July 6, 1953; William Rand to Flemming, July 21, 1953. For Flemming's response and firm endorsement of the strategic significance of the program, see Flemming to Rand, August 13, 1953. For Ezra Benson's support of the program, see Benson to Stassen, September 2, 1953. For industry continued support, see testimony and submissions of G. M. Tisdale, Paul Litchfield, Collyer, and other members of the Rubber Industry Advisory Panel, September 25, 1953. Stassen responded with a letter announcing the termination of the program as of June 30, 1954. See attachments to D. W. Figgis to D. A. Fitzgerald, October 22, 1953. See also R. M. Hill to D. W. Figgis, October 12, 1953. For the secret report that doomed the program, see The Rubber Research Development Program in Latin America: Evaluation Survey and Report, December 2, 1953. Harold Stassen's memo of December 9, 1953, officially ending the fourteen-year effort, has also been declassified and is on deposit at the U.S. National Archives, Washington, D.C.

For Rey Hill's suggestion that Schultes be assigned the task of preparing a monograph on *Hevea,* see R. M. Hill to R. F. Cook, July 10, 1953. Hill's telegram notifying Schultes of the assignment was received in Bogotá on July 17, 1953, and is preserved at the U.S. National Archives, Washington, D.C.

For Schultes on the threat of disease to the Eastern plantations, see Schultes, R. E., "Wild *Hevea*—an Untapped Source of Germ Plasm," *Journal of the Rubber Research Institute of Sri Lanka* 54, pt.1, No. 1, 1977, pp. 227–57.

Chapter Twelve. The Blue Orchid, 1947–48

Schultes described his encounter with the blue orchid in "*Aganisia cyanea*—Jewel of the Jungle," *American Orchid Society Bulletin* 30: 558–62, 1961. See also Schultes, R. E., "Orchidaceae Neotropicales V. Generis Aganisiae Synopsis," *Lloydia* 21(2): 88–99, 1958. Schultes's journal, covering his time on the Río Negro between September 4 and December 16, 1947, is among his personal papers.

While living on the Río Apaporis between October and December 1951, Schultes wrote a long and eloquent article about Richard Spruce. See Schultes, R. E., "Richard Spruce Still Lives," *The Northern Gardener* 7(1–4): 20–27, 55–61, 87–93, 121–25, 1953. For more of Schultes on Spruce, see "Some Impacts of Spruce's Amazon Explorations on Modern Phytochemical Research," *Rhodora,* 70: 313–39,1968; "Richard Spruce and the Ethnobotany of the Northwest Amazon," *Rhodora* 78: 65–72, 1976; "Richard Spruce: An Early Ethnobotanist and Explorer of the Northwest Amazon and Northern Andes," *Journal of Ethnobiology* 3(2): 139–47, 1983.

For Spruce's explorations, see Spruce, R. (ed. A. R. Wallace), *Notes of a Botanist on the Amazon and Andes,* 2 vols., Macmillan & Co., London, 1908. For the journals of Wallace and Bates, see Bates, H. W., *The Naturalist on the River Amazons,* John Murray, London, 1863; Wallace, A. R., *A Narrative of Travels on the Amazon and Río Negro,* Ward, Lock & Co., London, 1889. For a fascinating glimpse of the rivalry between Spruce and Wallace, see Balick, M. J., "Wallace, Spruce, and *Palm Trees of the Amazon:* A Historical Perspective," *Botanical Museum Leaflets* 28(3): 263–70, 1980.

Schultes published a number of papers on *Hevea* and its relatives based on his fieldwork on the Río Negro. Of particular interest are "Studies in the Genus *Hevea II:* The Rediscovery of *Hevea rigidifolia,*" *Botanical Museum Leaflets* 13(5): 97–132; "Studies in the Genus *Hevea IV:* Notes on the Range and Variability of *Hevea micro-*

phylla," Botanical Museum Leaflets 15(4): 111–38,1952. See also "Studies in the Genus *Hevea V," Botanical Museum Leaflets* 15(10): 247–54, 1952; "Studies in the Genus *Hevea VI," Botanical Museum Leaflets* 15(10): 255–72, 1952; "Studies in *Hevea VII," Botanical Museum Leaflets* 16(2): 21–44, 1953; "A New Infrageneric Classification of *Hevea," Botanical Museum Leaflets* 25(9): 243–57, 1977; "Studies in the Genus *Hevea VIII:* Notes on Infraspecific Variants of *Hevea brasiliensis," Economic Botany* 41(2): 125–47, 1987. For Schultes's studies of *Micrandra* and *Cunuria,* and descriptions of *Micrandra lopezii* and *M. rossiana,* see Schultes, R. E., "Studies in the Genus *Micrandra I:* The Relationship of the Genus *Cunuria* to *Micrandra," Botanical Museum Leaflets* 15(8): 201–22, 1952. For his description of *Vaupesia cataractarum,* see "A New Generic Concept in the Euphorbiaceae," *Botanical Museum Leaflets* 17(1): 27–36, 1955.

For Schultes's analysis of Spruce's *yagé* collection, see Schultes, R. E., et al., "De Plantis Toxicariis e Mundo Novo Tropicale Commentationes III: Phytochemical Examination of Spruce's Original Collection of *Banisteriopsis Caapi," Botanical Museum Leaflets* 22(4): 121–31, 1969. Schultes also analyzed Spruce's *yopo* collections. See Schultes, R. E., et al., "De Plantis Toxicariis e Mundo Novo Tropicale Commentationes XVIII: Phytochemical Examination of Spruce's Ethnobotanical Collection of *Anadenanthera peregrina," Botanical Museum Leaflets* 25(10): 273–88, 1977.

For his discovery and description of *Pouteria Ucuqui,* see Murça Pires, J., and R. E. Schultes, "The Identity of *Ucuquí," Botanical Museum Leaflets* 14(4): 87–96. For his description and account of the psychoactive liana *Tetrapterys methystica,* see Schultes, R. E., "Plantae Austro-americanae IX: Plantarum Novarum vel Notabilium Notae Diversae," *Botanical Museum Leaflets* 16(8): 179–228, 1954.

Chapter Thirteen. The Divine Leaf of Immortality

For the botany and ethnobotany of coca, see the articles of Tim Plowman cited above. The classic source is Mortimer, W. G., *History of Coca: The Divine Plant of the Incas,* reprint of 1901 edition, And/Or Press, San Francisco, 1974. For three excellent collections of essays and articles covering many aspects of the history, ethnobotany, medicinal potential, pharmacology, and politics of coca, see Pacini, D., and C. Franquemont (eds.), *Coca and Cocaine: Effects on People and Policy in Latin America,* Cultural Survival Report 23, Cambridge, 1986; Rivier, L. (ed.), "Coca and Cocaine, 1981," *Journal of Ethnopharmacology* 3(2–3): 106–379. For papers on coca and cocaine, see Andrews, G., and Soloman, D. (eds.), *The Coca Leaf and Cocaine Papers,* Harcourt Brace Jovanovich, New York, 1975. For an excellent review of the early history of cocaine, see Grinspoon, L., and Bakalar, J. B., *Cocaine: A Drug and Its Social Evolution,* Basic Books, 1985. For two superb books on the cultural importance of coca, see Allen, C. J., *The Hold Life Has: Coca and Cultural Identity in an Andean Community,* Smithsonian Institution Press, Washington, D.C., 1988; Antonil (A. R. Henman), *Mama Coca,* Hassle Free Press, London, 1978. See also Gagliano, J., *Coca Prohibition in Peru,* University of Arizona Press, Tucson, 1994.

For the results of the nutritional assay, see Duke, J. A., D. Aulik, and T. Plowman, "Nutritional Value of Coca," *Botanical Museum Leaflets* 24(6): 113–19, 1975. See also

Burchard, R. E., "Coca Chewing: A New Perspective," *in* Rubin, V. (ed.), *Cannabis and Culture,* Mouton, The Hague, pp. 463–84, 1975. For medicinal value of coca, see Weil, A. T., "The Therapeutic Value of Coca in Contemporary Medicine," *Journal of Ethnopharmacology,* 3: 367–76, 1981. For Tim's final solution to the puzzle of domesticated coca, see Bohm, B., F. Ganders, and T. Plowman, "Biosystematics and Evolution of Cultivated Coca *(Erythroxylaceae)," Systematic Botany* 7(2): 121–33, 1982.

For history of the Inca, see Baudin, L., *Daily Life in Peru Under the Last Incas,* Macmillan, New York, 1968; Brundage, B. C., *Lords of Cuzco* and *Empire of the Inca,* both University of Oklahoma, Norman, 1985; Conrad, G. W., and A. Demarest, *Religion and Empire,* Cambridge University Press, Cambridge, England, 1984; Hemming, J., *The Conquest of the Incas,* Harcourt Brace Jovanovich, New York, 1970; Métraux, A., *The History of the Incas,* Schocken Books, New York, 1976; Rowe, J. H., "Inca Culture at the Time of the Spanish Conquest," *Handbook of South American Indians,* Smithsonian Institution, Bureau of American Ethnology, Bulletin 143, vol. 2, pp. 183–330, 1946; Von Hagen, V. W., *The Incas of Pedro de Cieza de León,* University of Oklahoma Press, Norman, 1976; Zuidema, R. T., *Inca Civilization in Cuzco,* University of Texas Press, Austin, 1990.

For Inca roads, see Hyslop, J., *The Inka Road System,* Academic Press, 1984; Von Hagen, V. W., *Highway of the Sun,* Little, Brown & Co., Boston, 1955. See also Hyslop, J., *Inka Settlement Planning,* University of Texas Press, Austin, 1990. For Inca notions of land and sacred space, see Sallnow, M. J., *Pilgrims of the Andes,* Smithsonian Institution Press, Washington, D.C., 1987. Ronald Wright has written two superb books: *Stolen Continents,* Viking, New York, 1991, and *Cut Stones and Crossroads,* Penguin Books, New York, 1984.

For Andean botany and ethnobotany, see Franquemont, C., et al., *The Ethnobotany of Chinchero, an Andean Community in Southern Peru,* Fieldiana New Series, no. 24, 1990; Gade, D. W., "Plant Use and Folk Agriculture in the Vilcanota Valley of Peru," doctoral dissertation, University of Wisconsin, Madison, 1967; Goodspeed, T. H., *Plant Hunters in the Andes,* University of California Press, Berkeley, 1961.

Both Tim and I published papers on *chamairo,* the peculiar sweetening agent. See Plowman, T., "Chamairo: *Mussatia hyacinthina*—an Admixture to Coca from Amazonian Peru and Bolivia," *Botanical Museum Leaflets* 28(3): 253–61, 1980; Davis, E. W., "The Ethnobotany of Chamairo: *Mussatia hyacinthina," Journal of Ethnopharmacology* 9(2–3): 225–36, 1983. For a dramatic description of the Pongo de Mainique, see Matthiessen, P., *The Cloud Forest,* Ballantine Books, New York, 1961.

Chapter Fourteen. One River

For medicinal plants, especially of the Iquitos area, see Maxwell, N., *Witch Doctor's Apprentice,* Citadel Press, New York, 1990. For the folk use of *yagé* in the Iquitos region, see Luna, L. E., *Vegetalismo: Shamanism Among the Mestizo Population of the Peruvian Amazon,* Almqvist & Wiksell, Stockholm, Sweden, 1986. For Schultes on Amazonian coca, see Schultes, R. E., "Coca in the Northwest Amazon," *Botanical Museum Leaflets* 28(1): 47–59, 1980. For his discovery of a novel way of using resin to scent the powder, see Schultes, R. E., "A New Method of Coca Preparation in the Colombian Amazon," *Botanical Museum Leaflets* 17(90): 241–46, 1957.

Between May 1951 and his departure from South America in the fall of 1953, Schultes was accompanied in the field by Isidoro Cabrera, who kept a journal, scribbled by night in pencil in small notebooks. These long entries proved immensely helpful both as a record of their daily rounds, and as a means of verifying itineraries and localities.

For ethnography of the Vaupés, the starting point is once again the remarkable work of Gerardo Reichel-Dolmatoff. His book *Desana: Simbolismo de los Indios Tukano del Vaupés,* published in Bogotá by the Universidad de los Andes, 1968, appeared in English as *Amazonian Cosmos: The Sexual and Religious Symbolism of the Tukano Indians,* University of Chicago Press, Chicago, 1971. Among his many publications, all of them brilliant, I found these to be especially provocative: "The Cultural Context of an Aboriginal Hallucinogen: *Banisteriopsis Caapi,*" *in* Furst, P. T. (ed.), *Flesh of the Gods,* pp. 84–113, Praeger Publishers, New York, 1972; *Beyond the Milky Way,* UCLA Latin American Center Publications, Los Angeles, 1978; "Desana Shamans' Rock Crystals and the Hexagonal Universe," *Journal of Latin American Lore* 5(1): 117–28, 1979; "Brain and Mind in Desana Shamanism," *Journal of Latin American Lore* 7(1): 73–98, 1981.

Reichel-Dolmatoff inspired Brian Moser and Donald Tayler to descend the Río Piraparaná in 1960. For their time among the Tukano, see Moser, B., and D. Tayler, *The Cocaine Eaters,* Longmans, Green & Co., London, 1965. The husband and wife team of Stephen and Christine Hugh-Jones also were encouraged and assisted by the great Colombian anthropologist. For their superb publications, based on their fieldwork on the Piraparaná, see Hugh-Jones, C., *From the Milk River: Spatial and Temporal Processes in Northwest Amazonia,* Cambridge University Press, Cambridge, England, 1979; Hugh-Jones, S., *The Palm and the Pleiades: Initiation and Cosmology in Northwest Amazonia,* Cambridge University Press, Cambridge, England, 1979; Hugh-Jones, S., "Like the Leaves on the Forest Floor: Space and Time in Barasana Ritual," *Proceedings of the 42nd International Congress of Americanists,* vol. 2, pp. 205–15, Paris, 1977.

For Makuna and Cubeo ethnography, see Århem, K., *Makuna Social Organization,* Almqvist & Wiksell, Stockholm, Sweden, 1981; Goldman, I., *The Cubeo,* University of Illinois Press, Urbana, 1979. For a discussion of language, marriage rules, and settlement patterns, see Sorensen, A. P., "Multilingualism in the Northwest Amazon," *American Anthropologist* 69: 670–84, 1967. While living among the Yukuna on the Miriti-paraná, Schultes described the dance of the Cunuri in two unpublished manuscripts that are among his personal papers. See "The Evolution Dance of the Yukunas" and "A Folk Tale of the Cunuri Dance of the Yukunas of Colombia." He also referred to the Cunuri celebration in Schultes, R. E., "The Amazon Indian and Evolution in *Hevea* and Related Genera," *Journal of the Arnold Arboretum* 37(2): 123–48, 1956.

Schultes published numerous articles on psychoactive snuffs. For the first discovery, including the description of his self-experiment with yákee, see Schultes, R. E., "A New Narcotic Snuff from the Northwest Amazon," *Botanical Museum Leaflets* 16(9): 241–60, 1954. For his work among the Waika, see Schultes, R. E., and B. Holmstedt, "De Plantis Toxicariis e Mundo Novo Tropicale Commentationes II: The Vegetal Ingredients of the Myristicaceous Snuffs of the Northwest Amazon," *Rhodora* 70(781): 113–60, 1968. Review papers include Schultes, R. E., "The Botanical Origins of South American Snuffs," *in* Efron, D. H. (ed.), *Ethnopharmacological Search for Psychoactive Drugs,* Public Health Service Publication No. 1645, Government Printing Office, Washington, D.C., 1967; Schultes, R. E., "Evolution of the Identification of the Myristicaceous Hallucinogens of South America," *Journal of Ethnopharmacology* 1: 211–39, 1979.

For the first report of an oral preparation derived from *Virola,* see Schultes, R. E., "De Plantis Toxicariis e Mundo Novo Tropicale Commentationes V: *Virola* as an Orally Administered Hallucinogen," *Botanical Museum Leaflets* 22(6): 229–40, 1969. Follow-up studies were reported in Schultes, R. E., and T. Swain, "De Plantis Toxicariis e Mundo Novo Tropicale Commentationes XIII: Further Notes on *Virola* as an Orally Administered Hallucinogen," *Journal of Psychedelic Drugs* 8(4): 317–24, 1976; Schultes, R. E., T. Swain, and T. Plowman, "De Plantis Toxicariis e Mundo Novo Tropicale Commentationes XVII: *Virola* as an Oral Hallucinogen Among the Boras of Peru," *Botanical Museum Leaflets* 25(9): 259–272, 1977.

Schultes addressed the issue of admixtures in "De Plantis Toxicariis e Mundo Novo Tropicale Commentationes XI: The Ethnotoxicological Significance of Additives to New World Hallucinogens," *Plant Science Bulletin* 18: 34–40, 1972. For the best recent work on orally administered *Virola* preparations, see McKenna, D. J., G. H. N. Towers, and F. S. Abbott, "Monoamine Oxidase Inhibitors in South American Hallucinogenic Plants; Part 2: Constituents of Orally-Active Myristicaceous Hallucinogens," *Journal of Ethnopharmacology* 12: 179–211,1984.

Reichel-Dolmatoff recalled his initial meeting with Schultes in a letter to Tom Riedlinger dated May 10, 1993. At the memorial service for Tim, held at the Field Museum of Natural History in Chicago on January 19, 1989, I was asked to say a few words. This eulogy appeared later in a number of publications, including *Economic Botany* 43(3): 416–19, 1989.

Acknowledgments

THE WRITING OF this book was facilitated by a grant from the Canada Council. Research assistance was provided generously by archivists, reference librarians, and curators at numerous institutions including the U.S. National Archives, National Anthropological Archives of the Smithsonian Institution, National Library of Agriculture, National Library of Medicine, Library of Congress, Lauinger Library of Georgetown University, Tozzer Library and Economic Botany Library of Harvard University, and the Tina and Gordon Wasson Ethnomycological Collections, Harvard University. My fieldwork was supported in part by the Social Science and Humanities Research Council of Canada. Tim Plowman's coca research was conducted at the Botanical Museum of Harvard University under a contract with the U.S. Department of Agriculture, with Richard Evans Schultes directing the study. Heading up the effort at the USDA was Jim Duke, whose foresight and leadership made the project possible.

For correspondence, interviews, and assistance in locating specific research materials I am indebted to Edgar Asebey, Henrik Blohm, William Burroughs, Isidoro Cabrera, Alvaro Fernández-Pérez, Helen Floden, Peter Furst, Hernando García Barriga, Allen Ginsberg, Albert Hofmann, Paul Hurley, Jesús Idrobo, Ernest Imle, Roberto Jaramillo, Weston La Barre, Clara Loring, Mimi Marshall, Frederico Medem, Eunice Pike, Gerardo Reichel-Dolmatoff, Tom Riedlinger, Dave Russ, Rusty Russell, Russ Seibert, Calvin Sperling, Hans Ungar, Fernando Urrea, Bill Vickers, Gordon Wasson, Irmgard Weitlaner, Richard Wheaton, Wayne White, and Johannes Wilbert.

In South America I was assisted by many individuals, only a few of whom are mentioned in the book. Enrique Forero welcomed me on my arrival in Colombia and within a week led me to the rain forests of the Darien. In Medellín, Mariano Ospina provided a room at the Jardin Botanico "Joaquin Antonio Uribe." Doel Soejarto made possible several early botanical forays. Other scholars who offered guidance along the way include Ramon Ferreyra, Edgardo Machado, Cesar Vargas, Lucia Atehortua, Stephan Beck, Fernando Cabieses. Michael Hill introduced me to Sebastian Snow. Noel Prince guided me through the streets and history of Bogotá, and his family welcomed me with the generosity and kindness typical of Colombians. Jorge Fuerbringer and Frederico Medem received me in Mocoa and Villavicencio. Jim and Kathy Yost made possible my time among the Waorani. Dave Scoble saw me through a difficult moment in the Darien. Ñilda Callañaupa and the people of Chinchero introduced me to the plants of the Andes. Pedro Juajibioy brought Tim and me into the lives of the Kamsá and Ingano, Adalberto Villafañe and the Ika of Sogrome revealed the visionary world of the Sierra Nevada de Santa Marta. The Waorani of Quiwado shared their extraordinary knowledge of the forest. Juan Evangelista Rojas welcomed us on his farm.

A special word must go to Lee Jacobs. Lee introduced me to Juan's farm and on several occasions joined Tim and me in the field. He was with us on the drive south to Lima from southern Ecuador, accompanied Tim to the Apurimac, and joined us once more in Cuzco for the drive south to Bolivia. He had a gardener's love of plants, and his humor and wild spirit made him the finest of companions and friends. He is absent from this book only because by chance he was not with us during the expeditions that best tell the coca story.

For good company during some of my other travels I am grateful to Christine and Ed Franquemont, Steve King, Mike Madison, Nina Marshall, Dennis McKenna, Terence McKenna, Chuck Sheviak, Sebastian Snow, Arthur Sorensen, Calvin Sperling, John Tichenor, Etta Vendetta, and Jim Zarucchi. For friendship and moral support during the writing of this book I thank my sister Karen Davis, as well as Paul Burke, George and Cathy Cragg, Anna Gustafson, Shane Kennedy, Ian MacKenzie, Chuck and Loraine Percy, Tom Rafael, Chuck Savitt, and David and Tara Suzuki.

In my academic training and in the writing of this book I was inspired by many sources. One of Schultes's last graduate students, I was preceded at the Botanical Museum by a remarkable group of ethnobotanists, including Mel Bristol, Bob Bye, Tommie Lockwood, Mike Balick, Homer Pinkley, Andrew Weil, and, of course, Tim Plowman. At the Peabody Museum my undergraduate tutor, David Maybury-Lewis, introduced me to anthropology and the work of many of the scholars cited in this book. Those to whom I am especially indebted include Johannes Wilbert, Christine and Stephen Hugh-Jones, John Hemming, Peter Furst, Michael Taussig, and in particular Gerardo Reichel-Dolmatoff, whose influence is readily apparent in the text.

Several friends and colleagues reviewed all or portions of the manuscript. My thanks go to Michael Carlisle, Lavinia Currier, Simon Davies, Karen Davis, Luz Fandino, Joel McCleary, Corky and Scott McIntyre,

Gail Percy, Travis Price, Tom Riedlinger, Dorothy and Dick Schultes, Calvin Sperling, and Andrew Weil. I would like especially to thank Howard Boyer and Caroline Alexander for their sustained interest and invaluable critiques. My agent, Michael Carlisle, was behind the book from the start, as was Bob Bender, who proved to be both an insightful and a thorough editor, and a superb person in every way. As the project grew, ultimately consuming the better part of five years, Bob's support and faith in the book never wavered.

At Simon & Schuster thanks also to Rose Ann Ferrick, Theresa Czajkowska, Karolina Harris, Michael Accordino, and Johanna Li.

Naturally, I am deeply indebted to Dick Schultes and his wonderful wife, Dorothy, who so often welcomed me into their home and shared my enthusiasm for the project. I will never be able to repay them for all that they have done in the twenty or more years since I first walked into Professor Schultes's office.

My parents in their wisdom sent me to Colombia at the age of fourteen, a trip that instilled a love of Latin America that has never left me. To my mother, Gwendolyn, and my late father, Edmund, I owe the gratitude of a son whose life has been forged by opportunities made possible by their generosity. To my children, Tara and Raina, and my wife, Gail, I owe just about everything else. Gail has worked with me on every phase of the book, from proposal to final manuscript, reading and editing, offering advice and sharing her insights as an anthropologist and artist. Her patience, understanding, and love were the haven into which I retreated to write this book.

Finally, there is Tim, to whom the book is dedicated, as dear a friend as anyone could wish for and a man whose tragic death left an immense void in the lives of so many who knew and loved him.

Index

About the Author

WADE DAVIS holds degrees in anthropology and biology, as well as a Ph.D. in ethnobotany, from Harvard University. His previous books include *The Serpent and the Rainbow,* a firsthand account of his experiences in Haiti researching the zombi phenomenon and its place in Vodoun culture. That book was translated into ten languages and was later released as a film with the same title. Davis has research projects under way in Borneo, Colombia, and Tibet. A native of British Columbia, he and his family divide their time between Washington, D.C., and a fishing lodge in remote northern Canada.